S0-ARY-479

Earth and Earth-Rock Dams

Earth and

Serre-Ponçon Dam Electricité de France / *Photo Courtesy of Coyne and Bellier, Consulting Engineers*

Earth-Rock Dams

Engineering Problems of Design and Construction

6 2 7.8

JAMES L. SHERARD, Sc.D.

RICHARD J. WOODWARD, M.S.

STANLEY F. GIZIENSKI, M.S.

WILLIAM A. CLEVENGER, B.S.

Members of the firm:

Woodward-Clyde-Sherard and Associates

Consulting Engineers

John Wiley and Sons, Inc.

New York · London · Sydney

Library of Congress Catalog Card Number: 63-14068
Printed in the United States of America

Preface

Though earth dams have been constructed since the earliest times, the increase in our ability to build safe and economical structures has been greater during the last 30 years than in all previous history. Because of new confidence gained by developments in soil mechanics and hydrology, earth dams have been safely constructed at sites and with heights which would have been thought dangerous or impossible only a few years earlier. The momentum of this progress shows no signs of slackening. Each year higher dams with improved designs and more ingenious details are proposed and built.

There remains a wide variety of opinion and practice among engineers working in the field. Many aspects of designing and constructing large earth dams will probably always fall within that group of engineering problems for which there are no universally accepted or uniquely correct procedures. In spite of advances in related technology, it is likely that the building of earth dams will always remain an empirical process.

In addition to evaluating the lessons learned from his own experience, the earth dam engineer must familiarize himself with the practice and experience of his colleagues, and this is not an easy task. It is unfortunate, but quite true, that much of the valuable knowledge acquired on many interesting projects remains the property of the engineer in charge. This book has been prepared in an attempt to present the main elements of current practice, experience, and opinion. It is being published with full recognition that the content will not gracefully withstand even a short period of time.

Since the main emphasis has been directed toward the empirical aspects of the subject, in some places this has been at the expense of a thorough review of the theory. For example, seepage theory, which is well covered elsewhere in the literature, has been treated in a condensed form (Chapter 5). On the other hand, considerable discussion has been devoted to the mechanics and limitations of current methods of stability analysis, although we came close to omitting the topic altogether because it is in such a rudimentary state of development. The sections describing the causes and mechanics of failure (Chapter 2) and summarizing measurements of movements and pore pressures (Chapter 3) have been made as complete as the available information allowed.

The content is confined to consideration of the foundation and dam embankment after the site has been selected. Except for brief treatment of the relationship of spillways to the design of the embankment, problems of appurtenant structures are not covered.

The practices described are mainly those prevailing or coming into use in the United States, although many references are made to the work of engineers in other countries. The discussion includes not only conventional earth dams but also dams composed mainly of rockfill with rolled-earth cores. Earth and gravel dams with impervious membranes of concrete, steel and asphalt are covered in a special section (Chapter 9). Hydraulic fill dams are not treated, primarily because there has been no recent experience with them in most parts of the world.

One of the principal deterrents to the preparation of a book of this type, and probably the main reason why none has appeared in English for more than 20 years, is the certainty that soon after its publication the author will alter some of his opinions under the pressure of new experience. As authors of this book, though we have stopped far short of recommending the "best" method to proceed on many points and have instead described the conflicting views current among earth dam engineers, we have little doubt that during the time this book is in print we will change our minds about some of the opinions expressed.

Important contributions to the content were made by so many friends and colleagues that it is impossible to thank them individually here. Tables and drawings taken from other sources are acknowledged in the text. The organization and clarity of presentation have been improved greatly by the editorial efforts of Miss Mary F. Barnett. For both technical assistance and sympathetic forebearance during the preparation of the manuscript, finally, we owe the greatest debt to our partners.

July, 1963

JAMES L. SHERARD
New York, N.Y.

RICHARD J. WOODWARD
Oakland, California

STANLEY F. GIZIENSKI
San Diego, California

WILLIAM A. CLEVENGER
Denver, Colorado

Contents

one

Design Considerations

1.1 FACTORS INFLUENCING DESIGN

As with most civil engineering structures, the design of an earth dam is based both on precedent and on analytical studies. The personal experience and preferences of the individual designer, however, play a larger role in earth dams than in the design of most other structures. At a given site it is usually possible to build a variety of dams which would be both safe and economical, and there are many examples where competent engineers have proposed widely different designs for the same reservoir.

In addition, the characteristics of the particular site have a greater influence on the design of an earth dam than they do on many other engineering structures. The quantities, types, and locations of the soils available for the construction of the embankment may dominate the entire design, although in most cases the soils must be considered in conjunction with conditions such as the nature of the foundation, the climate of the region, and the size and shape of the valley, to name a few. Besides these conditions imposed by nature, the design may be affected by the length of time available for construction or the intended function of the reservoir. The main design considerations are treated individually in the following pages.

The fundamental steps in designing an earth dam are:

1. A thorough exploration of the foundation and abutments, and an evaluation of the quantities and characteristics of all the embankment construction materials available within a reasonable distance of the site.
2. A study of all the conditions which may influence the design as described below in this section.
3. The selection of possible trial designs.
4. An analysis of the safety of the trial designs.

5. The modification of the designs in order to meet minimum stability requirements.
6. The preparation of detailed cost estimates.
7. The final selection of the design which seems to offer the best combination of economy, safety, and convenience in construction.

The design has not been completed, however, until the dam is built and the reservoir is in successful operation. During construction on almost every job, problems arise which require design changes. For this reason it is highly desirable that the designer also have charge of building the dam. As new conditions are exposed during construction, no one better than he can make adaptations which are based on the original design concept.

When he cannot direct construction personally, the designer may be justified in using a different, and perhaps more conservative, design. Relatively conservative or simple designs may also be advisable for dams in parts of the world where experienced contractors and supervisory personnel are not available. Even in the United States, earth dams constructed by the U.S. Bureau of Reclamation just after World War II, when no dams had been built for a long period, were provided with somewhat flatter slopes than usual, mainly because it was feared that the personnel would not have the skill necessary to produce embankments of high quality (Ref. 147).

In some countries, currency restrictions or other circumstances may make it necessary to use only the equipment which is already available there. The design must be planned accordingly and will often be quite different from that which would have been chosen had other equipment been on hand. Similarly, in some parts of the world, the relative costs of work done by hand labor and by machinery will have a large influence on the selection of the most economical design.

Another special condition that may have a great effect on the design arises at a site where the reservoir is to be developed in two (or more) stages. This involves the construction of a low dam which will be raised in later years as more reservoir capacity is needed. In such circumstances a quite different embankment design and spillway scheme may be advisable from that which would be used if the higher dam were to be built at once (Sec. 1.3*f*).

1.1a Materials Available for Embankment Construction

One of the principal economic advantages of the earth dam is that nature has already fabricated the construction materials and placed them free of charge at the site. In some cases only one type of soil is readily available, and the principal design problem is to determine the most economical dam that can be made from it. If it is an impervious soil, the design will consist of a homogeneous embankment with only a small amount of pervious material to control internal seepage (Sec. 1.2*a*). If it is a pervious sand or gravel, a dam with a very thin earth core may be used where enough impervious soil is available to make a core (Sec. 1.2*b*); otherwise, an impervious membrane may be constructed of manufactured material such as reinforced concrete, steel plate, or asphaltic concrete (Chap. 9).

At the other extreme are sites with a great number of varied soils which can be incorporated in different combinations into the embankment. In such a case it is usually advisable to plan a zoned dam with the finer and more impervious materials placed in an internal core, and the coarser materials utilized to provide strength and drainage in upstream and downstream zones (Sec. 1.2*c*).

When the material which is most easily available at the site is so erratic that it cannot be relied upon to have the consistent minimum properties needed for an impervious core or other embankment zones of definite properties, it can often be utilized most economically in the form of "random" zones. In the design, these random zones are assumed to have the properties of the least desirable material in the excavation; hence, during construction anything can be put into the zone in any order. Depending on the soil types, the use of large random zones may or may not result in a larger embankment volume than would be required if the material were selectively excavated or processed and placed in a more tightly zoned dam. Random zones should be considered at any site where, for any of a number of reasons, an appreciable volume of waste excavation is contemplated. Since this waste material has to be hauled away and dumped in spoil piles somewhere if not used in the dam, it may be considerably cheaper to dump it into special zones in the dam. In fact, it is not infrequently the case that, because of limited area available for dumping spoil piles, the haul distance for the waste material may be considerably longer than if it were placed in the dam.

Where two types of soils occur in horizontal layers in the same borrow pit, they can be either excavated selectively and placed in different zones of the dam or excavated together and blended into a single material with intermediate properties (Fig. 1.1:1). The properties of such a blended mixture can be controlled to some extent by varying the excavation procedure to obtain different percentages of the two soils.

Fig. 1.1:1 Face of borrow excavation for Long Lake Dam, Washington (1948), in which power shovel mixed upper layer of silt with lower layer of sand. Shovel cut height was varied to obtain approximate 50:50 mixture of materials (Ref. 347). (Courtesy USBR)

It is also possible to prepare an embankment material of intermediate properties by mixing two materials from different borrow areas on the construction surface. Although this procedure has been used on a few dams, it has usually proved to be uneconomical and unsatisfactory. Generally speaking, although the properties of the natural soils at a given site may not be ideal and could be improved by blending, screening, or washing, it is usually more economical to utilize the available materials in their natural states without processing.

There are two main circumstances in which different materials have been mixed (other than by blending with a shovel in the excavation): (1) where no natural deposits of fine-grained and impervious soils are available and the impervious core is made from a mixture of a small amount of bentonite and the natural soil available; and (2) mixing of rock and soil to obtain an impervious material. Neither procedure has been used frequently. Although the use of bentonite has been studied at quite a few sites, it has been used only rarely, primarily because of the high cost. Bentonite is sometimes mixed on the construction surface with pulvimixers or disks, and sometimes in a central mixing plant (see Refs. 5, 122, 363, 373, 435, 440, 597, 640). Rock and soil have been mixed together usually on the construction surface but occasionally in central plants. (See for example the description of the construction of the Pirquitas Dam in Argentina in 1961, Ref. 576.) Soil and rock are mixed to provide a suitable means to use soft rock in the dam (Sec. 11.5*d*) and, occasionally, to construct a special zone which has the high strength of the rockfill and the low permeability of the fine-grained soil (Sec. 11.5*c*).

REMOVAL OF OVERSIZED COBBLES

The only processing operation now in common use is the removal of oversized cobbles from embankment materials. The largest cobble which can be placed in an embankment is customarily limited to a diameter a little less than the final thickness of the compacted layer. In the core of a dam constructed with sheepsfoot rollers, the maximum diameter of cobbles is customarily set at 5 or 6 in., while somewhat larger sizes are allowed in embankments constructed with rubber-tired rollers. In embankment sections of pervious soil and in zones of random properties, cobbles or rocks up to diameters of 24 in. or more have been used satisfactorily.

If the soil for the core of the dam contains only a few percent of oversized cobbles, the sole purpose of removing them is to improve the compaction and quality of the embankment. But if the soil contains a large quantity of oversized cobbles, two other benefits may be obtained: the embankment can be made more impervious, and the oversized cobbles can become an important source of material for the construction of free-draining zones.

Cobbles can be removed on the construction surface by hand labor or by rakes mounted on tractors, or they can be screened out in rock separation plants before the material is transported to the dam site. The mechanics of these operations are discussed in Sec. 11.2*c*. If the soil contains only a few percent of oversized cobbles, the contractor is commonly given the option to choose the method of removal. On jobs where the material contains a large quantity of cobbles, the designer specifies that a screening plant be used.

A coarse, rocky soil with a large percentage of fines can be incorporated into the dam in several ways:

1. Without screening, to construct pervious or semipervious zones, or random zones, in which the permeability is not important to the embankment design.
2. By screening on a coarse screen (6 to 12 in.), to obtain free-draining rock on the one hand and semipervious material on the other.
3. By screening on a fine screen (about 3 in.), to obtain free-draining rock on the one hand (Figs. 1.1:2, 1.1:3) and impervious material on the other.

Many successful and economical earth dams have been constructed from screened materials, the finer soil being used for the core of the dam and the coarser material for free-draining zones or riprap.

The unit cost of taking out the oversized material with hand labor or with rakes on the construction surface is directly related to the quantity of material removed. On the other hand, the percentage of oversized materials screened out at a separating plant does not have a

Fig. 1.1:2 Construction surface of rockfill embankment from material obtained by screening a dirty talus deposit, Caribou Dam, California (1958).

Fig. 1.1:3 Pervious zone constructed from angular cobbles obtained by separating on a 3-in. screen. Prineville Dam, Utah (1960). (Courtesy USBR)

large influence on the unit cost of the processed material. There is a wide range of opinion among contractors and engineers concerning the point at which the use of a screening plant is justified. Contractors who have had good experiences with them will automatically set up a plant when the material contains more than 2 to 4% of oversized material. Others will go to almost any extreme to avoid screening.

The use of a screening plant has a number of advantages over the removal of cobbles on the construction surface:

1. The material passing the screens is well mixed and can be compacted into a homogeneous embankment section with a minimum of blending required on the construction surface.
2. A screening plant is more effective in removing all of the oversized rock. Better compaction is obtained and more material is available for the construction of pervious embankment zones.
3. The water content of the finer fraction of the material can be adjusted very effectively by sprinkling in the screening plant, so that minimum time is required for water content control.
4. Less equipment and fewer men are required on the construction surface, and higher rates of embankment placement can be obtained. This may be particularly important when the area of the construction surface is small. Construction is delayed less often by the necessity of rerolling layers and digging out cobbles.

The main disadvantage of screening is the cost. This consists of the initial installation of the plant, the plant operation and maintenance, and the extra hauling, dumping, and reloading at the plant. The unit cost varies greatly depending on the total quantity of material processed, the haul distances and grades, and, to a lesser extent, the soil properties. At dams where the quantity exceeds 1,000,000 yd.^{3}, the cost commonly ranges between 5¢ and 15¢/yd.^{3} of material processed.

MATERIALS FROM REQUIRED EXCAVATIONS

The quantities and properties of the materials from the "required excavation" for the spillway, foundation cutoff trench, outlet works, and other appurtenant structures often have an important influence on the embankment design. These materials usually have less desirable properties and are more erratic and difficult to use than soils from borrow areas. It may also be inconvenient to arrange the construction schedule so that the materials can be placed directly into the dam without stockpiling. Nevertheless, for many major reservoirs, the volume of the required excavation has exceeded the volume of the dam. Since this material has to be excavated and hauled somewhere, economy dictates that as much as possible be used in the embankment. Of the more than 200,000,000 yd.^{3} of material which have been placed in the massive dams of the Army Engineers in the Missouri River Basin, about 70% has come from required excavations (Ref. 74).

Because the spillway often contains the bulk of required excavation, many large dams have been constructed almost completely of material from the spillway. In certain cases the spillway design is based largely on the fact that the excavated material can be used in the dam. As a result, a much larger cut is made than would be economical under other circumstances.

Required excavation may consist of soil or rock in any combination. The soil should be explored in the same way as the soils in prospective borrow areas. Because of the confined working area in required excavations, and because of the necessity of avoiding interference with the construction of the appurtenant structures, difficulties frequently arise in controlling the water content and the uniformity of the soil. For these reasons, the soil is frequently placed in random zones, which are located in the embankment in such a way that the soil properties are not too important to the safety or functioning of the dam.

The rock from required excavations has been used in many cases for the construction of free-draining rockfill zones. In other cases, it has been too soft or badly weathered to serve as pervious rockfill. Soft rock which may slake when wetted has been especially troublesome to use successfully and has often been placed in random zones. At Cachuma Dam in California (Ref. 47), for example, Zone 3 was a very light and suspect siltstone, Fig. 1.1:4. This zone was completely surrounded with pervious material and protected from any appreciable seepage from either the reservoir or rainfall. In addition to being placed in random zones, soil or rock of dubious properties can frequently be used to construct berms at the upstream and downstream toes of dams at sites where the foundations are soft and the extra stability is needed (Sec. 8.4).

It is often difficult in the design stage to estimate the quantities of required excavation which will be available for the embankment. Besides the fact that the quality is often not as good as expected and so necessitates more wastage, the relative construction schedules of the appurtenant structure and the embankment sometimes do not correspond as anticipated. For dams in which large amounts of required excavation are to be used, it is desirable to provide a flexible embankment cross section so that the dimensions of the zones can be varied during construction without the necessity for redesign.

1.1b Character of the Foundation

It is possible to build an earth dam on almost any kind of foundation. The foundation characteristics may or may not have a large in-

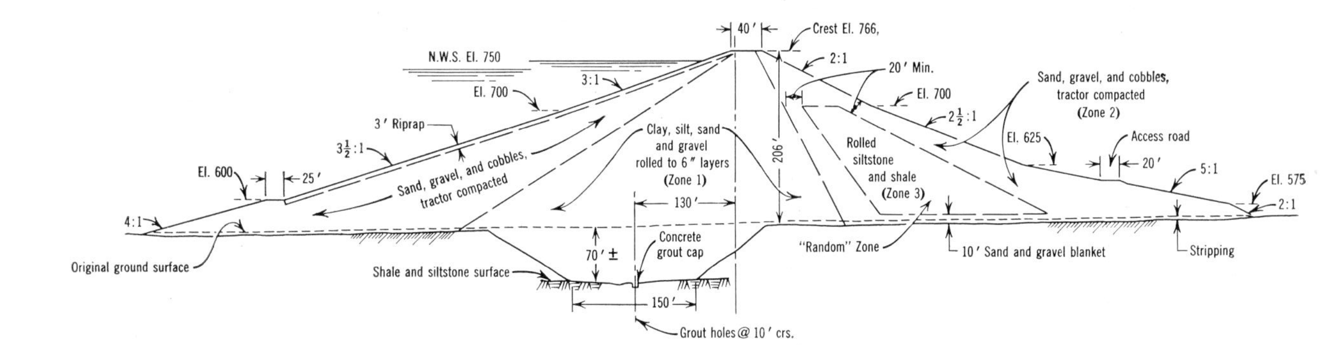

Fig. 1.1:4   Cross section of Cachuma Dam (after Walker, Ref. 147).

fluence on the embankment design. In some circumstances the problem of treating the foundation may be the most difficult and important element of the designer's work. The relationship of the foundation characteristics to the foundation treatment used and to the embankment design is covered in detail in several later sections, but the main aspects are discussed briefly below.

Although in some cases the foundation can be strengthened by one of several means (Sec. 8.4), the embankment that is on a foundation of soil with low shear strength must usually be made broader and with flatter slopes than would otherwise be required. For soft foundations, consideration must also be given to the embankment settlement from the standpoints of possible reduction in freeboard (Sec. 1.3*c*) and possible differential settlement cracks (Sec. 1.4*a*).

For pervious soil foundations, either a seepage cutoff must be provided or the underseepage must be controlled to avoid the danger of piping. The type of foundation cutoff used may have a large influence on the design of the embankment (Chap. 6). The existence of loose sands in the foundation and the possibility that a liquefaction failure could occur constitute one of the most difficult problems for the designer (Sec. 8.3). At sites where the depth or properties of the alluvium vary greatly under the upstream half of the dam, the cost and difficulty of making the foundation cutoff may be considerably less at one location than at others. In this way the nature of the material underlying the dam can determine the location of the foundation cutoff and thereby influence the location of the dam core.

In the case of rock foundations, special design details are used to seal the core of the dam to the rock (Chap. 10). If the rock must be grouted and the estimated time required for grouting is large with respect to the construction time available for the whole dam, a design must be chosen which will allow grouting to be carried out either as the embankment is being built or after it has been completed.

1.1c Climate

The design of a dam is often considerably affected by climate, mainly because of the importance of the weather during the construction period. For embankment sections of fine-grained soils, very little work can be done during wet weather or in weather where the temperature is well below freezing. But providing that it is possible to maintain the haul roads, embankment sections of pervious soil and rock can be placed as easily in wet weather as in dry. Earth dams in rainy localities should be designed with a minimum of fine-grained soil and a maximum of pervious material other things being equal.

In regions where construction must be carried on during a rainy season, it is often preferable to design the dam with a sloping upstream core instead of a central core. In the sloping core dam, the pervious material comprising the main part of the embankment can be placed during the wet weather and the earth core can be built later during the shorter dry season. In the central core dam, however, it is not practical to permit the construction surfaces of the impervious core and the pervious shells to have greatly different elevations, and it is more difficult to plan the work so that no shutdowns will be necessary during long wet periods.

In arid or semiarid regions it is much easier than in wet climates to control the construction water content of fine-grained soils. In very arid regions, however, except during a short rainy season, there may not be enough water available to raise the water content of the embankment material to a satisfactory value for compaction. An extra year of construction may be required to build a small reservoir for storing flood runoff for construction water. In rare cases involving off-channel reservoirs in very arid climates, the difficulty and expense of obtaining water for compaction may be so great as to dominate the whole economic picture.

Once completed, ordinary earth dams are not readily damaged by the weather, and it is usually not necessary to provide special design details to protect them. Occasionally, in very dry climates, it is difficult or impossible to grow a cover of protective grass on the downstream slope, and a layer of gravel or rock must be laid down. In very wet areas, masonry gutters may be required on the downstream slope to prevent erosion from surface runoff (Sec. 8.6). There are no records of damage to earth dams as a result of freezing temperatures, and no special design precautions are required.

1.1d Shape and Size of the Valley

The shape and size of the valley at the dam site may have an important influence on the design of the dam in a number of ways. In broad valleys with gently sloping abutments, the embankment section is not usually much influenced by either design or construction problems. In narrow valleys or in valleys with steep abutments, special design provisions may be necessary.

In very narrow valleys, the area of the construction surface is small, and large quantities of equipment cannot be used efficiently. It may be desirable to provide a simple design with as few internal zones and special construction provisions as possible. In narrow valleys with steep rock walls, the margin of safety of the embankment slopes against shear failure is increased by the confining influence of the valley walls. From the standpoint of shear strength, somewhat steeper slopes can be used (Sec. 7.4*e*), especially in cases where the valley curves appreciably (in plan) over the length of the dam.

Within a thin core dam, the core can be placed in the center of the embankment or sloped upstream. If it is placed in the center, the upstream and downstream slopes of the embankment are constructed with roughly the same steepness, but if it is sloped upstream, the upstream slope is made considerably flatter than the downstream. Thus where the width of the valley changes greatly upstream from the dam axis, the position of the earth core may have a large influence on the total embankment volume and cost.[1]

Materials are usually transported to the embankment construction surface on haul roads which are cut into the valley walls or which run up the slope of the embankment itself. For a high dam in a narrow valley with steep abutments, it becomes difficult and expensive to provide haul roads on the abutments at every elevation, and there is not enough room on the embankment slope to construct roads of reasonable grades. In such circumstances a rockfill embankment, which can be dumped in high lifts from relatively few haul roads, may have an economic advantage.

For dams in very narrow valleys with steep walls, it may be necessary to use piping-resistant cores, special filters, or other design details which will minimize the danger of embankment cracking due to differential settlement.

1.1e River Diversion

The necessity of handling the river during construction has a relatively more important influence on the design considerations for an earth dam than for a concrete dam. The difficulty of the problem at a given site depends largely on the width of the valley and the volume and variation of flow in the river.

In order to give the contractor as much flexibility and room for

[1] At the site of the 430-ft. earth-rockfill Ambuklao Dam constructed in the Philippines in 1956, the valley widened appreciably upstream from the dam axis. Largely owing to this fact, the dam was provided with a central core which would allow the construction of a steeper upstream slope and consequent smaller total embankment volume (Ref. 94).

ingenuity as possible, many engineers prefer not to specify a definite method of handling the water.[1] Leaving it to the contractor results in a lower cost for river diversion but increases the problems for the designer because he must anticipate all the possible methods the contractor might use and then try to make the design adaptable to each. On major rivers where the diversion problem is a large part of the total construction difficulty and cost, it may be advisable to design in detail and specify the method of diversion to be used.

In narrow, steep-walled canyons, the river must be diverted through a tunnel or conduit before foundation treatment can be completed over the whole river bottom. In wider valleys, portions of the embankment on the abutments can be constructed before the river is diverted. In this case the part constructed last, which is called the "closure" section, requires special design consideration from two standpoints—the provision of enough borrow materials and the prevention of piping through differential settlement cracks.

Suitable borrow materials must be set aside in advance, because it is often necessary to construct the closure section rapidly to avoid overtopping. It is not possible to stop work to look for additional construction materials or to process the water content of the material. Special sections of borrow pits containing ample quantities should be ready for immediate use.

Because the closure section is narrow and the material is placed rapidly, it is especially susceptible to the development of differential settlement cracks (Sec. 2.3). This is particularly true in cases where the foundation is compressible, although cracking has also occurred in dams which had negligible foundation settlements.[2] To prevent trouble, it may be desirable to provide extra filter drains in the closure section in order to control any leaks developing through cracks.[3] The closure section may also be compacted at a higher average water content so

[1] For example, it is the practice of the U.S. Bureau of Reclamation not to specify a method for diversion. The Bureau engineers do compute the height of cofferdam which would be required for a reasonable design flood, but the contractor is asked to take full responsibility. Another common procedure is to specify a minimum height of cofferdam while leaving the contractor free in all other decisions. In any event, the engineer must insist on power to approve the diversion plan finally used and must assure himself as to its adequacy and safety.

[2] See, for example, the case described by Casagrande in which substantial cracking developed in the closure section of a recent, well-constructed 80-ft. high, homogeneous dam of clay founded on shale (Sec. 2.3*b*, Ref. 15).

[3] At Garrison Dam on the Missouri River, where differential foundation settlement of 6 ft. in 500 was expected, the closure section was designed with special downstream filters in order to control leakage through possible differential settlement cracks (Ref. 183).

that the embankment will be better able to adjust to the differential settlement without cracking.

If a closure section is to be built where the river flow is large, there may be no way to cross the valley with earth-moving equipment until the closure is made. In such a case it is necessary to provide a design in which the embankment materials are obtained from the side of the river on which they are to be used. Thus it may be advisable to provide different embankment design sections employing different construction materials on opposite banks of the river.

Where diversion cofferdams of appreciable volume are required, economy demands that they be incorporated into the dam embankment if possible. The size and location of the upstream cofferdam frequently influence the position of the foundation cutoff trench, if one is used. In recent years cofferdams with heights of 150 to 180 ft. have been built at several major dams.

The pool formed behind the cofferdam as the river is being diverted may flood the most suitable available borrow areas. The designer then has several choices: (*a*) to use other, less suitable or more costly, construction materials; (*b*) to excavate the borrow materials before diversion and stockpile them above the high water; (*c*) to change the diversion scheme so that the water level is below the level of the borrow area; and (*d*) to provide dikes around the borrow area. Since large amounts of money may be involved, the choice requires careful study and judgment.

On occasion the cost of river diversion has been so great that the designer has elected to pass the flood waters over the partially completed earth embankment. This has been done frequently when the embankment has been very low or when only the foundation cutoff trench has been completed and backfilled, but less frequently when the floods have had to pass over the top of a high embankment during its construction. With proper design and planning, however, there is no reason why this method cannot be successfully used in many cases.[1]

Although little is known about the tolerable seepage through large rockfill (see, for example, Ref. 642), some flood water can be passed by simply letting the water percolate through partially completed rockfill embankment sections. At some rockfill dams as much as 2,000

[1] Any engineer contemplating such a procedure should study the excellent paper by A. Weiss which describes experiences with passing floods over several earth dams in Mexico (Ref. 162). Reference 642 contains a description of an interesting study of the possibility of passing floods over and through partially completed rockfill dams in Australia. Also see Ref. 331 for an account of passing large floods over the partially completed Brownlee Dam in Idaho in 1957.

ft.^{3}/sec. have been passed in this manner without damage. At the Sirinumu Dam in New Guinea, where this procedure was contemplated as a design measure, a steel mesh was used on the downstream slope to hold the surface rocks in place (Ref. 651).

1.1f Probable Wave Action

An earth dam must be protected from damage caused by the erosive action of waves beating on the upstream slope (Sec. 2.6). The severity of the wave action and the amount of protection needed for the embankment are related to the length of the reservoir surface over which the wind can blow and the wind velocities in the locality. In small reservoirs with only a few acres of surface area, waves only a few inches high will develop and no protection is required. In reservoirs with dozens of square miles of surface area in windy regions, waves with heights up to 8 ft. or more may rise and be driven against the upstream slope of the dam. The main damage is caused simply by the erosive action of the water driving repeatedly against the embankment, but occasionally sheets of ice and floating logs carried by the waves are hammered like battering rams against the dam. One has only to observe this activity to realize that the forces of impact must be enormous. In the springtime large sheets of ice break loose from the banks of some large reservoirs, and pieces 12 in. thick and as much as 100 acres in area may be smashed against the upstream face of the dam by a strong wind.

In many cases the cost of wave protection on the upstream slope may be a large fraction of the total cost of the embankment, and it may, consequently, have a great influence on the design. A layer of dumped rock riprap is the best, and usually the most economical, wave protection. When there is an adequate amount of suitable rock available for the purpose, the designer has no problem except to choose the thickness and the gradation of the riprap (Sec. 8.5). If rock for a riprap blanket is not available within easy hauling distance, there are several alternative solutions which may prove to be less costly.

One unusual method is to flatten the upstream slope to the point where the wave energy is dissipated on the slope without eroding the embankment material. This solution has not been used frequently and probably is economical only in the circumstance where a very flat upstream slope is needed for some other purpose. For example, at the Fort Randall and Gavin's Point Dams on the Missouri River (Fig. 1.2:20), the upstream slopes were flattened to 15:1, not only to avoid the necessity of providing wave protection but also, and primarily, because great quantities of soft rock (Niobrara chalk) were available from the required excavations (Ref. 561).

1.1g Time Available for Construction

For the design and construction of the most satisfactory and economical dam, an adequate amount of time is required. On the other hand, once a decision has been made to construct a reservoir, each year spent in planning, design, and construction is a year in which the reservoir is not producing income. There are cases where the annual income from the reservoir will be large compared with the cost of the dam or the possible savings to be obtained by additional studies and a longer construction period. Under these conditions simple economy may demand that the dam be completed as rapidly as possible.

When the construction time is limited, it is frequently necessary to use a somewhat different design from that which would be the most economical if a longer period were available. In a high dam which is constructed rapidly, for example, the construction pore pressures may be higher than those which would develop if the construction period were longer. Such a dam may require flatter side slopes or different interior drainage from a slowly constructed dam (Sec. 1.4*a*).

As discussed in Sec. 1.1*a*, the most economical dam is frequently built of material obtained from the excavation for the spillway and other appurtenant structures. If construction time is limited, it may not be possible to coordinate the construction of the various individual units so that the required excavation is ready at the time when it can be placed in the embankment. The engineer may have to design an embankment which either does not use the required excavation or uses only a fraction of it.

Although the design is somewhat limited by the tolerable water loss from seepage, the designer has many choices of methods to treat the foundation. These methods differ widely in the amount of time they require. At one extreme the designer can do everything possible to shut off the flow of seepage under the dam by using a complete foundation cutoff through the soil deposits and an extensive grout curtain in the rock. This decision adds considerably to the time required for the entire project, since most of the foundation treatment must be completed before any great amount of embankment material can be placed. At the other extreme the designer can provide little or nothing in the way of an underseepage barrier so long as the estimated seepage loss is not excessive. Since the quantity of underseepage usually cannot be predicted precisely, such a design carries a calculated risk that the quantity will be larger than anticipated.

Another saving in construction time is possible if the grouting can be carried out simultaneously with the embankment construction. The designer can provide for this in several ways: (*a*) by using a dam with an upstream sloping core of earth or an upstream membrane of manufactured material (Chap. 9); (*b*) by using an interior concrete gallery or tunnel from which the grouting can be carried out while the embankment is being constructed (Sec. 10.1*c*); or (*c*) by grouting through the embankment after it is completed (Sec. 10.2*e*).

As discussed in Sec. 1.1*c*, the time available for construction may be closely related to the weather. In wet regions the design features which can help reduce construction shutdowns due to rain include upstream sloping cores and embankments using minimum amounts of fine-grained impervious soils. Dams consisting of sand, gravel, or rock with impervious membranes of reinforced concrete, plate steel, or asphaltic concrete can be rapidly constructed with a minimum of delays during wet weather.

Often when a dam is built of fine-grained soils, the rate of construction may be severely restricted by the necessity for careful construction control. If the control is required primarily to assure the impermeability of the core, considerable time can be saved by providing a design with an impervious membrane of manufactured material and then relaxing the control.[1]

1.1h Function of the Reservoir

Depending on the use planned for the reservoir, different designs may be justified at a given site. The reservoir function determines in part the tolerable loss of water due to leakage through and under the dam. It also determines the reservoir hydrograph which influences the dam stability.

[1] One example is the dam for which the only materials available are gravel and fine, slightly silty sand. Both materials are located in the same deposit of stratified alluvium and can be excavated selectively with scrapers. A satisfactory and economical dam can be constructed with the fine, silty sand as a core, but the work cannot be carried out very rapidly because of the care needed in selecting the borrow materials and constructing the core. If a manufactured membrane were used as the water barrier, the dam could be constructed in a fraction of the time with the fine sands and gravels placed indiscriminately.

The tolerable seepage loss for some domestic water supply projects may be very low because of the scarcity or the cost of water. In these cases it may be desirable to go to great lengths to keep the loss as low as possible by using a highly impervious material for the core of the dam and a complete foundation cutoff (Chap. 6). For flood control dams, on the other hand, there is no economic limitation on the maximum quantity of leakage. Highly impervious embankment cores or foundation seepage barriers are not required; in fact, dams retaining large flood control reservoirs have even been constructed of relatively pervious sand. Some reservoirs have been built primarily for the purpose of storing flood water to replenish downstream supplies of ground water and consequently, were designed without foundation seepage barriers.

At some reservoirs the water level is never lowered and the upstream slope never subjected to "rapid drawdown" conditions. At the other extreme, the reservoirs for some hydroelectric power installations are kept full for long periods of time and then lowered in a span of a few hours to obtain "peaking" power. For dams constructed to retain reservoirs of such specialized types, especially conservative slopes and zonings for the upstream embankment portions are often justified.

Flood control reservoirs may be designed to remain empty for years, to fill only for a short time during a storm, and then to empty quickly as the water is discharged through large outlet works. For a dam with such a function it would theoretically not be necessary to provide as extensive an internal drainage system as would be required if the reservoir remained full for long periods of time. Since the future use of a reservoir cannot always be predicted reliably, however, some engineers design flood control dams on the assumption that the dams may have to serve other functions in future years. Dams retaining reservoirs which remain empty for long periods may require additional design attention to the problems of preventing damage from animal burrows (Sec. 2.7) and drying cracks (Sec. 2.11).

The design is sometimes influenced to a large degree by the rate at which the reservoir level can be drawn down and the frequency with which the reservoir will be emptied under normal operating conditions. At some sites—because of the large size of the reservoir basin or the absence of bottom outlets—it is practically impossible to lower the water level. At such sites, a much more conservative attitude to all design features is justified than for sites at which, because of a relatively small reservoir volume or a big outlet conduit, it is a simple matter to empty the reservoir. In the latter case, the engineer has more freedom for the use of new design details and materials which may result in large economies, but which may entail some calculated risk that repairs will have to be made if the dam does not perform in all aspects as anticipated.

Sec. 1.1i Earthquake Activity

Designing a dam to resist earthquake damage is probably the most difficult task which ever faces the earth dam engineer. The analytical methods available for approaching the problem are in a rudimentary state of development and have almost no practical value. The designer must rely heavily on his knowledge of dams which have been subjected to earthquake shocks in the past (Sec. 2.5) and on his best judgment concerning the type and magnitude of damage which could occur at his site (Sec. 8.2).

The design measures taken to increase safety in regions of seismic activity generally consist of a more conservative approach to the main elements of the dam. The designer may decide to use better filters, downstream drains of larger capacity, thicker cores of more piping-resistant materials, or flatter side slope than otherwise (see Sec. 8.2 for a longer discussion of this subject). In some cases such conservative provisions will result in a completely different type of dam than would have been considered most economical and satisfactory in an earthquake-free locality.

Extra cost and time in construction will, of course, be required for a more conservative embankment section and foundation treatment. In addition, a particularly detailed investigation of the geology and surface conditions is essential in seismic regions. Faults or other displacements of the foundation rock should be explored thoroughly, and more attention than usual should be given to the properties of soils underlying the dam.

No general rules can be stated as guides to the degree of conservatism necessary in a given case. For this the engineer has only his judgment and his conscience to direct him. The amount of extra money and time which should be spent on a dam in a seismic area will be governed to some degree by the size, location, and function of the reservoir. A dam retaining a small reservoir is potentially much less dangerous than a dam of the same height retaining a large reservoir. A dam located at a great distance from any inhabited area warrants less concern than a dam situated above a city. Similarly, dams for flood control reservoirs, which will remain empty except during river floods, can be treated less conservatively than dams for reservoirs which will

be full most of the time. The rate at which the re
emptied is another important factor when considerin
danger of an earthquake shock. If the water level can b
through the outlet works in a few hours or even days, the
safer than if weeks or months are needed.

1.2 TYPES OF EARTH DAMS

1.2a Homogeneous Dams

As the term implies, homogeneous dams are constructed entirely or almost entirely of a single embankment material. They are so named to distinguish them from zoned dams, which contain materials of different kinds in different portions of the embankment. Homogeneous dams have been built since the earliest times and are used today whenever only one type of material is economically available. Although some of the highest dams yet constructed are essentially homogeneous, homogeneous embankments are used most often in dams of low to moderate height. Very low dams are almost always made homogeneous, because their construction tends to become unduly complicated if they are zoned.

There is no strict or universally accepted definition of the term "homogeneous." It is applied to any dam constructed primarily of one material, even though the embankment may have an internal drainage system. Sometimes large sections of a dam classified as homogeneous are completely separated from each other by thin bands of more pervious material provided as internal drains. Such drains actually give the embankment many of the benefits of a zoned dam.

Some of the benefits of a zoned dam can also be obtained in a homogeneous structure when the soil is placed selectively or when different construction methods are employed in different portions of the embankment. In the first case, the finer materials in the borrow pit are selectively excavated and placed in a relatively impervious core zone. In the second case, zones of lower permeability are created in otherwise homogeneous embankments by using either more compaction or a higher construction water content.[1] Using more compaction will make a considerable difference in the permeability of some ma-

[1] See, for example, the description of the design of Três Marias Dam in Brazil, which is an essentially homogeneous clay dam with the central core compacted at a higher water content (Refs. 446, 486), Fig. 1.2:35.

terials, especially of residual soils and other materials which break down as they are being rolled.[1] Using a higher construction water content to create zones of lower permeability is a method chosen most frequently when the embankment materials are clayey soils. Additional water during compaction makes these soils considerably more impervious than they are when compacted on the dry side of optimum water content (at the same dry density), and the difference persists with time. The difference in permeability, which may be 10 to 100 times, is due to an actual difference in the structure of the compacted soil (Refs. 60, 61).

Homogeneous dams are usually composed of impervious or semi-pervious soils, but many successful embankments have been built of relatively pervious sands and sand-gravel mixtures. As part of the Petenwell Hydroelectric Project on the Wisconsin River, which was completed in 1950 (Ref. 96), a number of very long homogeneous earth dams with maximum heights of 60 ft. were made of uniform medium sand (average coefficient of permeability of compacted embankment equal to 70,000 ft./yr.), which was the only available construction material. Aside from the expected high leakage, the performance of these dams has been completely satisfactory.

DOWNSTREAM DRAINS

Any homogeneous dam with a height of more than about 20 to 25 ft. should be provided with some type of downstream drain constructed of material appreciably more pervious than the embankment soil. The purpose of such a drain is twofold: (1) to reduce the pore water pressures in the downstream portion of the dam and hence to increase the stability of the downstream slope against sliding, and (2) to control any seepage water as it exits at the downstream portion of the dam in such a way that the water does not carry away particles of the embankment soil, i.e., that "piping" does not develop (Sec. 2.2). The effectiveness of the drain in reducing pore pressures depends primarily on its location and extent. Piping, on the other hand, is controlled by making sure that the gradation of the pervious material from which the drain is constructed meets the filter requirements for the embankment material (Sec. 1.3*b*).

[1] Laboratory studies for the 200-ft. North Catamount Dam built in Colorado in 1956 indicated that the primary embankment material available (a decomposed granite in the form of a silty sand) could be made 100 times more impervious by an increase in the compactive effort associated with raising the density from 95 to 98% of Standard Maximum Proctor Dry Density.

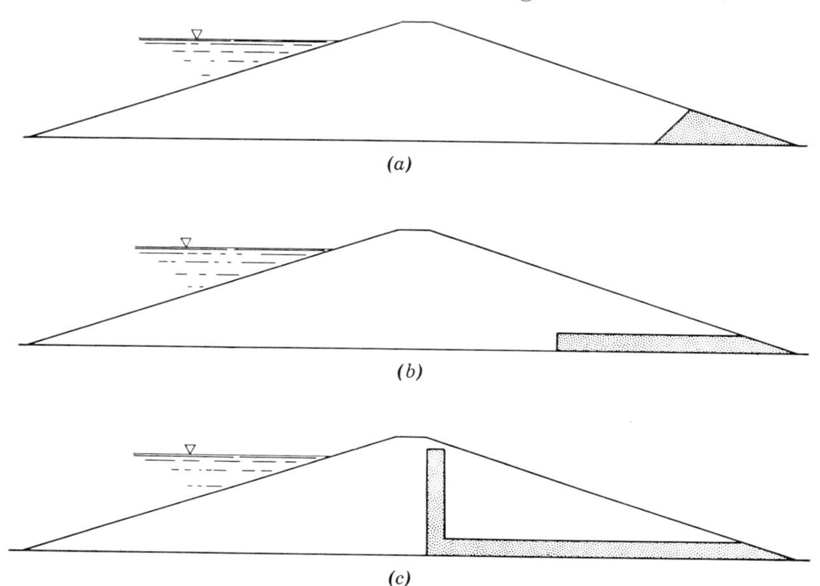

Fig. 1.2:1 Drains used in homogeneous dams. (a) Toe drains. (b) Horizontal blanket drains. (c) "Chimney" drain.

The design of the downstream drainage system is governed mainly by the height of the dam, the cost and availability of pervious material, and the permeability of the foundation. For low dams, a simple drain of the type shown in Fig. 1.2:1*a* can be used successfully. Toe drains such as this were installed in some of the oldest homogeneous dams in an effort to prevent the softening which was often observed at the downstream toe. For reservoir depths much greater than about 50 ft., most engineers prefer to place the drain further inside the embankment where it will be more effective in controlling the seepage and reducing the pore pressures.

Horizontal drainage blankets of the kind shown in Fig. 1.2:1*b* are widely used for dams of moderate height.[1] The Bureau of Reclamation's 150-ft. Vega Dam (Fig. 1.2:2) is one of the highest which has

[1] Although such blankets are frequently used over the downstream one-half or one-third of the foundation area, in a few cases where the drainage blanket has also been used to accelerate the time rate of the settlement of the foundation, they have been placed over nearly the whole base of the dam, stopping short of the upstream toe only a sufficient distance to provide a reasonable hydraulic gradient. (See, for example, Ref. 660 for the description of the Hulah Dam in Oklahoma, which was founded on a thick bed of soft clay and the drainage blanket was extended to within 50 ft. of the upstream toe.)

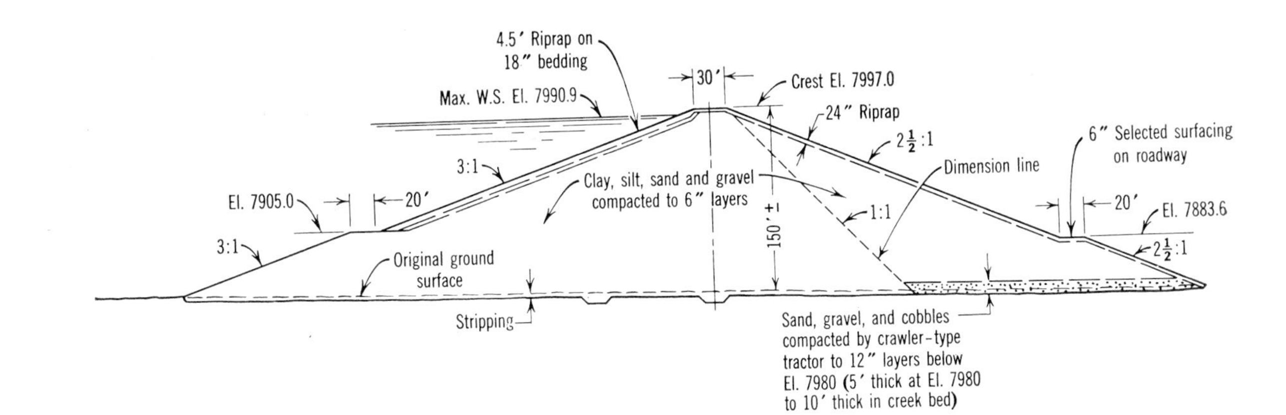

Fig. 1.2:2 Cross section of Vega Dam (after Walker, Ref. 147).

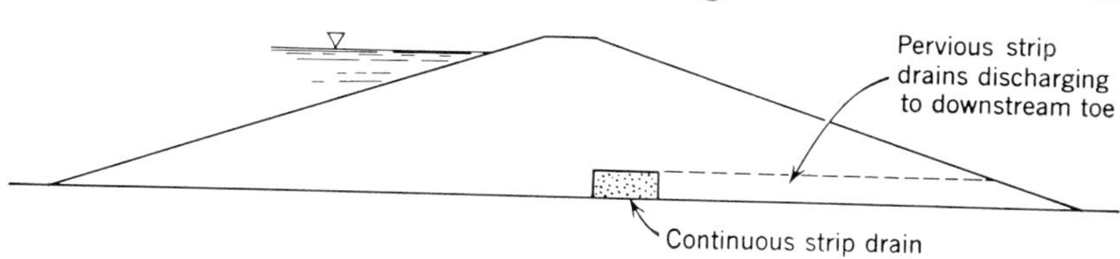

Fig. 1.2:3 Strip drains used in lieu of a continuous horizontal drainage blanket.

been constructed with a homogeneous section and a horizontal downstream drain. At sites where pervious material is very scarce, the general effect of the horizontal blanket can be obtained by placing internal strip drains (Fig. 1.2:3).

One of the principal disadvantages of horizontal drainage blankets results from the fact that earth dam embankments tend to be stratified and consequently more pervious in the horizontal direction than in the vertical. It can occasionally happen that horizontal layers much more impervious than the average material are constructed into the embankment, so that in spite of the horizontal drain the seepage water may flow horizontally on top of a relatively impervious layer and discharge on the face of the downstream slope. Surface sloughing due to this condition has caused concern at several major dams in recent years. Repairs have been made by installing pervious blankets on the downstream slopes or constructing vertical drains to connect with the horizontal blanket. The vertical drains, composed of sand and gravel, have usually been excavated in the form of a row of large, closely spaced holes by means of earth drilling rigs (see, for example, Ref. 93).

In an effort to prevent trouble with stratification and to intercept seepage water before it reaches the downstream slope, many of the higher homogeneous dams have been constructed with inclined or vertical "chimney" drains, Fig. 1.2:1*c*. Although frequently termed a recent innovation, chimney drains were used on some very old dams. Fig. 1.2:4 shows a section of the Sherburne Lake Dam, which was constructed in Montana between 1915 and 1918 with a vertical gravel drain varying in thickness from 5 to 10 ft. (Ref. 266).[1]

A typical example of the chimney drains used in recent years is the section adopted at the Mammoth Pool Dam completed in 1959 on the San Joaquin River in California (Ref. 654). This is one of the world's highest earth dams. As shown in Fig. 1.2:5, the chimney drain was

[1] See Refs. 204 and 270 for interesting early discussions of the use of drains inside earth dams.

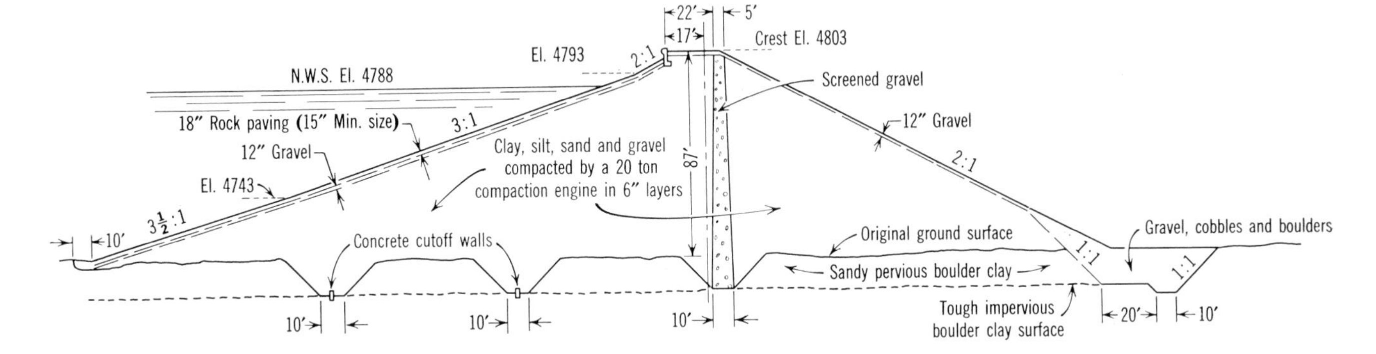

Fig. 1.2:4 Vertical chimney drain used at Sherburne Lakes Dam, Montana (1916) (after Walker, Ref. 147).

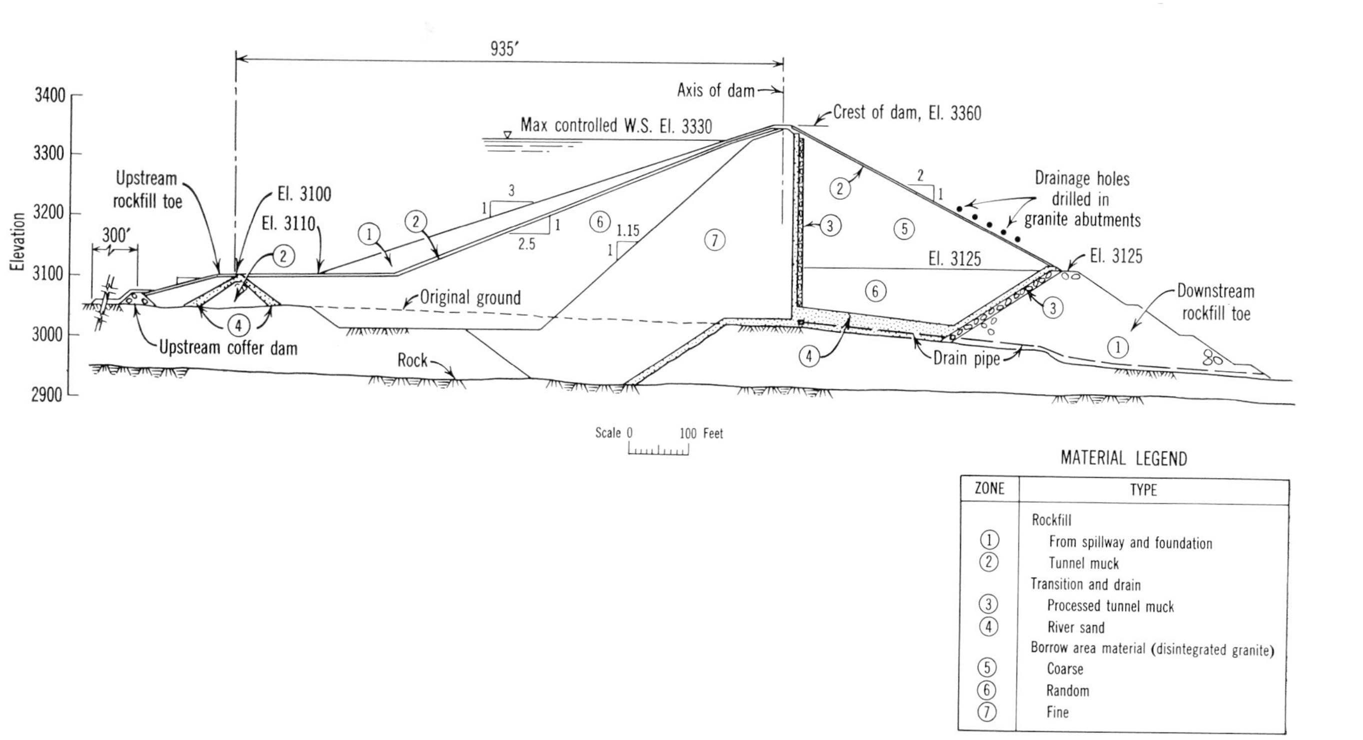

Fig. 1.2:5 Chimney drain use in Mammoth Pool Dam, California (1958) (after Terzaghi, Ref. 654).

constructed in two vertical bands. The more pervious downstream band consisted of crushed granite screened between ¼ in. and 1½ in., and the upstream band was well-graded river sand. In addition to controlling the normal seepage through the dam, this drain was intended to control concentrated leaks which might develop through settlement cracks.

In several recent major dams, the chimney drain has been inclined at a considerable slope, sometimes upstream and sometimes downstream. Cross sections of a number of chimney drains are shown in Fig. 1.2:6. With an upstream inclined drain such as that at the North Hartland Dam the upstream portion of the impervious embankment can be made to act as a relatively thin core. In addition to controlling the seepage through the embankment and increasing the stability of the downstream slope when the reservoir is full, the chimney drain is effective in reducing pore water pressures both during construction and following rapid reservoir drawdown (Sec. 1.4).

The dimensions and permeability of pervious drains must be chosen in such a way that the drainage system can carry away the anticipated flow with an ample margin of safety for unexpected leaks. If the dam and foundation are relatively impervious and the expected seepage is low, a drain can be constructed of material which is not highly pervious, although it should always have a coefficient of permeability at least 10 to 100 times greater than that of the average embankment material.

1.2b Thin Core Dams

There are several conditions which may make a thin core dam the best design for a given site. Where there is an ample supply of pervious embankment material but only a limited supply of soil suitable for an impervious core, the designer may have no other choice than a thin core dam. Where there are ample supplies of both pervious and impervious soil, a thin core dam may be more economical or more easily constructed for a combination of reasons.

a. The unit cost of placing pervious material may be less than that of placing the available impervious materials.
b. The embankment volume can be made smaller in a thin core dam than in any other.
c. The weather and the construction time available may not permit the use of a large zone of impervious material (Secs. 1.1*c*, 1.1*g*).

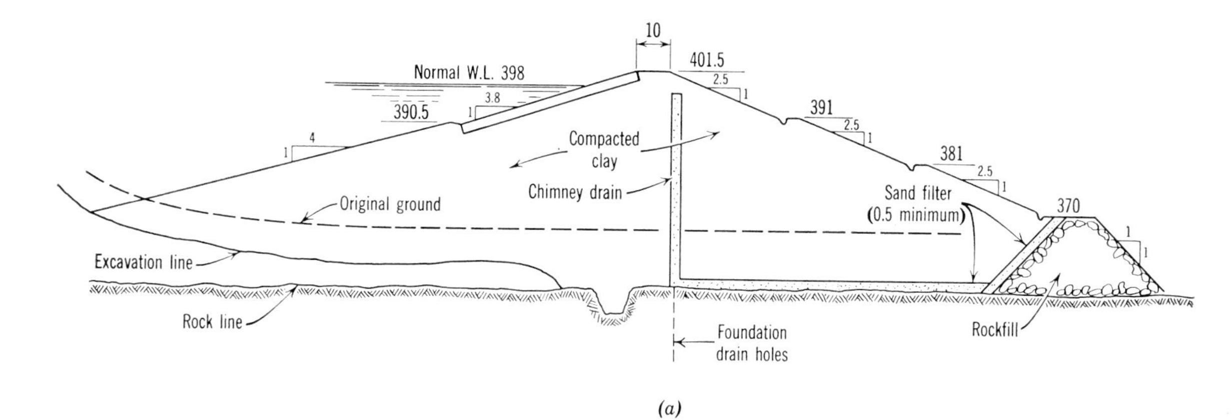

Fig. 1.2:6 Examples of recent dams with chimney drains. (a) Cross section of Vigario Dam, Brazil (after Ackerman et al., Ref. 103).

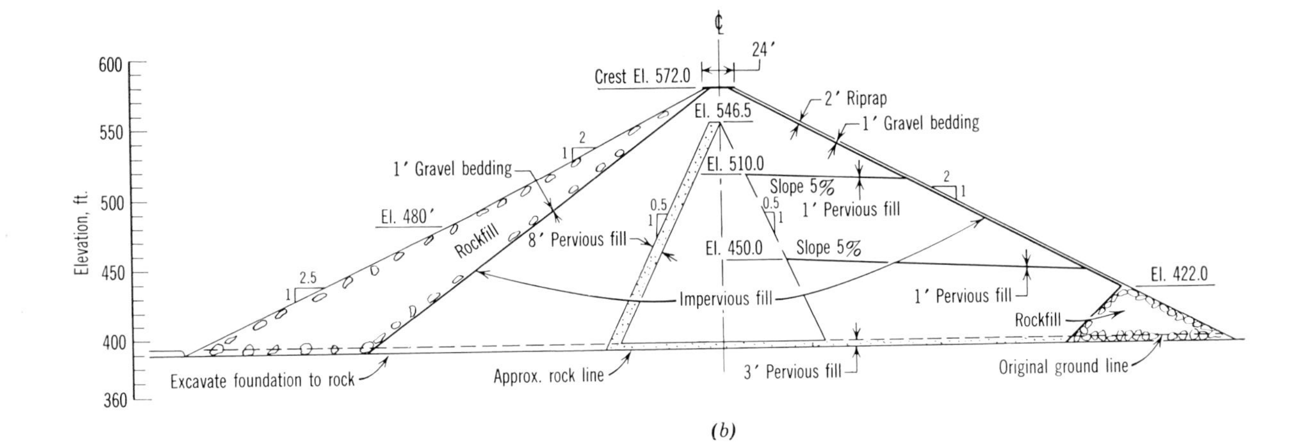

Fig. 1.2:6(b) *Cross section of North Hartland Dam, Vermont.* (Courtesy New England Division, U.S. Corps of Engineers)

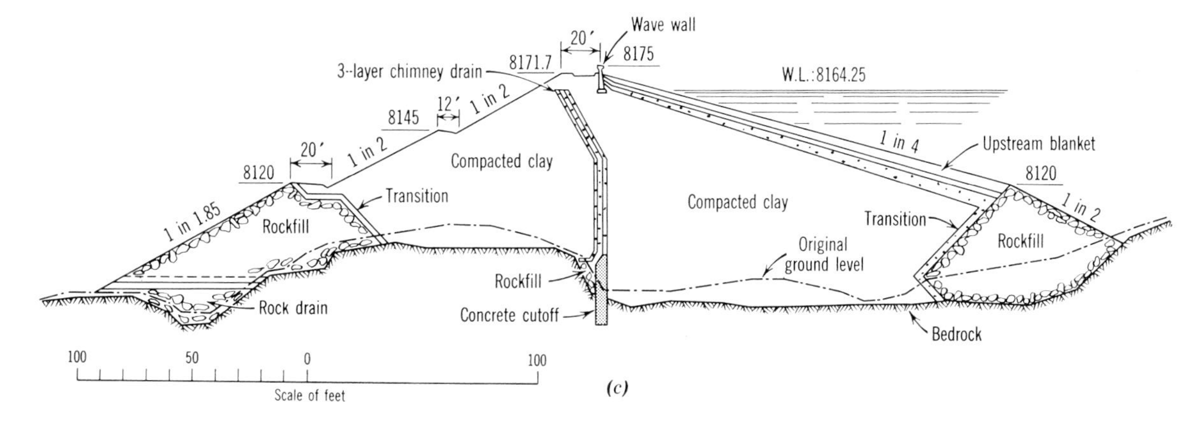

Fig. 1.2:6(c) *Cross section of Sasumua Dam, Kenya (after Terzaghi, Ref. 304).*

In this type of dam the core may be sloped upstream or placed in a vertical position near the center of the embankment, Figs. 1.2:7 and 1.2:8. The pervious zones are constructed either of sand and gravel or of rock obtained by quarrying or screening earth-rock mixtures. If rock is used, it may be placed in thin layers or dumped and sluiced in high lifts;[1] relative advantages of the two methods are discussed in Sec. 11.5.

MINIMUM THICKNESS OF CORE

At sites where impervious soil is scarce, the feasibility of constructing a dam with a thin earth core may depend to a large degree on the minimum thickness of impervious core which can be used safely. The problem of determining the minimum safe thickness is not amenable to theoretical treatment, and no definite rules can be given. It is governed for practical purposes by the following factors: (*a*) the tolerable seepage loss; (*b*) the minimum width which will permit proper construction; (*c*) the type of material available for the core and shells; (*d*) the design of the proposed filter layers; (*e*) precedent on similar projects.

Embankment sections constructed of most fine-grained soils are so impervious that even for very high dams only a few feet of core thickness is necessary to reduce the seepage to a negligible value. The core material normally has less shear strength than the rest of the embankment, and, consequently, from the standpoint of stability, the thinner the core is made the better. On the other hand, a thick core has more resistance to piping, especially to piping which may develop in differential settlement cracks. Because the piping resistance of compacted embankments is largely a result of the soil properties (Sec. 2.2), the minimum allowable core thickness should always depend to a great extent on the plasticity and gradation of the core material.

Although all the factors given above must be taken into consideration, the minimum safe core thickness must finally be selected very much on the basis of judgment. The following criteria represent a

[1] There is some debate whether a dam with a rolled earth core and large sections of rockfill embankment should be called an earth or a rockfill dam. The name is unimportant, since the design considerations are essentially the same for dams with pervious zones of soil or rock. In this book the term "earth-rock" is used to describe all dams with earth cores and large zones of rock; the name "rockfill" will be restricted to dams constructed entirely of rock.

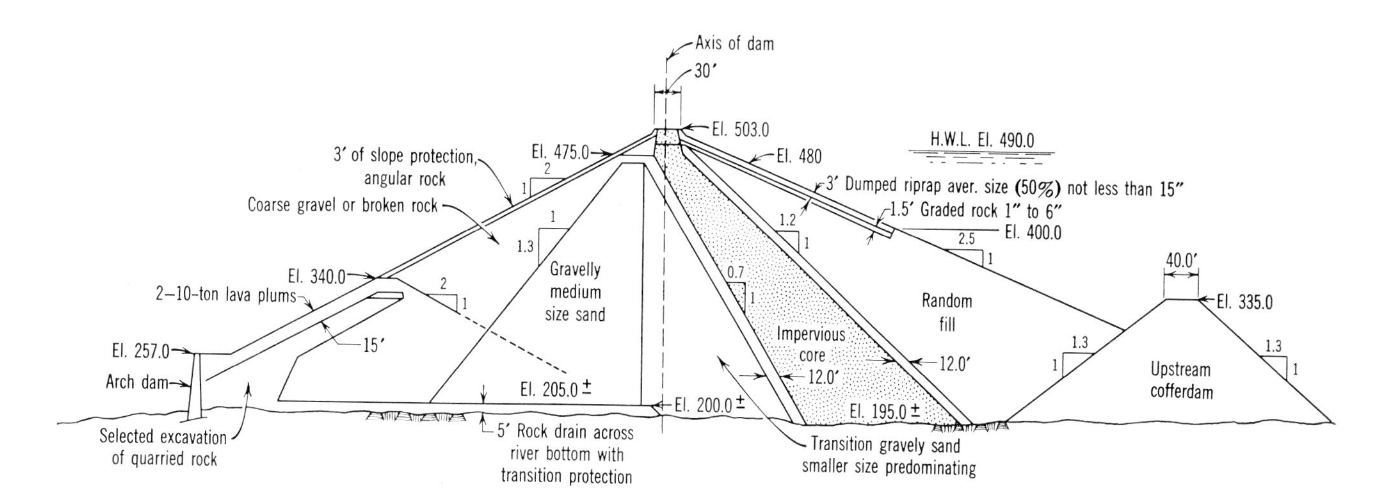

Fig. 1.2:7 Sloping core dam—Yale Dam, Washington (1952). (Courtesy Pacific Power and Light Company)

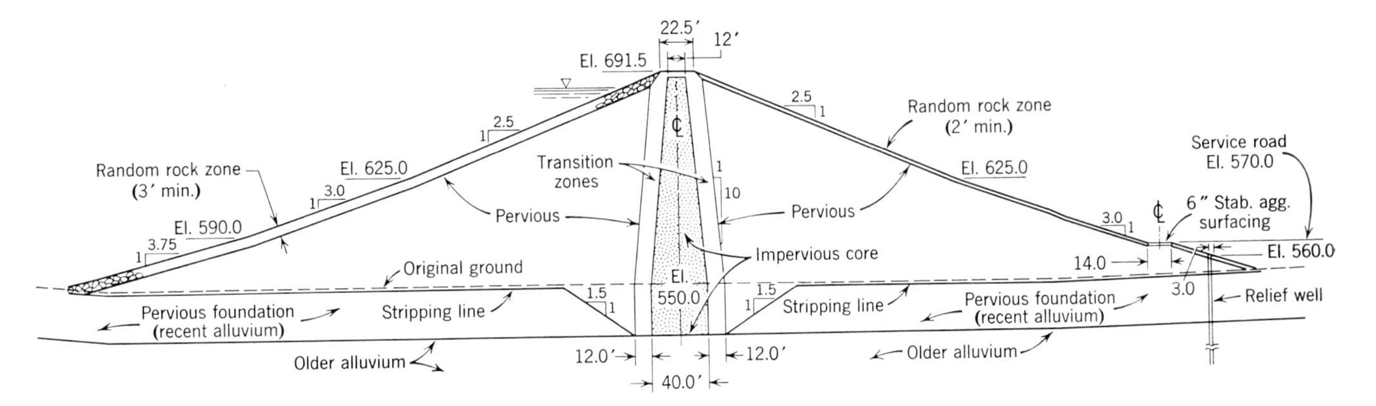

Fig. 1.2:8 *Central core dam—Success Dam, California (1960).* (Courtesy Sacramento District, U.S. Corps of Engineers)

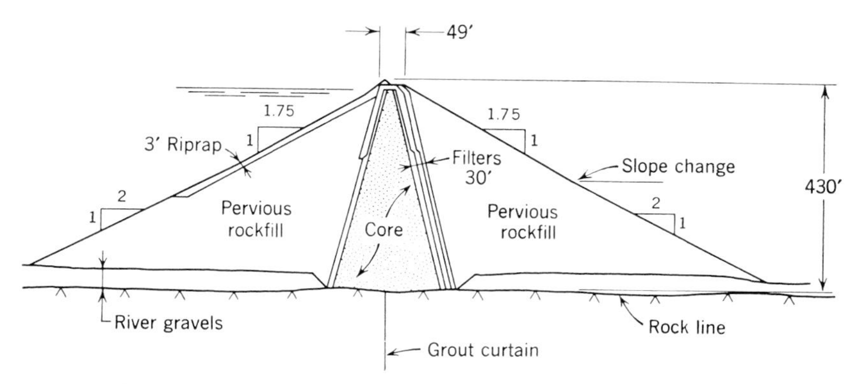

Fig. 1.2:9 Central core earth-rock dam—Ambuklao Dam, Philippine Islands (1956) (after Fucik and Edbrooke, Ref. 94).

rough cross-section of opinion among experienced earth dam engineers:[1]

1. Cores with a width of 30% to 50% of the water head have proved satisfactory on many dams under diverse conditions. Probably a core of this width is adequate for any soil type and dam height.
2. Cores with a width of 15% to 20% of the water head are considered thin but, if adequately designed and constructed filter layers are used, they are satisfactory under most circumstances.
3. Cores with widths of much less than 10% of the water head have not been used widely and probably should be considered only in circumstances where a large leak through the core would not lead to failure of the dam.

VERTICAL AND SLOPING CORES IN EARTH DAMS

Vertical and sloping cores both have points in their favor for the earth dam designer. One advantage of the vertical core is that higher pressures will exist on the contact between the core and the foundation and will provide more protection against the possibility of leakage along the contact.[2] Another advantage is that the thickness of a verti-

[1] See Ref. 373 for an interesting study of the properties and dimensions of some high thin core dams.
[2] This was a major consideration in the selection of a vertical core for the 430-ft. Ambuklao Dam in the Philippine Islands in 1956, Fig. 1.2:9.

cal core is slightly greater for a given quantity of impervious soil than the thickness of a sloping core.

The principal merit of the upstream sloping core is that the main downstream portion of the embankment can be constructed first and the core placed later. As discussed in Sec. 1.1*c*, this is a great advantage in areas which have only a short season of dry weather suitable for building a core from fine-grained soils. In addition, for a sloping core dam, the foundation grouting can be carried out while the embankment is being placed.

Another advantage of the sloping core is that the filter layers between the core and the upstream and downstream pervious zones can be made thinner with less difficulty in construction than the filter layers for vertical core dams (Sec. 1.3*b*).

One disadvantage of the sloping core which can be important at some sites results from the fact that the location of the area of contact between the core and the foundation depends on the depth of foundation excavation; i.e., when the excavation is carried deeper, the contact area moves upstream. In some cases the depth of excavation required to get down to a suitable contact between earth core and foundation cannot be determined reliably in advance and must be chosen during construction. Because of this fact and because it is frequently desirable to do the foundation grouting before the final excavation is made, it may be difficult to locate the grout curtain in the position desired relative to the core contact area.

For this and the other reasons discussed in the following, the sloping core dam may or may not be advantageous from the standpoints of speed and economy of foundation grouting. The advantage comes from the fact that grouting can be carried out while the main downstream pervious embankment section is being constructed. However, if it is anticipated that additional grouting may be required through the embankment after the dam is completed, the central core design is definitely preferable because the work can be done from the crest of the dam without lowering the reservoir. In addition, if the upstream shell of a sloping core dam is built of rock, it may be practically impossible to drill grout holes through it after construction.

There is little difference between the total embankment volumes of dams with vertical and sloping cores. The vertical core dam has a steeper upstream slope but a flatter downstream slope than the sloping-core dam, so that the total volumes are about the same. The theoretical minimum volume for a thin-core dam is obtained when the core is located in such a way that the upstream and downstream slopes have approximately the same safety factor against shear failure. When the core is sloped upstream, this condition is achieved with an average inclination of about 0.5:1 (Ref. 2).

In the debate on the relative desirability of using a sloping or a vertical core in a thin-core dam, it is sometimes speculated that the sloping-core dam may be less susceptible to cracking owing to differential settlement. The authors have no evidence to indicate that this is true nor do they have strong opinions on the subject. Furthermore, it has been suggested that the sloping-core dam is more resistant to cracking and damage during an earthquake shock. Here, again, the authors have no strong opinions except for rockfilled dams, for which they definitely believe the sloping-core dam is basically safer—not because the core is likely to crack less during an earthquake shock but because there is a larger body of stable rockfill downstream of the core (see Sec. 8.2*e*).

EARTH-ROCK DAMS WITH SLOPING CORES

A new type of dam was introduced in 1942 with the construction of the 250-ft.-high Nantahala Dam in North Carolina by the Aluminum Company of America (Ref. 334). As seen in Fig. 1.2:10, the main portion of the embankment consists of a large downstream zone of quarried rock, which was dumped in high lifts and sluiced, using the current construction practice for rockfill dams with upstream impervious decks of reinforced concrete. In lieu of the concrete slab, the water barrier at Nantahala Dam is a thin, sloping, rolled-earth core protected on both sides with filters of varying thickness and gradation.

Although the design of the Nantahala Dam was a considerable departure from the immediate previous practice, a number of dams similar in principle had been constructed in the western United States more than 40 years before. These consisted of downstream zones of dumped rock and upstream zones of impervious soil. The first were the Avalon and Lake McMillan Dams on the Pecos River in New Mexico, which were constructed in 1893 to a height of about 50 ft. (Refs. 207, 386). In the following years at least six other earth-rock dams with almost identical sections were built.

Since no filters were placed between the earth and rockfill sections, however, nearly all of these early dams suffered partial or complete piping failures. The most recent failure was that of the 62-ft. Schofield Dam constructed in 1926 in Utah, Fig. 1.2:11 (Ref. 385). When the reservoir was filled for the first time in 1927, cracks with a maximum width of 1½ in. opened on the upstream slope. In the spring of 1928,

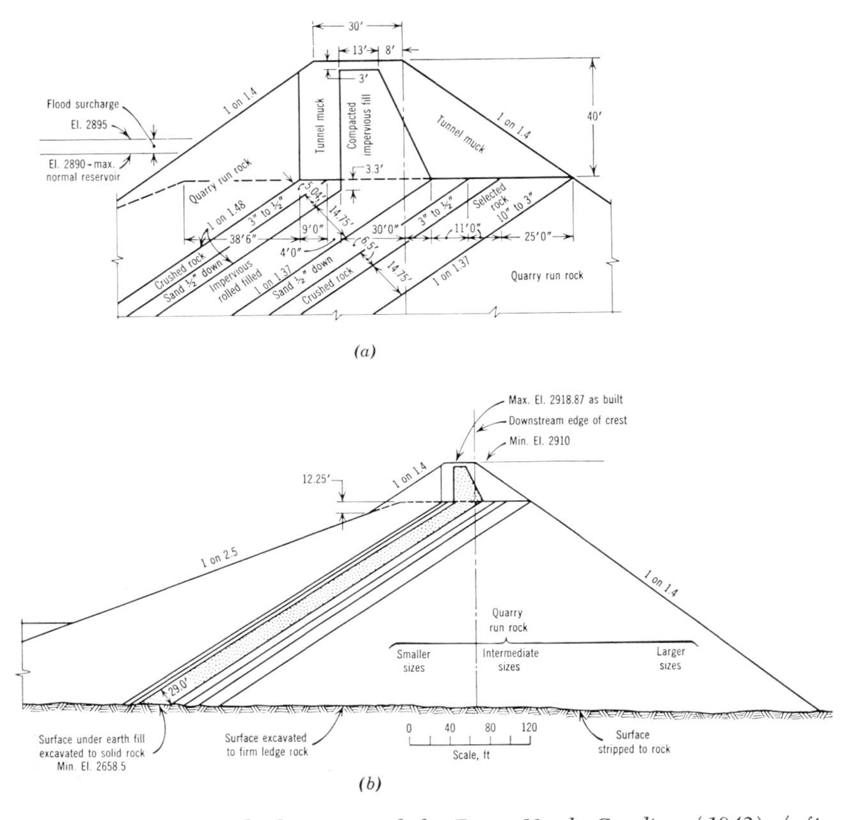

Fig. 1.2:10 Earth-rock dam—Nantahala Dam, North Carolina (1942) (after Growdon, Ref. 334). (a) Crest details. (b) Typical cross section.

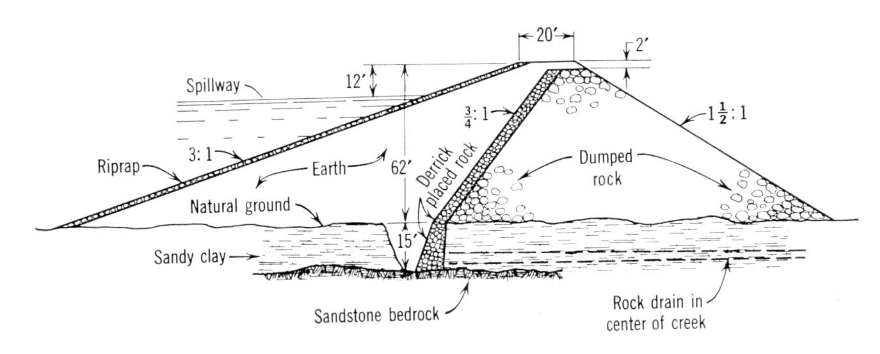

Fig. 1.2:11 Earth-rock dam—Schofield Dam, Utah (1926).

when the reservoir reached a new high level, large quantities of the upstream earth section were washed into the voids of the downstream rock zone. Complete failure was averted only by an emergency effort around the clock for several days, during which large quantities of earth and sand bags were dumped into the breach, Fig. 1.2:12.

For many years the failure of Schofield Dam discouraged designs combining large downstream zones of dumped rock and rolled-earth cores. During the planning of the Nantahala Dam, extensive laboratory investigations of the graded filter zones were carried out in order

Fig. 1.2:12 View of the failure of Schofield Dam, Utah (May 22, 1928). Soundings indicated that the cavity shown in the upstream embankment reached nearly to the foundation. (See cross section, p. 38.)

to provide a design that would prevent the migration of the fine-grained earth core material into the downstream rock zone. Compacted specimens of the core material were subjected to hydraulic gradients of more than 100 for several months. During this extreme test, no visible migration of the core material into the filter occurred and, as the result of many such tests (Sec. 1.3*b*), it was established conclusively that properly designed filters provide complete protection against piping failure of the type which occurred at Schofield Dam.

The performance of Nantahala Dam has been completely successful. Even though the crest settlement was higher than that of most rolled-earth dams (approximately 1.1% of the height in the first 15 yr.), the leakage has been negligible and no maintenance of any kind has been required. Because of the good experience with this dam, and because of its relative economy, six others with similar cross sections have subsequently been constructed by the Aluminum Company of America. All have given excellent performance (Ref. 336). Experience gained in their construction resulted in some slight design modifications, most of which are incorporated in the 215-ft. Bear Creek Dam completed in 1953, Fig. 1.2:13.

This type of dam is sometimes called the "Alcoa" dam or the "Growdon" dam after Mr. J. P. Growdon, the consulting engineer who was largely responsible for the design. Following the success achieved with these dams, a number of similar structures were built in the United States and abroad. Most of these are on rock foundations, but the Brownlee Dam, which was completed on the Snake River in Oregon in 1958, was founded on a 110-ft.-thick alluvial deposit of dense sand, gravel, and boulders, Fig. 1.2:14 (Ref. 331). For nearly all dams of this type, most of the rock for the embankment has been obtained from the spillway excavation. The downstream slopes have been made equal to the angle of repose of the dumped rock except at Kenney Dam in British Columbia, where the downstream slope was flattened with berms to an average slope of 1.75:1 in order to provide additional stability against earthquake shock, Fig. 1.2:15 (Ref. 328).

Figure 1.2:16 shows a cross section of the 270-ft. Hirfanli Dam (constructed in 1957–1960 in Turkey), which is an interesting variation of the typical sloping-core, earth-rock dam (Ref. 398). At the Hirfanli Dam a large quantity of the rock from the spillway excavation was too badly weathered to make a satisfactory dumped rock fill. This rock was used in a large zone just downstream from the core and was placed and compacted with a 50-ton rubber-tired roller in 12-in. layers.

The Dalles Closure Dam, a 2,000-ft.-long earth-rock dam completed in the Columbia River in 1957, is one of the most interesting ever constructed (Ref. 221). It was built without dewatering the foundation at a site where the water was 180 ft. deep and was flowing at a rate of 200,000 c.f.s. As seen in Fig. 1.2:17, the design consisted of a massive downstream rock section with upstream sloping zones of progressively finer rock spalls and gravels. The most impervious zone ("blanket material") consisted of a well-graded, sandy gravel with an estimated coefficient of permeability of about 8,000 ft./yr. The lower two-thirds of the high embankment were constructed by dumping under water. The project has been completely successful. Since there is always a tailwater on the dam, it is not possible to measure the leakage, but it

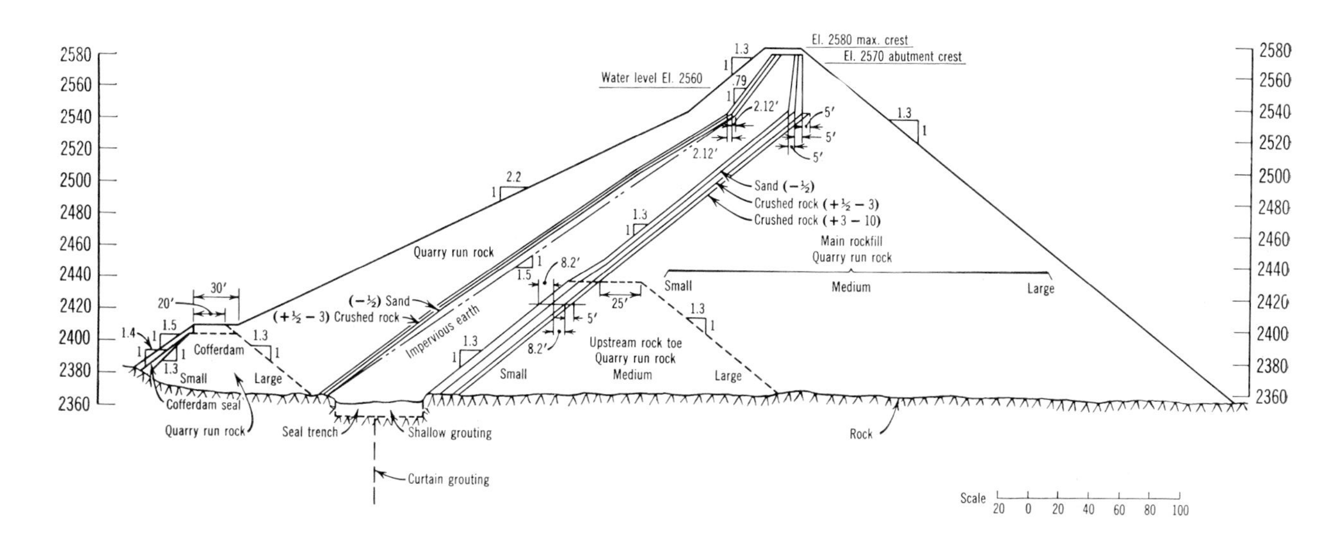

Fig. 1.2:13 Earth-rock dam—Bear Creek Dam (1953) (after Growdon, Ref. 335).

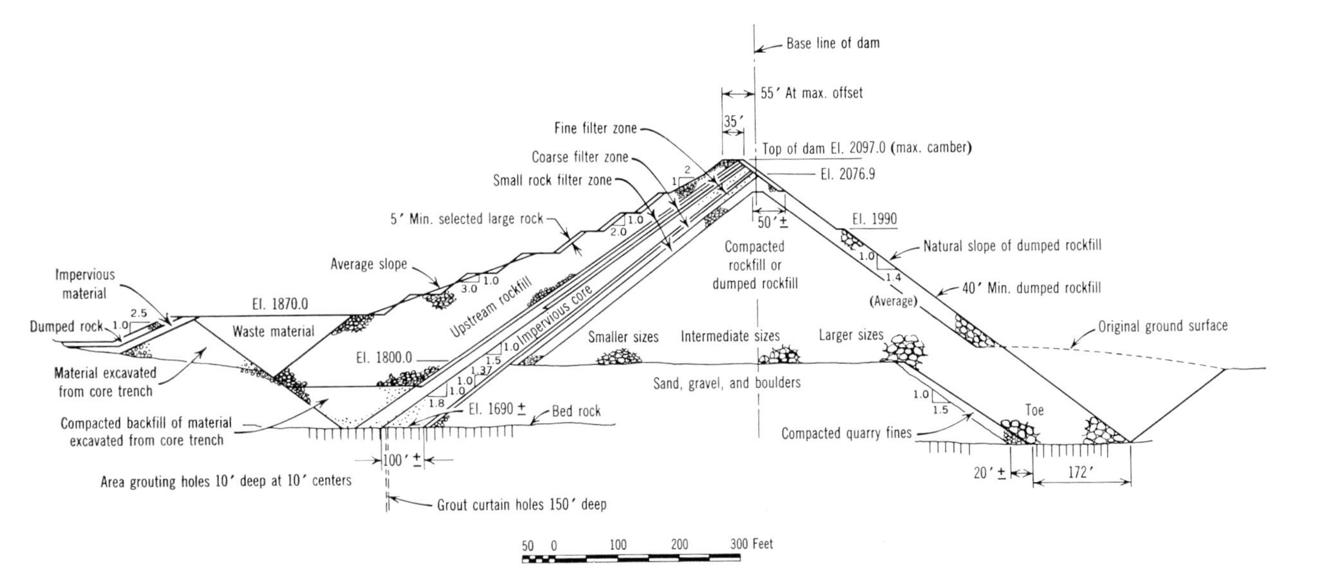

Fig. 1.2:14 Earth-rock dam—Brownlee Dam, Oregon (1958), constructed on alluvial foundation (after Mundal, Ref. 331).

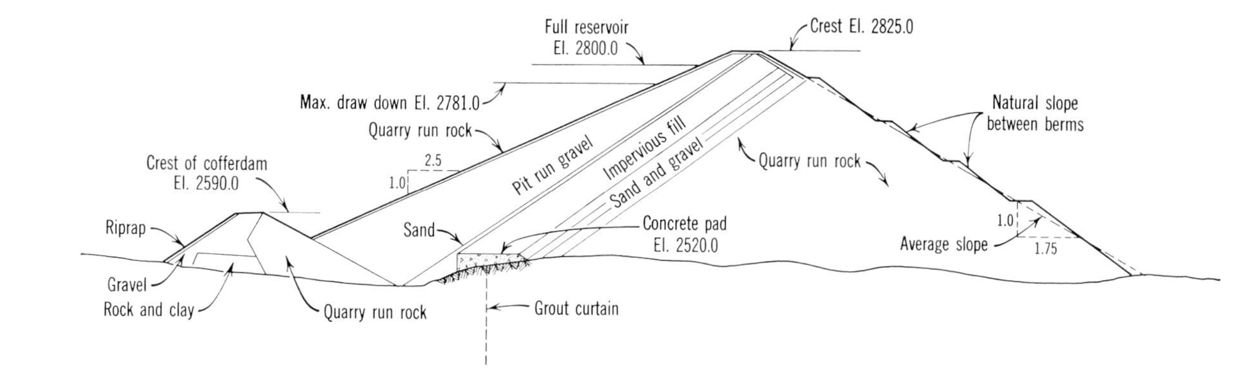

Fig. 1.2:15 Earth-rock dam—Kenney Dam, British Columbia (1958) (after Huber, Ref. 328).

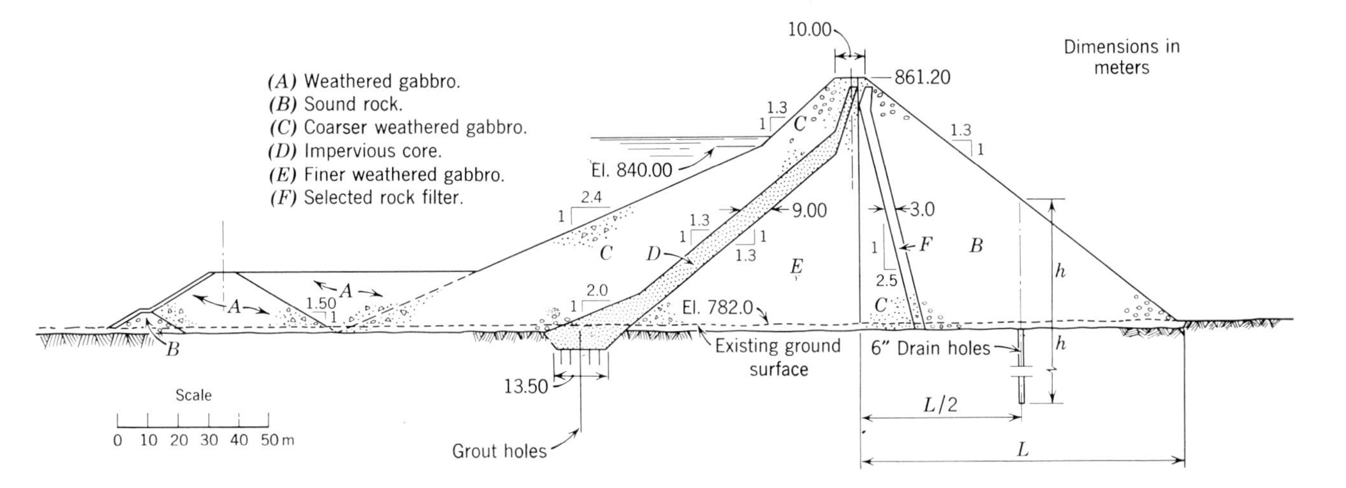

Fig. 1.2:16 Earth-rock dam—Hirfanli Dam, Turkey (1960) (after Noyan and Atatay, Ref. 298).

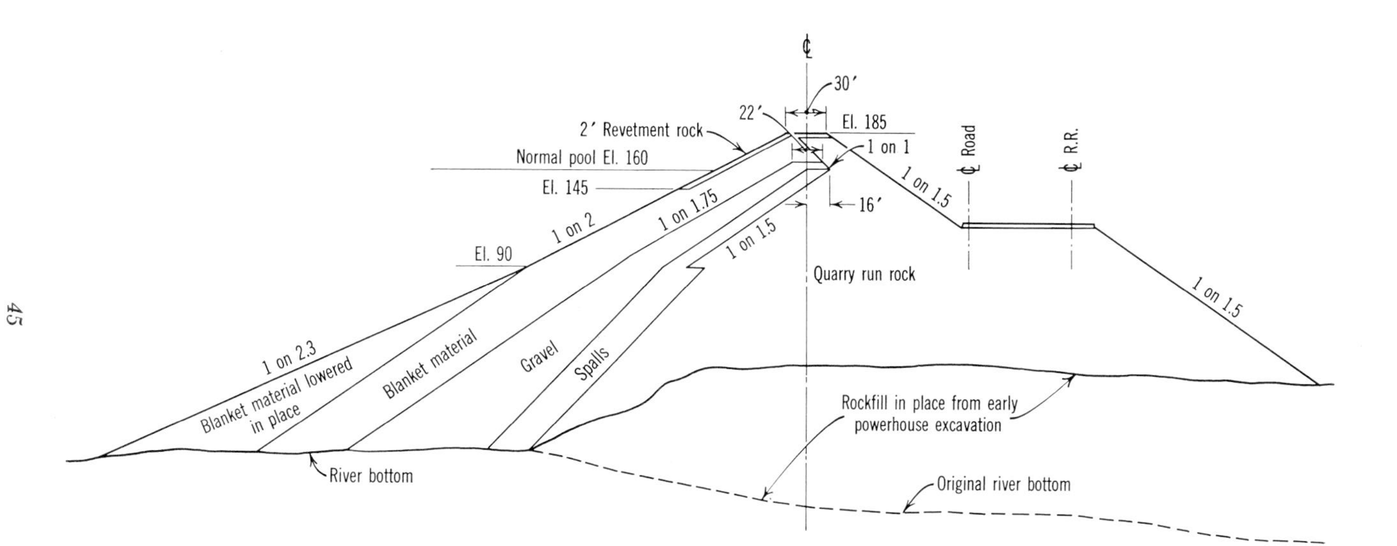

Fig. 1.2:17 The Dalles Closure Dam, Oregon (1957), constructed by dumping under water (after Pope, Ref. 221).

is not believed to be greater than the 170 c.f.s. estimated in the design stage.[1]

1.2c Typical Embankment Sections

The homogeneous dam and the dam with the thin impervious core can be considered as the extreme limits of possible zoning. Between these limits, earth dam engineers have used a great number of embankment designs with systems of internal zoning ranging from simple to quite intricate. Sections of some typical large earth dams constructed in the United States and abroad in the last 20 years are given in Figs. 1.2:18 through 1.2:55.

For the great majority of earth dams, the embankment is constructed with the same design section (the same zoning and slopes) over the whole length. At some sites, however, dams have been divided into two or more sections and a different design used in each. For dams of relatively short length there is usually little or no reason for changing the design, but for long dams several circumstances may make different sections advisable.[2] For example, different portions of the dam may have different foundation conditions or considerably different heights. Different portions may be located close to sources of different construction materials, so that the length of haul can be reduced if the design is varied. Or in the vicinity of appurtenant structures such as outlet works and spillways, the design may be changed to shorten the lengths of the more expensive concrete structures or to allow a better connection between earth and concrete.

1.3 DESIGN DETAILS

1.3a Embankment Side Slopes

No specific rules can be given for selecting the inclination of the outside slopes of the embankment. The general procedure is to make a first estimate on the basis of experience with similar dams and then

[1] One of the main indications that no large leaks are occurring is the fact that the spawning salmon are not attracted to the dam as they would be if any large flows came through.

[2] See, for example, the description of the factors controlling the design of the Army Engineers' Tuttle Creek Dam in Kansas in which seven different embankment sections were used in an 8,000-ft. long embankment (Ref. 420).

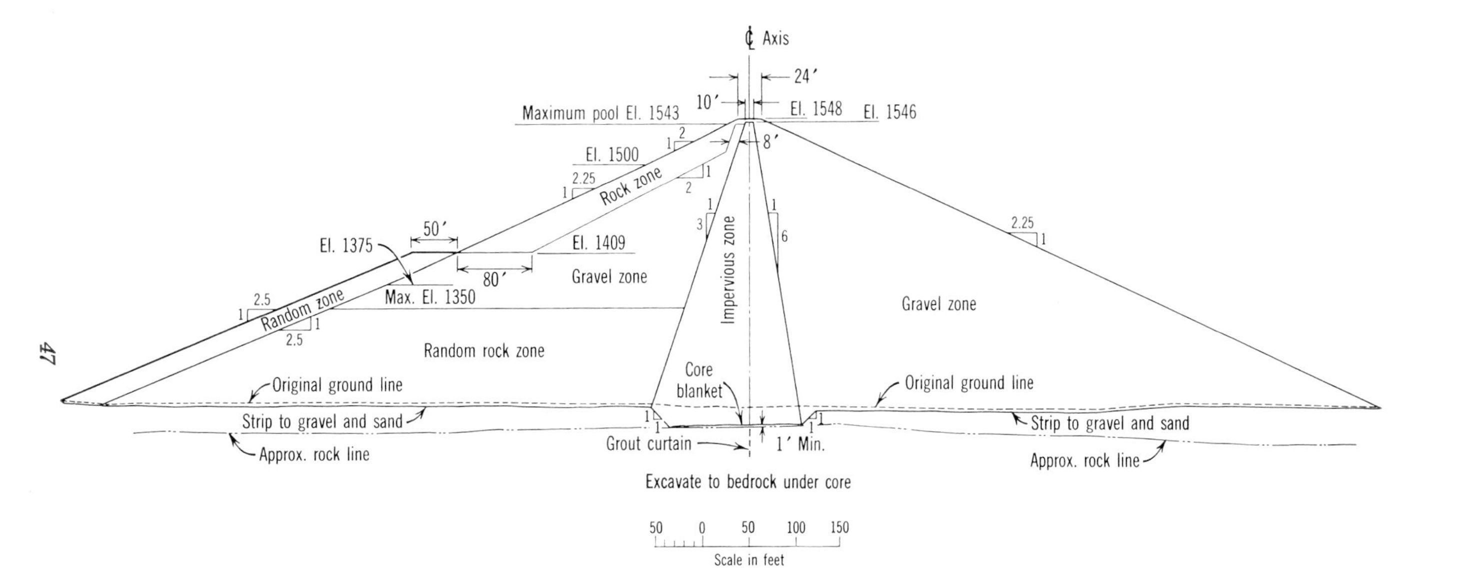

Hills Creek Dam, Oregon.

Fig. 1.2:18 Typical Designs.

Dams of the U.S. Army Corps of Engineers. (Courtesy, Portland District)

to modify the estimate as required after making theoretical analyses (Chap. 7). Except where there is a surplus of material available from required excavations, the most economical dam is obtained with the minimum embankment volume and therefore with the maximum slope steepness consistent with stability. For a given dam the allowable steepness depends generally on the internal zoning and on the strengths of the foundation and the embankment materials. In special cases it may also be influenced by the width of the valley and the rate of construction.

Average exterior slopes usually vary in the range between 2:1 and 4:1. Where the foundation is weak, however, much flatter slopes may be required (Sec. 8.4), and where the foundation is strong and the dam is to have large rockfill zones, the slopes may be made steeper than 2:1 (Sec. 1.2*b*).

Rock and gravel-fill dams with thin upstream sloping cores of earth on stable foundations are commonly designed with the downstream slope equal to the angle of repose of the material which varies approximately from 1.7:1 for rounded stream gravels to about 1.2:1 for angular quarried rock. If placed in thin, compacted layers, ordinary quarried rockfill can be constructed on slopes as steep as 1:1.

The downstream slope of a dam consisting of quarried rock or pervious granular soil with a central earth core on a rock foundation is commonly chosen on the basis of precedent in the range between 1.6:1 and 1.8:1. Precedent is the main factor for choosing the down-

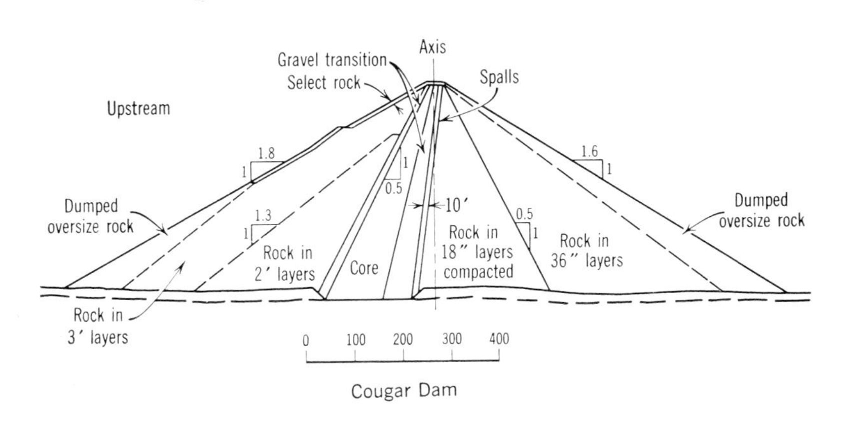

Cougar Dam, Oregon.

Fig. 1.2:19 Typical Designs.

Dams of the U.S. Army Corps of Engineers (after Thurber, Ref. 496).

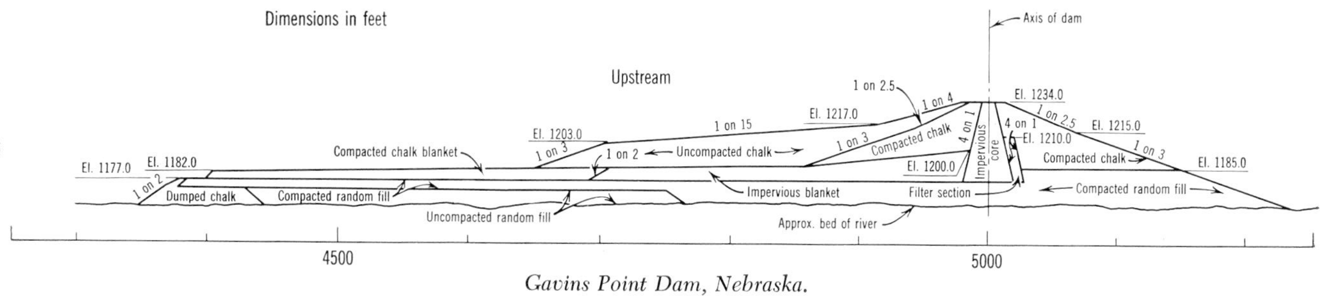

Gavins Point Dam, Nebraska.

Fig. 1.2:20 Typical Designs.

Dams of the U.S. Army Corps of Engineers. (Courtesy, Omaha District)

stream slope in such a dam because of the possibility that, while the dam may have ample computed safety factor against shearing failure, the crest may deflect by an excessive amount in the downstream direction (Sec. 2.10). Since we have no analytical method to predict this deflection, the downstream slope can only be chosen arbitrarily. Consequently, the compaction and the resultant rigidity of the downstream shell must be considered in connection with the selection of the slope. The Gepatsch Dam presently under construction in Austria (1962) at which the downstream rock shell is compacted with a heavy vibrating roller is probably the steepest dam of this type which has ever been contemplated, Figs. 1.2:45 and 11.5:2.

For rockfill slopes which are flatter than the angle of repose, the designer has two choices: (1) to use berms at intervals with the natural angle of repose between the berms; or (2) to make a smooth uniform slope. Both procedures can result in a dam face with an attractive appearance.[1] If the rockfill section is to be constructed in thin layers, either procedure can be used easily. If the dam is to be constructed in layers of more than 15 ft. in thickness, however, it is difficult to make a uniform slope since the material naturally falls on the angle of repose. Consequently, when the material is placed in thick layers, it is usually preferable to provide the berms. In some cases, the contractor may be given the choice within limits as to which procedure to use. For upstream rockfill zones consisting of relatively small rock, it may be desirable to use a uniform slope flatter than the angle of repose to provide better resistance to wave action; however, if the rock is large, it can be placed on the angle of repose even on the upstream face of

[1] See Sec. 11.5 for construction methods for rockfill embankment sections.

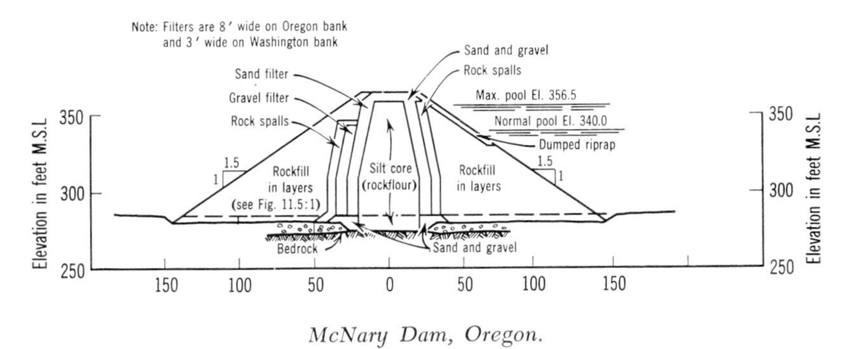

McNary Dam, Oregon.

Fig. 1.2:21 Typical Designs.

Dams of the U.S. Army Corps of Engineers. (after Drake et al., Ref. 323).

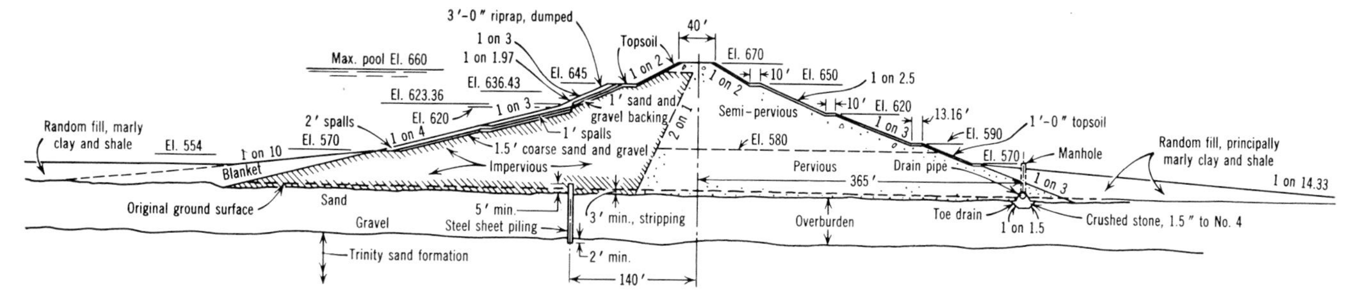

Denison Dam, Texas.

Fig. 1.2:22 Typical Designs.

Dams of the U.S. Army Corps of Engineers (Ref. 362).

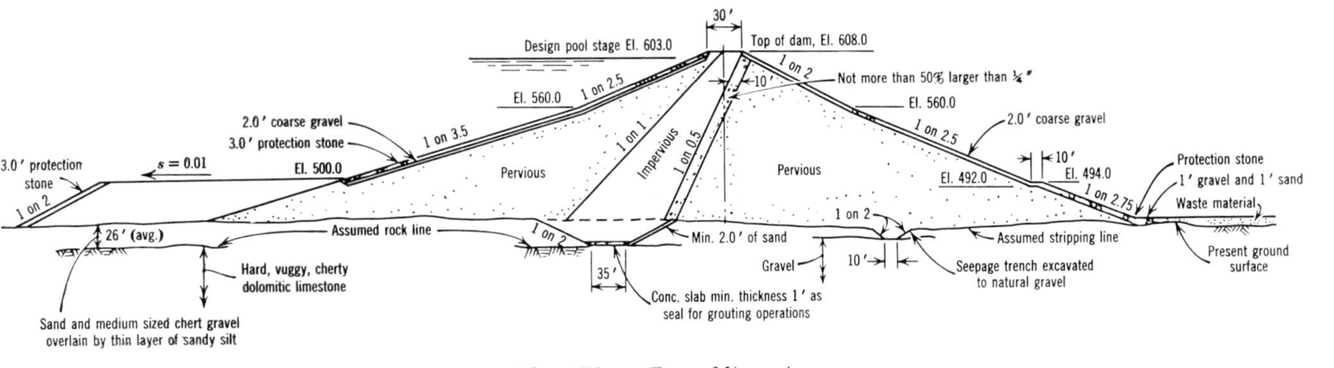

Clear Water Dam, Missouri.

Fig. 1.2:23 Typical Designs.

Dams of the U.S. Army Corps of Engineers (Ref. 362).

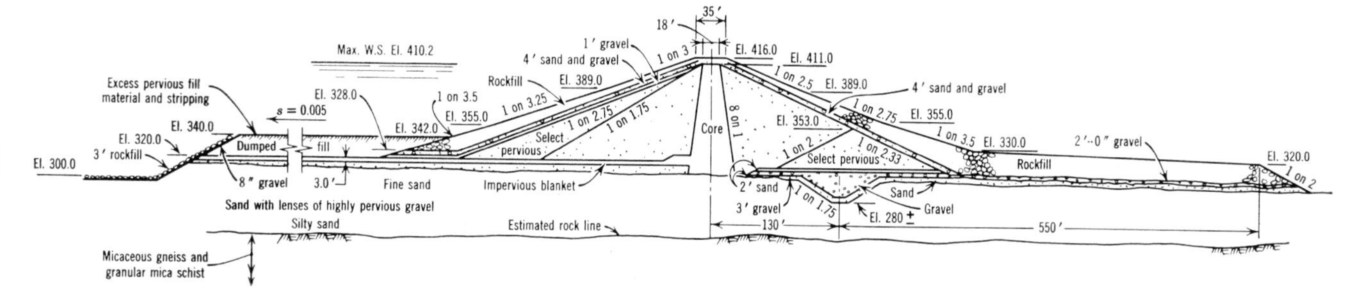

Franklin Falls Dam, New Hampshire.

Fig. 1.2:24 Typical Designs.

Dams of the U.S. Army Corps of Engineers. (Courtesy, New England Division)

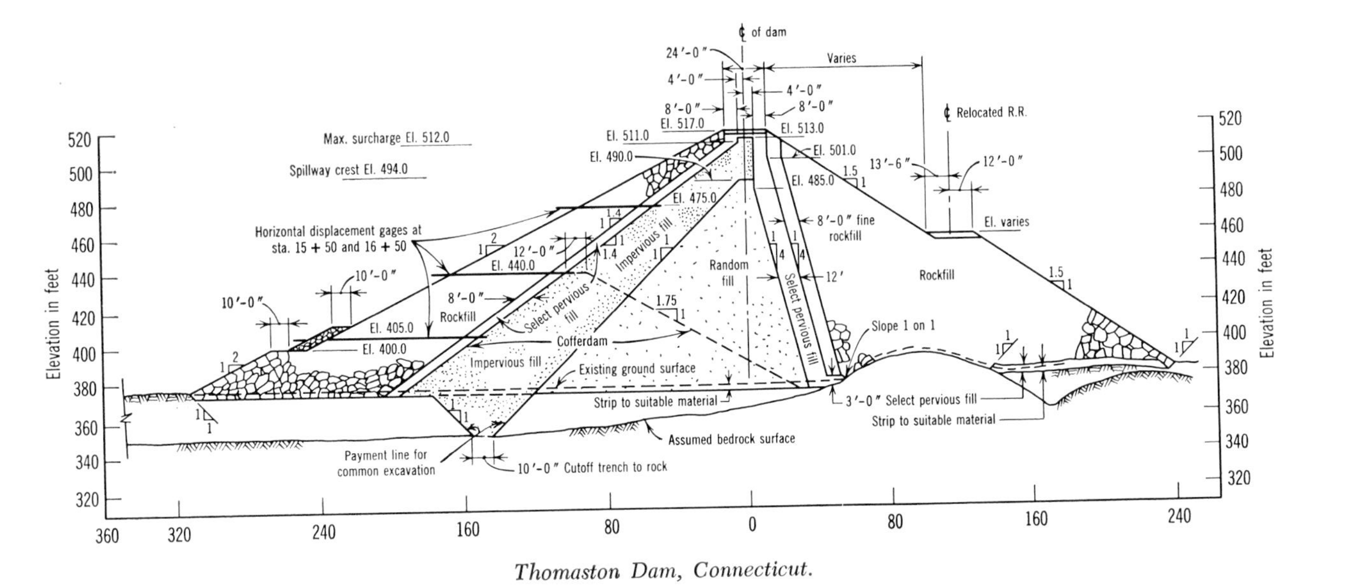

Thomaston Dam, Connecticut.

Fig. 1.2:25 Typical Designs.

Dams of the U.S. Army Corps of Engineers. (Courtesy, New England Division)

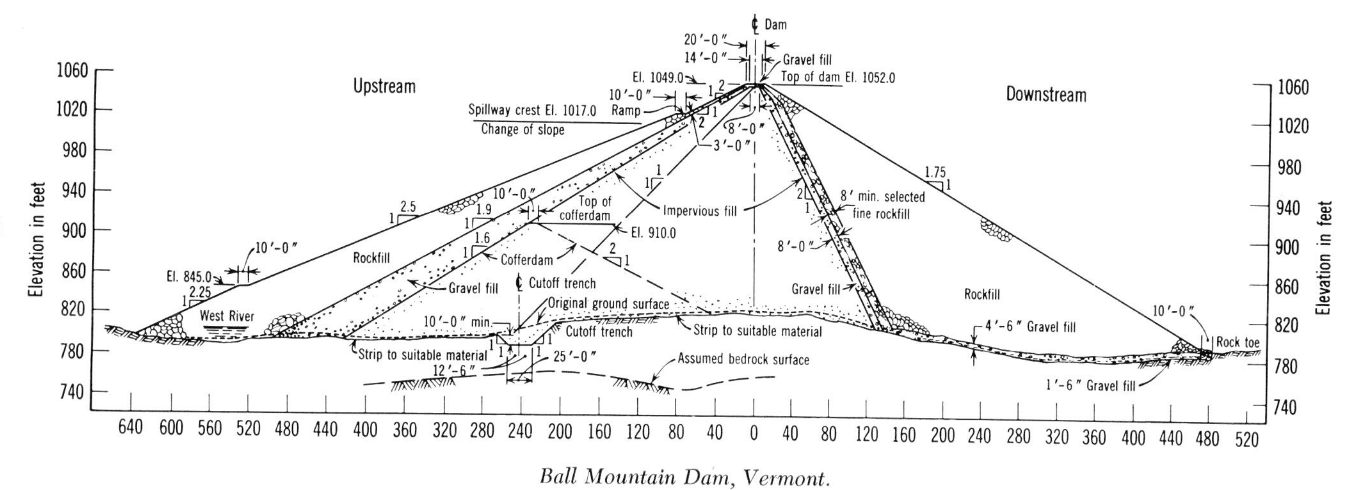

Ball Mountain Dam, Vermont.

Fig. 1.2:26 Typical Designs.

Dams of the U.S. Army Corps of Engineers. (Courtesy, New England Division)

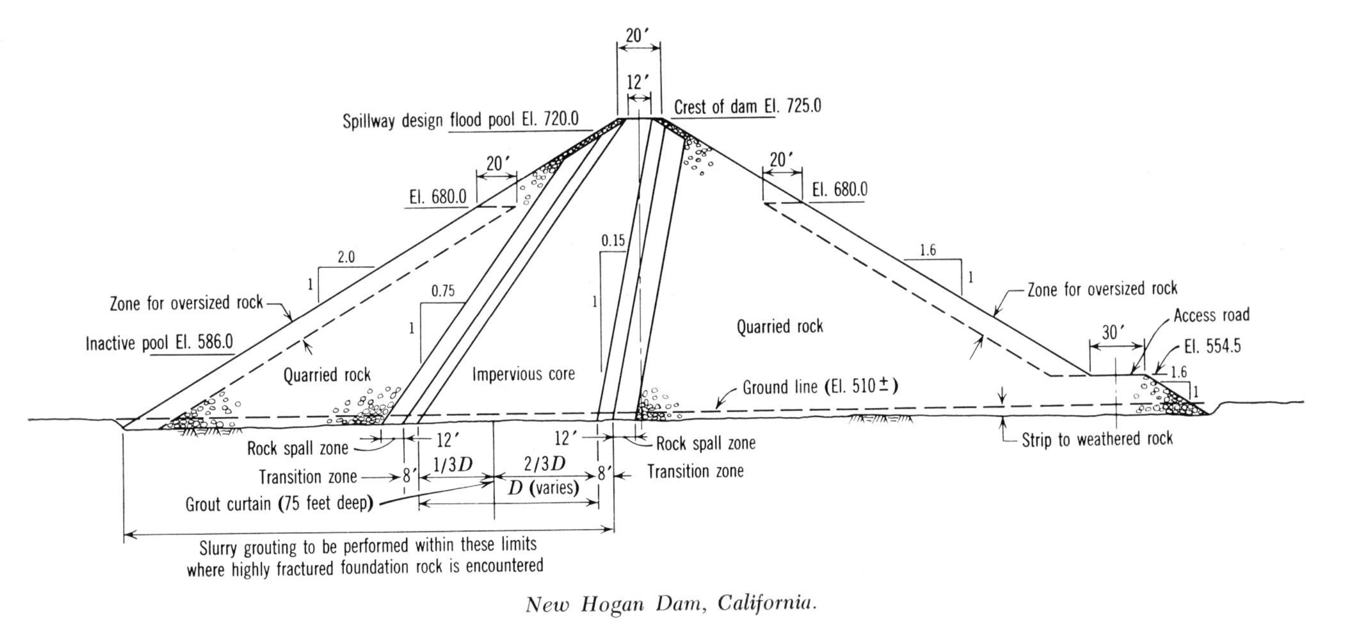

New Hogan Dam, California.

Fig. 1.2:27 Typical Designs.

Dams of the U.S. Army Corps of Engineers. (after Bird, Ref. 627).

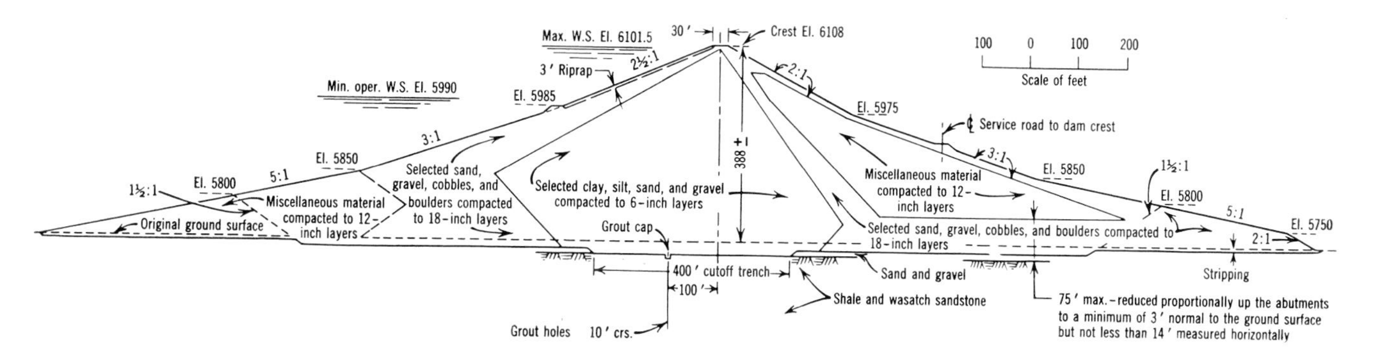

Navajo Dam, New Mexico.

Fig. 1.2:28 Typical Designs.

Dams of the U.S. Bureau of Reclamation (after Walker, Ref. 147).

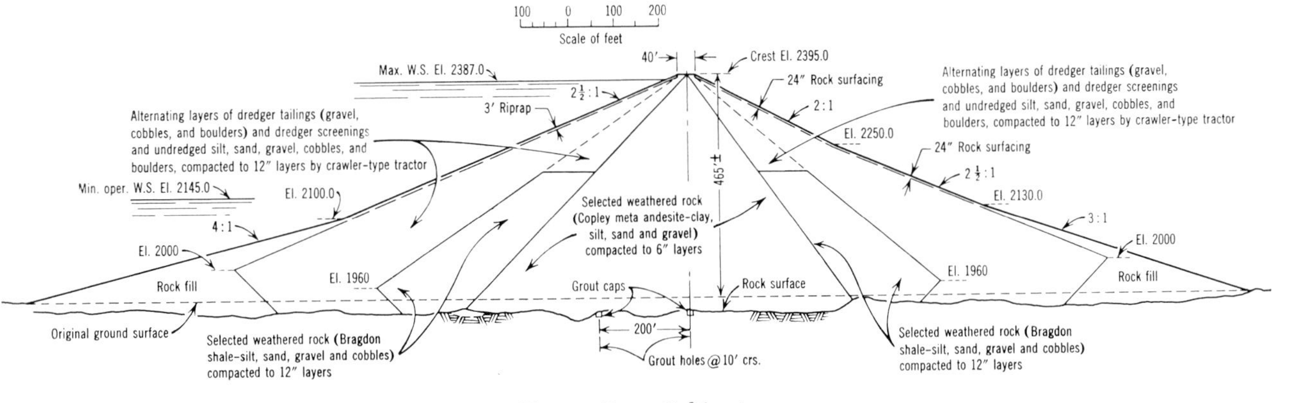

Trinity Dam, California.

Fig. 1.2:29 Typical Designs.

Dams of the U.S. Bureau of Reclamation (after Walker, Ref. 147).

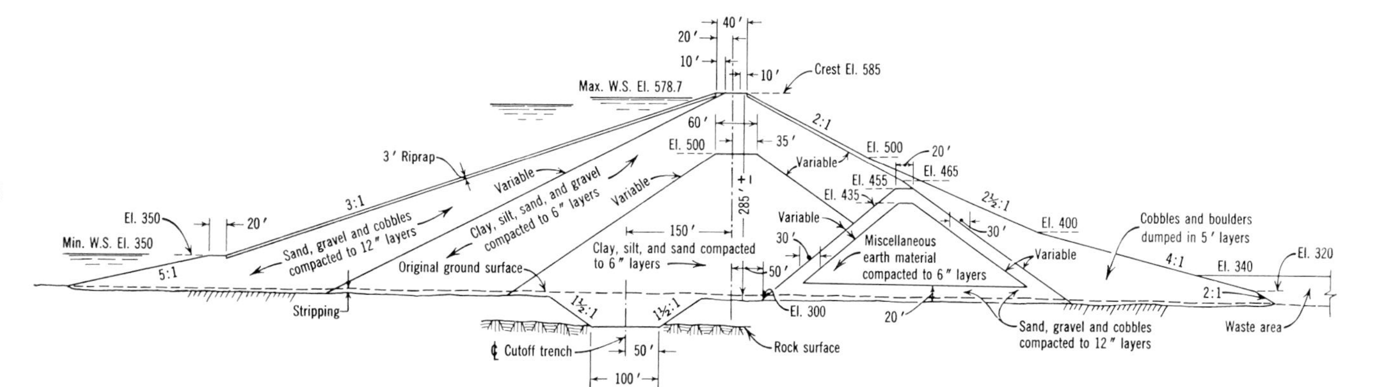

Casitas Dam, California.

Fig. 1.2:30 Typical Designs.

Dams of the U.S. Bureau of Reclamation (after Walker, Ref. 147).

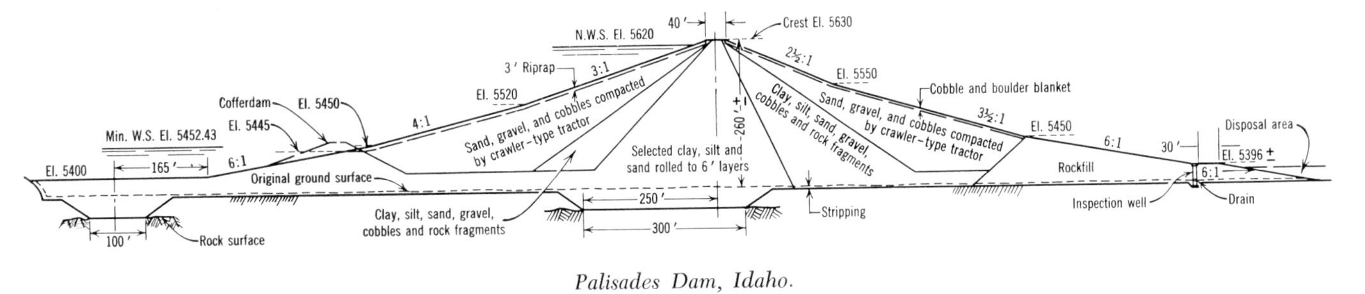

Palisades Dam, Idaho.

Fig. 1.2:31 Typical Designs.

Dams of the U.S. Bureau of Reclamation (after Walker, Ref. 147).

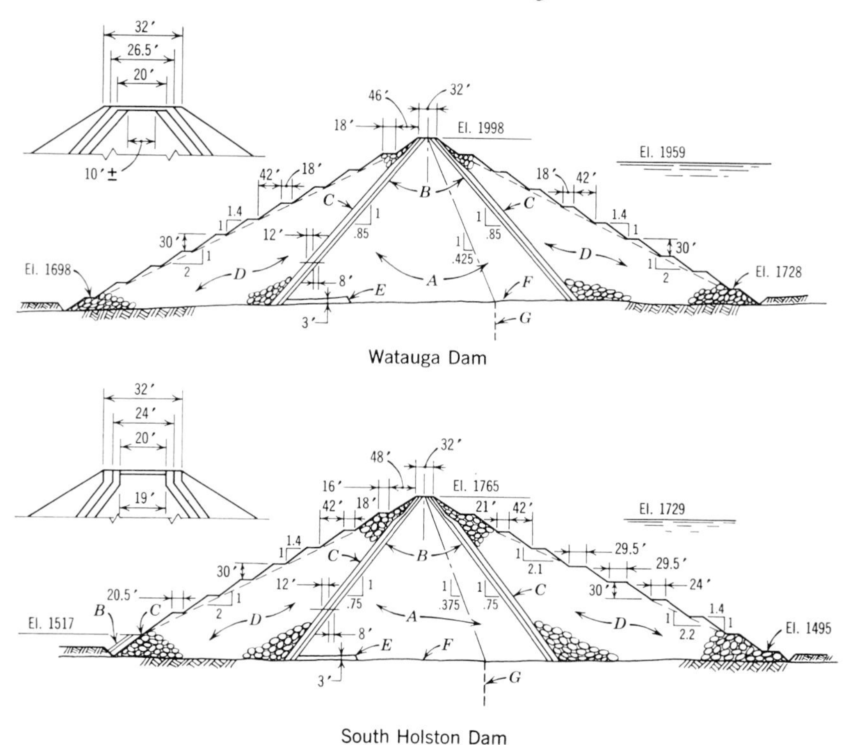

A: Impervious rolled fill. B: Fine filter. C: Coarse filter. D: Quarry run rock. E: Drain blanket. F: Weathered rock line. G: Grout curtain.

Watuaga and South Holston dams, North Carolina.

Fig. 1.2:32 Typical Designs.

Dams of the Tennessee Valley Authority (after Leonard and Raine, Ref. 392).

the dam, although it is often desirable to dump larger rocks immediately on the face.

As a general rule for homogeneous dams of fine-grained soil, the higher the dam, the flatter must be the slopes. For thin-core dams in which the bulk of the embankment consists of pervious granular soil or rock, the allowable slopes are almost independent of the height; specifically, they depend on it roughly to the degree by which the strength of the impervious core contributes to the stability of the embankment. In general, internal zoning in a dam permits the use of steeper slopes, both because the stronger materials are placed where

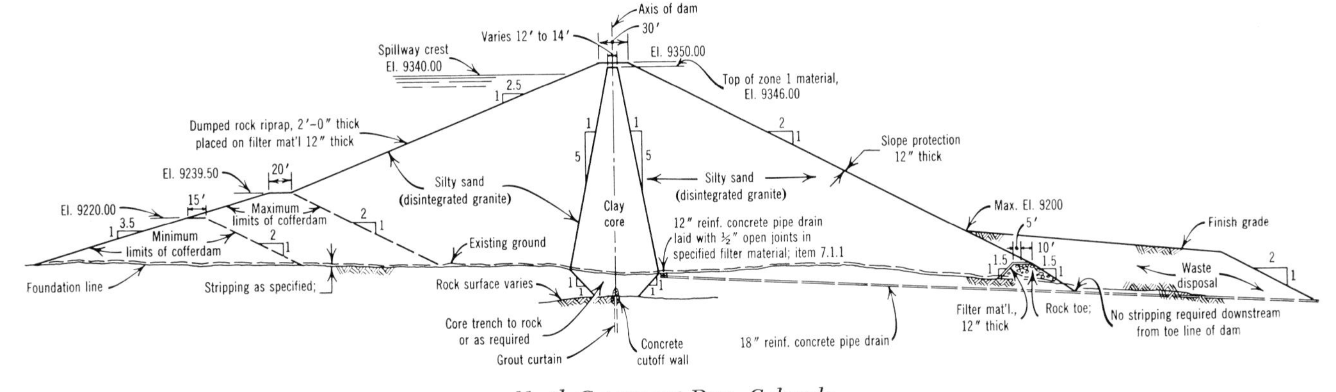

North Catamount Dam, Colorado.

Fig. 1.2:33 Typical Designs.

Dams of U.S. Consulting Engineers. (Courtesy, Black and Veatch)

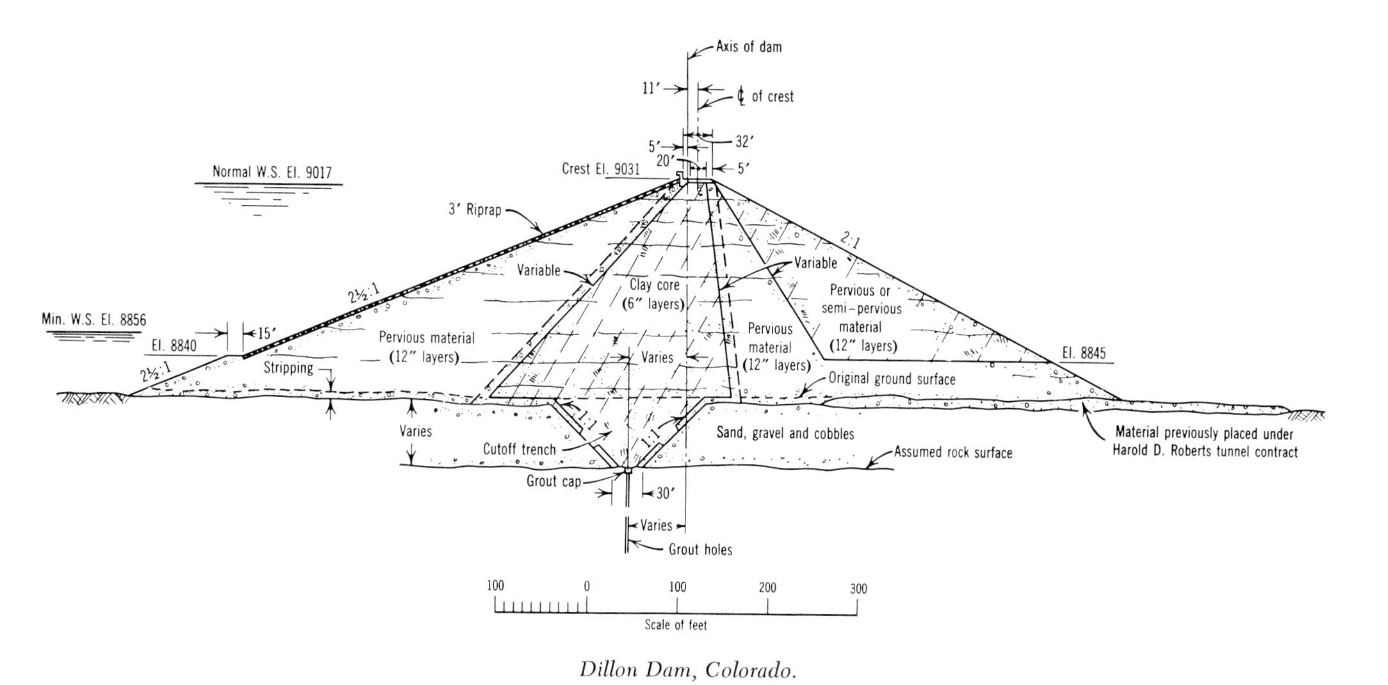

Dillon Dam, Colorado.

Fig. 1.2:34 Typical Designs.

Dams of U.S. Consulting Engineers. (Courtesy, Tipton and Kalmbach Inc.)

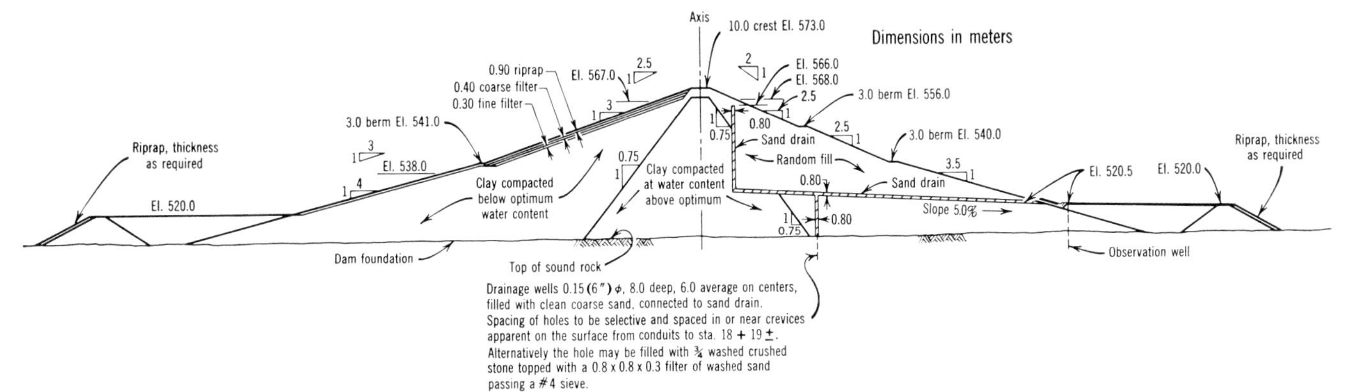

Três Marias, Brazil.

Fig. 1.2:35 Typical Designs.

Dams of U.S. Consulting Engineers. (Courtesy, International Engineering Co.)

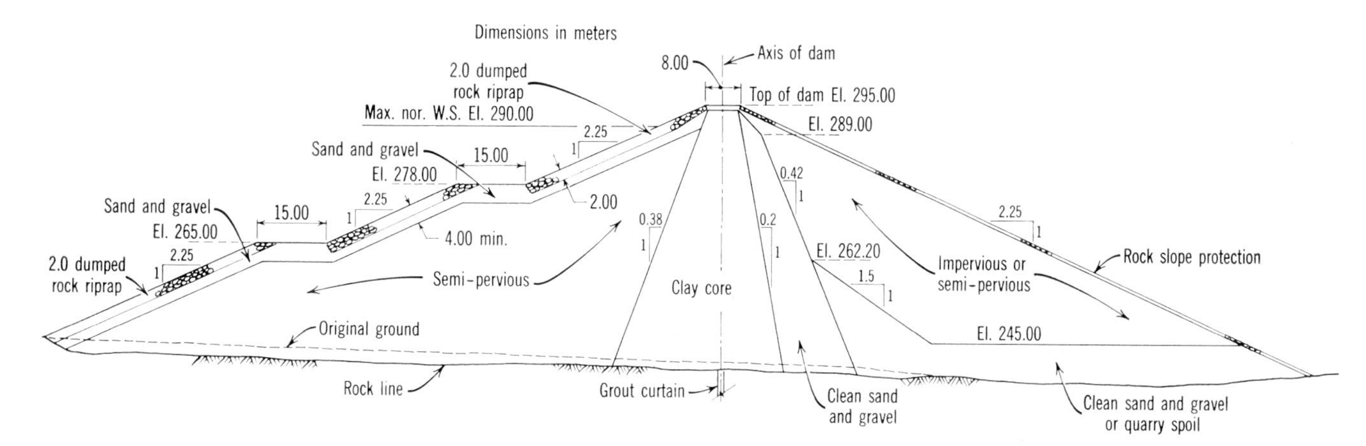

San Lorenzo Dam, Peru.

Fig. 1.2:36 Typical Designs.

Dams of U.S. Consulting Engineers. (Courtesy, International Engineering Co.)

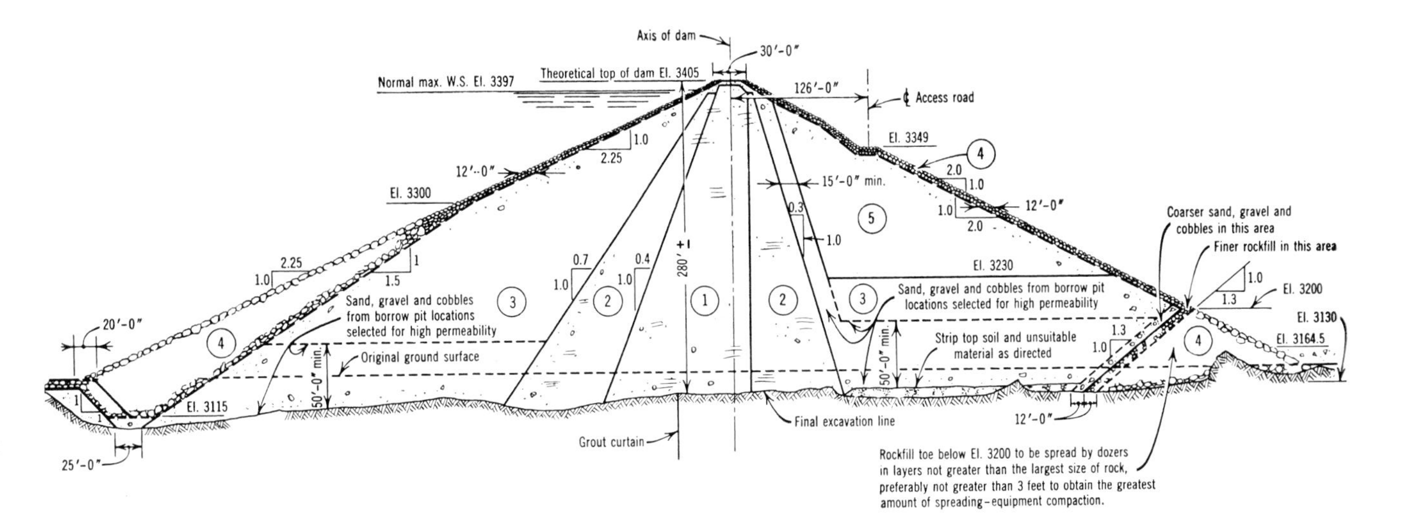

(1) *Impervious rolled earth—to consist of a mixture of silt, fine to medium sand, and gravel not exceeding 3 in. max. dimension.*
(2) *Semipervious rolled earth, transition zone—to consist of fine to coarse sand and gravel not exceeding 3 in. max. dimension, together with some silt sizes. The finest material to be adjacent to zone (1) with a gradual transition outwards.*
(3) *Rolled sand, gravel, and cobbles—the finest material to be adjacent to zone (2) and the coarsest adjacent to zone (4) and/or (5).*
(4) *Dumped rockfill—to consist of rock fragments from required rock excavations and primarily composed of well-graded fragments larger than 1/4 ft.³ in volume with only enough rock spalls and gravel to fill voids in the coarser material.*
(5) *Random fill—to have a min. dry weight = 115 lb./ft.³ and min. angle of internal friction = 35°*

Beardsley Dam, California.

Fig. 1.2:37 Typical Designs.

Dams of U.S. Consulting Engineers. (Courtesy, International Engineering Co.)

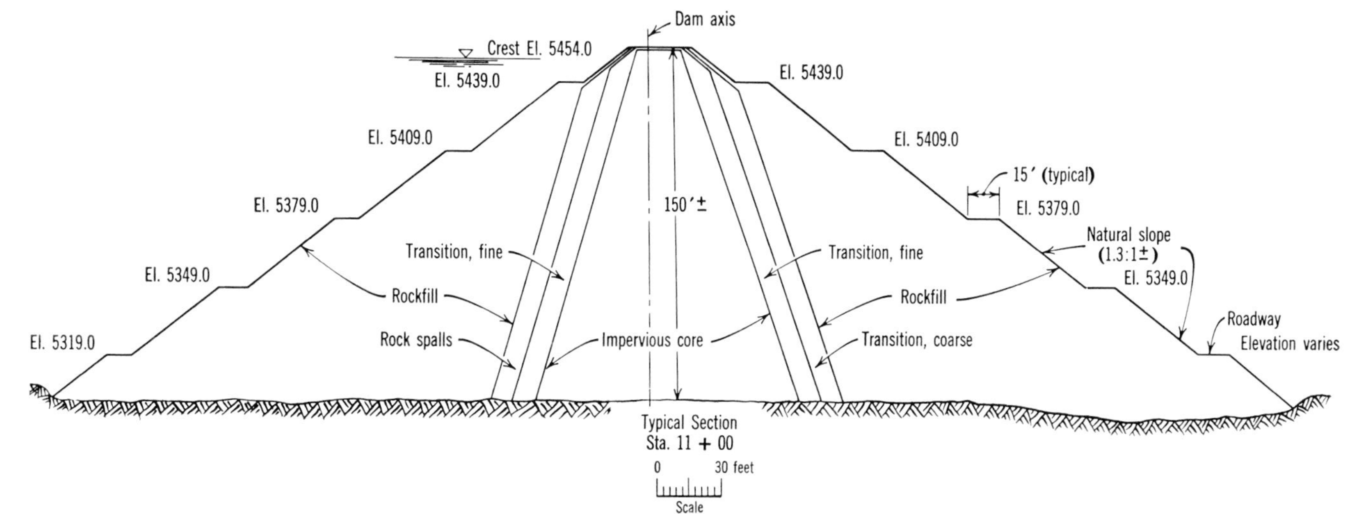

Ice House Dam, California.

Fig. 1.2:38 Typical Designs.

Dams of U.S. Consulting Engineers. (Courtesy, Sacramento Municipal Utility District)

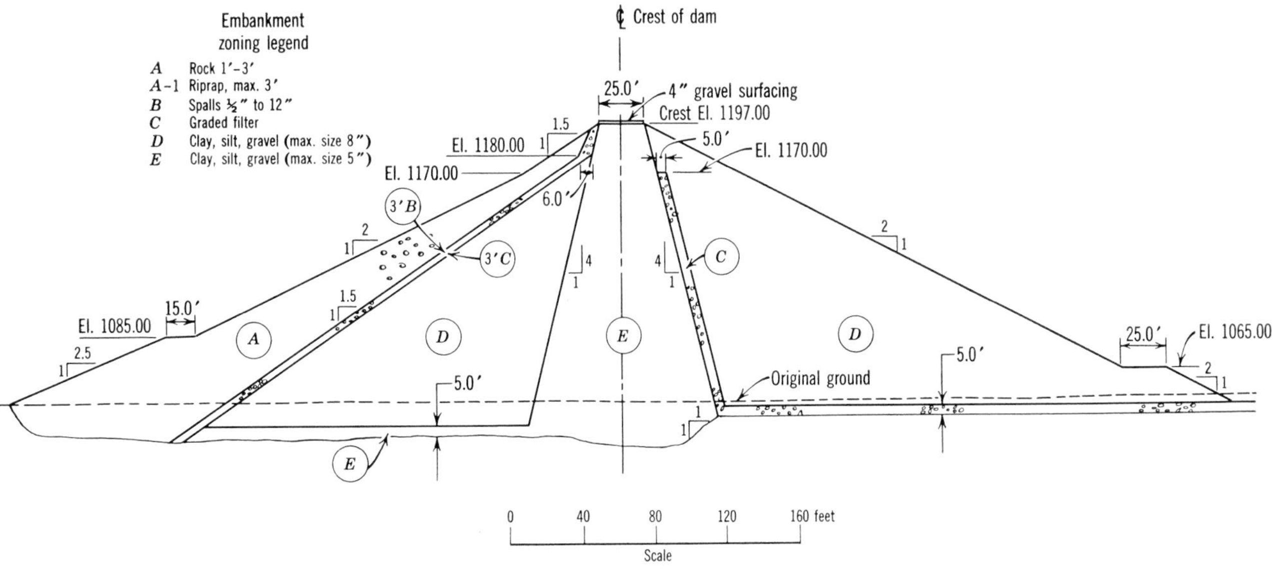

Virginia Ranch Dam, California.

Fig. 1.2:39 Typical Designs.

Dams of U.S. Consulting Engineers. (Courtesy, Browns Valley Irrigation District)

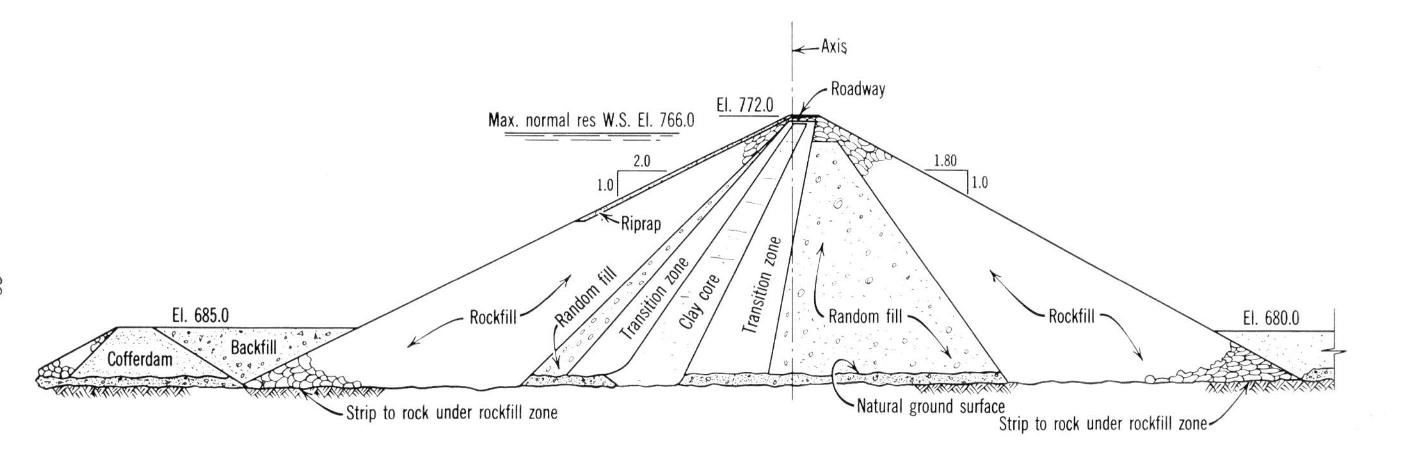

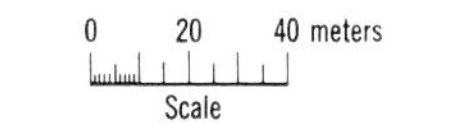

Furnas Dam, Brazil.

Fig. 1.2:40 Typical Designs.

Dams of U.S. Consulting Engineers (after Libby, Ref. 512).

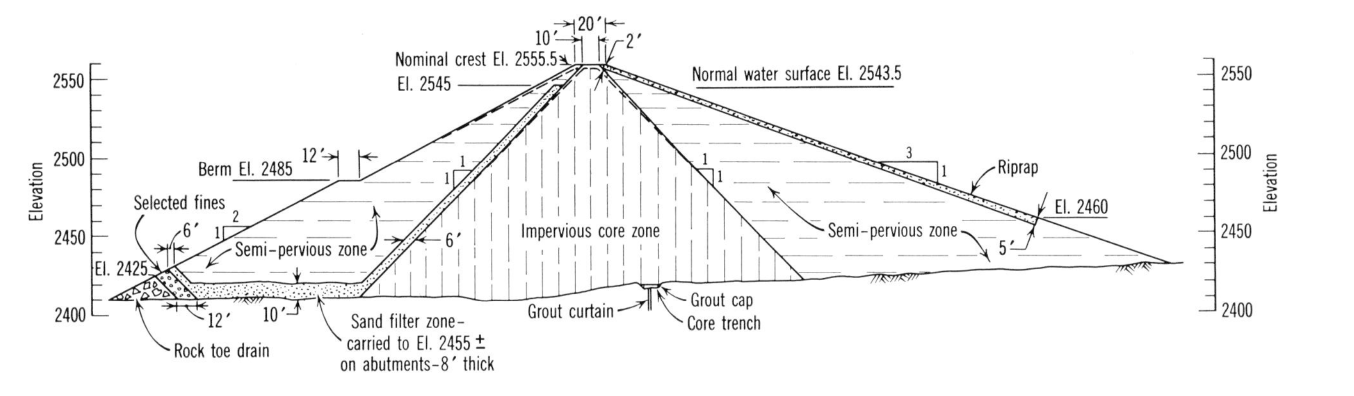

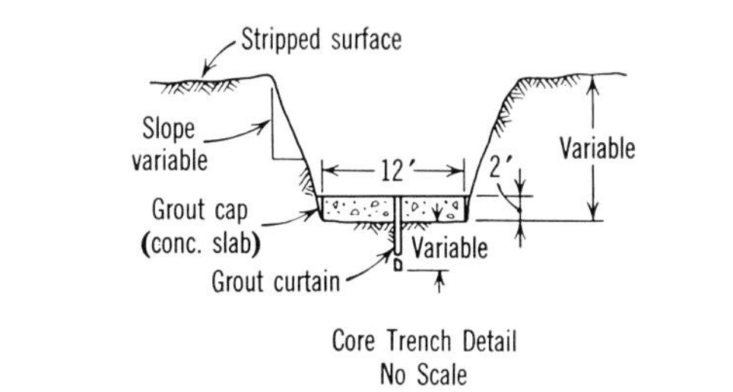

Mosquito Junction Dam.

Fig. 1.2:41 Typical Designs.

Dams of U.S. Consulting Engineers. (Courtesy, Kennedy Engineers)

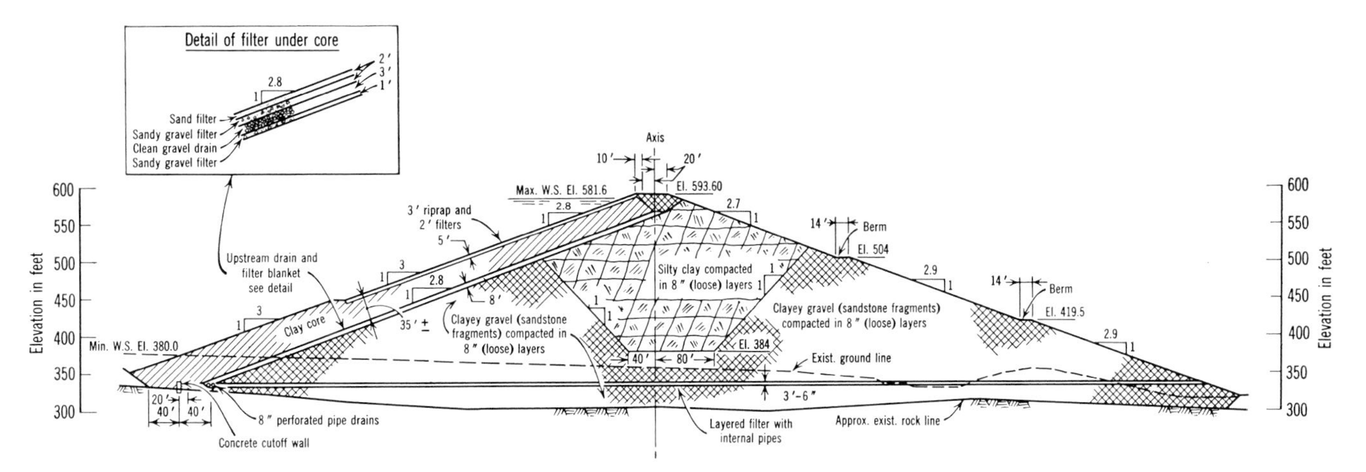

Briones Dam, California.

Fig. 1.2:42 Typical Designs.

Dams of U.S. Consulting Engineers. (Courtesy, East Bay Municipal Utility District)

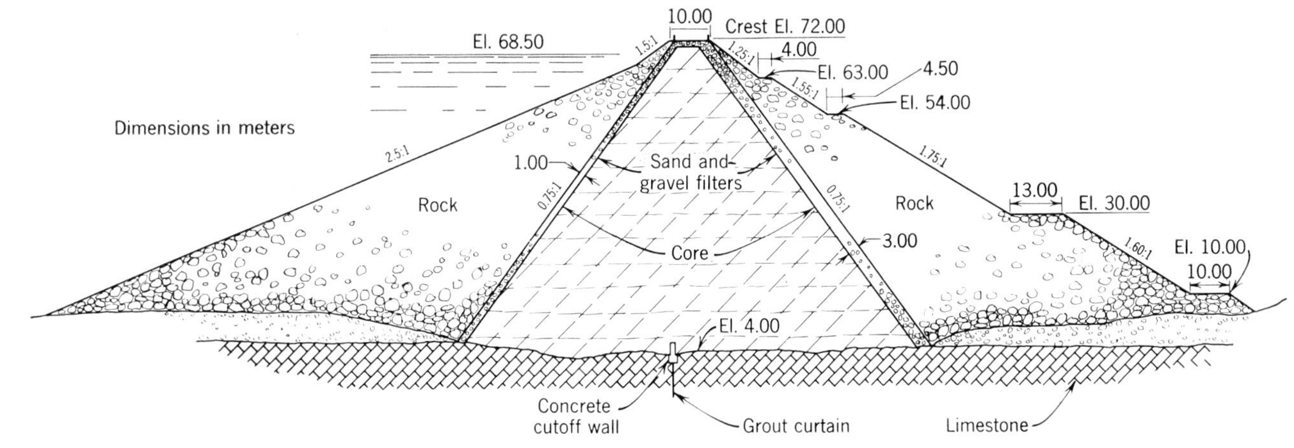

President Aleman Dam, Mexico.

Fig. 1.2:43 Typical Designs.

Dams of Foreign Engineers and Agencies (Ref. 344).

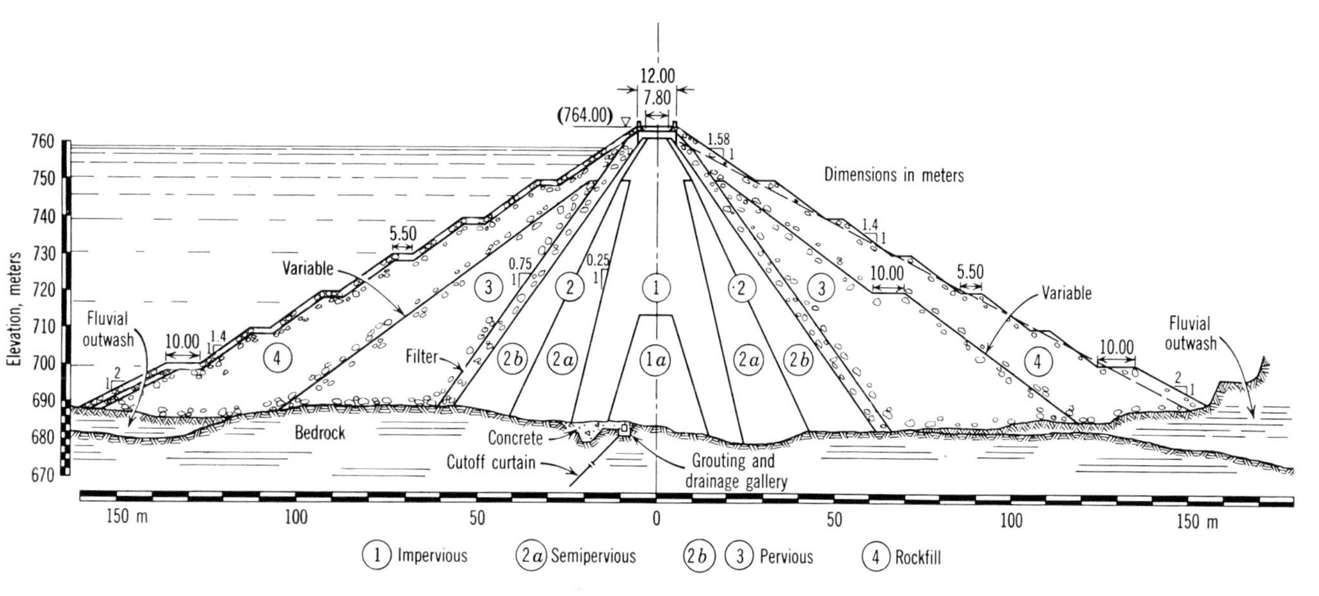

Las Pirquitas Dam, Argentina.

Fig. 1.2:44 Typical Designs.

Dams of Foreign Engineers and Agencies (after Grandi et al., Ref. 576).

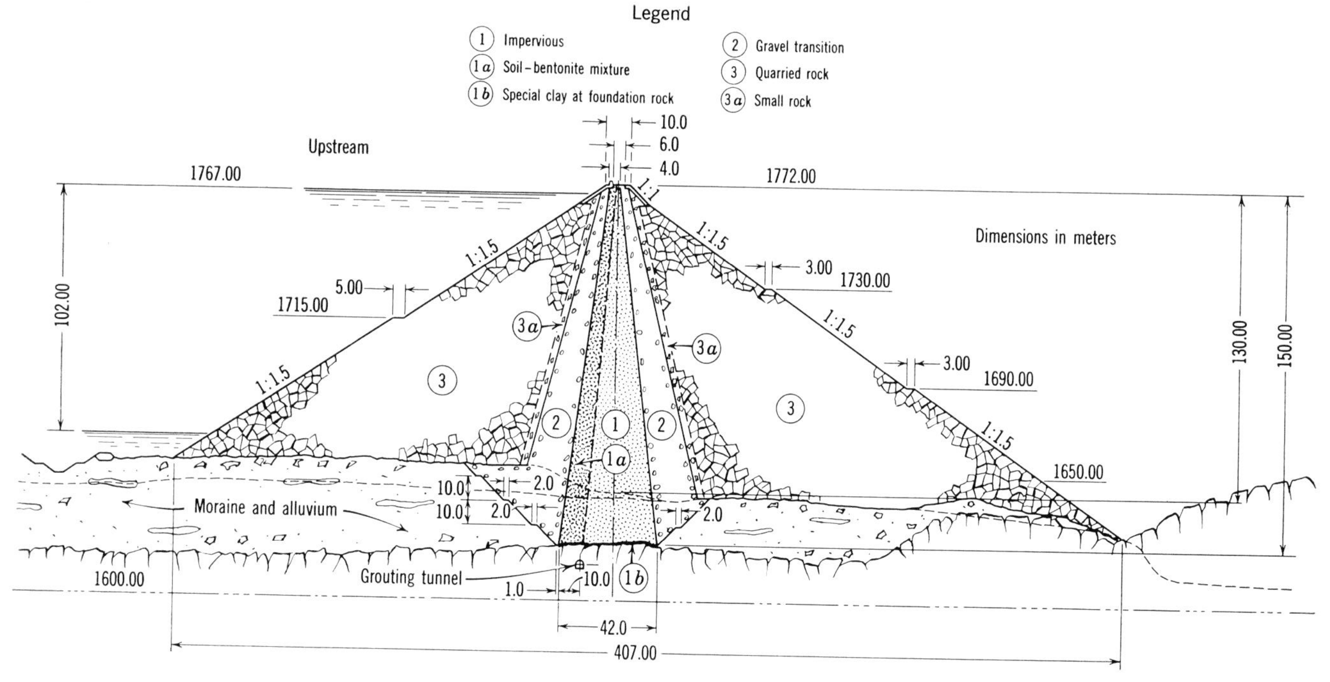

Gepatsch Dam, Austria. (under construction 1963)

Fig. 1.2:45 Typical Designs.

Dams of Foreign Engineers and Agencies (after Lauffer and Schober, Ref. 373).

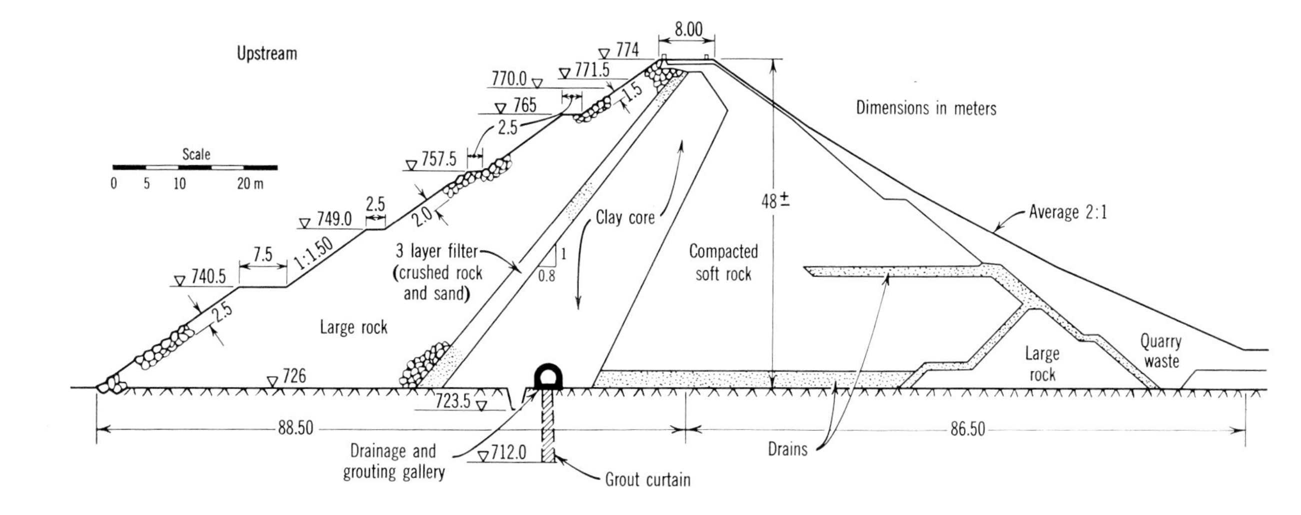

Lokvarka Dam, Yugoslavia.

Fig. 1.2:46 Typical Designs.

Dams of Foreign Engineers and Agencies (after Nonveiller, Ref. 303).

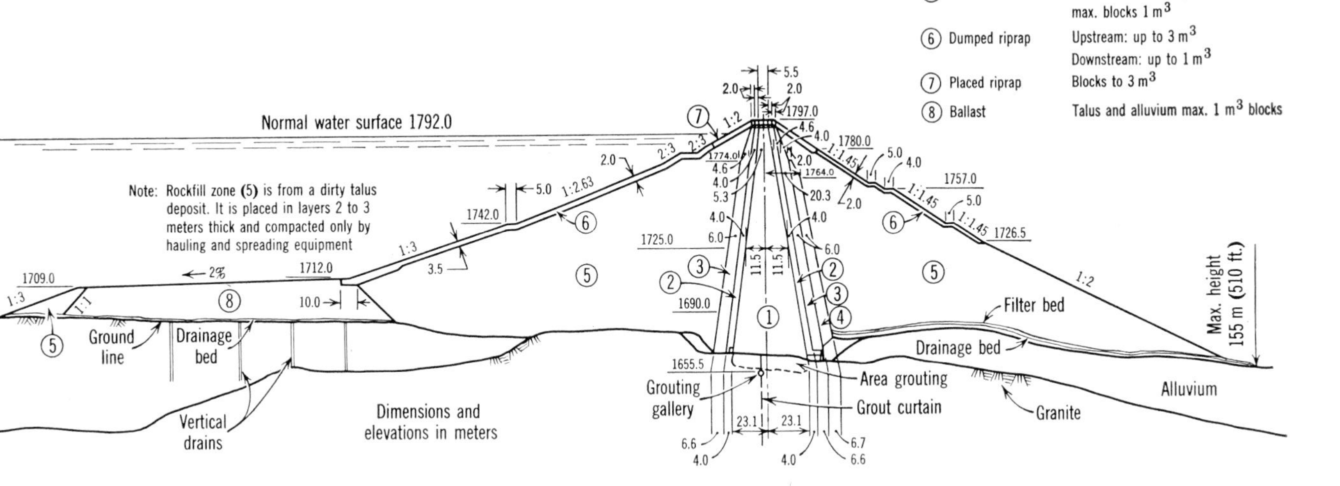

Goschenenalp Dam, Switzerland.

Fig. 1.2:47 Typical Designs.

Dams of Foreign Engineers and Agencies (after Cooke, Ref. 481).

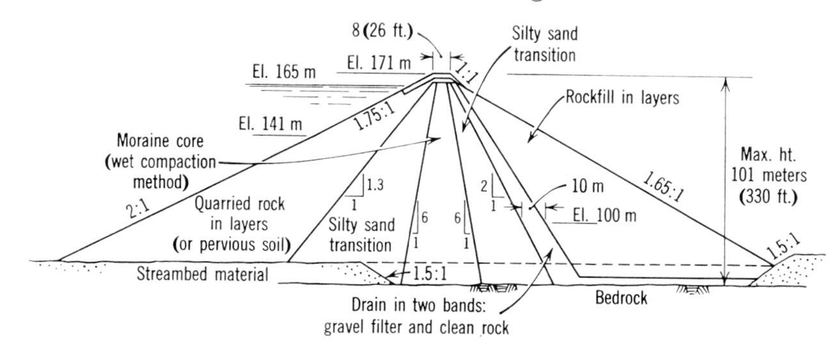

Messaure Dam, Sweden.

Fig. 1.2:48 Typical Designs.

Dams of Foreign Engineers and Agencies (after Cooke, Ref. 481).

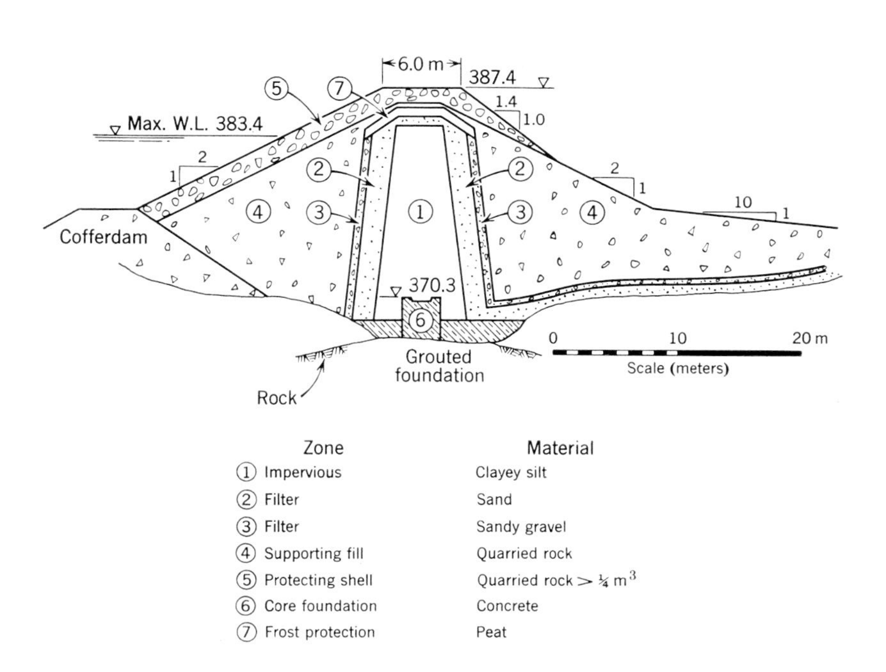

Zone	Material
① Impervious	Clayey silt
② Filter	Sand
③ Filter	Sandy gravel
④ Supporting fill	Quarried rock
⑤ Protecting shell	Quarried rock > ¼ m^3
⑥ Core foundation	Concrete
⑦ Frost protection	Peat

Tustervatn Dam, Norway.

Fig. 1.2:49 Typical Designs.

Dams of Foreign Engineers and Agencies (after Kjaernsli, Ref. 202).

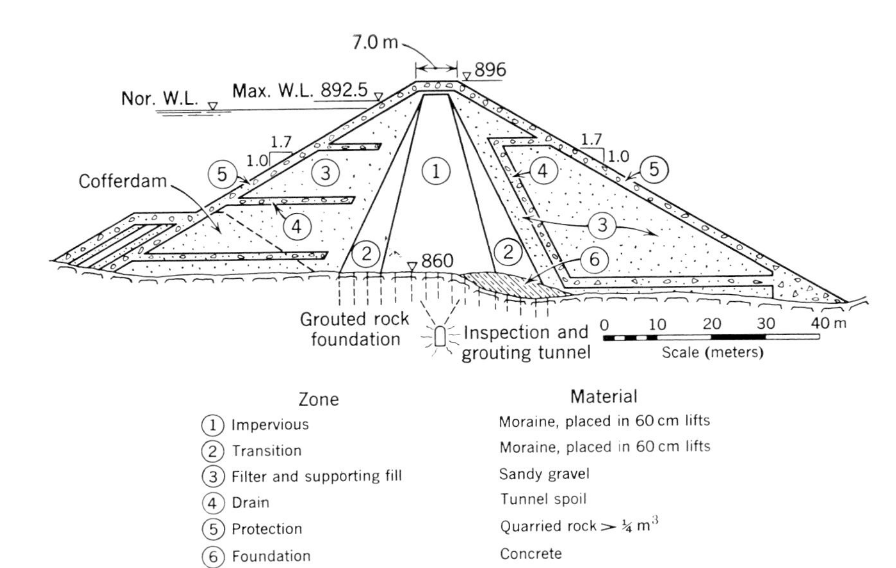

Bordal No. 1 Dam, Norway.

Fig. 1.2:50 Typical Designs.

Dams of Foreign Engineers and Agencies (after Kjaernsli, Ref. 202).

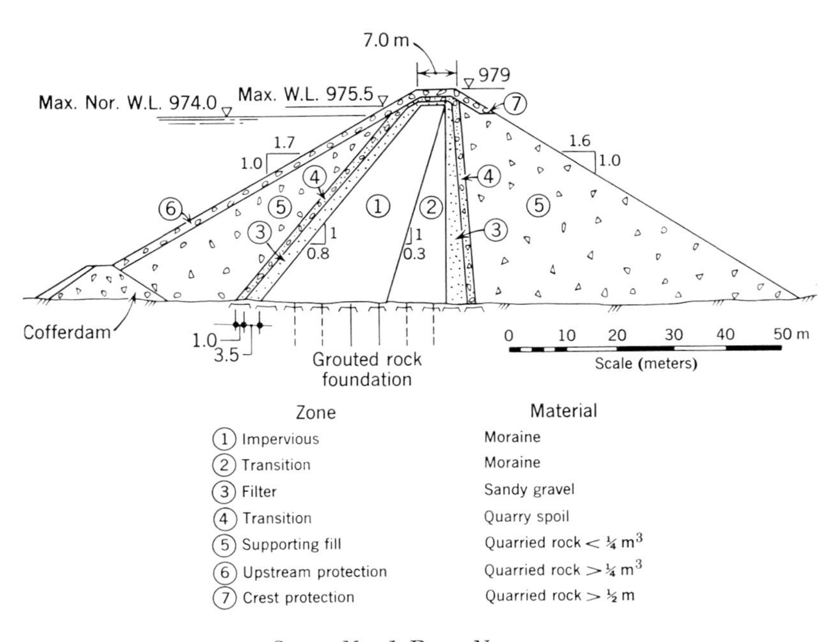

Songa No. 1 Dam, Norway.

Fig. 1.2:51 Typical Designs.

Dams of Foreign Engineers and Agencies (after Kjaernsli, Ref. 202).

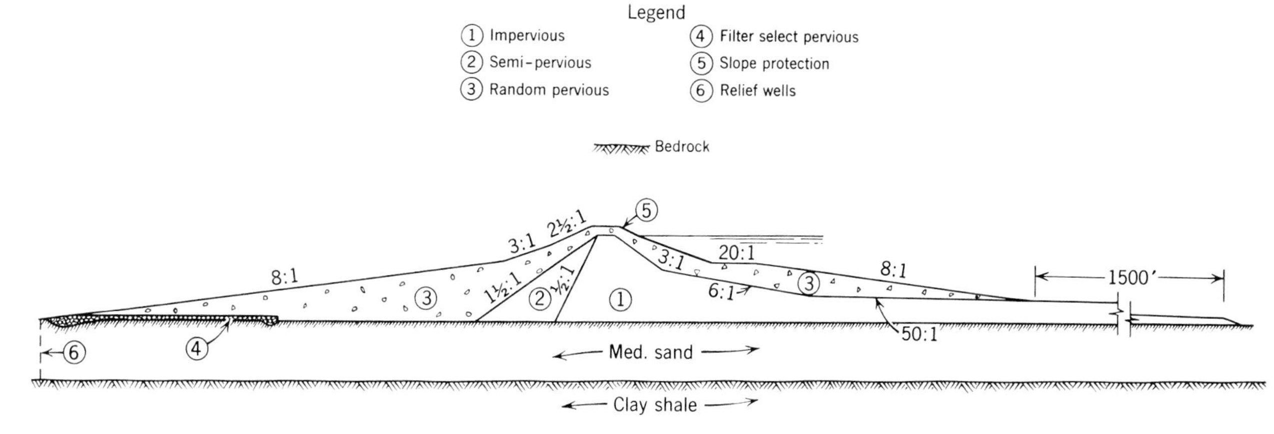

South Saskatchewan River Dam (proposed), Canada.

Fig. 1.2:52 Typical Designs.

Dams of Foreign Engineers and Agencies (after Peterson, Ref. 161).

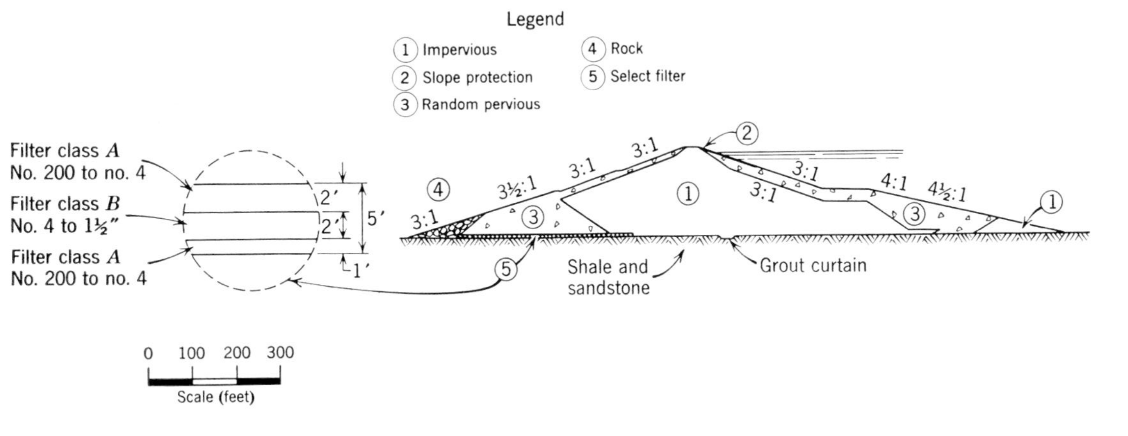

St Mary Dam, Canada.

Fig. 1.2:53 Typical Designs.

Dams of Foreign Engineers and Agencies (after Peterson, Ref. 161).

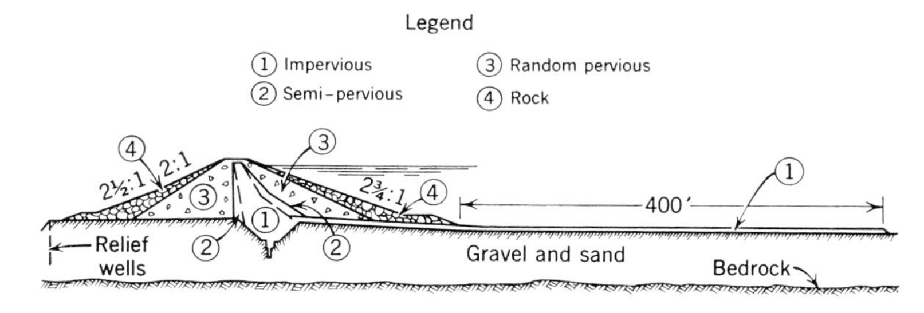

Rose Valley Dam, Canada

Fig. 1.2:54 Typical Designs.

Dams of Foreign Engineers and Agencies (after Peterson, Ref. 161).

they are most effective in resisting shear stresses and because the drainage systems control and reduce pore water pressures.

For any given safety factor against shear failure, an embankment with minimum volume is usually obtained when the side slopes are made steeper at the upper elevations and flatter near the bottom. Variable slopes should be considered for all earth dams higher than about 100 ft. The higher the dam and the softer the foundation, the greater is the economic advantage to be gained from varying the slopes.[1]

From the standpoint of stability, dams in narrow rock-walled canyons can be constructed with somewhat steeper slopes than otherwise because of the added stability given by the confining walls (Sec. 7.4*e*). In narrow valleys also, broad toe berms or very flat slopes at the toes of the dam can be provided relatively cheaply due to the small quantities of embankment material required.

1.3b Filter Zones

As water from the reservoir seeps through the pores of an earth dam, seepage forces are exerted on the soil particles in the direction

[1] See Refs. 662 and 663 for description of the Cheffia Dam currently under construction in Algeria, where the slopes were varied from steeper at the top to flatter at the bottom in the form of a continuous curve. The downstream slope varies from 1:1 at the top to 3:1 at the bottom.

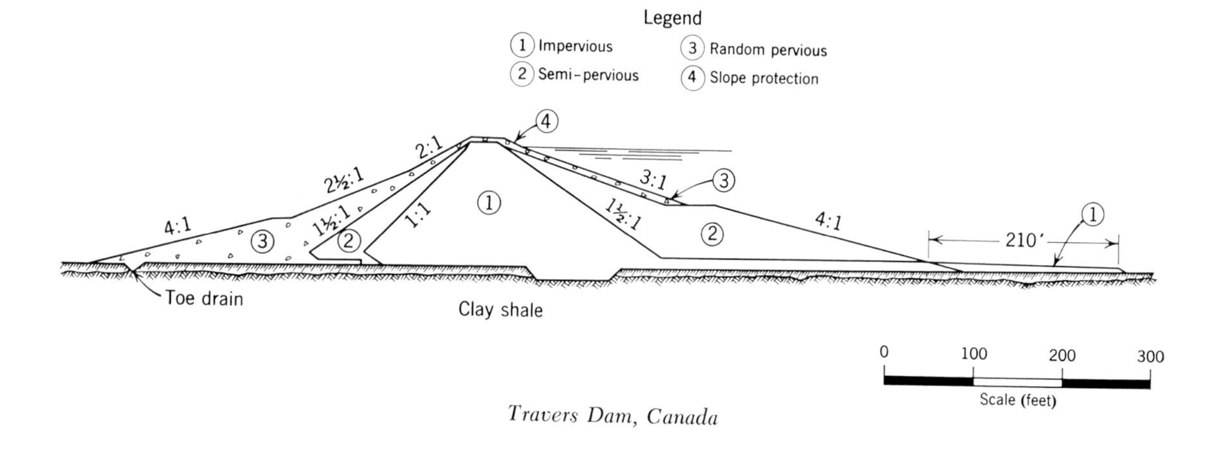

Travers Dam, Canada

Fig. 1.2:55 Typical Designs.

Dams of Foreign Engineers and Agencies (after Peterson, Ref. 161).

of the flow. Within a single embankment zone the individual particles acted on by the seepage forces cannot move because they are held in place by neighboring particles. At the point inside a zoned dam where the water discharges from fine material into coarse material, however, it is theoretically possible for the finer soil particles to be washed into the void spaces of the coarser material. Actually, there are no records available of failures caused by the migration of particles from a zone of fine soil to a zone of coarse soil (that is, sand or gravel). The only records of internal piping failures of this type have been at old dams where zones of fine-grained soils have been placed directly against zones of quarried rock, as was the case at the Schofield Dam (Sec. 1.2*b*). Nevertheless, it is generally accepted as good practice to require that the relative gradation of adjacent soil zones meet established "filter criteria" to prevent any possibility of appreciable migration of soil particles.

Where the difference in coarseness between the fine and coarse embankment zones is too great to meet filter criteria, zones of intermediate gradation must be provided. These are constructed of sands and gravels with special gradation characteristics and are called filters.

In a zoned dam with an internal impervious core and several upstream and downstream zones of increasingly pervious material, the transition from fine to coarse soil necessary to prevent piping is frequently achieved by the progressive zoning without any need for specially graded filter zones. Filters are seldom constructed, for example, for the dams of the U.S. Bureau of Reclamation (USBR), in which large zones of gradually varying materials are used. In homogeneous dams and dams with thin impervious cores, however, the difference between the impervious and pervious embankment zones is frequently so great that one or more layers of filter material with special gradation characteristics are necessary. This is particularly true for earth-rock dams (see discussion starting on p. 37).

While reliable quantitative criteria for the design of filters have been developed only in recent years, progressively graded filters have been used for more than 60 years to protect downstream drains in earth dams from internal erosion. In some of these old dams the filter designs selected on the basis of the engineer's "best judgment" meet all the modern design criteria for filters.

FILTER DESIGN CRITERIA

The two principal requirements for a satisfactory filter are that it must be more pervious than the protected soil in order to act as a drain, and that it must be fine enough to prevent particles of the protected soil from washing into its voids. While some early experimenters studied this problem,[1] the first rational approach to filter problems in modern times was the work of Terzaghi (Refs. 621, 630, 631). The earliest of the systematic laboratory investigations which led to our present criteria for analyzing filters were carried out at the Harvard University Graduate School of Engineering by Bertram working in 1938–9 with the assistance of Terzaghi and Casagrande (Ref. 387). Bertram's experiments, largely on very uniform sands, developed laboratory procedures for evaluating the efficacy of filters. His studies were subsequently confirmed and extended by researches carried out in the soil mechanics laboratories of the Army Corps of Engineers and the USBR (Refs. 390, 391). In all of these filter research programs specimens of soil of various types protected by filters of varying gradations were subjected to much higher hydraulic gradients than exist in earth dams. These results have demonstrated conclusively that properly designed filters provide complete protection against piping.

While there are some slight differences of opinion concerning quantitative criteria for satisfactory filters, the following rules are widely used:

1. The 15% size of the filter (i.e., the particle size which is coarser than the finest 15% of the soil, D_{15}) should be at least five times as large as the D_{15} size of the soil being protected by the filter.
2. The D_{15} size of the filter should not be larger than five times the D_{85} size of the protected soil.
3. The gradation curve of the filter should have roughly the same shape as the gradation curve of the protected soil.
4. Where the protected soil contains a large percentage of gravels, the filter should be designed on the basis of the gradation curve of the portion of the material which is finer than the 1-in. sieve.
5. Filters should not contain more than about 5% of fines passing the No. 200 sieve, and the fines should be cohesionless.

These rules are conservative and should be adequate for any soil type. They are too conservative for some types such as clays and coarse soils with clayey fines, which have inherent resistance to piping because of their cohesion. Many examples are available in which successful dams of coarse gravel or rock have been used adjacent to zones

[1] See, for example, the thoughts on filter design by F. P. Sterns about 1900 (Ref. 389).

of fine-grained cohesive soils. The criteria are probably too conservative also for well-graded coarse soils with silty fines, since these materials in themselves are natural filters. Additional research on many soils is needed in order to extend the rules.

At a site where extensive quantities of filters are necessary and where the easily available natural materials do not meet the criteria as given above, laboratory filter tests may demonstrate that the materials are adequate. The thicker the filter layer, the greater is the permissible deviation from the normal criteria.

DIMENSIONS OF FILTER LAYERS

Theoretically, individual protective layers of properly graded filter material can be very thin. From the practical standpoint, however, the minimum thickness of a layer is that which can be constructed without danger of gaps or of areas of segregated material being incorporated in it. Because they are easy to place, horizontal filter layers can safely be made thinner than steeply inclined or vertical filters. Minimum thicknesses for horizontal layers are about 6 in. for sand and 12 in. for gravel, although thicker layers are generally specified. For vertical or inclined filters, which are used for chimney drains in homogeneous dams or for protecting earth cores in earth-rock dams, a minimum horizontal width of 8 to 10 ft. is desirable for ease in construction, while 12 to 14 ft. is preferable. Figure 1.3:1 shows the placing of two 8-ft. wide vertical filter layers located between the rolled earth core of fine silt and the upstream zone of quarried granite rock at McNary Dam (see cross section, Fig. 1.2:21). This method of placing has been used at many dams. The filter materials were dumped in 6-in. layers from trucks which traveled longitudinally along the filter, and a minimum of hand labor was required to maintain the boundaries between the layers.

Though not too frequently done on earth dams, layers or bands of filter material which are on a slope of 1:1 or flatter can be spread in a single relatively thin layer if one zone of the embankment is constructed ahead of the other. However, this procedure contains the risk of erosion and contamination of the exposed filter. Also, when spreading with a tractor on a steep slope, great care is needed to prevent holes from being torn in the filter. At the Briones Dam, Fig. 1.2:42, the intricate filter was constructed with a specially designed and fabricated traveling bridge complete with a spreading blade and a vibrating plate tamper.

Fig. 1.3:1 Placing two vertical filter layers between upstream zone of quarried rock and central core of cohesionless silt, McNary Dam (*1950*). The *bulldozer is working on the silt core. The fine filter adjacent to the core is a medium to fine sand. The coarse filter on which the man is standing is a pit-run river gravel. The quarried rock zone is shown in the far right-hand side of the photograph.* (*See dam cross section, Fig. 1.2:21.*)

At a number of dams where the filter material is very scarce and expensive, vertical filter bands 3 to 5 ft. wide have been constructed, usually by means of wooden forms. However, much more hand work and supervision is required to ensure the integrity of narrow filters than is required for filters of normal width. For example, at Sasumua Dam (Kenya), Fig. 1.2:6, the 5-ft. wide vertical chimney drain was composed of three vertical bands of material with these gradations (Ref. 304):

BAND	WIDTH	GRADATION
Upstream	2 ft. 0 in.	25 mesh to 1 in.
Middle	2 ft. 0 in.	1 to 3 in.
Downstream	1 ft. 0 in.	1 in. to dust

Figure 1.3:2 shows typical construction of thin vertical filters with wood forms set up to minimize contamination.

PROCESSING FILTER MATERIALS

The design criteria cited above allow a considerable variation in the gradation of suitable filters so that pit-run deposits of sand and gravel can frequently be employed as filter materials. In recent years, however, it is becoming more and more the practice to process filters. Processing sometimes involves washing and screening the material in a manner similar to that used for the preparation of concrete aggregate.[1] In other cases natural or screened materials are blended. Crushed and screened rock is also occasionally used for filters.

Various washing methods have been devised to remove excess fines from natural sands and gravels intended to be used as filters. At a few dams it has been possible to wash fines from gravels by dumping the material into the river and re-excavating it. At the Kenney Dam in

[1] At the Swift Creek Dam in Washington, Fig. 3.3:3, the 12-ft. wide vertical filters contained a total of about 300,000 cu. yd. of stream gravels, which were screened on a ¼-in. sieve. The material was washed, screened on vibrating screens, and stockpiled under a separate contract, at a cost of approximately $2.00/yd^3.

Fig. 1.3:2 Construction of thin vertical filters at Tustervatn Dam (1958). Wooden forms are being used to keep individual bands separated. (See dam cross section, Fig. 1.2:49.) (Courtesy Norwegian Geotechnical Institute)

Fig. 1.3:3 Self-propelled vibrators operated by one man, used in Europe for compacting filters and other zones of cohesionless sands and gravels.

British Columbia, Fig. 1.2:15, sand for the fine filter was washed in a trough as described below (Ref. 48):

A flat bottomed steel trough was built on a 28 degree slope; at the lower end of this was placed a steel bin with a sand gate at the bottom, and provision for water flow at the top. Sand was pushed in the upper end of the trough by a bulldozer, numerous small jets of water played uphill against it and the sand and water flowed down into the bin. Sand with about 1% of silt settled to the bottom and the overflow water carried off the rest of the silt . . . it was a cheap and effective operation.

FILTER CONSTRUCTION

In the same manner as other cohesionless embankment materials, filter layers are usually compacted with excess water by means of tractors, vibratory rollers, or rubber-tired rollers. In Europe small self-propelled vibrators are often employed, Fig. 1.3:3 (Ref. 427). On some jobs, however, there has been trouble with the crushing of the filter materials while they were being handled, and it has been necessary to adopt special methods of placing and compaction to avoid creating too many fines. The grousers on the treads of tractors in particular exert very high unit pressures which crush soft or brittle filter materials. Filter material which is to be compacted should be sufficiently hard and durable to prevent much breakdown. Laboratory research appears to indicate that the efficiency of the filter is not greatly influenced by its density, and at a number of major dams limited quantities of filter materials have been placed without compaction in order to avoid crushing them. However, horizontal filter layers which extend over any appreciable fraction of the base of the dam should be well compacted to avoid the possibility of subsequent liquefaction (Sec. 2.9).

Where large areas of filter are exposed temporarily during construction, there is always a risk that they may be clogged by silt from surface water running into the drains. This happened, for example, in the early part of the construction of the Briones Dam which has an intricate series of inclined filters under an upstream sloping core, Fig. 1.2:42. At this site, when a considerable quantity of the filter was in place, and before it was covered with the rolled core, heavy rains backed water up against the partially completed embankment. This water was very dirty and deposited enough silt in the filters such that it was necessary to remove and replace an important quantity.

A number of construction methods have been used to keep filter layers intact and clean. For example, with vertical filters placed in chimney drains, it has been found helpful in preventing contamination to keep the construction surface several feet higher than the average surface of the embankment, Fig. 1.3:4.

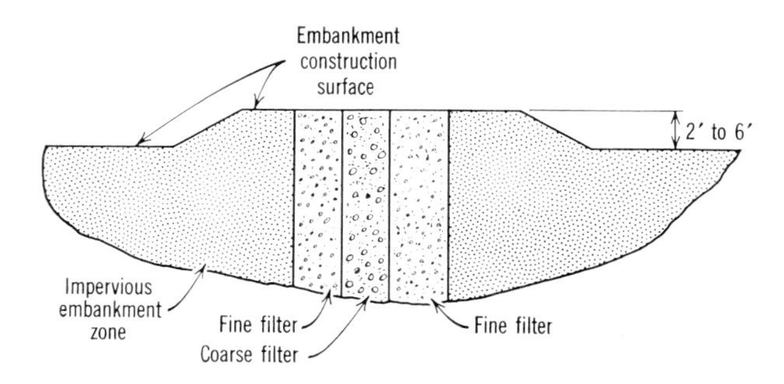

Fig. 1.3:4 Keeping the construction surface filters for chimney drains higher than the adjacent embankment surface assists the effort to keep the filter boundaries intact and well defined.

Fig. 1.3:5 Natural segregation of coarse material at abutments must be prevented in the construction of filters.

For filters constructed of well-graded sands and gravels, special precautions must be taken to prevent segregation of the coarser particles. As the material is spread into layers, the spreading equipment tends to separate the coarser gravels and push them to the end of the row (Fig. 1.3:5). For this reason it is advisable to limit filter material to a maximum size of about 3 in.

PIPES INSIDE FILTERS

In a number of old dams, porous pipes of metal or masonry were embedded in the drains to assist in carrying away leakage water. Because of bad experiences in these dams with pipes which were not surrounded by adequate filters, many engineers now will not place a pipe inside an earth dam under any circumstances.[1] Other engineers will use them only in the downstream third of the embankment in

[1] For example, The U.S. Army Engineers currently do not use pipes under any circumstances, even at the downstream toe.

such a way that complete failure of the pipe could not be serious. In a few modern earth dams, however, extensive use has been made of pipes (see, for example, Briones Dam, Fig. 1.2:42). In the design of the Mission Dam in British Columbia, large leakages were anticipated through differential embankment settlement cracks. To carry this water, a number of 24-in. drainage pipes surrounded by filters were embedded in the pervious foundation and were extended to within a few feet of the upstream face of the dam.

The main advantage of pipes is that they can carry large quantities of water and, in circumstances where large flows are anticipated and coarse, pervious material for drains is scarce, they may effect some economy. A secondary advantage is that leakage is channeled into a number of concentrated flows which can be more easily measured and analyzed. The main disadvantage of pipes is that they constitute a natural escape channel for internal erosion if the surrounding filter should become ineffective. Pipes inside earth dams have torn apart, crushed, filled up, and caused piping failures.

All pipes used must be surrounded by carefully constructed filters with a thickness of not less than 12 in. and preferably of more. The perforations in the pipe should not be larger than one-half of the D_{85} size of the filter.

1.3c Embankment Freeboard

Sufficient freeboard (the height of a dam above the spillway crest) must be provided so that there is no possibility whatever of the embankment being overtopped. The necessary freeboard is calculated by assuming that the maximum river flood will occur when the reservoir is full and that the highest possible waves will develop at the same time. The minimum freeboard equals the computed head on the spillway crest at maximum flood discharge, plus 1.5 times the wave height (for runup on riprapped slopes), plus a safety factor. The safety factor, which generally varies between 2 and 10 ft., is selected by considering the size of the reservoir, the dam height, the reliability of the data from which the flood computations are made, and customary practice. In very cold climates some engineers add a few extra feet of freeboard to compensate for the possibility that the upper part of the dam may be damaged by frost penetration and ice segregation, although no known damages from this cause have occurred.

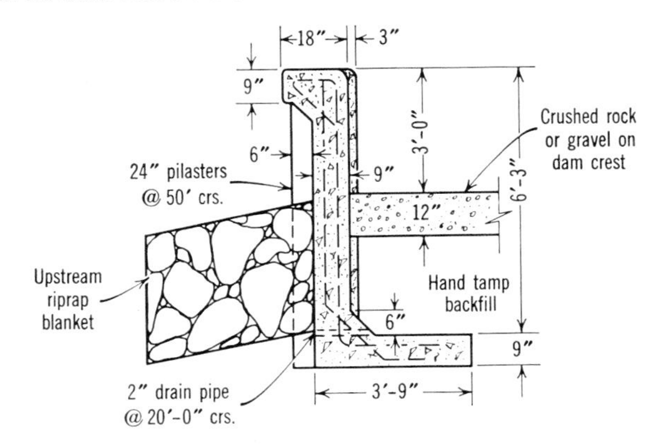

Fig. 1.3:6 Concrete parapet wall used at upstream edge of crest on many of the older USBR earth dams (Ref. 35).

The necessary freeboard is directly dependent on the spillway design, and the amount of freeboard provided affects the cost of both the spillway and the embankment. It is often more economical to add freeboard enough for storing a large part of the maximum flood temporarily in the reservoir, so that a spillway of lesser capacity can be used.[1]

Some additional freeboard can be obtained by constructing a parapet or wave wall of concrete or stone at the upstream edge of the crest. Such parapet walls are not used as frequently now as in past years. Before World War II, for example, most of the USBR earth dams had walls of the type shown in Fig. 1.3:6 added as extra freeboard, and their heights were not considered in the selection of the minimum freeboard. On the USBR dams constructed since 1945, the practice of using wave walls has been abandoned. For one reason, it was found that in cold climates they act as hindrances to snow cleaning.

In the special case where high waves are expected during extreme storms, however, such walls may provide a safe and economical means of keeping the freeboard to a minimum. Since the cost of raising the crest elevation a few feet is greater for a high dam than for a low (the

[1] Design studies for the Army Engineers' Buford Dam in Georgia showed that it was $5 million cheaper to construct a dam with 40 ft. of additional height and take advantage of a natural saddle in the abutment for the spillway rather than to provide a spillway with lower crest elevation and a capacity for the whole computed flood (Ref. 3).

increased height being added to the bottom of the embankment), parapet walls have a greater economic advantage on high dams.

1.3d Crest Width and Camber

The width of the dam crest has no appreciable influence on the embankment or foundation stability and does not have a large influence on the embankment volume.[1] It is determined only by the working room which will be needed on it. Most major dams have crest widths varying between 20 and 40 ft., with the wider dimensions being used for the higher and more important structures. Many successful small dams have been constructed with crests less than 20 ft. wide. No dam should have a crest width of less than 10 ft., because this is the minimum needed for an access road to permit maintenance work.

A crest which is to be traveled often should be provided with a pavement and a wearing surface constructed in accordance with highway practice. The crest should be drained with a slope toward the reservoir, and as a minimum topping should have a layer of well-graded granular material with a clay binder.

At the end of construction the crest should be given a sufficient camber to allow for the postconstruction compression of the embankment and foundation without a reduction in the freeboard. It is always preferable to provide an excess of camber since, in addition to the danger associated with loss of freeboard, an earth dam with a noticeable camber is esthetically pleasing. Moreover, even the slightest sag in the crest of a dam in a moderately wide valley can be easily recognized and always appears much larger than it actually is.

The camber is varied in any smooth curve or series of straight lines from zero at the abutments to a maximum at the center of the valley where the expected settlement is highest. Except for dams in which many feet of foundation settlement are anticipated, the extra height necessary for the camber can be obtained easily by a slight steepening of the slopes on the upper embankment.

The part of the postconstruction crest settlement which is due to foundation compression is estimated by the techniques of soil mechanics; the part which is caused by embankment compression, on the other hand, can best be estimated by measurements on other dams

[1] For a dam of 100 ft. or more, an increase in the crest width of 10 ft. will increase the volume of the embankment between 1 and 5% depending on the height, the side slopes, and the shape of the valley.

(Sec. 3.1). The long-time compression of well-constructed dams ranges between 0.2 and 0.4% of the embankment height, depending on the soil type. The compression of sluiced, dumped rockfill embankments can be predicted roughly from measurements made on other dams, Fig. 3.2:3. Though very little information is available concerning the postconstruction settlement of rockfills which are placed in layers (Sec. 3.2*a*), the designer can assume that they will compress more than rolled earth embankments but less than rockfills dumped in high lifts. Compression of clean rockfills placed in thin-rolled or vibrated layers, however, probably is generally less than that of rolled earth embankments.

1.3e Curving Earth Dams for Arch Action

Although it is far from a universal practice at the present time, many major earth dams constructed in narrow valleys during the last 10 years have been curved upstream. This curve is provided in an attempt to produce axial compression in the core of the embankment as the dam settles, thus counteracting the tendency for the development of axial tension and cracking (Sec. 2.3). The first modern earth dams to be arched were the Aluminum Company of America's thin, sloping core earth-rock dams (Sec. 1.2*b*), in which large post construction compression in the dumped rockfills was expected. All of the subsequent dams of this type have been constructed with an upstream curve. Subsequently a number of high earth dams in relatively narrow valleys have also been constructed with upstream curves.

The value of arching in earth dams is problematical. Many engineers are skeptical as to whether it has any benefit except possibly for an embankment with a very thin core in a narrow valley where considerable settlement or downstream movement is expected. However, since the cost of building a dam on a slight curve is negligible,[1] and since the curve certainly cannot harm the performance, it is to be ex-

[1] The extra cost of building the dam on a curve can be taken as negligible. Actually, of course, additional surveying is required; however, a surveying crew is needed in any event, and it is highly improbable that contractors will give different unit prices for a dam which has a curved axis than they would for a dam at the same site with a straight axis. The engineers charged with the responsibility for surveying and staking out the embankment in the field during construction of the Cougar Dam (Ref. 496), which is arched on a radius of 1,640 ft., concluded that their work was no more difficult than it would have been if the axis had been straight. All dimensions were computed and laid out along radial lines from the center of the circle, which was established on the ground downstream from the dam. At other dams, the curve has been laid out with chords from a straight axis.

Fig. 1.3:7 Construction surface of dam being constructed with an arch having a radius of about 2,000 ft. (Arstaddelen Dam, Norway, 1962).

pected that more high earth dams in the future will be arched upstream. The radius of curvature at the crest has generally varied in the range between 1,300 and 5,000 ft., Fig. 1.3:7.

With certain conditions of foundation or surface topography, the volume and the cost of a dam may be made much less if the axis is broken or curved in the opposite direction, i.e., with the concavity upstream. A few major dams have been built in this way in recent years. One example is the high Serre-Ponçon Dam in France (Ref. 615), which has an appreciable bend in the axis with the concavity upstream; however, the settlement of the foundation and the embankment was low and there was actually a final longitudinal compression in the crest instead of tension. Therefore, it was apparent that the curve in the axis was a perfectly satisfactory design.

In general, however, building earth and rockfill dams with an arch in the wrong direction should not be considered good practice. It should never be done unless there are relatively great potential savings to be made in the design and unless the anticipated settlement is not great. Also, if it is done, more attention should be paid to design measures which increase the safety against the possible development of tension cracks through the core.

1.3f Raising Earth Dams

Special design problems may be encountered where an old dam must be raised or where a new dam must be built in such a way that it can be raised at a later date. Concerning the first problem, each dam is such a special case that the experience of the past does not offer much assistance.[1] The considerations involved are essentially the same as for the construction of a new dam, with the exception that there should be less trouble with handling the river during construction. Where the height increase is small compared with the original height, economy will usually require that the core of the extension be tied to and made part of the existing dam. Where the height increase is great, it may prove more economical to build a completely new core and to use the existing dam only as a cofferdam during construction and as a means of increasing the stability of the upstream slope. It may be necessary to explore the properties of the existing dam in the same way as the foundation in order to determine its actual condition before deciding whether to use it as an integral part of the new dam core.

The design of an addition to an existing dam is governed to a large degree by the anticipated level of the reservoir during construction. If the reservoir can be emptied, the designer can build part of the new dam upstream from the old. In the more usual case where it is necessary to keep the reservoir in operation, all added embankment material must be placed above the crest and downstream from the old dam.

In the case where a dam is to be built in stages, the embankment for the first stage must be designed in such a way that it can be incorporated economically into the higher dam. In the simple case of the homogeneous dam, for example, the structure can be raised by adding to the existing embankment with the same material. A dam with an upstream impervious membrane of manufactured materials such as reinforced concrete can also be raised easily by building up the embankment downstream and simply extending the membrane.[2] A

[1] For some of the little material available in the engineering literature on this subject, see Refs. 81, 82, 208.

[2] One of the reasons for choosing a rockfill embankment with an impervious upstream membrane of asphaltic concrete at the Montgomery Dam in Colorado in 1957 was the fact that the structure could be rapidly and economically raised a planned 40 ft. (Ref. 84).

dam with a thin central earth core, on the other hand, is inconvenient to raise in stage construction because of the difficulty of extending the core.

Often the major problem both in extending old dams and in planning dams to be raised in stages occurs in connection with the spillways (Ref. 306). If a reservoir is to be elevated by only a few feet, the engineer can frequently retain the existing spillway by raising the crest through some relatively simple means and perhaps also by increasing the capacity of the energy dissipator. If the water level is to be raised a considerable distance, however, a new spillway may be required, and this can have a major influence on the cost and, when building a new dam, the desirability of raising the dam in stages.

In a case where stage construction is being considered, it may be desirable to prepare alternate plans for a high dam and for a low dam capable of being raised, and to ask contractors to quote on both schemes. When this was done for the Dillon Dam on the Blue River in Colorado in 1958, the bids for the high dam were so attractive that the owner decided to abandon plans for stage construction and to build the dam to the full height immediately.

1.4 DESIGN PROVISIONS TO CONTROL PORE PRESSURES

1.4a Control of Construction Pore Pressures

Under certain conditions during construction, embankment compression due to the weight of the overlying fill may cause high pore water pressures to develop in the impervious sections of earth dams. Such pressures may exceed any which occur later due to seepage from the reservoir, and, consequently, they may control the design of the dam from the standpoint of slope stability.

Most of the information available on construction pore pressures comes from piezometer measurements recorded at existing dams (Sec. 3.3). These measurements indicate that the magnitude and distribution of the pore pressures during construction depend primarily on the construction water content, the properties of the soil, the height of the dam, and the rate at which dissipation can occur from drainage.

Since high construction pore pressures exist only during the first few years of the life of the dam, it is desirable to avoid using a more conservative and expensive design than would otherwise be required. A number of design and construction procedures are available to eliminate the necessity for making the slopes flatter and the total volume of the dam greater because of the "during construction" stability condition:

1. The maximum construction pore pressures can be limited by compacting the impervious section of the embankment at an average water content a few percent below Standard Proctor Optimum.
2. The impervious section can be made thinner, so that high construction pore pressures will have less influence on the stability and will dissipate more rapidly.
3. Internal drains can be installed within the impervious section to accelerate the pore pressure dissipation.
4. Longer construction periods can be required to allow more time for pore pressure dissipation.
5. Lower safety factors can be tolerated against the possibility of slope slides during construction than would be permissible when the dam is retaining the reservoir.

CONTROL OF CONSTRUCTION WATER CONTENT

Both theory and experience indicate that a relatively small difference in the average construction water content has a large influence on the pore pressures which develop. The method of limiting them by compacting on the dry side of optimum water content was developed and is used extensively by engineers of the USBR (Refs. 18, 393). Their experience over the last 15 years has demonstrated that, at least in the semiarid geographical areas where most of their dams are built, it is practicable to hold construction pore pressures to tolerable maximum values.

In the USBR procedure the range of water content used for construction of the impervious embankment section is determined by laboratory tests. The lower limit is taken as the value at which no "compression on saturation" will occur in a compacted specimen in a consolidation test. The upper limit is determined by computation (Sec. 7.5) as the construction water content at which the maximum tolerable pore pressure will not be exceeded. These limits (especially the upper) vary as a function of the weight of the overlying fill, and a narrower range is used in the lower portions of high dams than in either the upper portions of high dams or low dams. The USBR methods of testing and presenting the results of the limiting moisture content studies are given in Ref. 393.

The experience of the USBR indicates that by placing impervious materials at an average water content between 1 and 3% below Standard Proctor Optimum, the pore pressures can be kept to reasonable values. These values are usually limited to a maximum of about 30% of the weight of the overlying embankment.

In the last 10 years many other major dams in addition to those of the USBR have been built with low water contents to reduce pore pressures. However, although this practice is accepted by many experienced engineers, many others argue that the dam built with low water content may be too brittle to follow differential settlement without cracking.

The influence of relatively small differences in the compaction water content on the ability of compacted soils to deform without cracking is great. Embankments of most fine-grained soils are stiff and brittle when compacted with an average water content several percent below Standard Proctor Optimum, whereas when compacted a few percent above Optimum, they are capable of appreciable deformation without cracking. As described in Sec. 2.3, serious cracks have developed in some dams, and it is known that the likelihood of cracking is greatest in dams composed of certain soil types, located at sites with steep abutments, and constructed with relatively dry compaction. However, not enough is known about this phenomenon to establish reliable criteria which will assure that a dam can safely be constructed at a given site with any predetermined low water content.

Many groups of engineers, including the Army Corps of Engineers, are so concerned about this danger that they prefer to compact impervious sections at or above optimum water content and deal with construction pore pressures by some other method. The arguments for wet and dry compaction as presented at an open meeting on the subject by Thomas A. Middlebrooks and Fred C. Walker in Switzerland in 1953 provide an excellent summary of current opinions.[1]

Thomas A. Middlebrooks, Chief, Soil Mechanics, Geology and Geo-Physical Section, U.S. Army Corps of Engineers, Washington, D.C.

"In addition to the regular criteria relating to stability and permeability we give special attention to the core material in order that it may satisfy the following criteria:

1. It must be placed at a density and at a moisture, which will not allow further consolidation on saturation.
2. It must be sufficiently plastic so that differential settlement will not cause cracks to develop through it.

[1] *Proceedings of the Third International Conference on Soil Mechanics and Foundation Engineering*, Switzerland (1953), vol. III. pp. 213–216.

Both of these criteria can be easily satisfied when the core material is placed well on the wet side of an optimum. In fact this is the only way criterion 2 can be fully satisfied.

You will certainly ask: what about pore pressure? My answer is that it is just another factor which must be considered in any design. It is not an overriding or controlling factor. My suggestion is that in cases where pore pressure will decrease the stability, that the previously mentioned criteria be applied to only the central portion of the core or impervious section, say at width of ⅓ to ¼ of the height of the dam.

It is my opinion that the danger of developing excessive pore pressures in modern rolled fill dams has been greatly overemphasized. I assure you that the danger of placing the fill on the dry side of optimum can be much greater. It is a simple matter to install piezometers and observe the pore pressure as it develops. The dangers inherent in a dry fill are not apparent in most cases until the reservoir is full.

In answer to the General Reporter's specific question, on what modifications in cross sections are contemplated by the Corps of Engineers, due to placement on the wet side, my answer is none. We will check the stability during construction more closely and install more piezometers. Since any pore pressure which develops will dissipate quickly after construction is completed, a lower factor of safety is tolerated during this period. Effective stresses are used in the stability analysis for all cases."

Fred C. Walker, Head, Earth Dam Section, Office of the Chief Engineer, U.S. Bureau of Reclamation, Denver, Colorado

"I am fully in agreement with most of Mr. Middlebrooks' comments. But there is a difference in our experiences that is largely connected with our design practices. If Middlebrooks had been in my position and I in his, our policy positions might well be reversed.

The Bureau of Reclamation has done the bulk of its work in the high dry plains and in the mountains of the Western part of the U.S. The Corps of Engineers, on the other hand, has concentrated most of its work on the wet lower alluvial valleys where the soft foundations are the rule. In consequence it is practical for us, in the Bureau of Reclamation, to take advantage of any possible saving in structure by consideration of the pore pressures to reduce the cost of the dam. In one instance we studied recently, we found that on a dam costing approximately 10 million dollars, we could reduce the costs of that dam by approximately 1 million dollars, by requiring control for the pore pressures. On the other hand, I must further emphasize Mr. Middlebrooks' point, that there is a great danger in constructing embankments too dry.

When I had to present papers for the 1948 High Dams Conference and the Rotterdam Conference, I was not yet in a position to say whether it was practicable to control pore pressures and prepare designs assuming such control. We had such plans under way but did not have the evidence of performance. I did not think that we had sufficient new information to present to this group at this time, to make the prepared paper worthwhile. However, on the numerous discussions I have heard here, I see that I have been mistaken. We have found that the pore pressures can be controlled under construction and kept to very low values, under the climatic conditions with which we have had to work in the Bureau of Reclamation. I will readily admit that in England and in Sweden and in many tropical countries such procedures are impossible. But if there is money to be saved and the climate permits then I think that it is well worthwhile to go after pore pressure control. Our experience leads me to believe that a structure of less than 100 ft. in height will not show a great saving in cost, whether pore pressures are considered at high values or at low, because other factors in the design predominate.

But for a dam of about 200 ft. height, costing approximately 3 million dollars, we can, if we can keep the pore pressures down to about 30% uplift, reduce the costs of that structure by about 500,000 dollars."

During the placing of an embankment at a high water content, the construction surface undulates directly in front of the equipment. Under certain conditions the shear stresses associated with this movement have caused shear planes on horizontal surfaces to develop in the cores of dams, Fig. 1.4:1. The factors which influence the development of these shear planes are not well established; however, since they have not occurred frequently and have not been a large source of

Fig. 1.4:1 Shear surface developed in fine-grained embankment compacted at high-water content (after Justin, Craeger, and Hinds, Ref. 134).

construction difficulty on many jobs, the problem is relatively unimportant.[1] Research by the USBR indicates that the shear planes can be produced in clayey soils in the laboratory when the material is compacted with a tamper at a water content above Standard Proctor Optimum (Ref. 100). Probably the planes develop only in embankments of fine-grained soils. On some projects where the engineer was at first concerned over the possible influence of these shear surfaces on the performance of the dam, tests indicated that the strength and permeability of the material along the surfaces were not appreciably different from those on other planes through the embankment (see, for example, the closing discussion of Ref. 415).

CORE WIDTH AND DRAINAGE

Measurements have demonstrated (Sec. 3.3) that during construction periods of normal length, little pore pressure dissipation from drainage occurs in the central portions of thick core dams composed of most fine soils. On the other hand, in the peripheral portions of the cores adjacent to pervious embankment zones, the pore pressures are completely dissipated, Fig. 3.3:1. The rate of dissipation in the center is approximately inversely proportional to the square of the core width, so that the construction pore pressures can be reduced considerably by reducing the thickness of the core. Any impervious embankment section in which high construction pore pressures can be maintained without appreciable dissipation for several years is sufficiently impervious that a thick core is not required to reduce the seepage loss from the reservoir.

In homogeneous dams or other dams at sites where only small quantities of pervious embankment material are available, dissipation of construction pore pressures can be accelerated by judicious installation of the internal system of downstream drains. The chimney drain, Fig. 1.2:1, because it provides a vertical strip of embankment with zero pore pressure in the central portion of the dam where the maximum pressures would otherwise develop, has a large beneficial influence on the during-construction stability condition. The fact that the downstream slope of Otter Brook Dam remained stable while the upstream slope bulged was undoubtedly due to the action of the inclined chimney drain in controlling pore pressures in the downstream portion (Sec. 3.3*b*).

[1] For example, see Refs. 655 and 675 for descriptions of situations at which these shear planes were observed and considered unimportant.

Special drains to control construction pore pressures have been built into a few dams with little cost and considerable success. Since such drains need only carry very small flows of water from the soil pores as the embankment compresses, they do not have to be large or exceptionally pervious. At the Army Corps of Engineers' North Hartland Dam in Vermont, Fig. 1.2:6, two horizontal blankets each 1 ft. thick were placed in the main downstream portion of the embankment. At the Usk Dam in England, in 1953, Fig. 1.4:2, similar drainage blankets were added to the design during construction when piezometer measurements indicated that unexpectedly high construction pore pressures were threatening the stability of the embankment. At both dams these drains had no function during the subsequent operation of the reservoirs. At the Selset Dam in England (built 1957–1959), high rainfall during the construction season threatened to cause high pore pressures, and extensive horizontal drains were included in the design as shown in Fig. 1.4:3. These drains performed very satisfactorily in controlling the pore pressures (Ref. 531).[1]

Figure 1.4:4 shows the projected design (1962) for the Las Palmas Dam on the Rio Cojedes in Venezuela. The bulk of the embankment is constructed of fine-grained soil, which is available in large quantities near the dam. Sand and gravel, available at a considerably greater haul distance and cost, are used in small quantities in horizontal drainage blankets in the upstream and downstream slopes to assure that no high pore pressures will develop in these zones during construction. These drains are also intended to control drawndown pore pressures in the upstream slope and to prevent any possibility that high pore pressures will develop in the downstream slope as the result of prolonged rainfall or leakage through the abutments.

LOW SAFETY FACTORS

Another reasonable procedure when high construction pore pressures are anticipated is to accept a lower safety factor for the during-construction condition. Since a construction slide in a compacted

[1] Drains of this type were originally planned in the high Swift Creek Dam in Washington (Ref. 77), but they were eliminated when piezometer readings during construction showed that negligible pore pressures developed. At the Troneras Dam in Colombia, in which unexpectedly high pore pressures developed during construction, a single drainage blanket in the upstream and downstream slopes was less effective than had been hoped (Ref. 543).

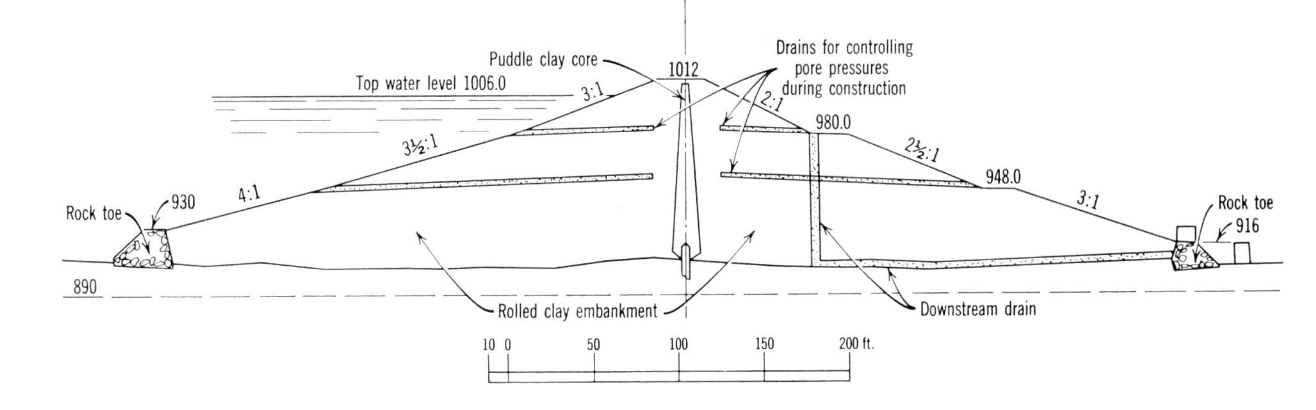

Fig. 1.4:2 Cross section of the Usk Dam, England (1953), showing horizontal drainage blankets installed within the embankment for the purpose of controlling construction pore pressure (after Little, Ref. 72).

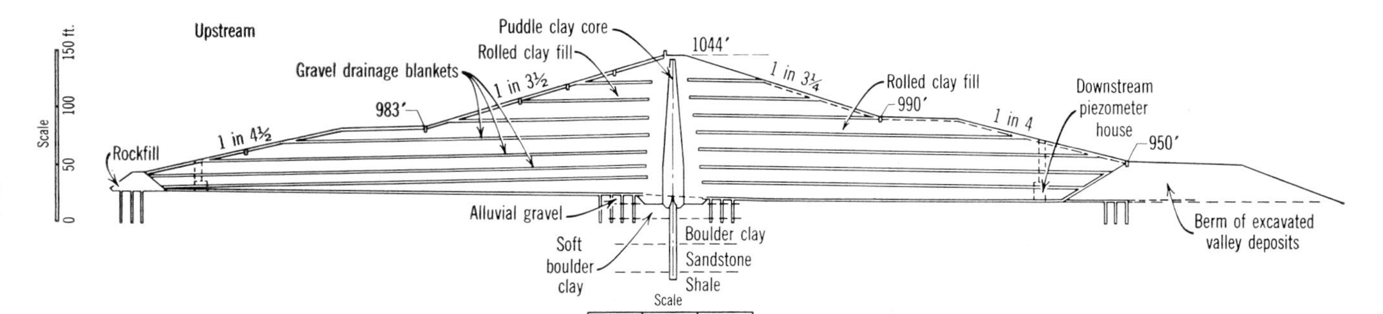

Fig. 1.4:3 ***Internal embankment drains*** *used to control construction pore pressures at Selset Dam (after Bishop et al., Ref. 531).*

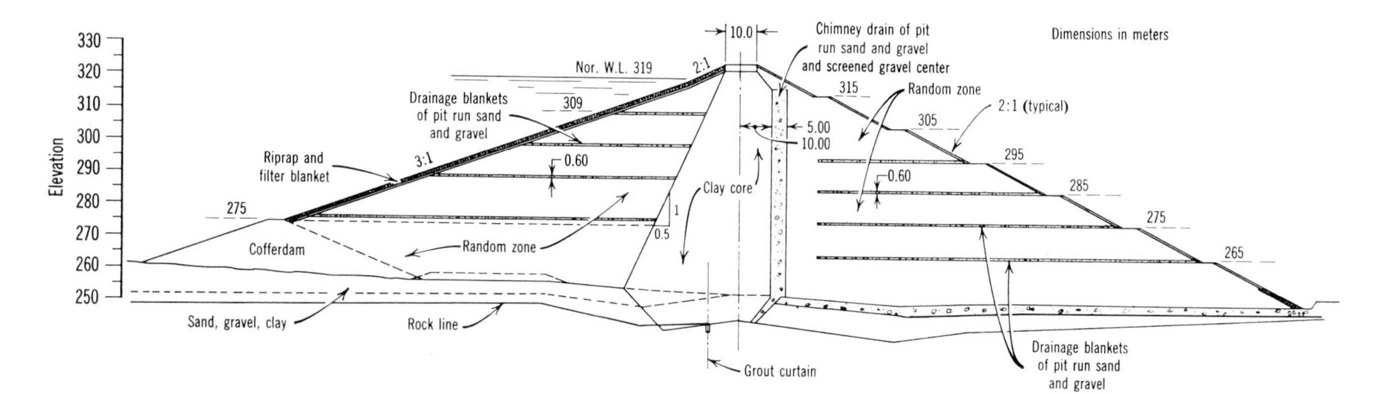

Fig. 1.4:4 Cross section of Las Palmas Dam (projected design). (Courtesy Ministry of Public Works, Caracas)

earth dam does not result in a failure of the catastrophic type, which involves loss of life and great property damage (Sec. 2.4), it is within the realm of activity for which a calculated risk is permissible.[1] Except when the dam is founded on a deposit of sensitive clay or silt (where the movement can be larger), the worst results of a construction slide due to high pore pressures can only be: (1) a few feet of relatively slow movement in a portion of the embankment; (2) repair costs which are a fraction of the original cost of the dam; (3) some delay in the completion of the construction; and (4) embarrassment for the engineer if the client has not been properly warned in advance about the calculated risk.

The selection of minimum allowable safety factors against shear failure at all times in the life of an earth dam is a difficult problem (Chap. 7). A computed safety factor of 1.5 is usually specified for the full reservoir condition, and a somewhat lower value is often considered acceptable for the condition of reservoir drawdown. For the during-construction condition, if realistic estimates of the embankment strength and pore water pressure are made, a computed safety factor only slightly greater than unity can reasonably be used provided that the construction pore pressures and embankment strains are carefully observed.

When low computed safety factors are accepted for the "during construction" condition, piezometers must be installed to measure the pore pressures which develop (Sec. 8.1), and monuments for measuring horizontal and vertical movement must be placed in those areas on the embankment slope and valley floor which would be the first to move if a slide occurred. Study of the measured pore water pressures and the movements of the monuments during construction will enable the engineer to anticipate an imminent slide in time to take preventive action before any large damage can occur.

[1] More than 40 years ago, following a number of construction slides on hydraulic fill dams which were much more spectacular than any possible slide of a rolled-earth dam, a similar opinion on the relative importance of construction slides was expressed by the late John E. Fields, eminent dam engineer: "The troubles at Necaxa and Calaveras Dams (slides during construction) were not properly dam failures, and to the general public were of little interest, as they involved neither its life nor property; the owner and the engineer were intensely interested and the latter has given considerable prominence to the phenomena presented. It is better to have one dam in ten fail during construction, than to incur the enormous expense necessary for all in order to insure against one failure. On the other hand, the failure of one dam in use involves such great and widespread consequences that any extra precaution and expense is justified to insure the safety of all." [*Transactions, American Society of Civil Engineers,* vol. 83 (1920) p. 1801.]

1.4b Control of Drawdown Pore Pressures

The critical stability condition for the upstream slope of an earth dam occurs as the reservoir is lowered after being full for some length of time. The reduced stability results from the removal of the supporting reservoir load and from the fact that the pore pressures in the slope may not dissipate immediately. If the material comprising the upstream portion of the dam is free-draining rock or sand and gravel. the water will flow out of the pores as rapidly as the reservoir is lowered. If the embankment material is less than free-draining, however, there will be some time lag during which the stability will be lessened.

At present the magnitude of the drawdown pore pressures cannot be estimated with any reliability (Sec. 3.4*b*), and consequently it is common practice to make the upstream slope flat enough to be stable under the maximum possible pore pressures. Except for embankments of very fine-grained compressible clays and silts, this is probably a very conservative procedure (Sec. 7.7).

While drains have not been widely used to control drawdown pore pressures, they can perform this function very reliably. Upstream sloping chimney drains in an otherwise homogeneous dam will change the direction of pore water flow following rapid drawdown and thereby can have an important beneficial influence on the magnitude of the pore pressures, Fig. 5.7:3. At a few dams where the designer has doubted whether the upstream slope was pervious enough to prevent high residual drawdown pressures, one or more horizontal layers of pervious material have been installed. Figure 1.4:5 shows a cross sec-

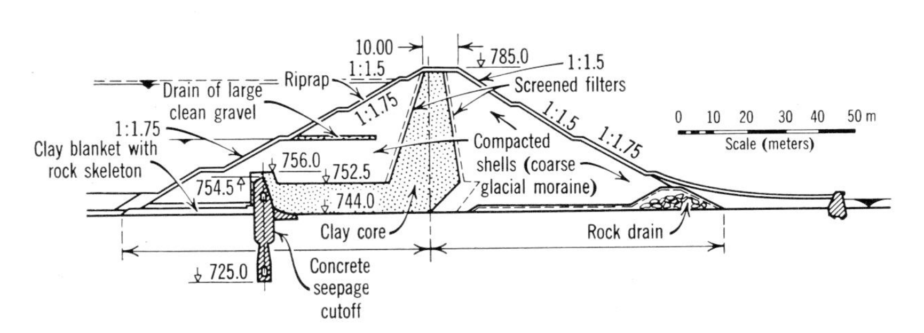

Fig. 1.4:5 Horizontal gravel drain used in the upstream slope of Rosshaupten Dam, Germany (1953) for the purpose of controlling drawdown pore pressures (after Treiber, Ref. 69).

tion of the 130-ft. Rosshaupten Dam (completed in 1953 in Germany), in which a horizontal drainage blanket was placed in the semipervious upstream slope at the low water level (Ref. 69). Figures 5.7:4 and 5.7:5 show the theoretical influence of drains in any such upstream slope on the pore pressures following reservoir drawdown.

Horizontal drainage blankets of the type frequently used under the downstream slope of homogeneous dams (Sec. 1.2*a*) have also been placed under the upstream slopes of a number of dams. Examples of these are the Arkabutla Dam in Mississippi (1943) and the Pomme de Terre Dam in Missouri (1960) where the blankets served the dual purpose of controlling drawdown pore pressures in the embankment and accelerating the consolidation of clay foundations, Figs. 1.4:6, 1.4:7. Drains placed in the upstream slope primarily for the control of drawdown pore pressures may also be of value in controlling construction pore pressures.

The problem of the permeability of upstream embankment zones which are designed to be free-draining frequently arises during construction, because the sand or gravel borrow materials intended for these upstream zones often are found to be dirtier and less pervious than anticipated. The stability of the upstream slope then comes into question, and in some cases during construction this has necessitated major design changes with added costs and delays.[1]

When this situation arose during construction of the Marmorea Dam in Switzerland in 1953, the problem was solved only by washing the fines from all of the material in the large upstream and downstream pervious zones (Refs. 111, 397). Another such case developed in Peru in 1956 during the construction of the 260-ft. San Lorenzo Dam, which has a large upstream zone of sand and gravel, Fig. 1.2:18. To eliminate the necessity for considering drawdown pore pressures in the stability analyses, the designers specified that the sand and gravel comprising the upstream slope should have a coefficient of permeability not less than 10,000 ft./yr. It was decided on the basis of the field explorations that an ample supply of sand and gravel meeting the specifications was available. In the early part of construction, the material obtained from the borrow pits was very clean and was placed by rolling in layers and sluicing with water from a 3-in. hose. However, the borrow pit became dirty, and, it was no longer possible to sluice the material, which then had to be placed as a semipervious fill with compaction water content near optimum. Since the material

[1] See also Sec. 11.4*b* for discussion of the construction problems with sand and gravel embankment materials which contain small quantities of fines.

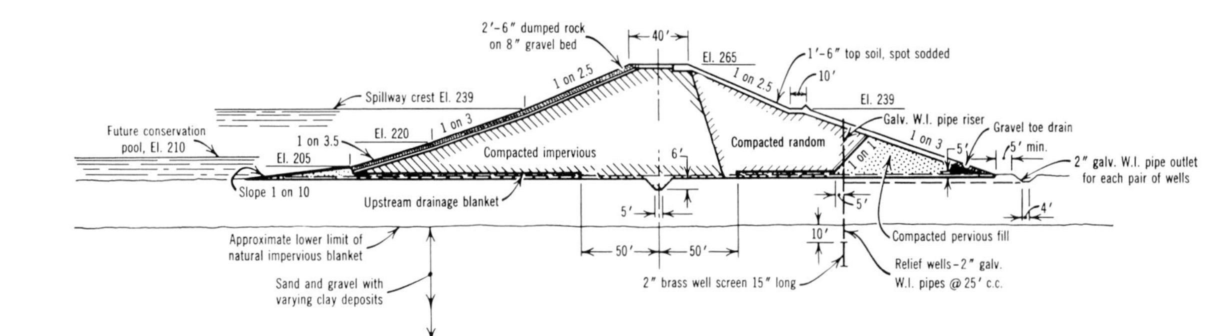

Fig. 1.4:6 Cross section of Arkabutla Dam (1943), Mississippi, showing pervious horizontal drain used under upstream slope (Ref. 362).

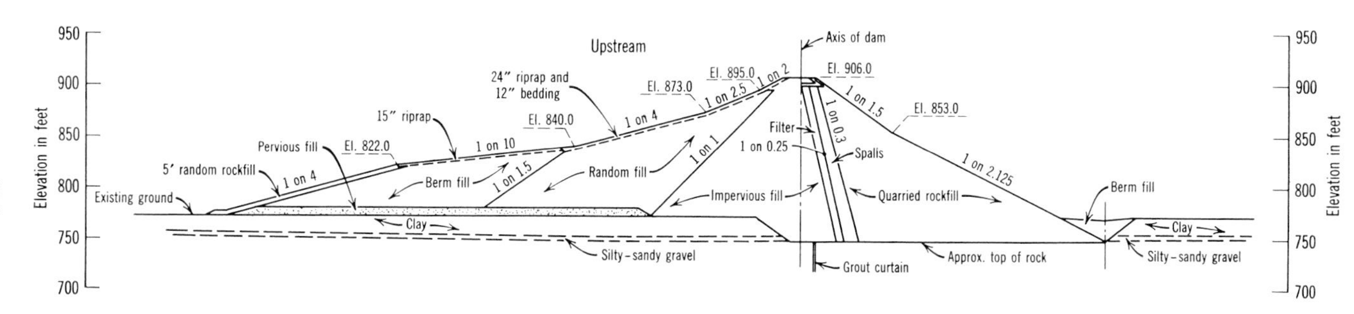

Fig. 1.4:7 Cross section of Pomme de Terre Dam (1961), Missouri, showing pervious horizontal drain used under upstream slope. (Courtesy Kansas City Districts, U.S. Army Engineers)

had ceased to be free-draining, the engineers were forced to re-evaluate the stability of the upstream slope during construction.

Suggested General Reading Covering Design Considerations and Practice

	Reference No.
General summaries of opinion and practice (books and comprehensive articles)	3, 15, 19, 34, 44, 88, 101, 118, 134, 217, 218, 316
USBR practice	31–33, 38, 41, 42, 99, 147, 194, 233, 342, 345–51, 396, 425, 426
U.S. Corps of Engineers practice	11, 12, 20, 27, 74, 81, 87, 301, 315, 362, 420, 584–586, 627
Discussions at conferences	66, 169, 173–180, 198, 203, 232
Mexican dams	344, 628
Prairie Farm Rehabilitation Administration (Canada) practice	161

two

Failures and Damages

2.1 NATURE AND IMPORTANCE OF FAILURES

The development of soil mechanics has given the engineer powerful analytical tools and rational procedures which have made obsolete many of the older, empirically developed "rules of thumb" formerly used for earth dams. Nevertheless, most of the procedures which constitute modern practice for design and construction have evolved from practical efforts to eliminate weaknesses exposed by unsatisfactory performances of older dams. A knowledge of the principal lessons learned from failures and damages in the past is an essential part of the training of the earth dam designer.

In this chapter the authors have attempted to provide a summary of the most instructive experiences with failures and damages. Of necessity such a summary prepared by a small group cannot draw upon all the experience which exists. Many factors in addition to their own natural reluctance to publicize their troubles cause owners and engineers to withhold the details about unsatisfactory performances of their dams. Experiences with some failures and damages remain the exclusive knowledge of a few people, and in other cases the information given to the profession is not complete or wholly correct. As a consequence, there are many misconceptions about the frequency, details, and importance of the failures which have occurred.

The magnitude of recorded damages to earth dams ranges from complete catastrophic failure, resulting in large property damage and loss of life, to relatively minor deterioration which may or may not necessitate remedial work. The worst type of complete failure occurs when the reservoir water suddenly breaks through the embankment and surges downstream in one devastating flood wave. Lesser damages fall into several categories, some of which can lead to complete failure if left unattended and some of which require only maintenance work even under the most extreme conditions.

As can be seen from Middlebrooks' comprehensive list of earth dams with unsatisfactory performance (Table 2.1:1), the most common cause of complete catastrophic failure has been that of water flowing over the tops of earth dams during great river floods when the spillway capacities were inadequate. Although there have been a few cases where overtopping did not result in complete failure, it must be assumed that earth dams cannot be designed safely to withstand the erosive action of water passing over the crest. Failures due to overtopping are not deficiencies in the design of the earth dam itself but rather the result of inadequate hydraulic design, a subject which is outside the province of this book.

The two other principal causes of catastrophic failure are piping (the progressive erosion of leaks which develop under or through the dam) and earth slides in the downstream portion of the embankment or foundation. Upstream slides in the embankment have not often threatened to cause complete failure of the dam since they usually happen after the reservoir has dropped below a dangerous level. Embankment or foundation slides occurring during construction never threaten a catastrophic failure unless water is retained in the reservoir while the dam is being built. Neither do slope and crest erosion by waves, wind, and rain lead to danger of complete failure except in very special circumstances.

The designer must make every conceivable effort to eliminate any possibility of unsatisfactory performance which could lead to a catastrophic failure. Design details which are provided against damages that would not lead to catastrophic failure, on the other hand, fall within the sphere of decision where a calculated risk may properly be taken. Because of the uncertainties in some aspects of earth dam design, the engineer is often faced with a situation where he can make large savings in the cost of the structure by risking certain damages. In some cases the cost of repairing the worst conceivable damage may be less than the extra cost of a design which would eliminate all possibility of damage. When the engineer chooses a design involving such a calculated risk, he must make sure that the owner and everyone else involved is thoroughly cognizant of it.

2.2 EMBANKMENT AND FOUNDATION PIPING

Piping, or progressive erosion of concentrated leaks, has caused a larger number of catastrophic failures than any other action except overtopping, and many of the modern techniques of earth dam design and construction have been developed to prevent it. For example, the present stringent requirements for uniformly compacted embankments with emphasis on control of construction water content and density have been developed to provide dense and homogeneous cores which reduce the incidence of concentrated leaks and resist piping when leaks do develop. Because of such requirements, and because of the introduction of graded filters in the downstream portions of dams, there have been extremely few piping failures in important modern dams.

The experience of piping failures available to the profession has come largely from older dams and from poorly constructed small dams. In addition, since the damage resulting from failures frequently obliterates the embankment or foundation zone in which the piping developed, we must often depend on construction and design records or on eyewitness accounts, if there were any, to reconstruct the causes and mechanics of the failure. As a consequence, the reliability of the records of a failure is frequently a matter for speculation.

2.2a Mechanics of Piping

As water seeps through the compacted soil of an embankment or the natural soil of a foundation, the pressure head is dissipated in overcoming the viscous drag forces which resist the flow through the small soil pores. Conversely, the seeping water generates erosive forces which tend to pull the soil particles with it in its travel through and under the dam. If the forces resisting erosion are less than those which tend to cause it, the soil particles are washed away and piping commences. The resisting forces depend on the cohesion, the interlocking effect, and the weight of the soil particles, as well as on the action of the downstream filter, if any.

If the embankment and foundation were completely homogeneous, the erosive forces would be evenly distributed; but actually the embankment and foundation are not uniform seepage media and the flow is not uniform. Concentrations of seepage quantity and velocity inevitably develop even though the total seepage may be small, and at the places where these concentrations emerge on the downstream side of the dam, the erosive forces on the soil particles are greatest. The removal of a small portion of the embankment or foundation by erosive action at any point accentuates the subsequent concentrations of seepage and erosive forces there.

Table 2.1:1 Unsatisfactory Performance of Earth Dams (After Middlebrooks, Ref., 3)

Dam and Location	Date		Reference[a]	Height	Type	Reason for Unsatisfactory Performance
	Built	Failed				
Alexander, Hawaii	1932	1932	ENR, Vol. 104	140	Hydraulic	Core pressure slide before completion of dam
Anaconda, Mont.	1898	1938	ENR, Vol. 121	72	Earth and concrete core	Seepage slide
Ansonia, Conn.		1894	ER, Vol. 30; EN, Vol. 47		Earth, rolled	Piping along outlet
Apishapa, Colo.	1920	1923	ENR, Vol. 91	115	Earth, rolled	Piping through settlement cracks
Ashti, India		1883	ASCE Proc., Vol. 49	58	Earth, rolled	Seepage through foundation
Avalon (old dam), N. Mex.	1893	1893; 1904	EN, Vol. 54	58	Earth and rock	Overtopped 1893, piping into rock 1904
Avoca, Pa.		1892	EN, Vol. 47			Overtopped
Baker City, Ore.		1896	EN, Vol. 47			
Balsam, N. H.	1927	1929	ENR, Vol. 54	60	Earth and concrete core	Erosion from spillway
Barton, Idaho	1910	1922	Sherard	40	Earth, rolled	Seepage slide
Bear Gulch, Calif.	1896	1914	Sherard	63	Earth, rolled	Drawdown slide
Beaver Park, Colo.	1914	1914	ASCE Trans., Vol. 65	87	Rock	Foundation leakage, repaired
Belle Fourche, S. Dak.	1911	1933	ENR, Vol. 111	122	Earth, rolled	Drawdown slide, concrete slope paving failure
Blairtown, Wyo.		1888	EN, Vol. 47		Earth, rolled	Piping along outlet
Blue Water, N. Mex.	1908	1909	EN, Vol. 62	35	Rock	Overtopped
Bolton, Conn.		1938			Earth, rolled	Overtopped
Bonney Reservoir, Colo.	1901	1903	EN, Vol. 47	34	Earth, rolled	Break in dam
Bradford, England		1896	ASCE Proc., Vol. 49	90	Earth, rolled	Piping along outlet
Breakneck, Pa.		1902	EN, Vol. 47		Earth, rolled	Overtopped
Bridgeport, Conn.	1855	1905	ER, Vol. 52		Earth, rolled	Overtopped
Brooklyn, N. Y.	1893	1893	EN, Vol. 47			Foundation seepage
Brush Hollow, Colo.	1925	1923; 1928	Sherard		Earth, rolled	Conduit break; Drawdown slide
Calaveras, Calif.		1918	ENR, Vol. 80	240	Hydraulic	Excessive core pressure
Castlewood, Colo.	1890	1933	ASCE, Trans., Vol. 65	70	Rock	Spillway over dam failed

Cobden, Ont., Canada		1894	EN, Vol. 48	35	Earth, rolled	
Cold Springs, Colo.		1912	ER, Vol. 66	50	Earth, rolled	Embankment seepage
Cold Springs, Ore.		1931	Corps of Engrs.	98	Earth, rolled	Riprap displaced by waves
Conshohocken Hill, Pa.		1873	ASCE Proc., Vol. 49; ENR, Vol. 18		Earth, rolled	Piping
Costilia, N. Mex.	1920	1924; 1941	ER, Vol. 75	125	Earth, rolled	Embankment seepage; Sloughs
Crane Creek, Idaho	1910	1928	Sherard	63	Earth and puddle core	Piping into tunnel outlet
Crane Valley, Calif.			Corps of Engrs.		Hydraulic	Riprap displaced by waves
Credit River, Ont., Canada		1910	EN, Vol. 63	50	Earth and concrete core	Overtopping
Dale Dyke, England		1864	Wegmann	95	Earth and puddle core	Probable piping along outlet
Dallas, Tex.		1891	EN, Vol. 47	29	Earth, rolled	Settlement
Dalton, N. Y.		1912	EN, Vol. 67	29	Earth and concrete core	Foundation piping
Davis Reservoir, Calif.		1914	EN, Vol. 72	39	Earth, rolled	Piping around outlet
Debris Barrier No. 1, Calif.		1904	EN, Vol. 53; EN, Vol. 58		Hydraulic	Overtopped
Dells and Hatfield, Wis.	1910	1911	EN, Vol. 66	34	Earth and concrete core	Overtopped
Desabia Forebay, Calif.	1903	1932	Sherard	53		Piping through embankment
Dry Creek, Mont.	1938	1939	ENR, Vol. 122	46	Earth, rolled	Piping
East Liverpool, Ohio	1901	1901	ER, Vol. 44		Earth, rolled	Piping along outlet
Elk City, Okla.	1925	1936	ENR, Vol. 116	30	Earth, rolled	Overtopped
Ellington, Conn.		1890	EN, Vol. 47		Earth, rolled	
Empire, Colo.	1906	1909	ASCE Proc., Vol. 49	30	Earth, rolled	Piping along outlet
English, Calif.		1883	Schuyler			Overtopped
Fairview, Mass.		1922	ENR, Vol. 89	30	Earth, rolled	Piping
Forsythe, Utah	1920	1921	Sherard	65	Earth, rolled	Piping under spillway; drawdown slide
Fort Collins, Colo.		1902	EN, Vol. 57			
Fort Peck, Mont.		1938	ENR, Vol. 121	250	Hydraulic	Foundation slide
Frazier, Idaho	1915	1935	Sherard	25	Earth, rolled	Seepage slide
Frazier Valley, B. C., Canada		1948	ENR, Vol. 140	12	Earth, rolled	Overtopped
Frenchmans Creek, Mont.	1951	1952			Earth, rolled	Overtopped
Fruit Growers Reservoir, Colo.	1898	1937	ENR, Vol. 118	36	Earth, rolled	Seepage slide
Garza, Tex.	1926	1926	ENR, Vol. 94; ENR, Vol. 100	80	Hydraulic	Core pressure slide
Gatum, Panama		1912	ER, Vol. 66	115	Hydraulic	Core pressure slide
Goose Creek, S. C.	1903	1916	EN, Vol. 76	22	Earth, rolled	Overtopping
Grand Rapids, Mich.	1874	1909	ER, Vol. 42	25	Earth, rolled	Overtopping

Table 2.1:1 Unsatisfactory Performance of Earth Dams (After Middlebrooks, Ref., 3) (Continued)

Dam and Location	Date		Reference[a]	Height	Type	Reason for Unsatisfactory Performance
	Built	Failed				
Greenlick, Pa.	1901	1904	EN, Vol. 52	60	Earth, rolled	Seepage
Greenville, S. C.		1927	ENR, Vol. 100; ENR, Vol. 103	140		Cast-iron pipe failed
Gunnison, Calif.		1890	EPG Journal, Vol. 44	20	Earth, rolled	Piping along outlet
Half Moon Bay, Calif.		1926	ENR, Vol. 96	80	Earth, rolled	Overtopping
Hatchtown, Utah	1908	1910; 1914	Sherard; EN, Vol. 75	65	Earth, rolled	Seepage sloughs; Complete failure due to seepage along conduit
Hatfield, Wis.	1908	1911	EN, Vol. 66	24	Earth and concrete core	Overtopping
Hebron, N. Mex.	1913	1914; 1942	ER, Vol. 69	56	Earth, rolled	Piping through dam; Overtopped
Holmes Creek, Utah	1903	1924	Sherard	65	Earth, rolled	Upstream slope slide, full pool
Hope Reservoir, R. I.	1882	1907	ER, Vol. 53; ER, Vol. 56	23	Earth, rolled	Seepage
Hornell, N. Y.	1912	1912	EN, Vol. 58		Earth and concrete core	Seepage
Horse Creek, Colo.	1911	1914	ER, Vol. 69; EN, Vol. 71	56	Earth, rolled	Piping
Horton, Kans.	1924	1925	ENR, Vol. 95	34	Earth, rolled	
Jeanette, Pa.		1903	ER, Vol. 48	20	Earth, rolled	Overtopped
Jefferson County, Colo.		1897	EN, Vol. 47			
Johnson, Nebr.	1940	1942–1945	Corps of Engrs.	47	Earth, rolled	Loss of filter through riprap
Johnstown, Pa.	1852	1889	ASCE Trans., Vol. 24	70	Earth, rolled	Overtopping
Julesburg (Jumbo), Colo.	1905	1907–1910	ER, Vol. 63	70	Earth, rolled	Serious leakage started in 1907, dam failed with 24-ft head in 1910
Kern, Ore.	1948	1949	Sherard	52	Earth, rolled	Excessive settlement of fill
Kettering, England		1905	ER, Vol. 52	46	Earth, rolled	Slide
Killingsworth, Conn.		1938	ENR, Vol. 121	18	Earth and concrete core	Overtopping
Kingsley, Nebr.	1941	1942	CE, Vol. 15	170	Hydraulic	Loss of fill through concrete blocks no filter

Knolbrook, Pa.		1894	EN, Vol. 32		Earth, rolled	
Knoxville Reservoir, Tenn.		1883	EN, Vol. 47		Earth, rolled	Foundation seepage
Lafayette, Calif.	1929	1928	ENR, Vol. 54	140	Earth, rolled	Foundation slide during construction
La Fruta, Tex.	1930	1930	ENR, Vol. 105; ENR, Vol. 106; ENR, Vol. 107	61	Earth, rolled	Foundation piping
Lake, N. Mex.		1893	EN, Vol. 47	52	Earth and rock	Overtopped
Lake Avalon, N. Mex.		1894	EN, Vol. 23; EN, Vols. 35 and 36; EN, Vol. 54	48	Earth, rolled	Overtopped
Lake Avalon, N. Mex.	1894	1904	EN, Vol. 54	48	Earth, rolled	Seepage
Lake Coedty, Wales		1925	ENR, Vol. 96		Earth, rolled	
Lake Dixie, Tex.		1940	ENR, Vol. 125		Earth, rolled	Overtopped
Lake Francis (old dam), Calif.	1899	1899	ASCE Trans., Vols. 58 and 59	50	Earth, rolled	Piping along outlet
Lake George, Colo.		1914	ASCE Proc., Vol. 49		Earth, rolled	Piping
Lake Malloya, N. Mex.	1914	1942	Sherard	50	Earth, rolled	Overtopped for 6 hour did not fail
Lake Toxaway, N. C.	1902	1916	ENR, Vol. 94; ER, Vol. 74	62	Earth, rolled	Seepage
Lake Yosemite, Calif.	1884	1943	Sherard; Schuyler	53	Earth, rolled	
Lancaster, Pa.		1894	ER, Vol. 39	21	Earth, rolled	Piping along outlet
Lebanon, Ohio		1882	EN, Vol. 9	30	Earth, rolled	Overtopping
Lebanon, Pa.		1893	ER, Vol. 27	40	Earth, rolled	Piping between fill and foundation
Leroux Creek, Colo.		1905	EN, Vol. 54	25	Earth, rolled	Overtopping
Lidderdale, Colo.		1909	ASCE Proc., Vol. 49	19	Earth, rolled	Overtopping
Lima, Mont.		1894	EN, Vol. 31; EN, Vol. 47	40	Earth, rolled	Erosion at spillway
Linville, N. C.	1919	1919	ASCE Trans., Vol. 84	160	Hydraulic	Core too flat
Lock Alpine, Mich.		1926	ENR, Vol. 96	25	Earth, rolled	Settlement on being saturated
Long Tom, Idaho	1906	1915	Sherard	50	Earth and puddle core	Piping into tunnel
Longwalds Pond, Mass.		1922	ENR, Vol. 89	30	Earth and concrete core	Piping
Lower Otay, Calif.	1897	1916	EN, Vol. 75	130	Rock and concrete core	Overtopping
Lyman, Ariz.	1913	1915	EN, Vol. 73	65	Earth, rolled	Piping
Lynde Brook, Mass.		1876	EPG Journal, Vol. 44	27	Earth, rolled	Piping along outlet
Magic, Idaho	1910	1911	Sherard; ER, Vol. 60	130	Earth, rolled	Piping through dam

Table 2.1:1 Unsatisfactory Performance of Earth Dams (After Middlebrooks, Ref., 3) (Continued)

Dam and Location	Date		Reference[a]	Height	Type	Reason for Unsatisfactory Performance
	Built	Failed				
Mahonoy City, Pa.		1892	EN, Vol. 27; EN, Vol. 47; ER, Vol. 26		Earth, rolled	Piping
Mammoth, Utah	1912	1917	ENR, Vol. 79; ER, Vol. 66	70	Earth and hydraulic	Overtopped during construction
Maquoketa, Iowa	1924	1927	ENR, Vol. 98	20	Earth, rolled	Piping at junction with concrete spillway
Marshall Creek, Kans.		1937	ENR, Vol. 119	80	Earth, rolled	Foundation failure during construction
Marshall Lake, Colo.	1908	1909	ER, Vol. 62	70		Seepage
Martin Davey Dam, Tex.		1940	ENR, Vol. 125		Earth, rolled	Overtopped
Masterson, Ore.	1950	1951	Sherard	60	Earth, rolled	Piping, dry fill
McMillan, N. Mex.	1894	1915; 1937	Sherard; Schuyler		Earth and hydraulic	Upstream earth piped into rock downstream
Mellville, Utah	1907	1909		36	Earth, rolled	Piping through foundation
Melzingah, N. Y.		1897	ER, Vol. 36	24	Earth, rolled	Overtopping
Middlefield, Mass.		1901	ER, Vol. 43	20	Earth, rolled	Overtopping
Mill River, Mass.	1865	1874	ASCE Trans., Vols. 3 and 4	43	Earth and concrete core	Seepage
Mission Lake, Kans.	1924	1925	ENR, Vol. 95		Earth, rolled	Settlement with overtopping
Mohawk, Ohio		1913	EN, Vol. 73		Earth, rolled	Settlement and seepage
		1915	EN, Vol. 73	18	Earth, rolled	Seepage
Montreal, Que. Canada		1896	EN, Vol. 47	18	Earth and rock	Seepage
Morena, Calif.	1912		ASCE Trans., Vol. 65	167	Rock	Overtopped, did not fail
Mountain Creek, Tex.	1931		Corps of Engrs.	36	Earth, rolled	Loss of filter through riprap
Mount Lake State Park, Minn.	1937	1938	ENR, Vol. 120			Overtopped
Mount Pisgah, Colo.	1910	1928	Sherard	76	Earth, rolled	Upstream slope sloughed on draw-down to 1 on 3
Mud Pont, Mass.	1873	1886	ER, Vol. 13	15	Earth and rock	Piping
Narragninep, Colo.	1908	1928–1951	Sherard	79	Earth, rolled	Continued sloughing of upstream slope and abutment leakage

Nebraska City, Nebr.	1890	1890	EN, Vol. 47	17	Earth and rock	Seepage
Necomah, Wis.		1905	ER, Vol. 52		Earth and concrete core	Core settled
New Bedford, Mass	1866	1868	ASCE Trans., Vols. 1 and 2	25	Earth, rolled	Piping along outlet conduit
New Bowman, Calif.	1927	1928	ENR, Vol. 54	170	Rock	Break in outlet tunnel, repaired
Nezaxa, Mex.		1909	EN, Vol. 62	190	Hydraulic	Slide in embankment during construction
North Dike, Wachusett, Mass.	1905	1907	Merriman	82	Earth, rolled	Slide in upstream slope
North Scituate, R.I.		1926	ENR, Vol. 96	6	Earth, rolled	Overtopped
Norwich, N.Y.	1892	1905	EN, Vol. 54	34	Earth, rolled	Overtopped
Peapack Brook, N.J.		1928	ENR, Vol. 100	32	Earth, rolled	Overtopped
Piedmont No. 1, Calif.	1903	1905	Sherard	50	Earth and concrete core	Outlet pipe sheared off at core wall, replaced
Pleasant Valley, Utah		1928	ENR, Vol. 100	63	Earth and rock	Piping through settlement cracks
Point of Rocks, Colo.	1911	1915	Sherard	86	Earth, rolled	Concrete placed on $1\frac{1}{2}$ upstream slope failed because of 5-ft waves, near failure in 1927 as a result of wave erosion
Portland, Me.	1889	1893	ER, Vol. 28	45	Earth, rolled	Piping along drain pipe
Portneuf, Idaho	1911	1950	Sherard	55	Earth, rolled	Concrete conduit disintegrated, and was replaced in 1950
Prairie River, Wis.		1912	EN, Vol. 68		Earth, rolled	Overtopped
Pratts Fork, Ohio	1934	1938	OCE files	21	Earth, rolled	Overtopped
Providence, R.I.	1816	1916	EN, Vol. 45	17	Earth and rock	
Puddingstone, Calif.		1926	ENR, Vol. 96		Hydraulic	Overtopped during construction because of a clogged outlet
Rector, Calif.	1946	1947	Sherard	150	Earth, rolled	Transverse cracking
Rocky Ford, Utah	1914	1915–1950	Sherard	70	Earth, rolled	High saturated line, reservoir level limit in 1950
Roxborough Reservoir, Pa.	1894	1894	ER, Vol. 30		Earth, rolled	Piping
Saluda, S.C.	1930	1930	ENR, Vol. 104	208	Hydraulic	Core pool lost during construction
San Pablo, Calif.	1921		Corps of Engrs.	220	Hydraulic	Fill loss, through riprap, no filter
Santee Cooper, S.C.	1942	1942–1946	CE, Vol. 18	80	Earth and hydraulic	Disintegration of porous concrete slope protection
Santo Amaro, Brazil		1907	Merriman	63	Hydraulic	Failed during construction due to slide
Schenectady, N.Y.		1916	EN, Vol. 76	30	Earth, rolled	Overtopped
Schofield, Utah	1926	1927	ENR, Vol. 100	62	Earth and rock	Transverse cracking and piping into rock

Table 2.1:1 Unsatisfactory Performance of Earth Dams (After Middlebrooks, Ref., 3) (Continued)

Dam and Location	Date		Reference[a]	Height	Type	Reason for Unsatisfactory Performance
	Built	Failed				
Scottdale, Pa.		1904	EN, Vol. 52	60		Piping
Seefield, Utah		1925	ENR, Vol. 79	130	Earth, rolled	Overtopped
Sepulveda, Calif.		1914	ER, Vol. 74	65	Earth and concrete core	Overtopped
Sheffield, Calif.		1925	ENR, Vol. 95	30	Earth, rolled	Earthquake slide
Shelton, Conn.		1903	ER, Vol. 47	20	Earth and rock	Piping
Sherburne, N.Y.	1892	1905	EN, Vol. 54	34	Earth, rolled	Overtopped
Sherburne Lake, Mont.	1916		Corps of Engrs.		Earth, rolled	Floating logs displaced hand-placed riprap
Short Creek, Ark.		1939	ENR, Vol. 122	57	Earth and rock	Overtopped during construction
Sinker Creek, Idaho	1910	1943	Sherard	70	Hydraulic and earth	Seepage slide
Six Mill Creek, N.Y.		1905	EN, Vol. 53	15	Earth, rolled	Overtopped
Snake Ravine, Calif.	1893	1898	EN, Vol. 40	64	Hydraulic	Poor compaction
Spartanburg, Pa.		1892	EN, Vol. 27	10	Earth and rock	Overtopped
Spring Lake, R.I.	1887	1889	EN, Vol. 20	18	Earth and rock	Piping along outlet
Staffordville, Conn.		1887	EN, Vol. 4	20	Earth and rock	Piping along outlet
Standley Lake, Colo.	1911	1916	ENR, Vol. 78	113	Hydraulic	Core too large, slides during and after construction
Stockton Creek, Calif.	1950	1950	Sherard	80	Earth, rolled	Failed at abutment, probably along contact or crack
Sublette, Idaho	1915	1916	Sherard	40	Earth, rolled	Conduit cracked as a result of settlement
Summer Lake, Ore.			ASCE Trans., Vol. 94	60	Earth, rolled	Foundation slide
Suputrida Canyon, Calif.		1914		65	Earth and concrete core	Overtopped
Swansen, Wales	1867	1879	SE, Vol. 3	80	Earth and rock	Piping
Table Rock Cave, S.C.	1927	1928	ENR, Vol. 100	140	Earth, rolled	Broken outlet pipe
Tappan, Ohio	1935	1934		52	Earth, rolled	Slide in foundation
Tecumseh, Ala.		1894	EN, Vol. 32			
Telluride, Colo.		1909	Colorado State Engr.	30	Earth and rock	Overtopping
Throttle, N. Mex.	1912	1942	Sherard	65	Earth, rolled	Overtopped, but did not fail
Torranto, Ont., Canada		1912	ER, Vol. 65	35	Earth, rolled	Overtopped

Trout Lake, Colo.	1894	1909	ER, Vol. 60	25	Earth, rolled	Overtopped
Tupper Lake, N.Y.	1906	1906	EN, Vol. 57	18	Earth, rolled	Piping along outlet
Turkey Creek, Colo.		1910	Colorado State Engr.	22	Earth, rolled	Slide during construction
Turlock Irrigation, Calif.		1914		56		Leakage around outlet
Turtle Creek, Tex.		1891	EN, Vol. 25	29	Earth, rolled	Foundation settlement
Union Bay, B.C., Canada		1912	EN, Vol. 67	20	Earth and timber core	Overtopped
Utica Reservoir, N.Y.	1874	1902	ER, Vol. 46	70	Earth, rolled	Insufficient compaction
Valentine, Nebr.	1911	1911	ER, Vol. 63	30	Earth, rolled	Settlement of spillway
Val Marie Dam, Sask., Canada		1952	ENR, Vol. 148		Earth, rolled	Overtopped
Valparaiso, Chile		1888	E & BR, Vol. 18	56	Earth, rolled	Slopes too steep
Victor, Colo.		1901	ER, Vol. 43	25	Earth, rolled	Overtopped
Virgin River, Nev.		1929	ENR, Vol. 103	120	Earth, rolled	Poor design and construction
Wachusett (new dike), Mass.		1907	Schuyler		Earth, rolled	Upstream slope slide, water surface 40 ft below top
Wagner, Wash.	1918	1938	ENR, Vol. 120	50	Hydraulic	Spillway failure—erosion
Walnut Grove, Ariz.	1888	1890	Wegmann	110	Rock	Overtopped
Weisse, Czechoslovakia		1916		42	Earth, rolled	Piping along outlet
West Julesburg, Colo.	1905	1910	ER, Vol. 63	55	Earth, rolled	Piping
Wilmington, Del.	1864–1887	1900	ER, Vol. 42	12	Earth, rolled	Piping along outlet
Winston, N.C.	1902	1912	EN, Vol. 67	24	Earth, rolled	Overtopped
Wise River, Mont.		1927	ENR, Vol. 99		Earth, rolled	Overtopping
Wister, Okla.	1951			90	Earth, rolled	Piping
Worcester, Colo.	1912	1951	Sherard	68	Earth, rolled	Concentrated seepage
Worcester, Mass.	1871	1876	ASCE Trans., Vols. 5 and 6	41	Earth, rolled	Leakage in culvert
Yuba (old dam), Calif.		1907	EN, Vol. 58		Earth, rolled	Overtopped
Yuba (new dam), Calif.	1949	1951	Sherard	25	Earth, rolled	Seepage slide, downstream slope enlargement
Zuni, N. Mex.	1907	1909	EN, Vol. 62	70	Earth and rock	Piping through abutment

[a] ENR = *Engineering News Record;* ER = *Engineering Record;* EN = *Engineering News;* ASCE Proc. = *ASCE Proceedings;* Sherard = "Influence of Soil Properties and Construction Methods on Performance of Homogeneous Earth Dams," by James L. Sherard, thesis presented to Harvard University at Cambridge, Mass., in 1952, in partial fulfilment of the requirement for the degree of Doctor of Science, Civil Engineering; ASCE Trans. = *ASCE Transactions;* Corps of Engrs. = "Slope Protection for Earth Dams," Corps of Engrs. Dept. (Waterways Experiment Station), March 1949; Wegemann = *Design and Construction of Dams*, by Edward Wegemann, Chapman and Hall, Ltd., London, 8th Ed., 1927; Schuyler = *Reservoirs for Irrigation, Water-Power, and Domestic Water Supply*, by J. D. Schuyler, John Wiley and Sons, Inc., New York, 2nd Ed., 1908; EPG Journal = *Journal of Electric Power and Gas;* CE = *Civil Engineering;* Merriman = *American Civil Engineering Handbook*, by Mansfield Merriman, Chapman & Hall, Ltd., London, 1930; OCE files = Files of Office of Chief of Engineers, Department of the Army, Washington, D.C.; SE = *Sanitary Engineer;* Colorado State Engr. = Annual Report, "Colorado State Engr., Denver, Colo., 1910; and E & BR = Engineering and Building Record.

Most of the serious trouble from piping has resulted from progressive backward erosion of concentrated leaks which develop through or under the dam. The erosion starts at the point where the seepage water discharges and works toward the reservoir, gradually enlarging the seepage channel until, in the final stage, complete failure occurs. Theoretically, piping can commence inside a zoned dam where the seeping water discharges from the finer material of the core into the coarser adjoining pervious zone. Actually there have been no documented records available of failures being caused by migration of soil particles from a zone of fine soil to a zone of coarse soil (sand or gravel). The only records of internal piping failures of this type have come from old dams where zones of fine-grained soil were placed directly against zones of large quarried rock (see the description of the Schofield Dam, Sec. 1.2b).

2.2b Leaks Which Have Led to Piping

When first observed, the leaks which have led to piping failure have varied considerably in size, and the rates of development have been widely different. At some dams the leak was seen after the first filling of the reservoir; in others it appeared only after many years of leak-free operation. In some cases the leakage water first emerged as a small seep which, to the naked eye, ran clear for years and then increased gradually until rapid failure occurred. In other cases, a large and muddy leak preceded complete failure by only a few days or hours.[1] Figure 2.2:1 shows the rapidly progressing piping failure of a small earth dam from a leak which developed at the abutment contact when the reservoir was first filled.

There have been many origins of the leaks which have led to piping failures. Probably the most common cause of embankment leaks has been poor construction control, which can result in inadequately compacted or pervious layers in the embankment, inferior compaction adjacent to concrete outlet pipes or other structures, or poor compaction and bond between the embankment and the foundation or abutments. Embankment leaks through differential settlement cracks (Sec. 2.3) have also been a major source of trouble.

[1] Muddy seepage water emerging at the downstream side of an earth dam is usually justifiable cause for alarm; however, in the case of the first reservoir filling where the mud may be coming from mudfilled cracks in the bedrock, the flow of muddy water may be temporary and not dangerous (Sec. 8.1*d*).

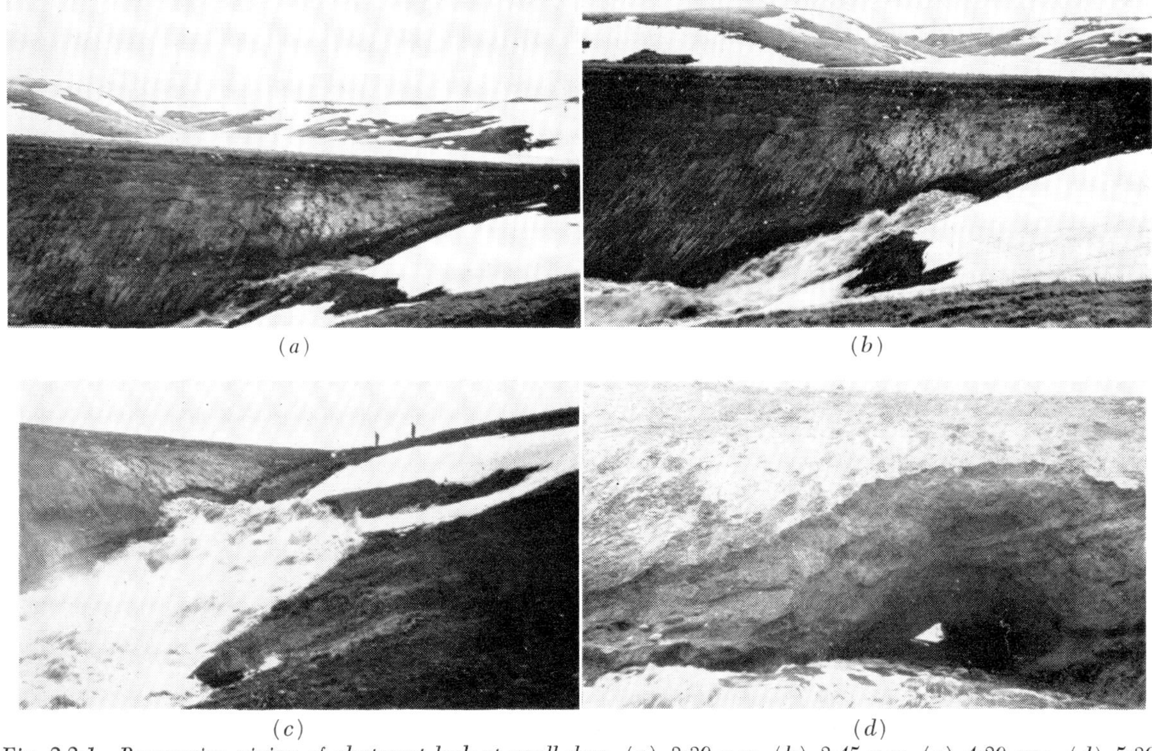

Fig. 2.2:1 Progressive piping of abutment leak at small dam. (*a*) 3:30 P.M. (*b*) 3:45 P.M. (*c*) 4:30 P.M. (*d*) 5:30 P.M.

Another cause of embankment leaks has been cracking in outlet pipes, which is caused by foundation settlement, spreading of the base of the dam, or deterioration of the pipe itself. In such cases either the eroded soil may be lost into the pipe, or the water in the pipe, if under pressure, may be forced out into the embankment downstream. In addition to these sources of embankment leaks, animal burrows and drying cracks have sometimes caused difficulty (Secs. 2.7, 2.11).

Leakage through the natural foundation soils under dams is more common than leaks through the embankment because the natural soils are more erratic and often less dense than embankment materials. Any uncontrolled seepage erupting in the form of springs in the natural ground downstream from a dam is potentially dangerous. It must be carefully observed and if necessary controlled with filters (Secs. 6.4*b* and 8.1*d*).[1]

Where the natural surface layer of the foundation is sandy and the upward flow of seepage water is strong enough to carry sand particles with it, it is common for the sand to be deposited around the springs emerging on the ground surface in the form of a ring. "Sand boils," if unobserved and unattended, can lead to complete failure by piping.

The dimensions and spacing of sand boils vary with the foundation materials and water pressures. Individual boils frequently originate at the locations of exploratory borings.[2] At some dams, isolated boils several feet in diameter have appeared, Fig. 2.2:2. At other sites the whole foundation area downstream from the dam has developed numerous boils with diameters of only a few inches spaced 2 to 3 ft. apart. Reference 464 summarizes measurements of foundation pore water pressure along Mississippi River dikes where large underseepage and boils developed. These indicate that, while the conditions required to form boils differ considerably from site to site, boils should be expected in fine cohesionless soils whenever the upward seepage gradient exceeds 0.5 to 0.8.

[1] In addition to piping, underseepage sometimes reduces the margin of safety of the downstream slope against shear slides because high pressures develop in the foundation pore water near the downstream toe. It has been considered advisable at several major dams to limit the maximum reservoir storage after unexpectedly high pressures at the downstream toe were measured in piezometers. (Methods for reducing underseepage pressure are discussed in Sec. 6.4) Moreover, if the natural surface layer is relatively impervious, and if the pore pressures exceed the weight of the overburden, it is theoretically possible for the layer to "heave." The authors know of no actual cases where the foundation pressures became so high that large sections of the soil downstream were known to have been lifted.

[2] At the first reservoir filling of the Fort Peck Dam on the Missouri River, a boil about 25 ft. in diameter and 35 ft. deep developed from seepage up the side of the casing of a test well (Ref. 569).

Fig. 2.2:2 Typical sand boil. (Courtesy W. J. Turnbull)

Individual large boils can be temporarily controlled by ringing them with dikes of sandbags or other materials. The dike allows the development of a back pressure on the boil and thus decreases the upward hydraulic gradient and prevents the loss of sand on the surface (Fig. 2.2:3). Boils can be permanently controlled with drainage relief wells (Sec. 6.4*c*) or blankets of filter material. Where numerous small boils are closely spaced over large areas, for example, they have been checked with inverted filters consisting of 3- to 5-ft.-thick blankets of pervious materials varying from sand at the bottom to coarse gravels at the top.[1]

2.2c Resistance to Piping

Records of dams which have developed concentrated leaks demonstrate a very wide range of susceptibility to piping. In one study of

[1] For example, at the Petenwell Project (Ref. 96), where the foundation for the earth dam consisted of uniform medium clean sand, boils of about 12 in. in diameter spaced approximately 6 ft. on center developed generally over an area of several hundred feet below the dam. The boils were of classic sand-ring shape, and a rod could easily be pushed down into them by hand for 5 ft. or more. This situation was satisfactorily treated by covering the ground surface with a blanket of progressively coarser gravel which had a total thickness of about 5 ft. The gravel blanket was graded from coarse sand and pea gravel at the bottom to about 3 in. in diameter at the top.

Fig. 2.2:3 Sandbag protection around boil which developed downstream from major earth dam. Standpipes indicate the elevations to which the pressure head in underlying foundation soil rose before installation of the relief well (pipe in foreground).

leaks in 31 dams (Ref. 340) the influence of the soil properties and the embankment construction method on the piping resistance (that is, on resistance to piping after a condition of leakage exists) was analyzed as carefully as the rather sketchy information available permitted. From this it was concluded that the embankment soil properties, and particularly the plasticity of the fines, had a larger influence on piping resistance than the method by which the embankment had been compacted. As seen in Table 2.2:1, which summarizes the results of the study, embankments constructed of clay with plasticity index greater than 15 demonstrated the highest resistance to piping, while embankments constructed of fine uniform cohesionless sand had the lowest resistance. Laboratory research is urgently needed to extend knowledge concerning the influence of soil type and density on the piping resistance of compacted soil.

Table 2.2:1 Rough Empirical Relationship Observed between Piping Resistance in Earth Dam Embankments and Soil Types and Construction Methods (In Order of Decreasing Piping Resistance)

Greatest piping resistance	1. Clay of high plasticity (P.I. greater than 15%). Well compacted.
	2. Clay of high plasticity (P.I. greater than 15). Poorly compacted.
Intermediate piping resistance	3. Well-graded coarse sand or sand-gravel mixtures with binder of clay of medium plasticity (P.I. greater than 6). Well compacted.
	4. Well-graded coarse sand or sand-gravel mixtures with binder of clay of medium plasticity (P.I. greater than 6). Poorly compacted.
	5. Well-graded cohesionless gravel-sand-silt mixtures (P.I. less than 6). Well compacted.
Least piping resistance	6. Well-graded cohesionless gravel-sand-silt mixtures (P.I. less than 6). Poorly compacted.
	7. Very uniform fine cohesionless sand. (P.I. less than 6). Well compacted.
	8. Very uniform fine cohesionless sand. (P.I. less than 6). Poorly compacted.

Experience indicates that graded filters are very effective in controlling piping, and many abutment or foundation leaks strong enough to cause serious erosion have been made safe by the installation of drains or blankets of pervious material graded in such a manner that the eroded soil cannot be washed away.

Current lack of knowledge on this point is a ridiculous anachronism considering the general advance of modern soil mechanics and the great need for the information. At the present time, when a designer compares the relative desirability of several alternative soils for the core of a dam, he generally does not even consider the inherent "piping" resistance. This is in direct contrast to the fact that this property of the core has probably more influence on the safety of the dam than any other.

2.2d Sloughing

Progressive sloughing (or raveling) is a type of damage closely related to piping which has occurred on a few older homogeneous dams. The process begins when a small amount of material at the downstream toe erodes and produces a small slump or miniature slide. It leaves a relatively steep face, which becomes saturated by seepage from the reservoir and slumps again, forming a slightly higher and more unstable face. This raveling process can continue until the remaining portion of the dam is too thin to withstand the water pressure and complete failure occurs suddenly as the reservoir breaks through.

Failure of this type has taken place only when the whole downstream portion of the dam has been saturated. In sloughing failures concentrated leaks of measurable size may or may not develop, but it is possible for the total quantity of leakage to remain small until just before failure.

The failure of the Sinker Creek Dam in Idaho in 1943 is a good illustration of the mechanics of progressive backward sloughing (Ref. 95). This homogeneous dam was constructed in 1910 and raised in 1918 to a height of 70 ft. It was composed primarily of silty, sandy gravel and retained an irrigation reservoir which was filled for only a part of the year. During the first 25 years of operation large sections of the downstream slope became saturated whenever the reservoir was full for any length of time, and concern over this condition prompted efforts to keep the period of maximum reservoir storage as short as possible. The state of the dam was not considered critical, however.

Because of high rainfall in the spring of 1943, the reservoir filled several months earlier than usual and the State Engineer's Office inspected it frequently. Nothing unusual was noticed except that larger areas of the downstream slope were saturated than in previous years.

On the evening of June 19, a small amount of the softened downstream toe sloughed and left a short, nearly vertical face. Under the action of relatively small concentrated leaks, this saturated face was unstable, and after standing for a short period it sloughed again, producing a higher face. This process continued until the reservoir broke through the thin remaining embankment. The time between the first indication of sloughing at the downstream toe and the complete failure was approximately 8 hours. Until the reservoir broke through the dam there was little leakage, so that the whole content of the reservoir was released to form one large flood wave downstream.

2.3 DIFFERENTIAL SETTLEMENT CRACKS

While the danger of cracking has not been widely publicized or understood by earth dam engineers,[1] it is possible that a larger number of leaks which have led to piping failures have originated from embankment cracks than from any other source. Although many of these failures have been in small and cheaply constructed dams, a considerable number of large well-constructed dams have developed alarming cracks in recent years.

There are two reasons why the subject of cracking has not been given more attention. First, there is always an understandable desire on the part of both the engineer and the owner to conceal serious defects which develop in a dam structure. When a slope slide occurs in an embankment its presence is obvious even to the casual observer and it cannot be easily hidden; but an open crack, which may be potentially more dangerous than a slide, often cannot be discovered except by very close inspection. Consequently it is a simple matter to hide surface manifestations of serious embankment cracks, so that while many individual engineers have had experience with them, the details have not been made widely known and the profession has had no reliable aggregate of experience on the subject.

The second reason why cracking in earth dams has not received the consideration it deserves is that the true cause of failure often has not been identified. A large group of failures which have occurred when reservoirs were filled for the first time have been attributed to piping through leaks along the outlet conduit or at the abutment or foundation contacts. Actually, piping in many of these cases undoubtedly started in embankment cracks. Many such failures took place without witnesses, but even if there had been reliable records of the events leading to failure, it would still have been difficult in many cases to establish the cause with certainty. In some circumstances severe and dangerous cracking may develop but never be visible on the surface of the dam.

[1] The total number of references to the problem of earth dam cracking in the literature is surprisingly small. The following include the bulk of the important published experience and discussions of the subject: Refs. 15, 95, 161, 626, 628, 629.

2.3a Mechanics of Cracking

Cracking of the type discussed in this section develops because portions of the embankment are subjected to tensile strains when the dam is deformed by differential settlement. Depending upon the geometry and relative compressibility of the foundation, abutments, and embankment, earth dams may be twisted in different ways which result in quite different cracking patterns. Cracks may open parallel or transverse to the axis of the dam and may form in vertical or horizontal planes or in any intermediate direction. They may be either localized or continuous for great distances through the impervious core. They may open as much as 6 in. in width, though widths of an inch or two have been more common.

Cracking has developed most commonly in dams of moderate height (less than 100 ft.) and in the upper portions of high dams. Obviously, the great earth pressures which exist deep within high dams will work against the development of cracks; however, we have no quantitative information on the subject.

TRANSVERSE CRACKS

The most dangerous cracks are those which run transversely, creating a path for concentrated seepage through the core. They are caused by differential settlement between adjacent lengths of the embankment, usually between the portion located at the abutment and the portion in the center of the valley. The worst cracking develops when the foundation under the higher portions is compressible and the abutments consist of steep and relatively incompressible rock. Cracks of the type shown in Fig. 2.3:2 frequently open over the midheight of the abutment and, depending on the geometry of the strains imposed, may be vertical or inclined in either direction, Fig. 2.3:1.[1]

In narrow valleys with rock abutments, arching of the upper portion of the embankment can prevent the crest from settling as much as the foundation, and in severe cases roughly horizontal cracks open at the bottom of the arch, Fig. 2.3:1*b*. Cracking of this type was responsible for the failure of the 112-ft. Apishapa Dam, Colorado, in 1923 (Refs.

[1] This type of cracking is sometimes associated with surprisingly high tensile strains of the embankment in the direction parallel to the crest of the dam, though very few measurements of this movement have been made. (See record of measurements at Rector Creek Dam, Sec. 3.2*b*.)

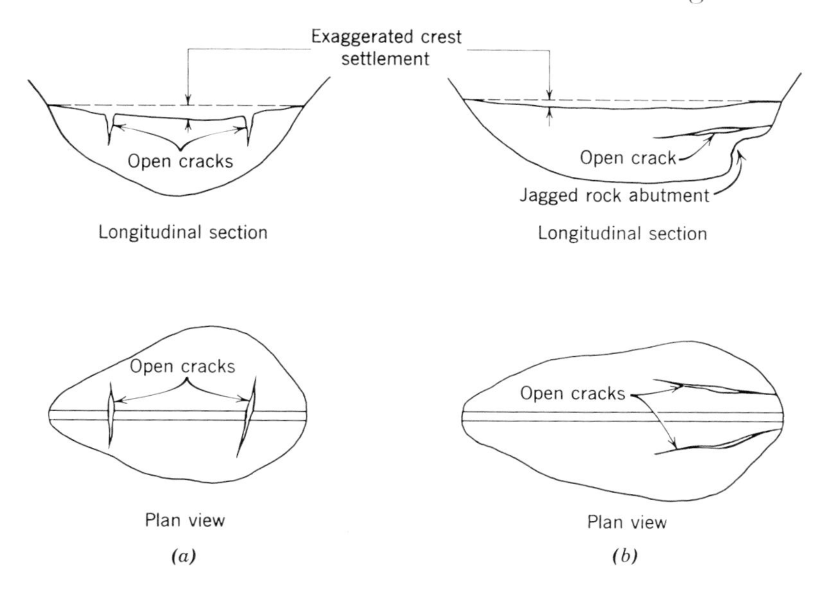

Fig. 2.3:1 Typical transverse differential settlement cracks.

95, 104, 105). The foundation consisted of compressible soil and the abutments of steep hard rock. When the reservoir filled for the first time, the foundation and the lower part of the embankment settled, the upper part of the embankment arched between the abutments, water broke through a large horizontal crack, and the dam failed completely in a few hours.

Although the most critical conditions develop at sites where the foundation is compressible, compression of the embankment, even when it is well constructed, can also cause dangerous transverse cracking. Casagrande (Ref. 15) describes a case of this kind in which nearly horizontal cracks opened in an 80-ft. dam. It was a homogeneous structure of clay of low to medium plasticity built with good modern construction techniques. The cracks appeared in a length of the embankment which had been completed last as a closure section (Sec. 1.1*e*). Since the foundation was shale, and so could not have contributed materially to the settlement, the cracking must have been caused by settlement from compression of the embankment itself.[1]

[1] See also Ref. 627 for a description of the discovery of a crack in the thin core of a recently constructed 200-ft. dam. It was evidently caused by differential settlement in the vicinity of a haul road which had been left on the abutment.

(*a*)

Fig. 2.3:1 Typical transverse cracks on two small dams due to larger settlement in the center of the valley than on abutments. (a) Cracks in well constructed dam due to foundation settlement. (b) Cracks due to settlement of lower part of dam that was constructed too dry.

Crack on
upstream
slope
6″x 9″
Notebook
Close-up
of crack
above

(b)

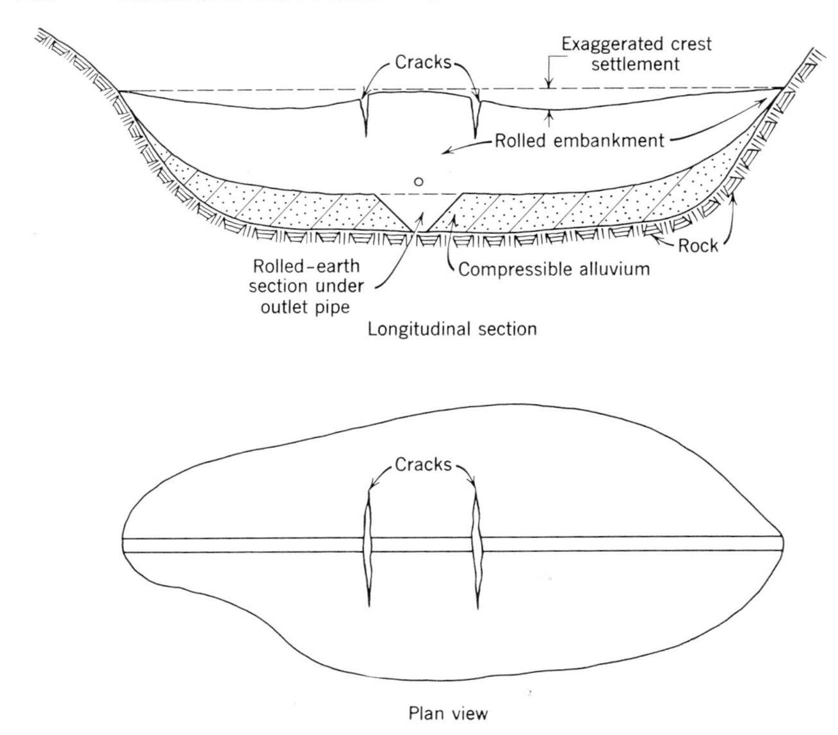

Fig. 2.3:3 Cracking due to differential settlement between natural foundation soil and rolled-earth support under outlet pipe (or other discontinuity in the foundation).

Another kind of transverse cracking has developed at a few small dams under the conditions shown in Fig. 2.3:3, when sections of rolled-earth embankment have been placed in trenches through compressible natural soil foundations for the purpose of supporting outlet conduits. Differential settlement between the portion of the dam at the outlet conduit and the adjacent portions has resulted in severe cracks.

LONGITUDINAL CRACKS

Although they are not normally dangerous, longitudinal cracks (Fig. 2.3:4) occur frequently and have developed in some of the largest dams built in recent years. Although some have opened to depths of more than 20 ft., they usually are only a few feet deep.

Longitudinal cracks are produced by several types of differential movement. In dams with rolled-earth cutoffs which are much less compressible than the natural foundation soil underlying the slopes, longitudinal cracking may be caused by the tendency of the slopes to settle more than the crest, Fig. 2.3:4*a*. Dams with central cores of rolled earth and upstream and downstream shells of dumped quarried rock frequently develop the cracks near the crest at the juncture between the core and the dumped rock sections, Fig. 2.3:4*b*, because the rock shells continue to compress appreciably after construction while the rolled core does not.[1] Whenever longitudinal cracks appear at the crest of a dam, they are probably due in part to the tensile stresses which develop at the tops of earth slopes.

The main danger associated with longitudinal cracks is that they may occur in conjunction with other unseen cracks running trans-

[1] Longitudinal cracks of this type opened at the upstream and downstream edges of the crests in the Cherry Valley Dam in California (Ref. 26) and the Mud Mountain Dam in Washington (Ref. 484), both of which have the rolled-earth cores and rock shells.

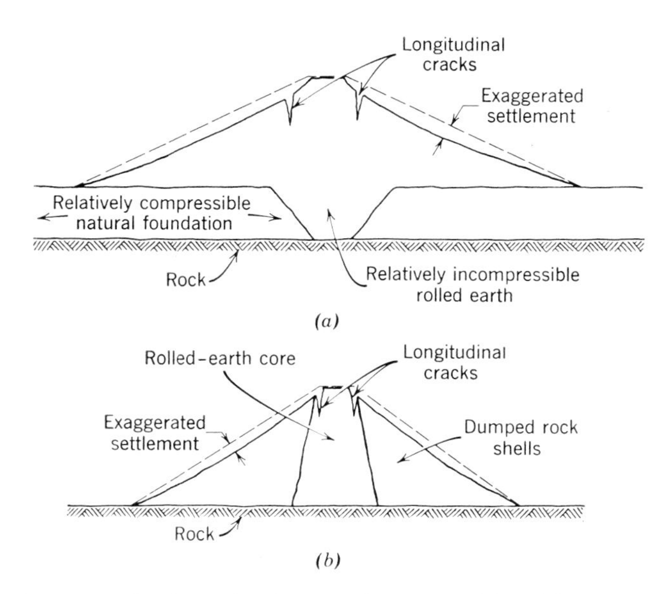

Fig. 2.3:4 Longitudinal cracks. (a) Cracking caused by differential foundation settlement. (b) Cracking caused by differential settlement between embankment sections of dumped rock and rolled earth.

versely through the core. Longitudinal cracks should be carefully observed and, when movement stops, filled by trenching and backfilling with compacted impervious soil. It is always advisable to put down test pits to be sure that a crack is actually vertical and longitudinal and not a surface manifestation of an inclined crack cutting through the dam core. If a crack is of appreciable depth and length, the engineer should try to prevent it from filling with surface water while it is under observation and before it is sealed.

INTERIOR CRACKING NOT VISIBLE ON THE DAM SURFACE

Because cracking of the types described in the previous sections cannot be observed if the exterior slopes of the dam are blanketed with cohesionless sand, gravel, or rock riprap, such cracks have probably developed unrecognized in many dams. But in addition to the types of cracks previously discussed, there is another category of cracks which are never observed on the surface of the dam and whose existence can only be inferred.

One crack of this type, shown in Fig. 2.3:5*a*, may occur in dams with narrow vertical central cores of compressible impervious material. During construction the core tends to compress more under the weight of the overlying fill than the shells do, so that a part of the weight of the core is transferred to the shells by shear stresses and arching. Within the core the vertical stresses on horizontal planes at various elevations may be so low that they are approaching zero. This arching effect, combined with variable shear strengths developing on the vertical planes separating the core and shells, can conceivably result in horizontal cracks in the core.[1]

During the construction of several dams with narrow cores it has been suspected that this action was taking place. For one structure in Sweden (Ref. 106) it had been estimated that the post-construction settlement would be 2 ft., but the actual settlement was measured at only 2 in. It was the opinion of the engineers that the core had probably arched between the less compressible upstream and downstream zones and that a hidden crack or series of cracks having a combined thickness of as much as 12 in. could possibly have opened. This hypothesis created enough concern to cause the engineers to drive a row of steel sheet piling through the core before filling the reservoir.

The records of a number of dams have indicated that cracks of this type may have developed, although their possible existence is dis-

[1] A theoretical approach to the analysis of this problem is suggested in Ref. 577.

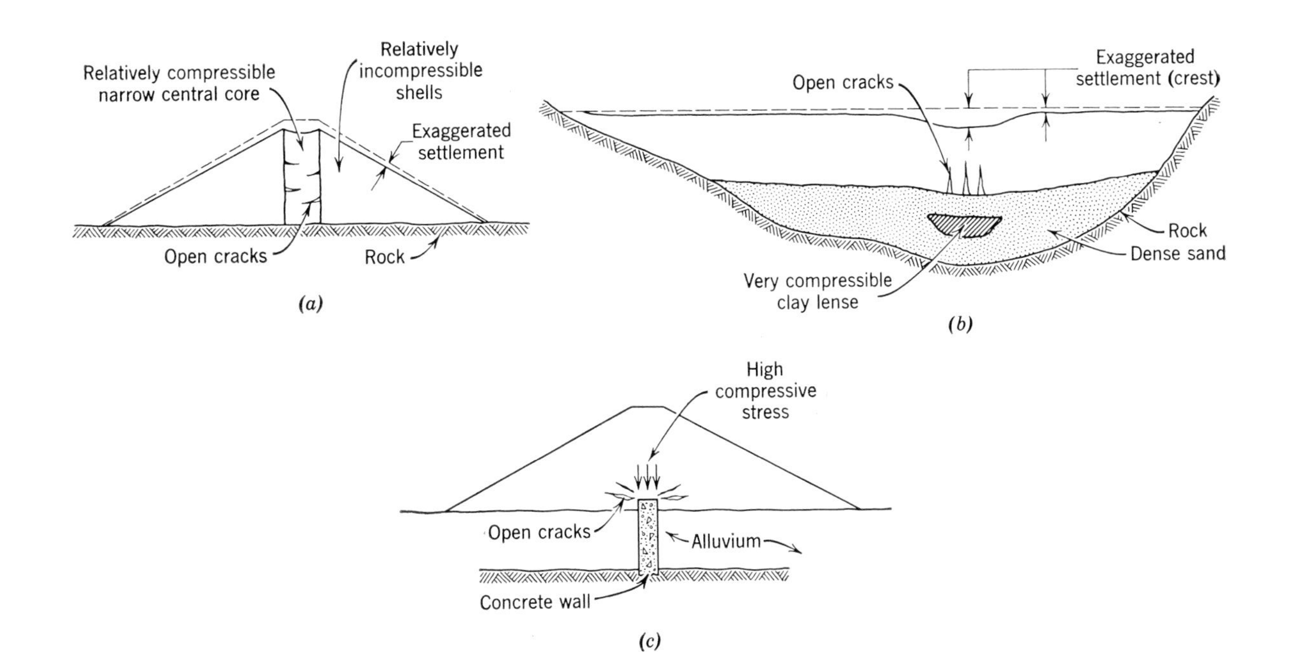

Fig. 2.3:5 Internal embankment cracking.

covered only by accident, as when drill water is lost in bore holes put down in the core after construction. On the other hand, the cracks and water losses in drill holes can always be explained as well by other phenomena, and the authors know of no experience which can be unquestionably attributed to cracks of this kind.

Also suspected when water is lost through drill holes or when leaks develop through the dam is a second category of internal cracks which seem to occur where a relatively short length of embankment is underlain by a more compressible foundation material than that which exists under the rest of the dam, Fig. 2.3:5*b*. Tension may develop in the lower part of the dam in the same way that it occurs in the lower fibers of a simple structural beam, and if the tensile extension is severe, cracks may open through the core. While this action has been suspected at several structures (see, for example, Ref. 15), it is always difficult to diagnose the trouble with complete certainty.

However, it is definitely known that localized internal cracks open in impervious zones adjacent to interior concrete structures, Fig. 2.3:5*c*. Since the concrete is much less compressible than the surrounding soil, concentrations of high compressive stress develop near the top of the concrete structure. These stress concentrations are associated with relief of stress in other areas and they sometimes cause tension and cracking. At Neversink Dam, (Ref. 587) which had a rigid concrete foundation cutoff wall extending well up into the embankment, cracking of this type produced a pervious zone about 50 ft. above the foundation in the otherwise very impervious core. Similar localized cracks can open in the embankments adjacent to reinforced concrete outlet conduits and even in the vicinity of steep, jagged, hard rock abutments.[1]

2.3b Influence of Embankment Properties

The amount of cracking which will develop at a given dam depends on the magnitude of the strain imposed and on the deformability of the embankment. There exist no reliable guides, either from field observations or laboratory tests, for estimating the maximum amounts of embankment settlement which can take place at a given site without the development of cracks. Cumulative records of embankment cracking, perhaps supplemented by laboratory research on the stiffness of compacted impervious soils, may in the future provide some definite criteria for the designer.

[1] See the experience described in footnote on p. 135.

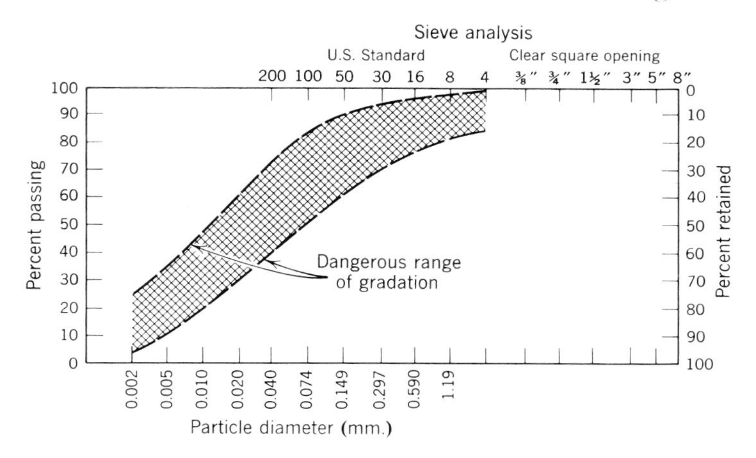

Fig. 2.3:6 Range of gradation of soils suspected to be most critical embankment materials from standpoint of cracking (Ref. 95).

One study of the performance of 17 dams, some of which cracked and some of which were subjected to large strains without cracking, furnishes a little information about the relationship between embankment soil properties and susceptibility to cracking (Ref. 95). Although the evidence on which this study was based was sketchy, it indicates that embankments of inorganic clays of low to medium plasticity (plasticity index less than 15) with gradation curves falling within the range shown in Fig. 2.3:6 are probably more susceptible to cracking when compacted dry than either finer or coarser materials. It also shows that clays of higher plasticity (plasticity index more than 20) which are finer than the gradation range in Fig. 2.3:6 will withstand much larger deformations without cracking.

The study provides some evidence that susceptibility to cracking is high in embankments of residual soils containing coarse particles of soft rock, which break down and become appreciably finer when they are being placed and compacted. It is difficult to get enough water into such soils during construction, so that they are often compacted at a lower water content than intended (Sec. 11.3*e*). Moreover, embankments constructed of soft, decomposing residual soils are often made more brittle by the cementing action of the partially decomposed particles, and therefore are more likely to crack with small deformations than embankments of ordinary soils.

Most of the embankments studied which developed severe cracking were compacted at low construction water contents (usually 5% or more drier than Standard Proctor Optimum), but dangerous cracks have also opened in embankments which were not compacted excessively dry. The cracks shown in Figs. 2.3:1*a* and 2.3:1*b* occurred in a 45-ft. dam built with an average construction water content not lower than 2% below Standard Proctor Optimum. The material was a clay of low plasticity falling in the range of "most susceptible materials," Fig. 2.3:6, and the site had steep rock abutments. The cracking in this case was caused by a settlement of roughly 2 ft., which resulted from the saturation of underlying dry foundation soils when the reservoir was filled for the first time.

A recent study of the factors influencing cracking has been carried out by Narain, who subjected beams of compacted soil to bending tests (Ref. 626). He developed theory to allow the use of such tests for the prediction of cracking potential in earth dams and then checked the theory with a few cases of dams which cracked. The results of the study give promise of future developments. Some of the specific conclusions of value to the earth dam designer are:

1. Increasing the compaction water content from 2 or 3% below Standard Proctor optimum water content to the optimum substantially increases the flexibility of a compacted soil. Further increases in the water content have less influence.
2. The tensile strain at failure is not related to the strain at failure obtained in a typical laboratory compression test, and therefore the compression test cannot give a good index of susceptibility to cracking.
3. At a given water content, increased compaction effort substantially decreases the flexibility of the compacted soil.

2.4 EMBANKMENT AND FOUNDATION SLIDES

Slides, which are one of the frequent causes of failure, occur in earth dams in the same way that landslides develop in natural earth slopes—when the average stress along any potential sliding surface becomes greater than the average strength. Because earth movements are spectacular phenomena, and because they lend themselves to analytical treatment, the mechanics of this type of failure have received considerable attention by the profession. Present methods of stability analysis (Chap. 7) have been developed largely as the result of studies of actual slides, and therefore the designer must understand the mechanics of failures which have occurred in order to evaluate the reliability of his analytical procedures.

Tables 2.4:1 and 2.4:2, taken from Middlebrooks' summary of records of earth dams with unsatisfactory performance (Ref. 3), give an instructive statistical view of the chronology of slides.

Table 2.4:1 Relation of Occurrence of Slide to Age of Dam (In Percentage of Total Number of Dams on Which Slides Occurred)

NO. OF YEARS AFTER COMPLETION	PERCENTAGE OF SLIDES OCCURRING
0–1	29
1–5	24
5–10	12
10–20	12
20–30	12
30–40	11
40–50	0
50–100	0

Table 2.4:2 Chronological Distribution of Failures (In Percentage of Total Number of Dams on Which Slides Occurred)

CALENDAR YEAR (DURING WHICH SLIDE OCCURRED)	PERCENTAGE OF THE TOTAL NUMBER OF SLIDES RECORDED
1850–1860	0
1860–1870	0
1870–1880	0
1880–1890	3
1890–1900	3
1900–1910	16
1910–1920	23
1920–1930	26
1930–1940	23
1940–1950	3
1950–	3

As seen in Table 2.4:1, nearly a third of all the slides developed within the first year after construction, and more than a half took place within 5 years. Table 2.4:2 shows that 94% of the slide failures in the study occurred before 1940. Because a large proportion of existing earth dams (including nearly all of the very high ones) have been built since 1940, it is apparent that present design and construction practices have almost eliminated the likelihood of slides. Most of the histories available are at least 20 years old, although there have been a few recent failures by sliding in small and poorly constructed dams, dams with very soft foundations, and dams built a number of years ago.

Slides can be grouped into three categories:

1. Slides during construction involving the upstream or downstream slope (or both);
2. Slides on the downstream slope during reservoir operation;
3. Slides on the upstream slope after reservoir drawdown.

These are discussed individually in the following subsections.

2.4a Slides During Construction

Relatively few slides have occurred on rolled-earth dams during construction compared with the number which have developed during the operation of the reservoir.[1] Of these few, none have threatened loss of life or damage to property other than to the dam itself. The greatest damage from construction slides is usually that to the relationship between the engineer and the client when the client has not been properly informed about the calculated risk being taken to make economies in the design.

In every case of construction slide known by the authors, the dam was underlain by a foundation of either soft, brittle, or sensitive clay, usually of high plasticity, and a large portion of the sliding surface passed through the foundation.[2] Experience indicates that there are two types of construction slides, which differ primarily in the speed and

[1] In the past a number of spectacular failures took place during construction of hydraulic fill dams. Probably the best known of these was at Fort Peck Dam in Montana, where in 1938 about 5,000,000 yd.^{3} of earth slid rapidly and killed eight men (Ref. 133). This failure and at least a dozen similar failures of major dams which took place between 1905 and 1930 occurred because the hydraulic fill method of construction resulted in an embankment which was, at least in part, composed of very loosely compacted saturated sand which "liquefied." There are no records of similar trouble during construction of compacted earth dams. Since it is unlikely that a rolled-filled dam would ever be placed in such a loose state that liquefaction of the embankment could occur, the lessons learned from the slides on hydraulic fill embankments have no direct bearing on the analysis of rolled earth dams.

[2] See also the footnote on p. 369 concerning the recent Waco Dam failure through a shale formation.

magnitude of the movement. In the first type (slow slide), the movement starts gradually and continues at a uniform rate for a period usually of one to two weeks. The total horizontal and vertical components of movement are usually a small percentage of the height of the dam (often between 5 and 15%). While the sliding never completely stops, the movement slows down at the end of the initial period to an unimportant rate in a creeplike action.

The second type of slide (rapid slide) takes place suddenly, and the magnitude of the movement is usually equal to one-half or more of the total height of the dam. The major part of the sliding is over in a few minutes, and the movement either stops or slows to a creeplike rate within several hours.

The difference between the two types is the result of differences in the foundation clay. In both cases the movement starts when the average stress imposed on the potential failure surface by the weight of the embankment approaches the average strength available to resist movement. Slow slides occur when the foundation is a more or less homogeneous deposit of soft clay which is not sensitive; that is, which does not lose its strength appreciably under the shearing movement.[1] Rapid slides result when the foundation clay contains horizontal bedding planes, lenses, or layers of coarse silt or fine sand through which the high pore water pressures developing under the center of the dam can be transmitted outward toward the more lightly loaded areas under the toe. The failure comes abruptly, since much of the weight of the sliding mass is temporarily transferred to the pore water.[2] Actually, rapid failure is possible even where the clay is not soft and where, if it were not for the spread of high pore pressures in horizontal layers, the foundation would be adequate to support the dam.

A typical slow slide during construction occurred at the North Ridge Dam in Canada (Ref. 109) in September 1953, when the dam

[1] In the late 1930's, slow-moving construction slides developed in the Tappan and Clendening Dams, both of them built for the Muskingum Flood Control Project in Ohio and both designed with large central clay cores and heavy rock toes on soft clay foundations (Ref. 134). The designers concluded that the failures were caused by incompatibility between the strain characteristics of the well-compacted embankments and the underlying soft clay foundations; i.e., that the embankments were so "overcompacted" that the bases of the dams expanded and pushed the toes upstream. This novel explanation provoked much discussion, but no failures have been attributed since then to overcompaction. Consequently, while there is undoubtedly good reason for not constructing heavily compacted, rigid embankments on soft clay foundations, it is probable that the failures of Tappan and Clendening Dams can be most easily understood as typical slow slides.

[2] See Ref. 126, pp. 401–404 for a discussion of the pore water action in this type of failure.

was nearly completed. The first indications of the slide were the appearance of cracks, a slight bulging of the fill, and an overthrust at the downstream toe. The widest cracks, which opened to a maximum of about 6 in. at the top, were 11 ft. deep along the upstream slope and 22 ft. deep along the downstream berm at the locations shown in Fig. 2.4:1. The ground surface for a distance of 60 ft. beyond the downstream toe heaved and moved outward about 3 ft., the movement of the downstream berm was 4.4 ft. horizontally and 0.7 ft. downward. As soon as the slope movement was noticed, fill construction was immediately stopped and berms were added at both toes. The dam was successfully completed in 1956 to its full height with no further modi-
mediately stopped and berms were added at both toes. The dam was largely responsible for the slide, had a liquid limit of approximately 72%, plasticity index of 51%, and natural water content equal to 36%.[1]

The failure at Marshall Creek Dam in Kansas (Refs. 128, 129), which occurred in 1937 when the embankment was within 10 ft. of the crest, is a typical example of a rapid slide. The following description was taken from the Report of the Board of Engineers.[2]

> The watchman, riding on the running board of one of the cars, noticed a large crack in the roadway near the center of the dam. A hasty inspection revealed that the crack was increasing in width. The cars were ordered from the dam and immediately thereafter the general movement began. *The major movement covered a period of 15 to 20 minutes and was accompanied by a rumbling noise and dense clouds of dust.* (Italics by the authors.)

As Fig. 2.4:2 illustrates, the movement was such that the height of the slide scarp was greater than half the height of the dam. Marshall Creek Dam was underlain by 20 to 30 ft. of "non-continuous layers of soil composed of silt and clay in every possible mixture" (Ref. 129), which must have been capable of transmitting pore pressures horizontally.

REPAIR OF CONSTRUCTION SLIDES

For dams at which slow moving slides have occurred, the foundation clay has always been of high plasticity and high water content. Since thick deposits of soft clay consolidate and gain strength very slowly, it has usually not been practicable to do much more than stabilize the

[1] For descriptions of other typical slow moving slides, see the experiences at Chingford Dam in England in 1937 (Ref. 123) and Lafayette Dam in California in 1928 (Ref. 130).

[2] From private correspondence with Mr. R. V. Smrha, Chief Engineer, Kansas State Division of Water Resources.

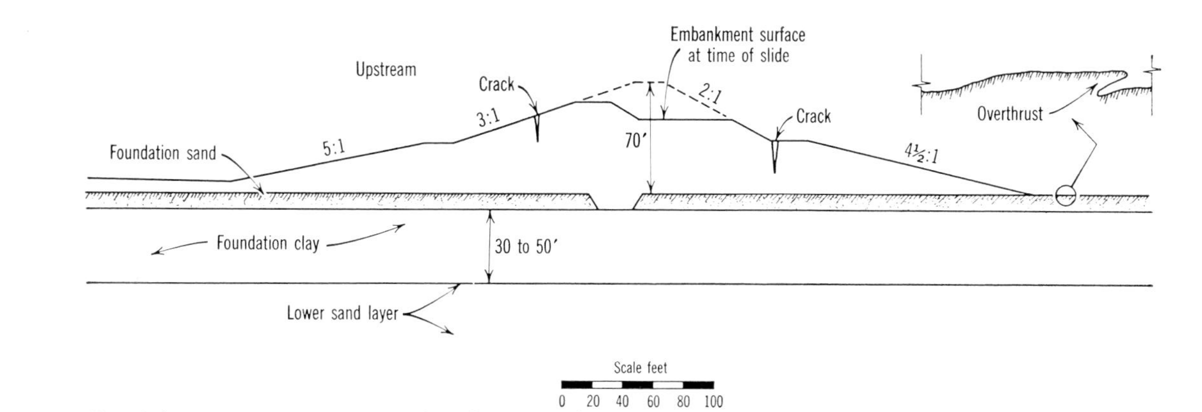

Fig. 2.4:1 Cross section of North Ridge Dam showing surface manifestations of slide during construction (after Peterson et al., Ref. 109).

(*a*)

(*b*)

Fig. 2.4:2 Views of construction slide at Marshall Creek Dam. (a) Looking along crest from right abutment. (b) Aerial view looking upstream. (Courtesy Sverdrup and Parcel, Consulting Engineers)

slide. In most cases the designers have been satisfied to leave the dam at the height which had been reached when the failure commenced rather than to try and build to the originally planned height.[1]

North Ridge Dam was built in the days of modern soil mechanics and was fitted with piezometers in the foundation clay. After the slow slide in 1953, further construction was postponed until the excess pore pressure in the foundation clay had dissipated, and it was then possible to finish the small remaining portion of the embankment as planned. The dam was completed to its originally proposed height without excessive movement.

Since failures of the rapid type are caused largely by the existence of high pore water pressures in horizontal foundation layers, dams which have suffered rapid failures can usually be reconstructed to the originally planned height without a long wait for foundation consolidation.

2.4b Downstream Slope Slides During Reservoir Operation

SERIOUS HAZARD OF DEEP DOWNSTREAM SLIDES

Two distinct types of downstream slides have occurred: deep slides, which generally pass through the clay foundation, and shallow surface slides. Deep slides nearly always take place during full or almost full reservoir and frequently reduce the freeboard by extending further upstream than the upstream edge of the crest. The internal pore water pressures which cause deep slides are the result of seepage from the reservoir through or under the dam. After a slide takes place, there is no relief in this pressure. The unstable vertical slide scarp left standing often sloughs or slides again until it breaches the dam and releases the reservoir water in one great flood wave. Many dams have been saved from complete failure after downstream slides only by around-the-clock emergency action.

Deep downstream slides generally move at about the same rate, or somewhat faster, than slow slides during construction. A typical downstream movement might be 3 to 4 ft./day in the first day or so and then approximately 1 ft./day for several weeks.

[1] At Lafayette Dam in California the slopes were flattened considerably after the failure, and no attempt was made to complete the embankment to its originally planned height. Even so, a slow, uniform movement of the slopes has continued during the 25 years since completion. For another typical example, see the description of the Muirhead Dam slide in Scotland in 1940 (Ref. 119).

TYPICAL DEEP DOWNSTREAM SLIDES

Fruit Growers Dam. At the Fruit Growers Dam in western Colorado (Ref. 136), which was constructed to a height of 32 ft. in 1910, a deep downstream slide occurred in 1937 when the reservoir was at the maximum elevation. The sliding soil buried the downstream end of the outlet conduit, making it impossible to lower the reservoir. As is frequently the case, a stream of water poured from the slide scarp at a point about 14 ft. below the dam crest, softening the whole downstream slope and threatening to cause a second slide which would cut back through the dam and release a flood wave. Complete failure was averted by a cut made by a bulldozer through the embankment and the natural abutment where the dam was only 5 ft. high. Once the flow started through this trench, it widened and deepened so rapidly that the entire reservoir was emptied in about 9 hours.

Great Western Dam. A similar experience occurred in 1958 at the 50-ft. high Great Western Dam near Denver, Colorado. This dam had been raised slightly and strengthened the previous year by the addition of granular material on the downstream slope. The slide began after the reservoir had been at the new maximum elevation for about two weeks. The sliding surface passed into the natural clay foundation and extended well upstream of the upstream edge of the crest, actually cutting below the water level in the reservoir, Fig. 2.4:3. As a consequence, water from the reservoir had free access to the sliding surface. Figure 2.4:4 shows photographs taken about 4 days after the slide commenced.

The sliding surface passed through the steel outlet conduit, crushing it and rendering it useless. In order to draw the reservoir down before the crest level sank lower than the water in the reservoir, a syphon pipe was placed over the top of the embankment and was used to lower the water level by approximately 1 ft./day. At the time the authors

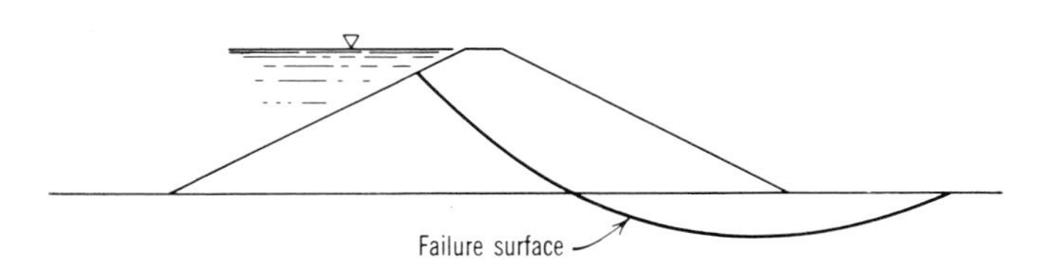

Fig. 2.4:3 Deep downstream slides frequently extend to the upstream slope below the reservoir water level.

(*a*)

(*b*)

Fig. 2.4:4 Views of downstream slope at Great Western Dam, Colorado (1958), four days after movement commenced. (a) View of dam from left abutment with slide in foreground. Note curve in dam crest, subsidence on upstream slope and heaved area on downstream slope. Backhoe in immediate foreground is working to deepen the spillway. (b) Closeup view of the upstream slope at location where it was intersected by the slide looking from dam crest.

inspected this slide, approximately 3 ft. of freeboard remained. Since the slide was also moving at about 1 ft./day, with all the emergency methods available, the repair crews were just able to preserve their slender freeboard. They could not increase it by sandbagging the crest because any added weight near the top of the sliding mass caused the movement to accelerate. The dam was saved when, by good fortune and hard work, the reservoir level was at last pulled down at a more rapid rate than the slide moved. Subsequently the dam was completely rebuilt.

SHALLOW DOWNSTREAM SLIDES

Shallow slides, most of which follow heavy rainstorms, do not as a rule extend into the embankment in a direction normal to the slope more than 4 or 5 ft.[1] Some take place soon after construction, while others occur after many years of reservoir operation.

At Costilla Dam, a 125-ft. embankment constructed in New Mexico, in 1920, a downstream slide developed in June of 1942. The crack at the upper edge of the slide was about 20 ft. below the crest. It ran parallel to the crest for about 80 ft. and then dropped downward into the heavy rock riprap which covered the lower portion of the slope. The sliding mass was found by observation in a test pit to be only a few feet thick, and the maximum vertical movement was only about 1 ft. In discussing the cause of the slide, the New Mexico State Engineer said: "The slide was the result of saturation of the downstream face by shallow percolation of excessive precipitation which fell on the surface of the dam, rather than from deep percolation through the dam by water impounded in the reservoir." (Ref. 137.)

Shallow slides following rainstorms have often occurred where there are heavy layers of downstream rock riprap, evidently because the riprap layer retains the rainwater in many small puddles and produces greater saturation of the downstream slope. Poorly drained downstream berms or poorly drained roads on the downstream slope also allow accumulation of surface puddles which saturate the embankment directly below the berm or road and lead to slides.

Shallow surface slips involving only the upper few inches of the embankment have sometimes occurred when the embankment slopes have been poorly compacted. This is a frequent difficulty in small, cheaply constructed dams, where the construction forces often do not make the determined effort necessary to prevent a loose condition in the outer slope (Sec. 11.6). The outer few feet then soften during the first rainy season, and shallow slides result. Such a condition is normally not serious, but it can often be embarrassing to the engineer.

[1] However, it is often difficult to determine from visual examination whether the slide is only superficial. Engineers at several dams have found it necessary to make test pits for observing the actual thickness of embankment affected.

It may also prove to be rather expensive to bring the slope back to a stable condition.

2.4c Upstream Slope Slides Following Reservoir Drawdown

Although it is conceivable that upstream slides could take place during full reservoir, all of the cases (except upstream slides during construction) known to the authors have occurred following reservoir drawdown. Upstream slides have not caused complete failure or loss of water from the reservoir, although they have occasionally blocked the entrances to outlet conduits and made these useless for further lowering of the reservoir, sometimes creating a very awkward and dangerous situation. Following an upstream slope slide caused by reservoir drawdown, the excess pore pressures within the embankment soil adjacent to the surface of sliding are dissipated to a large extent. Consequently, there is a lesser tendency for continued sloughing and sliding than there is in the case of downstream slides, in which the pore pressures are not likely to be diminished. Since the slide comes to equilibrium at a stage of low reservoir, there is small likelihood of catastrophic failure even though a large earth movement has taken place.

Extremely rapid rates of reservoir drawdown are not necessarily involved. A study of upstream slides in 12 dams (Ref. 95) indicates that the majority were caused by a drawdown approximately between maximum water surface and mid-height of the dam at average rates varying between 0.3 and 0.5 ft./day. Most drawdown slides have developed when the reservoir was lowered for the first time, though a few have occurred after many years of successful operation. In some of the latter the delay may have been due to a decrease in the shear strength of the clay embankment or foundation with time. In every case studied by the authors, however, the slide was caused by a drawdown which was either faster or over a greater range than had occurred previously.

Upstream slides *usually* move at the same rather slow rates as deep downstream slides. The majority have been deep-seated, with the sliding surface passing well into the clay foundation. The upper edge of the sliding surface normally extends only to the upstream edge of the crest. Some upstream slides have cut as far as the middle of the crest, and a few have reached to the downstream edge of the crest, Fig. 2.4:5. The authors know of none, however, which has caused the whole crest to sink and reduce the freeboard.

Fig. 2.4:5 Upstream slope slide at Mount Pisgah Dam in Colorado (1928). Note *that slide extends just to the downstream edge of the crest and the reservoir is at a low level. The upstream slope of the dam was 1½:1 and faced with a thin concrete slab.*

There have been a few shallow upstream slides, such as that at Belle Fourche Dam in Montana (Ref. 135). Built in 1909 as a homogeneous embankment of highly plastic clay, the dam has a 2:1 upstream slope and a maximum height of 115 ft. In 1931, more than 20 years after construction, the reservoir was drawn down at an unprecedented rate and a shallow layer of soil slid down the upstream slope. The slide had a maximum depth normal to the sliding surface of about 10 ft., and the difference in elevation between the top and the bottom edges of the slide was approximately 45 ft. A movement of this type is similar to a shallow downstream slide resulting from heavy rain.

In some dams constructed for hydroelectric power, where the reservoir remains essentially full all the time, the riprap wave protection is placed only at the top of the dam in the range of elevation over which the reservoir will operate. If the upstream slope is relatively steep, troublesome surface slides may develop and become progressive as the reservoir is being filled and the water level is still below the bottom of the riprap. These slides are apparently first caused by erosion from wave action on the unprotected lower slope of the dam.

Even if the embankment material is basically of grav
slides may develop if the slope is steep. This condi
if the slope is not originally dressed to a smooth uni
especially, if localized steeper slopes from haul road
The authors know of one recently constructed m
general type where the upstream portion of the dam consisted of coarse, sandy gravel on a 2:1 slope where considerable surface sloughing took place during the first reservoir filling even though there was no lowering of the water level at any time.

2.4d Influence of Soil Types

Almost all slides during construction and all deep upstream and downstream slides after construction have occurred in dams underlain by foundations of clay relatively high in plasticity and natural water content. In addition, a strong correlation exists between the incidence of slides and the use of fine-grained and highly plastic soil in the embankment.

In a study of 65 old homogeneous dams in the western United States (Ref. 95), the 14 slides which occurred all involved embankments of clay of at least medium plasticity. The D_{50} values for the soils from which these 65 dams were constructed covered the whole range of soil types used for the impervious sections of earth dams, varying from finer than 0.005 mm. to coarser than 2.0 mm. As seen in Fig. 2.4:6, the correlation between the fineness of soil and the susceptibility to sliding was strong enough to outweigh the influence of all other factors, including steepness of slopes, construction methods, and reservoir activity, on the susceptibility to sliding. Of the embankments included in this study, all that were constructed of clay having an average grain size (D_{50}) finer than 0.006 mm. failed by sliding; of those made from clays with median grain size ranging between size 0.02 mm. and 0.006 mm., approximately half failed by sliding; and of those made from soils with median grain size in the range between 0.02 and 0.06 mm., only a small fraction failed by sliding. None of the dams constructed of soils with an average grain size coarser than 0.06 mm. failed by sliding, even though many had steep slopes and were poorly compacted. From these statistics and from experiences common to many engineers, there seems to be justification for the statement that rolled-earth dam embankments have not failed by sliding unless the embankments or the foundations consisted of relatively fine-grained soils.

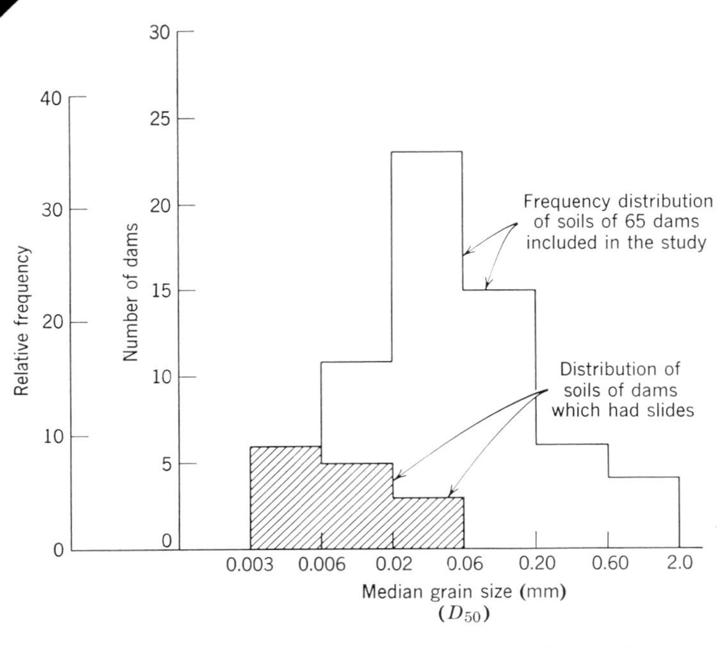

Fig. 2.4:6 Statistical relationship between occurrence of slides and average grain size of embankment soil from study of performance of 65 old dams (Ref. 95).

2.5 EARTHQUAKE DAMAGE

The information available concerning the performance of earth dams during earthquakes is so meager that it offers little assistance to the designer planning a dam in a region of seismic activity. Observations of damage to similar structures such as highway and railroad embankments, levees, and tailing piles indicate clearly that loose, uncompacted earth embankments subjected to severe earthquake shocks are literally shaken to pieces. They settle as much as 50% of their height, spread at the base, and develop large cracks in all directions. Most of the earth dams which have been damaged were constructed before 1920 with design details and construction methods which would be considered less than adequate at the present time.[1]

No large well-constructed modern dams have as yet been subjected to severe shocks. Only a few dams as high as 100 ft. have been badly shaken, and in most cases neither the intensity of the shock nor the

[1] The first comprehensive collection of records of earthquake damage from various parts of the world was recently completed by Ambraseys (Ref. 435).

resulting damage is known in reliable detail. A few of the more instructive and better documented cases are described below.

SAN FRANCISCO EARTHQUAKE—1906

The 1906 earthquake in San Francisco was one of the most severe on record in the United States. At that time there were three relatively large rolled-earth dams retaining municipal water supply reservoirs located a few miles south of San Francisco—San Andreas Dam, Pilarcitos Dam, and Upper Crystal Springs Dam (Ref. 141). The intensity of the shock on the San Francisco Peninsula in the area of the dams was at least 10 on the Modified Mercalli scale.

The San Andreas and Upper Crystal Springs Dams were constructed in the main valley of the San Andreas Fault. During the earthquake this fault had a horizontal displacement as large as 21 ft. at some points. The Pilarcitos Dam is located a short distance off the main fault line.

San Andreas Dam. At the time of the shock the San Andreas Dam was 90 ft. high, having been built in the 1860's and raised in 1875. It was constructed of well-graded gravelly, clayey sand compacted in thin layers with a 3-ton roller. The embankment has conservative slopes of 3½ to 1 upstream and 3 to 1 downstream and a relatively narrow central core of wet puddled clay.

The dam was evidently subjected to an extremely severe shock. The main fault, though it did not cross the foundation, passed through the east abutment directly adjacent to the embankment. The concrete-lined outlet tunnel in the east abutment was twisted 10 ft. out of line. The distinguished consulting engineer C. Derleth of the University of California inspected the dam several days after the earthquake and wrote (Ref. 143): "As an eyewitness, I am convinced that this dam was subjected to the most severe earthquake shock." The natural ground around the dam was everywhere crisscrossed with cracks evidently caused by the tensile stresses in the rarefactions between the earthquake shocks. A single longitudinal crack 2 to 3 in. wide opened parallel to the axis of the dam near the center of the crest. Even though the reservoir was almost full at the time of the shock, no leakage or other alarming symptoms developed.

Upper Crystal Springs Dam. The Upper Crystal Springs Dam was built in 1887 with approximately the same construction method and design section as San Andreas Dam. It is probably the only dam of any size that has actually been displaced by an earthquake fault. In 1906, however, it was being used merely as a highway embankment across the reservoir. The water was at the same level on both sides, and the outlet conduit through the dam was open.

The fault passed through the reservoir, intersecting the dam at a point where the crest was about 20 ft. above the bedrock foundation (Ref. 435). The crest was offset approximately 8 ft. during the earthquake, and both longitudinal and transverse cracks developed on it. Some longitudinal cracks with a maximum width of 6 in. had an observed depth of only 4 ft. The transverse cracks were closed by the general compression which appeared to develop in the direction normal to the fault plane in the immediate vicinity of the fault.

Since there was no water head on the Upper Crystal Springs Dam, the experience provided no information on leakage. The tightness of the reservoir was not influenced noticeably even though the fault ran through the middle of it. Engineers who inspected the dam after the earthquake came to various conclusions about whether it would have failed if it had been retaining a reservoir (Ref. 436). Some decided that the core was sufficiently plastic so that no leakage would have occurred at all; others believed that considerable leakage would have developed through cracks in the clay core and would possibly have led to piping failure.

Pilarcitos Dam. Built in 1874, the Pilarcitos Dam was 95 ft. high at the time of the earthquake. The upstream slope was 2½ to 1 and the downstream 2 to 1. Like the San Andreas Dam, it had been constructed in thin compacted layers with a puddle core of clay. When it was inspected soon after the earthquake, it showed no cracking or other visible evidences of damage (Ref. 143).

The records of the San Andreas and Pilarcitos Dams, which are now approximately 90 years old, constitute two of the few reliable guides to the action of severe earthquake shocks on earth dams. Constructed of clay with methods similar to those used for modern dams, both were a little less than 100 ft. high at the time of the shock. Both were underlain by approximately 50 ft. of alluvial deposits, and both were subjected to extremely severe shocks during the 1906 earthquake, as well as to numerous lesser shocks in other years. Both have performed very well, and no damage has developed in either from earthquakes except for the longitudinal cracking in the center of the crest in San Andreas Dam.

KWANTO EARTHQUAKE, 1923—ONO AND LOWER MURAYAMA DAMS

The 120-ft. Ono and 80-ft. Lower Murayama Dams near Tokyo, Japan (Refs. 435, 437) were subjected on September 1, 1923, to very severe shocks, probably of about the same magnitude as the 1906 shocks near San Francisco. Both the Japanese dams had been compacted in layers and provided with central clay cores.

At the time of the earthquake the water in the Ono Reservoir was 19 ft. below the top of the dam. The crest settled approximately one foot as a result of the shock. The embankment developed serious cracks varying in width from 2 to 10 in. on the surface and extending to depths of 35 to 70 ft. Two local slides about 60 ft. long also occurred. No leakage is mentioned in the reports. Ono Dam was evidently very badly shaken even though it was located 60 miles from the epicenter. The cracks were filled with a clay-sand grout, however, and the dam continued in use.

At the Lower Murayama Dam, where the water level was 30 ft. below the crest at the time of the earthquake, the reports stated that the crest settled about 8 in., longitudinal cracks opened in the upper portion of the embankment, and the downstream slope bulged about 6 ft. No leakage was mentioned in these reports either, and the dam continued in use.

SANTA BARBARA EARTHQUAKE, 1925—SHEFFIELD DAM

Sheffield Dam, which was built by the city of Santa Barbara, California for water supply storage in 1918, is the only dam known to have failed completely as the result of an earthquake. The shock, which moved wooden houses nearby as much as 4 in. on their foundations, was estimated at about 9 on the Rossi-Forel scale. (Refs. 144, 435).

The dam failed early in the morning of June 29, 1925, with no witnesses, and in the subsequent investigations there developed some controversy over the methods by which the embankment had been constructed. It is believed that the construction was in relatively thick layers with a minimum of compactive effort, probably consisting only of the travel of the hauling equipment. Good photographs of the walls of the breach show that the material contains cobbles and boulders up to 2 ft. in diameter embedded in fine-grained soil. The upstream slope was faced with a 5-in.-thick slab of reinforced concrete, the lower edge of which was extended with a vertical wall down to bedrock as a foundation cutoff.

Before the failure a considerable quantity of leakage had occurred at the downstream toe for several years, and the lower portion of the downstream slope had been saturated continuously when the reservoir was full. The water level was about 15 ft. below the crest of the 35 ft. dam at the time of the earthquake.

Photographs of the failure are shown in Figs. 2.5:1 and 2.5:2. In several ways this failure was unique. First, because the whole central portion of the dam moved in a more or less intact condition about 100 ft. downstream. The site was inspected the week after the failure by a number of eminent dam engineers, and almost all independently used the phrasing that the central portion had moved downstream like a door on hinges. Second, the failure was unique because the concrete slab facing remained intact in the lower portion of the embankment across the full length of the dam, Fig. 2.5:1.

It seems probable from the appearance of the remaining embankment and the descriptions of the portions which moved downstream that the failure was caused by liquefaction of the lower portion of the embankment or the upper part of the foundation. Evidently the earthquake in shaking the relatively loose embankment tended to vibrate it into a denser condition. Since the lower embankment was saturated, the vibration caused the weight of the dam to be carried largely by the pore water. When the load was transferred to the pore water, however, the shearing strength of the base was severely reduced, and the force of the reservoir water acting on the upstream slope pushed

Fig. 2.5:1 View of upstream face of Sheffield Dam several days after failure. (Courtesy State of California, Dept. of Water Resources)

Fig. 2.5:2 View of failure of Sheffield Dam from right abutment. (Courtesy of State of California, Dept. of Water Resources)

the central portion of the dam bodily downstream. It is difficult for the authors to conceive of any other phenomenon which could explain the type of failure which occurred.[1]

CONTRA COSTA, 1955—ST. MARY'S DAM

Constructed in Contra Costa County, California in 1928, St. Mary's Dam is a small water supply dam 50 ft. high with a 2½ to 1 upstream slope and a 2 to 1 slope downstream. It has an upstream face of reinforced concrete and a central concrete vertical core wall which extends to within 5 ft. of the crest. Over the years the dam had performed very well, and there had been no cause for concern of any kind (Ref. 145).

A local earthquake shock on October 23, 1955 produced an estimated intensity of 7 on the Modified Mercalli scale (Ref. 435). On

[1] As in the case of most failures of old dams, the facts are obscured by time, and it is not unanimously agreed that the failure was due to a flow-slide. In 1949, the Los Angeles District of the U.S. Corps of Engineers carried out an investigation of the failure and concluded that the movement was probably not a flow slide but was due to shear failure on a more or less horizontal plane through the base (Ref. 3).

October 25, an engineer of the California Department of Water Resources inspected the dam as part of a routine check of all the water-retaining structures in the area after the earthquake. He found a longitudinal crack varying approximately between 0.1 and 0.2 ft. in width running the full length of the crest near the center just above the concrete cutoff wall. Subsequently a trench 7 ft. deep was opened on the crest as an observation pit, and it was found that the crack extended down 5 ft. to the top of the cutoff wall. A ½-in. space or crack was also discovered between the upstream portion of the earth dam and the concrete core wall. In the walls of the exploratory trench, which was located just above and upstream from the top of the concrete wall, the compacted embankment was badly fractured and cracked, evidently from localized high compressive stresses, Fig. 2.5:3.

This evidence makes it appear that there was a relative vibration between the concrete wall and the embankment. Cracking of this type could have developed if the wall had remained rigid and stationary and the earth dam had been vibrated several inches in the upstream-downstream direction at the crest, or vice versa. No damage could be found in the reinforced concrete slab on the upstream slope.

SOUTHERN CALIFORNIA, 1952—SOUTH HAIWEE, DRY CANYON, AND BUENA VISTA DAMS

The Tehachipi earthquake of July 21, 1952, had an estimated maximum intensity of 10 on the Modified Mercalli scale and caused cracking in several low earth dams in Southern California (Refs. 145, 146).

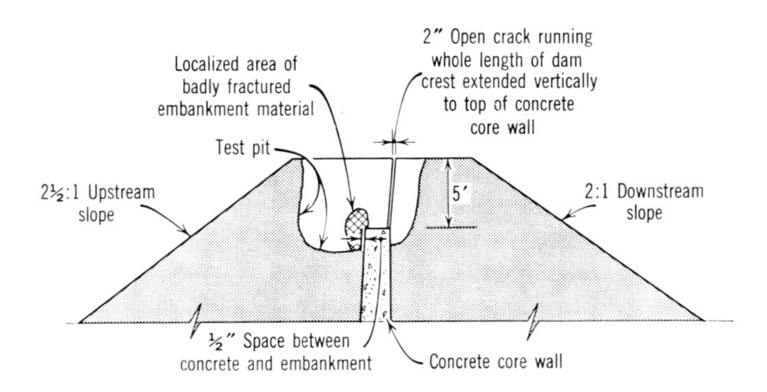

Fig. 2.5:3 Cracking as seen in walls of exploratory test pit following 1955 earthquake shock—Saint Mary's Dam.

One was the South Haiwee Dam in the Owens Valley of Inyo County, which had been built in 1912 by the hydraulic fill method. It has a height of 90 ft. and a crest length of 1500 ft. Following the earthquake, a number of ¼-in.-wide longitudinal cracks were found over a long part of the dam, generally near the upstream edge of the crest. One of these was approximately 1 in. wide for a short distance.

Test pits indicated that the cracks disappeared 5 to 6 ft. below the crest, and that at the bottom they dipped 6 to 14 in. toward the reservoir. A series of levels run on the crest for several weeks showed a maximum settlement of approximately ¼ in. during the first day and a total of 1 in. in 2 weeks. These measurements left no doubt that the earthquake had shaken the embankment considerably and had initiated new settlements again 40 years after construction.

The Dry Canyon Dam in Los Angeles County was built in 1912 to a maximum height of 55 ft. and a crest length of about 600 ft., with upstream and downstream slopes of 2½ to 1 and 2¼ to 1 respectively. Construction was partially by the hydraulic-fill and partially by the rolled-fill methods. Solution of gypsum in the abutments by seepage water caused a problem during the 1930's, but it was resolved without serious trouble (Sec. 2.8a).

After the earthquake a number of longitudinal cracks opened on the crest. The largest, which was located at the downstream edge, was about 2 in. wide. When a test pit was sunk 16 ft. to the top of the old hydraulic fill section, this crack went down in the walls of the pit almost vertically with a slight slope toward the reservoir. It stopped at the hydraulic fill, probably because of the saturated condition there. All the cracks were subsequently sealed, and no further trouble developed.

Also during the 1952 shocks the crest of the Buena Vista Dam in Kern County settled about 2½ ft. over a 75-ft. length where the embankment was only about 8 ft. high. The great settlement was undoubtedly due to the fact that for many years large quantities of gypsum had been leached from the foundation by seeping water (Sec. 2.8a).

LOS ANGELES COUNTY, 1930—CHATSWORTH DAM NO. 2

Chatsworth Dam No. 2 in Los Angeles County was constructed in 1918, probably by the hydraulic-fill method, to a height of 35 ft. It has a crest length of about 2,000 ft. and 2½ to 1 upstream and downstream slopes. Before the earthquake the dam had been leaking for a number of years, and repairs and an extension of its height were under consideration.

An earthquake estimated to have an intensity of at least 7 on the Modified Mercalli scale occurred in the afternoon of August 30, 1930. At the dam the caretaker's automobile, though its brakes were set, reportedly moved 4 ft. in one direction and then reversed for 6 ft., throwing the driver from the running board. In houses nearby, windows were broken, big chimneys toppled, and loose articles were thrown from shelves and mantles.

The shock caused a number of troubles at the dam:

1. A loose, dumped earthfill, which had been deposited on the upstream slope that spring when the reservoir was half full, slid into the reservoir. The fill had been dumped in an effort to decrease the saturated condition of the lower downstream slope, and it was not bonded to the dam. At the time of the earthquake it was on a fairly steep slope and was thoroughly saturated.
2. A system of longitudinal crest cracks appeared over the full length of the dam. Of four very long ones, two were in the upstream half of the crest and two were in the downstream half. These varied in width from $\frac{1}{4}$ to 3 in. Observations in test pits indicated that they ranged from 5 to 12 ft. deep and that their widths decreased uniformly.
3. Seepage developed through the embankment and discharged on the downstream face about 400 ft. from the left abutment and 10 ft. above the embankment toe. When observed approximately 2 hr. after the shocks, the measured flow was 8 gal./min. and the water was highly turbid. At 8:00 the next morning the flow had decreased to 4 gal./min. and was less turbid, and in two days it was only 1 gal./min. and had become clear. The leak caused no further trouble.
4. In the weeks following the earthquake, a new area of seepage developed in the valley floor downstream.
5. A number of surveyors who were working on the dam at the time of the shock gave the opinion that the crest had settled between 1 and 3 in. and had moved downstream a similar distance.

WEST YELLOWSTONE, 1959—HEBGEN DAM

Hebgen Dam on the Madison River near West Yellowstone, Montana had always been a satisfactory structure before the earthquake of 1959. It was constructed in 1914 to a height of 90 ft. with a crest 700 ft.

long. The embankment was built as a rolled fill from a gravelly clay of medium plasticity. As seen in Fig. 2.5:4, the dam has a thick central core of concrete, which is founded on bedrock over most of its length. The reservoir has a length of about 17 miles and a capacity of 350,000 acre-ft.

A relatively severe shock occurred at 11:35 P.M., August 17, 1959 (Refs. 570, 571, 572, 573). One of the main faults passed along the north shore of the reservoir within about 700 ft. of the dam, Fig. 2.5:5. Figure 2.5:6 shows a closeup of the fault at a point near the dam where there was a maximum vertical displacement of the ground surface of 15 to 18 ft. with no appreciable horizontal displacement. The earthquake caused large waves on the reservoir, open cracks and subsidence of the ground, a number of landslides around the reservoir edge, and considerable damage to wood frame buildings in the area. In addition, a massive rock slide estimated at 50,000,000 tons of rock dammed up the river 7 miles downstream and created a lake which backed up nearly to the downstream toe of Hebgen Dam. From every visual indication, the dam was in the area of greatest surface movement. The shock intensity at the site was probably at least 10 on the Modified Mercalli scale,[1] but the dam did not fail even though it was badly shaken.

A few minutes after the first shock, the caretaker rushed to the dam and in the moonlight observed the reservoir action from the high ground above the right abutment. The reservoir was full, and the first waves created had already overtopped the dam before he arrived. A few minutes later another wave struck the dam with such momentum that water 3 ft. deep ran uniformly over the crest for 10 minutes. Subsequently the wave receded and seemed to travel to the other end of the reservoir. After 10 minutes it returned, and water flowed over the crest for another 10-minute period. This action was repeated, and although the estimates of depths and durations are rough, there can be no doubt that the water flowed over the dam at least four times.

The crest and the downstream slope were surprisingly resistant to erosion. Much of the vegetation on the downstream slope was swept away, but there were no concentrations of erosion which in any way threatened the dam. The deepest erosion gullies were about 3 ft.

Both the upstream and downstream portions of the embankment settled over the whole length of the dam with respect to the concrete core wall. The amount decreased almost linearly with the distance

[1] The magnitude was measured at 7.5 to 7.8 on the Richter Scale, which by comparison gave readings of 8.2 and 7.7 for the great San Francisco and Tokyo earthquakes (Ref. 570).

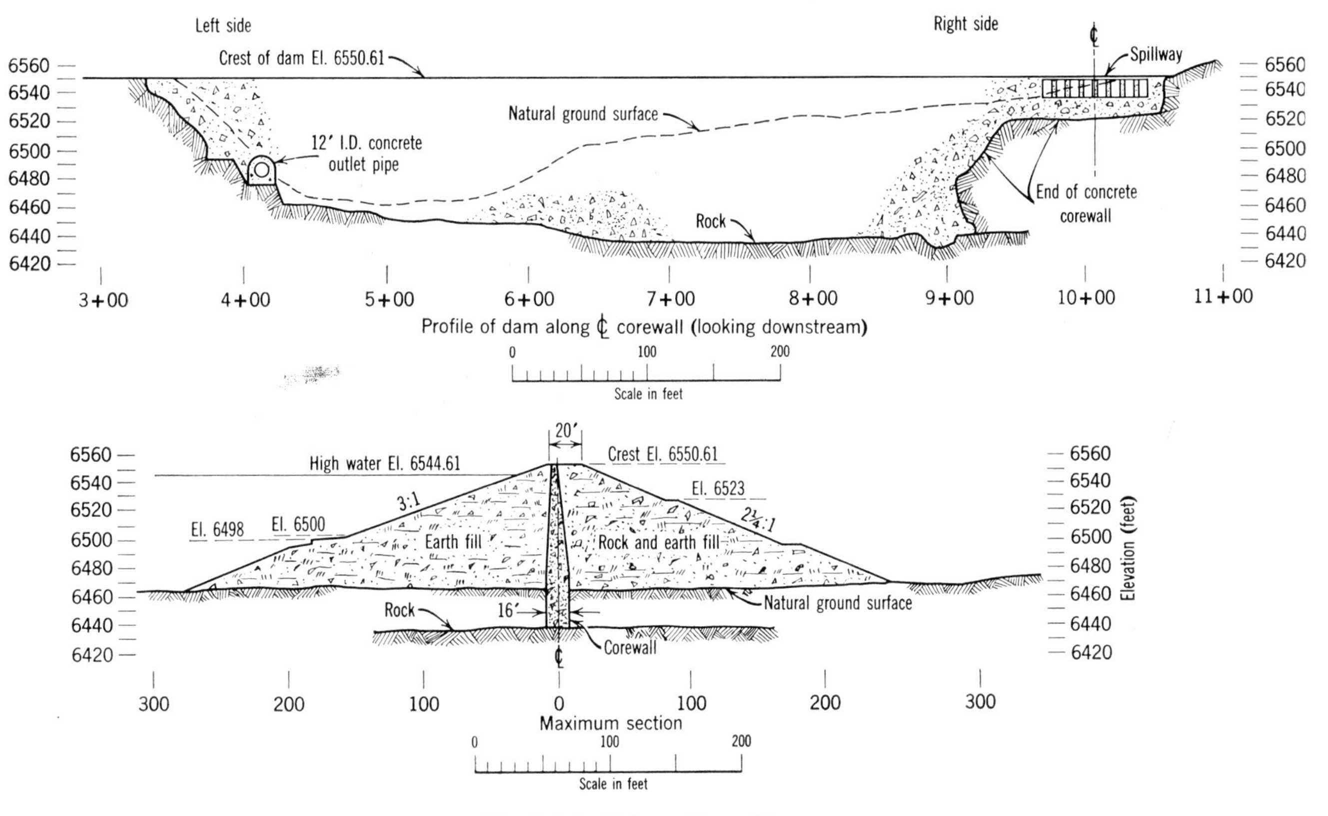

Fig. 2.5:4 Hebgen Dam, Montana.

Fig. 2.5:5 Aerial view of Hebgen Dam taken four days after the earthquake of August 17, 1959. The main earthquake fault can be seen on the hillside of the left side of the photo running parallel with the road along the north shore of the reservoir.

Fig. 2.5:6 Closeup view of face of main fault scarp near Hebgen Dam.

from the right abutment, where the maximum settlement downstream from the core wall was about 4 ft. at a point adjacent to the concrete spillway. At this point the dam was lowest but the thickness of alluvium under it was greatest. The settlement at the left abutment was only about 1 ft. The upstream slope slumped more than the downstream, the maximum settlement upstream with respect to the core wall being approximately 6 ft. near the middle of the dam.

The core wall, which was exposed when the upstream and downstream portions slumped (Figs. 2.5:7, 2.5:8), was in perfect condition over most of its length, although four vertical cracks about equally spaced over some 85 ft. developed at the extreme right end where the wall joined the spillway structure. The maximum displacement of these cracks was about 3 in. as shown in Fig. 2.5:9.

Longitudinal cracks, the widest being about 12 in. and the average 2 to 3 in., opened in the embankment on the crest downstream from the concrete core wall. These were most severe at the right end of the dam in the immediate vicinity of the spillway, but they extended

Fig. 2.5:7 Crest of Hebgen Dam three days after earthquake looking from right abutment. Exposed concrete core wall shown to left side of photo was completely buried before earthquake. Note that cap of soil remains on top of concrete wall in spite of serious overtopping. Cracking in embankment crest is wider at this point than the average crest cracks.

Fig. 2.5:8 View of upstream slope of Hebgen Dam three days after earthquake looking from center of dam toward left abutment. Note uncracked condition of newly exposed concrete core wall at the upstream edge of crest.

over the whole crest. The downstream slope was free from cracking except at the right end, where a number of cracks developed. No bulging was visible at the toe. A number of severe longitudinal cracks appeared on the portion of the upstream slope exposed above the high water line. The maximum width of these was about 6 in., and some were several hundred feet long.

Gaps also opened on both sides of the core wall between the concrete and the earth embankment. Apparently they were caused by relative upstream and downstream vibration between embankment and wall, a phenomenon similar to that which seems to have occurred at St. Mary's Dam described in the foregoing. Near the central portion of the dam the spaces between the embankment and the wall were as much as 6 in. wide.

A number of leaks developed near the abutment contact at the right end of the dam in the general vicinity of the core wall cracks. Seepage water was clear from the beginning and not plentiful, and it finally stopped when leaks in the badly cracked spillway structure were sealed.

Fig. 2.5:9 Typical crack which developed in concrete core wall at extreme right end, Hebgen Dam. (Author S. F. Gizienski.)

SUMMARY

From the incidents described above and from other records, the following observations are warranted:

1. In the majority of dams shaken by severe earthquakes, two primary types of damage have occurred: longitudinal cracks at the top of the embankment, and crest settlement. The crest settlement was not usually great enough to threaten failure by overtopping.
2. Only in the case of the Sheffield Dam did a complete failure occur in a structure which could properly be called a rolled-earth dam. The Sheffield failure was probably due to liquefaction of the very loose and saturated lower portion of the embankment.

3. The damage to the typical dam appears to have been caused primarily by the horizontal component of the earthquake movement in the upstream-downstream direction; that is, the direction transverse to the longitudinal axis of the dam. Probably the amplitude and the acceleration of the horizontal component of the movement of the crest in this direction is much larger than the movement of the foundation, and this results in a "whipping" action of the thinner top of the dam and causes longitudinal cracks.[1]
4. Earthquake shocks have caused remarkably few slope slides in earth dams even though some of the embankments must have been very poorly compacted.
5. There is some evidence for the speculation that earth dams are damaged more severely by ground movements with longer periods (lower frequencies) than those which cause greatest damage to building structures. For this reason dams located very near the earthquakes may not be as badly cracked as dams many miles distant. For example, the Ono Dam in Japan, which was very badly shaken and damaged in the Kwanto Earthquake, was located approximately 60 miles from the epicenter.
6. There is strong evidence for the conclusion that earth dams with central concrete core walls are more badly cracked by earthquakes than embankments without core walls. The records of both the St. Mary's and the Hebgen Dams indicate that the embankment and the concrete wall did not vibrate together when the dam foundation was shaken. From the damage to the Ono and Murayama Dams there is even some indication that cracking may be accentuated by the existence of central clay cores.

2.6 RESERVOIR WAVE ACTION AND UPSTREAM SLOPE PROTECTION

The erosive wave action, which causes most of the trouble at earth dams occurs only at relatively infrequent intervals during unusually bad storms. It normally lasts for short periods, and since considerable

[1] In addition to the evidence for this conclusion offered by the appearance of the cracked dams, measurements at the Cachuma Dam in California, which has seismographs both on the crest and in the foundation of the dam, indicate that the ground acceleration measured in a small shock in 1957 was amplified from 0.010 g at the foundation to 0.025 g at the crest (Ref. 435).

time would be required for waves to erode completely through an earth dam even if there were no slope protection, damage from wave action has not caused a serious threat of complete failure except in rare cases. Usually it has necessitated repairs rather than emergency action. Only a few poorly constructed dams with completely inadequate freeboard or excessively steep upstream slopes have been in danger of failure from wave erosion.

The near failure of Point-Of-Rocks Dam in Northeastern Colorado illustrates the worst trouble that has occurred (Refs. 95, 149). It was constructed in 1911 with a maximum height of 90 ft. and a length of 1 mile. The homogeneous clay embankment originally had an upstream slope of 1.5 to 1 (Fig. 2.6:1), and the original protection consisted of a 4-in.-thick concrete slab reinforced by wire fencing, with construction joints every 20 ft.[1]

The dam is located in a flat part of the great plains of Colorado where high wind storms are common, and the reservoir is so situated that the wind can blow over it for 2 miles toward the dam. On the first filling, waves 5 to 6 ft. high developed. The steep upstream slope and concrete facing proved unsatisfactory from the beginning. In 1915 the waves cracked the slab and eroded the embankment to a depth of 2 to 3 ft. over long sections. A layer of rock riprap with a thickness varying from 2 ft. at the crest to 5 ft. at the toe was finally placed over the concrete slab. At the upper elevations the 2-ft.-thick riprap layer was hand-placed on a 12-in. layer of local sand and gravel. Over the following years, however, the slope proved too steep for the riprap repairs to stay permanently in place, and many thousands of tons of additional rock were dumped on the face.

On May 8, 1927, when the reservoir was 17 ft. below the top, a 2-day wind storm of hurricane proportions began. As shown in Fig. 2.6:1, the crest was nearly eroded through and a disastrous failure was averted only narrowly. During the latter part of the storm, when it appeared that the dam might fail completely, over 1,000 men worked night and day, sand-bagging at the high water line.

The history of this dam indicates that under certain circumstances wave action could cause a catastrophic failure. It should be noted, however, that the conditions resulting in near disaster at Point-Of-Rocks Dam would not be duplicated in modern design practice, since a 4-in. reinforced concrete slab on a 1.5 to 1 upstream slope would no longer be considered adequate for a dam retaining a large reservoir.

[1] It was the common practice in the early 1900's in the Rocky Mountain States to make the upstream slope considerably steeper than the downstream.

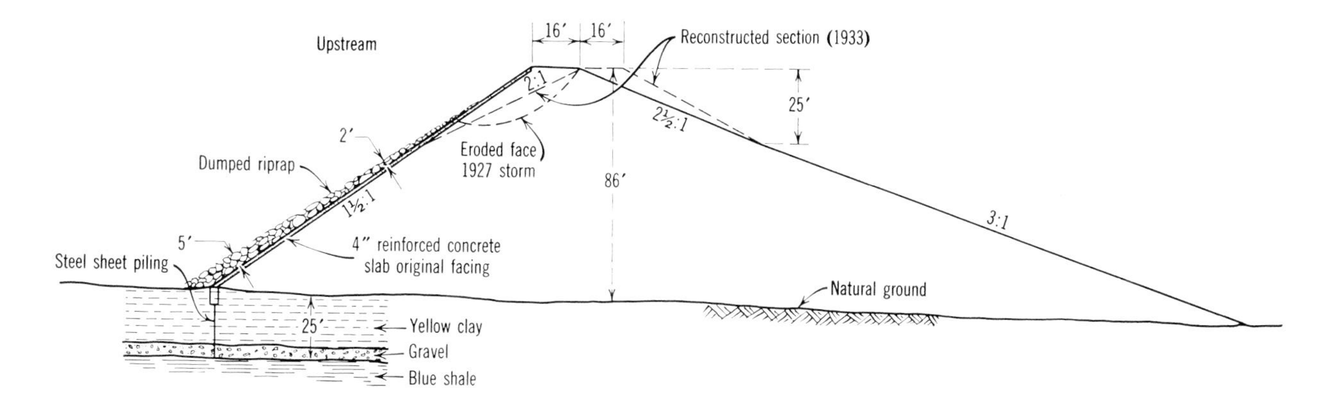

Fig. 2.6:1 Cross section of Point-of-Rocks Dam showing damage which occurred during the 1927 storm and the reconstructed section.

2.6a Slope Protection Failures

The upstream slopes of most earth dams have been protected with one of the following materials (in decreasing order of frequency):

1. Dumped rock riprap.
2. Hand-placed rock riprap.
3. Articulated concrete pavement consisting of individual slabs.
4. Monolithic reinforced concrete pavement.

A few dams have been faced with asphalt layers of various types or protected with floating log-booms, but such dams retain small reservoirs which have little or no wave action.[1] The few dams constructed with steel plate on the upstream face have been completely resistant to wave action.

One of the most comprehensive and reliable collections of data on slope protection performance was gathered by the U. S. Corps of Engineers during 1946 and 1947 in a survey of approximately 100 dams in the United States (Ref. 193). These dams varied in age from 5 to 50 years and were in localities with widely different climatic, topographic, and geologic conditions. Some of the main conclusions from the survey were:

1. Forty of the dams studied had layers of dumped rock riprap varying in thickness between 9 and 96 in. as upstream slope protection. In only two of these, or 5% of the dams protected with dumped riprap, were there failures which resulted in displacement of the rock and loss of slope protection. The average rock size on the two dams which failed was only 5 in., and the estimated wave heights were 2.5 and 4.0 ft., respectively. In addition to the two dams with displaced riprap, loss of the underlying gravel filter layer caused damage to three others in this group.
2. Twenty of the dams had hand-placed rock riprap with thickness between 12 and 48 in. In six of these, or 30% of the total, displacement of the rock occurred. The damaged dams had riprap blankets between 12 and 36 in. thick, and the maximum wave heights were estimated to be between 1.5 and 8.0 ft. In four of the six dams which failed, the primary damage was attributed to floating trees or ice.

[1] See Sec. 9.4 for further discussion of the use of asphaltic concrete on the upstream slope.

3. Concrete slabs varying in thickness between 4 and 12 in. were used on 14 of the dams, and five of these were damaged from wave action. Two dams had 8-in.-thick monolithic reinforced concrete slabs, which were not damaged.

2.6b Dumped and Hand-Placed Riprap

The fact that layers of dumped rock riprap are more successful than equivalent thicknesses of carefully hand-placed rock has been suspected for a number of years, but it was not confirmed definitely until the U.S. Corps of Engineers' comprehensive study. The primary reason for the superiority of the dumped riprap is that a moderate movement of any individual rock has little influence on the integrity of the protective layer. In contrast, if one large rock in a tightly knit, hand-placed blanket is moved, the filter is exposed and progressive erosion starts to undercut adjacent rocks. Hand-placed riprap is particularly vulnerable to damage by floating trees and ice layers, which can gouge one or two rocks out of place. Individual rocks in a dumped rock layer are only slightly jostled when rammed by trees or ice, and the layer remains intact.

During a heavy storm the waves on the surface of a reservoir beat repeatedly against the slope just above the reservoir water level, and their energy is dissipated in turbulent action on and within the rocks of a riprap layer. As a wave strikes the slope, the water rushes upward into the riprap and filter layer and then, in the lull before the next wave strikes, tumbles back downward. This action may damage dumped riprap layers in two main ways. First, if the filter material is too fine, the wave water moving in and out of the riprap may gradually wash the filter out; in an extreme case where the filter is completely removed, the individual rocks in the riprap layer settle and expose the embankment to wave erosion. Second, if the average size of rock comprising the riprap is not heavy enough to resist the hydraulic forces generated by the waves, rocks may be literally washed out of the layer.

Failure of the filter layer is more common. The Army study indicated that only very small rocks in graded dumped riprap layers are actually moved very far by waves. The larger rocks are more likely to be moved out of place because they have been undermined when the filter layer was washed out.[1]

[1] Even though the average rocks are not pulled out of place by the wave action, they may be set in slight rocking motion by strong waves. At some dams with quite adequate dumped riprap protective layers, creaking noises from the rock blanket can be clearly heard during severe storms.

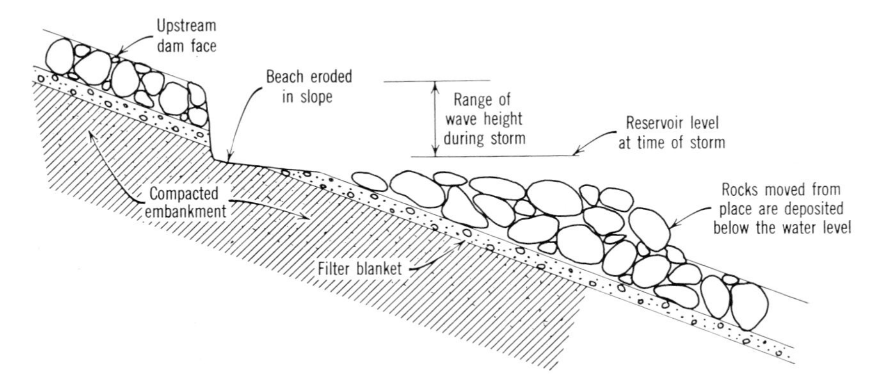

Fig. 2.6:2 Typical failure of rock riprap.

Beaching of the type shown in Fig. 2.6:2 is the typical result of failure of dumped riprap. After the first few rocks are moved out of place, waves cut a notch in the riprap layer just above the still water level and create a relatively flat beach. The rocks cut from the niche are then deposited on the embankment slope below the beached zone. Over a period of time the waves may cut many feet into the compacted soil of the slope, but except in an extreme case such as that of Point-Of-Rocks Dam, beaching failures do not threaten complete failure of the dam. However, repairs may cost hundreds of thousands of dollars, Fig. 2.6:3.

Another kind of damage occurs when riprap layers are constructed with very large rocks of uniform size and without enough small rocks to make a well-graded blanket. Such layers have been used on the slopes of several major earth dams where severe wave action was expected and where small rocks were not easily available. Since there were not enough small rocks to fill the spaces between the large ones, the blanket had holes in it and the filter material washed out. With the support for the large rocks gone, some of them sagged and moved slightly down the slope, creating an unsightly appearance and making it easier for the waves to erode the slope between the rocks. This type of failure is awkward and expensive to repair, because it is difficult not only to move the large rocks but also to use construction equipment on the slope.

Such experience indicates that if trouble with retaining the filter layer is to be avoided, dumped riprap should be well graded with an appreciable content of sizes smaller than 4 in. At some dams where the riprap has been obtained by screening on a 6-in. grizzly, trouble has

Fig. 2.6:3 Views showing typical beaching failure of dumped riprap in major earth dam during severe wave storm.

developed with the filters. If rocks deficient in sizes smaller than 6 in. must be used, the riprap blankets should be thicker or the filters coarser than would ordinarily be considered necessary.

The possibility that the whole riprap layer might slip down the slope in a mass movement has caused some concern, particularly in view of the fact that the soil on the outer slope is often less adequately compacted than that of the main embankment (see Sec. 11.6). Some designers even make an effort to analyze the factor of safety of the riprap, but the authors know of no case of such a riprap slide on a major earth dam.

Where the material comprising the upstream slope directly under the riprap is clay, some engineers have arbitrarily used 3 to 1 or 3.5 to 1 as the steepest allowable slope on which riprap can be safely maintained. This precaution is not justified by experience, however, because the performance of the riprap appears to be independent of the characteristics of the soil in the upstream slope.

Theory shows that rock of a given size will more readily be displaced by wave action when resting on a steep slope than on a flat slope; but here again the experience available does not indicate a great difference in the performances of well-constructed riprap layers over the common range of upstream slopes.

2.6c Articulated Concrete Pavement

Concrete slabs for wave protection have been used with a frequency second only to rock riprap. Most of them have been designed and constructed in some form of articulated pavement consisting of a continuous checkerboard of individual slabs with dimensions varying between 5 and 50 ft. and with thickness between 4 and 12 in. In most cases the slabs have been cast in place on the upstream slope, but a few designers have used precast slabs or units. The individual slabs often are not connected. In some cases, however, the slab reinforcing, which is usually wire mesh, has been extended through the construction joints, and in other cases tongue and groove or dowelled joints have been used. The main purpose of constructing the slope protection as an articulated pavement is to make it flexible enough to adjust to the expected embankment settlement without severe cracking. Approximately one-half of the dams with this type of protection have a gravel filter layer provided under the pavement.

Slope protection of this type has been the least successful of all the principal methods used. Almost invariably the failures have occurred because wave water has washed large quantities of the filter or embankment material through the cracks between the individual slabs. The loss of material has caused serious settlement of the slab and the development of more cracks through which the waves can wash the soil. A few failures have also occurred from deterioration of the concrete, especially where thin, unreinforced slabs were constructed of poor quality concrete. Because of their bad performance record, articulated concrete slabs should not be considered satisfactory upstream facing for dams which must resist wave action. Some examples of the mechanics of typical failures are given below:

BELLE FOURCHE DAM

Belle Fourche Dam in South Dakota was completed in 1908 with a maximum height of 115 ft. and a length of 1 mile (Refs. 95, 566, 567). The dam was well constructed in thin layers with a heavy roller. It has a homogeneous section consisting completely of highly plastic clay, and the upstream slope is 2 to 1. The slope protection is composed of 8-in.-thick precast concrete slabs, 6×5 ft. in plan dimensions, underlain by 1 to 2 ft. of gravel filter. The individual slabs are not connected in any way, and a space of ½ in. was allowed between them for drainage.

Wave action between 1912 and 1915 caused minor damage to the upstream face, displacing some of the slabs each year. In 1916 a serious break brought about the dislocation of more than 350 of the slabs and made extensive repairs necessary. In 1922 a similar large displacement occurred and was repaired. This problem still continues, with damage to the upstream slab following every major storm.

Waves striking the upstream face of the dam at an oblique angle probably cause the most destruction. As a wave recedes, the water head may be 8 or 10 ft. lower in the reservoir than in the filter behind the concrete slab. Jets of water have been seen spouting from the joints between the slabs as the waves pass, and in one case an individual concrete slab was observed to shoot out about 10 ft. from the dam face.

MINATARE DAM

Minatare Dam in western Nebraska, which was constructed in 1915 to a height of 60 ft., retains a reservoir that develops wave heights of approximately 6 ft. (Ref. 55). The upstream face is protected by an

8-in.-thick concrete slab, which was placed in pieces 20 ft. long and 10 ft. wide. The individual slabs are not connected except by flat tongue and groove joints; the tongue in the shape of an arc protrudes into the groove of the adjacent slab about 1 in. The slab rests on a 2.5 to 1 slope and is underlain by a 12-in. layer of natural sand and gravel.

Throughout the life of the dam there has been trouble with the material beneath the slabs washing out through the joints. The resulting settlement has caused cracking of the slabs and an accelerated washing of the underlying soil, and a great quantity of cement grout has been used to fill the cavities under the slabs. The maintenance forces have found that the parts of the slab which had been completely grouted to form a monolithic structure were the areas which most successfully resisted future trouble.[1]

KINGSLEY DAM

At Kingsley Dam, also in Nebraska, the articulated concrete slope protection was constructed in a manner different from the cases cited above, but it suffered the same type of damage. The dam has a maximum height of 160 ft. and was completed in 1941. The upstream slope is 3 to 1 except above the high water line, where it is 2 to 1. The length of the reservoir is approximately 23 miles, and because of the flat topography of the site and the strong winds which develop, the wave action is probably as severe as that at any reservoir in the United States. Maximum wave heights of 8 ft. are conservative estimates.

Since there was no supply of hard rock available within economical hauling distance from the site, the original slope protection (Refs. 55, 150) consisted of precast concrete blocks 4 ft. 6 in. long, 9 in. wide, and 12 to 18 in. thick normal to the slope. The blocks were placed in staggered rows and bonded with iron rods, which ran the full height of the slope and which were tied together by field welding. An opening 1 in. wide was left between the individual concrete elements for drainage.

In the first year of operation a large area of the concrete blocks was undermined and settled into the face of the dam. In the following two years most of the blocks were removed and replaced with a 3-ft.-thick layer of dumped rock riprap, which was transported from a quarry in Wyoming more than 200 miles away. The cost of reconstructing the

[1] This has been a frequent observation of maintenance crews on dams of this type.

(a) (b)

Fig. 2.6:4 Precast concrete elements being placed on upstream slope of Kingsley Dam in an attempt to reduce damage due to wave action (1960). (a) Looking down slope from crest. Wire rope in foreground used to tie elements together. (b) Standing on the slope looking from left to right at wave protection.

upstream face after the failure was estimated at more than $3,000,000.[1] As in the case of the other dams with articulated concrete faces, the failure was due to the soils washing through the openings between the individual units.

HAROLD DAM

This dam was constructed in Los Angeles County in 1924 to a height of about 35 ft. (Ref. 145). The slab on the upstream face was 4 to 5 in. thick and had continuous wire mesh reinforcement. There were construction joints in both directions at approximately 12-ft. centers. The slab remained in fairly good condition for more than 30 years, until June of 1955, when a storm with gale winds which lasted for three days undermined the slab and caused severe settlement and cracking.

[1] The maximum rock size for the riprap weighs about 1,000 lb., and the average size is estimated at about 100 lb. (Ref. 55). Because of the extremely heavy wave action, the maintenance crews have had trouble with moving of the rock and beaching of the slope after almost every major storm. In an effort to reduce the maintenance costs and provide better protection, the zone of worst erosion on the upstream face is being covered at the present time (1960) with a layer of specially designed concrete tetrapods tied together with steel wire rope, Fig. 2.6:4.

A description by the engineer of the State Inspection Agency could have been written for many similar failures:

> It is the writer's belief that destruction started after the embankment soil had worked out through the weep holes and slightly opened construction joints. The present condition took many years to develop to the point where slight settlement and cracking of the slabs has allowed an increased rate of erosion.

2.6d Monolithic Reinforced Concrete Slabs

The number of dams with continuously reinforced concrete slabs on the upstream slopes is relatively small, but this type of wave protection has been very successful. Monolithic slabs have been constructed from 8 to 12 in. thick, with continuous steel equal to 0.3 to 0.5% of the concrete area running in both directions at the center of the slab. As a rule no gravel drain has been provided under these slabs. In some dams the slab has been designed to serve as an impervious membrane as well as for wave protection and so has been extended down to bedrock (Chap. 9). More usually the slab has been intended only for erosion resistance and has been placed only on the parts of the slope needing protection.

The authors know of 12 earth dams with faces of this type which are subjected to moderate or heavy wave action, and none have failed. Although it might be expected that ice lenses developing in the embankments behind the slabs at the water level would cause bulging and cracking, this has not happened although several of the dams are located in regions with cold winters (for example, Sutherland Dam in Nebraska, Jumbo Dam in Colorado, McKay Dam in Oregon). Nor have any of the slabs been damaged by ice layers in the reservoirs, even when huge floating ice sheets have been dashed against the upstream slope during spring wind storms. From the experience available it can be concluded that a well-constructed monolithic concrete slab used for wave protection will have a life of at least 30 years and probably many more.

One of the disadvantages of concrete slab faces is that waves run up the smooth slope further than they do on a rough rock surface, Fig. 2.6:5, and a considerable amount of water can be thrown over the dam crest. Steps have been constructed in the concrete slabs at several dams to act as "wave breakers" near the high water elevation, but these have been spectacularly unsuccessful. For example, at the Suther-

Fig. 2.6:5 Views showing runup of waves on upstream concrete face with stepped "wave breakers" at Don Martin Dam. (Courtesy Mexican Department of Hydraulic Works)

land Dam in Nebraska (Fig. 2.6:6), water and spray is thrown 30 to 40 ft. in the air when waves hit the steps. This water is blown over the crest, wetting the downstream slope and the area below the dam. An appreciable amount of water is lost from the reservoir in this way, and unpaved roads are muddied as far as an eighth of a mile below the dam. In contrast to the poor performances of concrete steps, curved wave walls at the tops of the slopes have been very effective in turning back the wave water and keeping the crest dry.

Fig. 2.6:6 Monolithic concrete upstream slab at Sutherland Dam in Nebraska after 25 years of reservoir operation. Note steps constructed in the slab near the high-water line as wave breakers.

2.6e Test Section at Bonny Dam

At the USBR's Bonny Dam in eastern Colorado, where it was necessary to haul suitable rock for riprap from the Rocky Mountains 220 miles away, a test embankment was constructed in 1951 to determine the suitability of soil-cement and asphaltic concrete as upstream slope protection (Refs. 14, 345, 428, 594). The embankment was placed away from the dam on the side of the reservoir where greatest wave action was anticipated. It was built to a height of about 25 ft. with an upstream slope of 2 to 1 and was located approximately at the normal water level.

The upstream face was divided into two lengths of 200 ft. each, of

Fig. 2.6:7 View of Bonny test section at end of construction (asphalt paving to the left and soil-cement facing to right (Courtesy USBR)

which one was faced with asphaltic concrete and the other with soil-cement, Fig. 2.6:7. The soil-cement layer was constructed with a thickness of 7 ft. measured in the horizontal direction, while the thickness of the asphaltic concrete, facing was varied from 9 to 18 in. normal to the embankment face. The construction procedures used are described in Ref. 345.

Since completion of the test embankment, waves with a height of about 6 ft. have developed frequently. The soil-cement section has remained essentially intact, although the erosion pattern which developed indicated that the material had not been uniformly compacted and that poor bond existed between horizontal layers, Fig. 2.6:8. The

Fig. 2.6:8 Views of the condition of soil cement portion of the Bonny test section after eight years of reservoir activity (May 1960). Note that the soil-cement has provided excellent protection and is completely intact. The erosion pattern clearly indicates that the construction method used did not result in a perfectly uniform mass of soil-cement.

Fig. 2.6:9 View of condition of the 18-in.-thick asphaltic concrete facing at Bonny test section after eight years of reservoir activity (May 1960).

success of this test section has led the USBR engineers to adopt soil-cement wave protection on their Merritt Dam currently under construction (Refs. 666, 683, 689).

The asphalt section stood up fairly well for a number of years in spite of the fact that it soon developed hairline cracks which gradually increased in size until they were about ½ in. wide and several inches deep. Unfortunately, the section with the asphalt facing was undermined by the waves which attacked the end of the embankment, and much of the facing was lost through no inadequacy of the asphaltic concrete. The present condition of the asphalt in the short length of test section remaining is shown in Fig. 2.6:9.[1]

2.7 DAMAGE DUE TO BURROWING ANIMALS

Burrowing animals have been responsible for piping failures in a number of small earth dams and dikes but have not caused trouble in

[1] The asphaltic concrete on this test section was not as well designed or constructed as that which is commonly used for the impervious upstream membranes of asphaltic concrete. See experience on these in Sec. 9.4.

major dams because animal holes do not penetrate to a great depth. In the United States the worst pests have been muskrats and ground squirrels.

Muskrats burrow into embankments either to make homes or to dig passages from one pond to another, Fig. 2.7:1. The entrance to a muskrat hole, as seen in Fig. 2.7:2, is usually under the water surface. If willows or other bushes grow at the water line, the muskrat will dig under them because the roots reinforce the opening of the hole and also hide its existence. Unless there is water on both sides of a dam, a muskrat will dig only far enough to get under the upstream slope, Fig. 2.7:2*a*. If many muskrats are involved, however, their holes may dan-

Fig. 2.7:1 Animal burrow exposed in wall of trench cut in the upstream face of an old dam for the construction of a new conduit.

gerously honeycomb a small earth dam embankment even though each hole is short. Muskrats become the worst hazard for low earth dams when they make tunnels between the reservoir and ponds of water at the downstream toe, Fig. 2.7:2*b*. In some parts of Saskatchewan, highway embankments with swamps or sloughs on both sides of the road often grow so riddled with muskrat holes that they must be rebuilt every third or fourth year (Ref. 142).

Ground squirrels normally dig only in dry soil and will stop when they encounter seeping water or moist soil but they can be dangerous during periods when the reservoir is low for a number of years. Under such conditions ground squirrels have been known to enter the upstream slope below the high water level and dig completely through a small earth dam. The holes are usually 3 to 4 in. in diameter, but badgers will enlarge the holes to about 12 in. in order to follow and kill the squirrels.

Experience indicates that special designs to deter digging animals are not necessary except in regions which have unusually large numbers of them. At the Throttle Dam in northern New Mexico, which was constructed in a locality with considerable evidence of burrowing rodents, the designers used a 12-gauge galvanized iron central core wall running the full height of the dam from abutment to abutment (Refs. 151, 152). At Heart Lake Dam, built by the Colorado Game and Fish Department in 1952, an asbestos-bonded, bolted and corrugated sheet metal core was included in the central portion primarily for control of rodent holes in the upper section of the dam, Fig. 2.7:3.

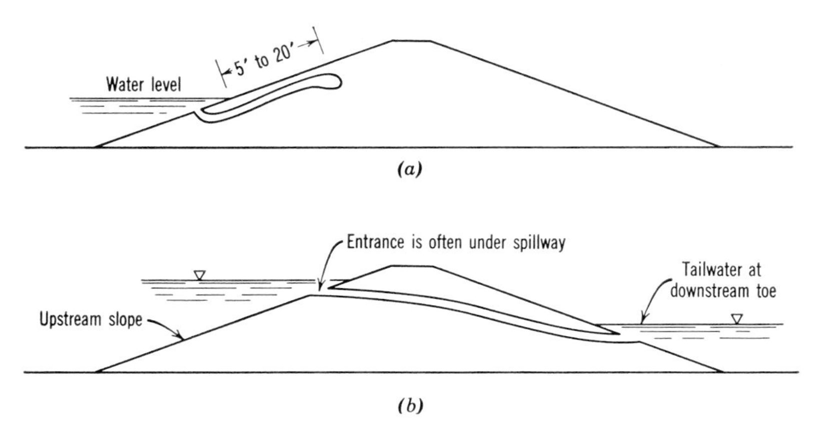

Fig. 2.7:2 Holes by burrowing animals in earth dams. (a) Typical muskrat hole for nest. (b) Typical muskrat hole used as passage between the reservoir and the downstream pond.

Fig. 2.7:3 Sheet-metal core being installed in Heart Lake Dam (Colorado) for purpose of preventing rodents from burrowing all the way through the embankment. (Courtesy Damon Runyon, Consulting Engineer)

All earth dams in regions with many muskrats should be carefully inspected at frequent intervals. If the dam becomes infested, trapping and poisoning may keep the burrowing under control. An interesting study at the University of Saskatchewan (Ref. 142) demonstrated that creosote floating on the surfaces of ponds is sufficient to drive out all muskrats. The study also showed that if the outer portions of an earth dam embankment were compacted as a mixture of creosote and soil, muskrats would be much less likely to burrow in.

2.8 DAMAGE CAUSED BY WATER SOLUBLE MATERIALS

2.8a Soluble Materials in Foundations and Abutments

The leaching of natural deposits of water soluble materials from abutments and foundations has caused difficulty at some dams. Gypsum which is gradually dissolved by seepage water from the reservoir has been particularly troublesome in this respect.

At the Dry Canyon Dam in southern California, which was constructed to 55 ft. in 1912, leakage through the abutments and foundations increased until laboratory tests were performed in 1933 to analyze the condition. They showed that approximately 5 ft.^{3} of solid material was being removed per day, and that most of it was gypsum (Ref. 145). By 1935 the rate of seepage had approximately doubled, and it was estimated that the annual loss of solids from the abutments varied between 40 and 100 yd.^{3} per year. The leakage continued to increase until extensive cement grouting was carried out in the abutments during 1936 and 1937.

At the Buena Vista Dam in Kern County, California, which is an old and relatively low dam retaining a large reservoir, considerable seepage occurred for many years through the foundation, which contained gypsum seams. In 1938 the amount of dissolved gypsum in the reservoir water was analyzed and compared with that in the water exiting downstream as seepage. From this study it was estimated that the seepage might be removing as much as 100 ft.^{3} of solids from the foundation each day.

As a result of the 1952 earthquake, which was fairly severe at the site of Buena Vista Dam, a portion of the embankment settled approximately 2½ ft. (Ref. 146). Because this occurred in a section where the dam was only 8 ft. high and where considerable foundation leakage had been taking place for years, the large settlement can only be attributed to the fact that quantities of gypsum had been leached from the foundation. Measurements of seepage loss from the McMillan Reservoir in New Mexico over a number of years have also led to the conclusion that gypsum in the rock formation is being gradually dissolved by seeping water (Ref. 225).

In addition to their propensity for creating cavities and flow channels, water soluble materials may cause trouble from the opposite effect in modern earth dams. The deposition of soluble material previously leached from the natural soil may tend to plug specially designed filters and pervious drains. At the Vermillion Dam in California (Ref. 93), for example, the gravel toe drain became inoperative soon after the reservoir was filled for the first time, and it was believed that iron oxide leached from the natural soils in the vicinity had filled the void spaces in the drain.

2.8b Soluble Materials in Embankment Soils

It was thought in previous years that a small percentage of water soluble salts in a prospective embankment material was potentially dangerous, but there has never been a record of failure or damage from this cause. Most engineers today do not consider it necessary to test embankment soils for salts except in extreme cases when the soil has an odd light color or when some other suspicious characteristic indicates that large fractions may be water soluble.

There are several reasons for the fact that small quantities of soluble salts are not as troublesome in embankment materials as they are in foundations and abutments. First, the water in a river running through a region where the soils contain much soluble salt has a high salt content itself, and consequently the reservoir water will show no great tendency to leach salts out of an embankment composed of such soils. Second, in the construction of modern rolled earth dams the water which is added to the borrow material to bring the construction water content up to optimum probably dissolves much of the soluble salt.

Solution of salts has been the suspected cause in the failures of a few old dams constructed in arid regions where the soils had a high percentage of soluble materials. However, these failures occurred many years ago when earth dam embankments were built with what would now be considered rudimentary construction control. It seems probable that many of the dams actually failed from piping which developed in leaks resulting from poor construction methods.

The failure of the Apishapa Dam in southern Colorado in 1923 was attributed, at least in part, by some engineers to solution of salts from the embankment material (Ref. 155). After the failure a sample of the soil taken from the borrow pit lost 6% of its weight when allowed to soak in pure water for 2 hr. (Ref. 104). However, a description of the portions of the embankment which remained after the failure strongly indicates that the presence of the salts was not an important factor:

> This material apparently gave a very impervious fill. On the upstream slope of each exposed cross-section, the line to which the water had penetrated could be followed for a distance down of 20 to 25 ft. in elevation from the water surface. At the lower depth, the water had gone into the fill only 5 or 6 ft. (Ref. 154).

This interesting observation leads to the conclusion that seepage had penetrated the upstream portion of the embankment uniformly and that the soluble salt in the soil did not have an immediate radical effect on the permeability. The probable cause of failure was piping originating in differential settlement cracks (Sec. 2.3).

2.9 FLOW SLIDES DUE TO SPONTANEOUS LIQUEFACTION

One of the most difficult problems facing the earth dam designer is the analysis of the stability of loose sand foundations against the possibility of liquefaction or flow slides, Sec. 8.3. The performances of existing dams give practically no assistance, since few major dams have been founded on loose sand foundations and no failures of rolled earth dams have occurred from liquefaction.

A number of major earth dams constructed by the hydraulic fill method have developed construction flow slides which were due primarily to liquefaction of the outer granular shells of the embankment.[1] In all of these failures the mechanics of movement were similar. The outer shells were in a loose state because they had been deposited hydraulically and not compacted. When the high fluid pressures acting in the upstream-downstream direction caused sufficient shear strains in the outer shells of the embankments during construction, flow slides resulted.

The failure of the Sheffield Dam in Santa Barbara in 1925 following an earthquake shock probably was caused by liquefaction of the lower portion of the dam or the foundation (see Sec. 2.5). In this case the records are vague about the manner of construction and the construction material, but it is known that the embankment was poorly compacted and that the material was basically granular and fully saturated.

A flow slide of the internal downstream sand drain at Red Mountain Dam, constructed near San Diego, California in 1949, provides interesting evidence that liquefaction of localized embankment zones can occur. This dam had a maximum height of about 55 ft., a homogeneous section of clayey soil, and a horizontal drainage blanket under the downstream slope. The drainage blanket consisted of a layer of sand about 3 ft. thick, which had been placed in one layer without compaction. Some minor abutment leakage developed following the first reservoir filling. Subsequently the exit of the drain was inadvertently blocked by a pile of waste stripping material, which prevented water from discharging freely. In the fall of 1950, after the reservoir had been filled for approximately one year, a large portion of the sand drain flowed out from under the downstream slope. This caused the downstream toe overlying the drain to slump. The scarp of the slump extended up the downstream slope to approximately 15 ft. above the toe.

The failures described above leave no doubt that flow slides can occur in the lower portions of embankments consisting of loosely compacted granular materials, however, these histories provide no indication as to whether loose sand foundations under well-compacted rolled-earth dams could liquify and flow out from under the dam.

[1] See, for example, the descriptions of the failures of Calaveras Dam in California, 1925 (Refs. 359, 644); Necaxa Dam in Mexico, 1909 (Ref. 360); Alexander Dam in the Hawaiian Islands, 1930 (Ref. 361).

2.10 DAMAGE CAUSED BY DOWNSTREAM DEFLECTION IN ROCKFILL DAMS WITH CENTRAL CORES

No matter how steeply the slopes are constructed, rockfill dams with thin central cores of earth or concrete on rock foundations never develop slope slides of the type which occur in earth dams. If the downstream slope is too steep, however, the dam crest deflects an excessive amount when the reservoir is filled for the first time, and it may continue to move gradually downstream with a creeplike action. No theoretical methods are available for analysis of this phenomenon, and the few records of movement available give little guidance in the problem of determining the critical downstream slope for a dam of this type. The following examples describe the kind of trouble which develops when the slope is made very steep.

CRANE VALLEY DAM, CALIFORNIA

The record of the Crane Valley Dam in Madera County, California provides an example of excessive and long-time movement. The dam was completed in 1910 with a section as shown in Fig. 2.10:1. The earth embankment upstream from the central concrete core wall was placed by the hydraulic fill method, with the core pool located just upstream from the wall. The whole downstream half of the dam below the wall was constructed of quarried rock dumped in high lifts. Construction photographs indicate that the rock is primarily large, angular, and hard, with a large content of dirty fines.

When the reservoir was filled for the first time, the crest of the dam started to move downstream. In 1914 after considerable movement had already taken place, measuring points were established on the top of the core wall, which protruded from the crest, and measurements of the horizontal movement were started. These indicated that between 1914 and 1952, the crest moved downstream at a regular but decreasing rate, with a total movement of about 5 ft. It is estimated that an additional 9 ft. of movement occurred between 1910 and 1914 before measurements were begun. To retard the downstream movement, the engineers placed large quantities of additional rock on the downstream slope by dumping from the crest. As a result, the crest presently has a width of about 100 ft., and the downstream slope is 1.3 to 1.

In 1928, a vertical exploratory shaft was sunk at the downstream edge of the concrete core wall in the rockfill section near the center of

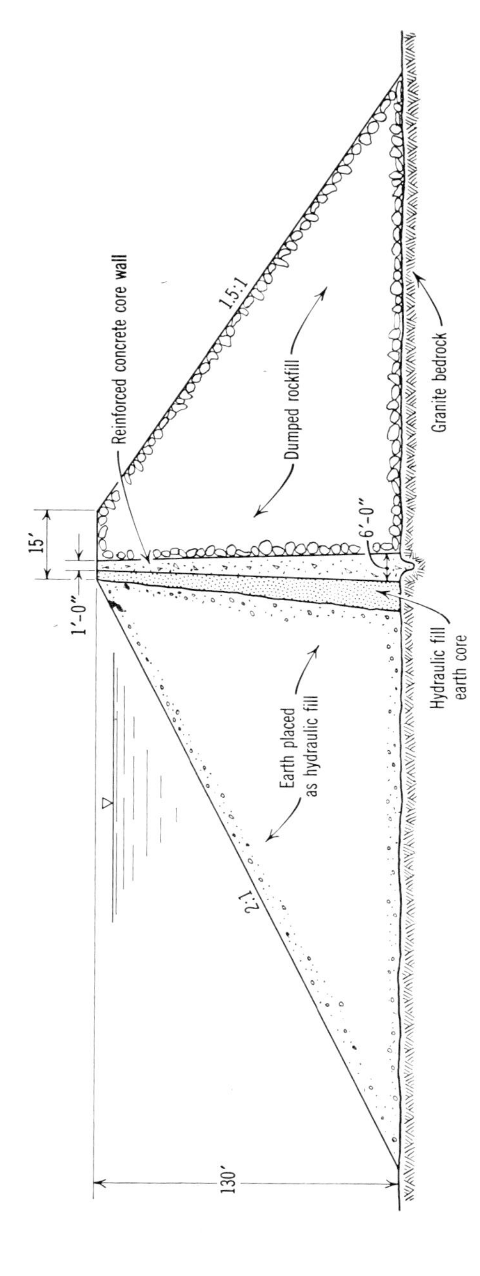

Fig. 2.10:1   Cross section of Crane Valley Dam as constructed (1909).

the valley. It was found that the bottom of the wall, which had been keyed into the granite bedrock of the foundation, had not moved. At a height of about 6 ft. above the bedrock, however, there was a horizontal crack and a change in the slope of the wall. Careful observations indicated that the wall was essentially intact above the crack (at least at the location of the test pit) and that most of the crest deflection resulted from rotation at the crack. The engineers who made the investigation concluded that the wall had been badly cracked both vertically and horizontally but that fine soil from the upstream hydraulic fill section had sealed the cracks and prevented any appreciable seepage.

The engineers also made an effort to sink an exploratory shaft on the upstream side of the core wall. They were able to get down approximately 80 ft. but could not go any deeper because the material of the hydraulic fill core was practically a liquid mud "showing little signs of consolidation and having the appearance of quicksand." No doubt the high lateral pressures exerted by this nearly fluid earth zone upstream from the concrete wall had a great influence on the high downstream movements which occurred.

OUED KÉBIR DAM, NORTH AFRICA

Similar excessive downstream movement occurred in 1929 during the first filling of the reservoir behind the 115-ft. Oued Kébir Dam for the water supply system of the city of Tunis (Ref. 264). This dam had a hollow, rigid central concrete core 30 ft. thick at the bottom and 6 ft. thick at the top, Fig. 2.10:2. Downstream from the core the embankment consisted of quarried rock, about half of which was hand-placed while the other half was dumped without sluicing. The average downstream slope was 1:1, and the foundation consisted of hard rock.

When the reservior filled, the crest settled more than 2 ft. and moved downstream 2 ft. 8 in. The bottom of the concrete core wall sheared off and moved 5 in. downstream.

2.11 DAMAGE DUE TO SURFACE DRYING

Surface drying cracks have caused a constant maintenance problem on a few low dams constructed with homogeneous sections of clayey soil. Usually the main cracks, which in extreme cases have been several inches wide, develop near the top of the dam parallel with the

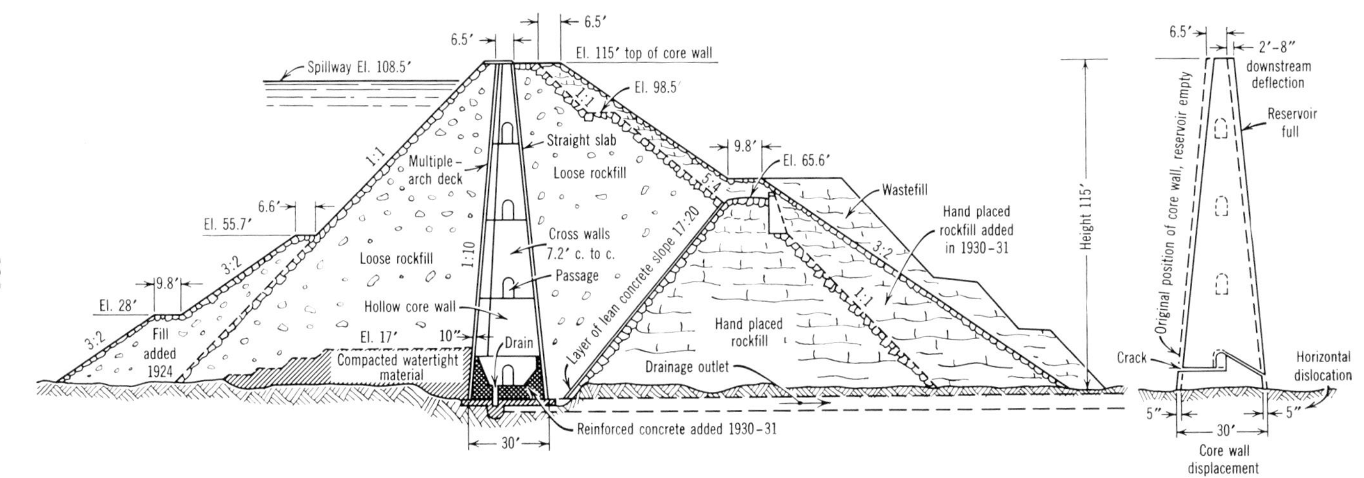

Fig. 2.10:2 Qued Kébir Dam, Tunisia, showing embankment section and core wall movement (after Noetzli, Ref. 264).

crest. They appear to be aided by the tensile stresses at the top of the embankment slope. The worst conditions develop when some combination of the following three factors occurs: (1) hot, dry climates during which the reservoir remains empty for long periods; (2) embankment construction materials of highly plastic or extremely fine silty soil; and (3) embankments not compacted to high densities.

Excessive drying cracks frequently develop on low clay dams constructed for flood control purposes. The designer in such cases may reason that the reservoir will have to retain water only for short periods and at infrequent intervals, and therefore he specifies less embankment compaction than he would normally require. Surface erosion concentrates in the embankment cracks which develop. In many cases it has been necessary to refinish the slopes and crests of such dams completely and to add a layer of cohesionless material to minimize the surface drying and erosion. In extreme cases it has been necessary to provide continuous irrigation of the embankment through a ditch on the crest in order to avoid development of shrinkage cracks (Ref. 34).

The higher the content of clayey fines in a poorly compacted embankment, the more the embankment can be expected to shrink and crack. On the other hand, it is the cohesionless, silty materials which are most susceptible to erosion. Some of the worst examples of trouble from this source have been in the arid southwestern part of the United States where, in some cases, homogeneous dams of very fine clayey silt and silty clay of low plasticity have been so badly eroded with concentrated gullies, starting in drying cracks, that they have had to be almost completely reconstructed. This type of reconstruction is very awkward and expensive work.

DRYING CRACKS DURING CONSTRUCTION

If the construction surface of an embankment of fine-grained soil is allowed to dry in the sun, drying cracks can greatly increase the overall permeability of the material. This has happened even on dams constructed in accordance with good modern practice.

The Bureau of Reclamation's Lovewell Dam in Nebraska illustrates the kind of trouble which can arise (Refs. 430, 682). The dam is a homogeneous embankment of clay of low plasticity with a downstream drainage blanket, Fig. 2.11:1. During construction in 1956, earth placement was suspended for three weeks at a time when the partially completed embankment was 6 ft. high. Although the work stoppage occurred during a period of no rainfall and dry winds, no unusual

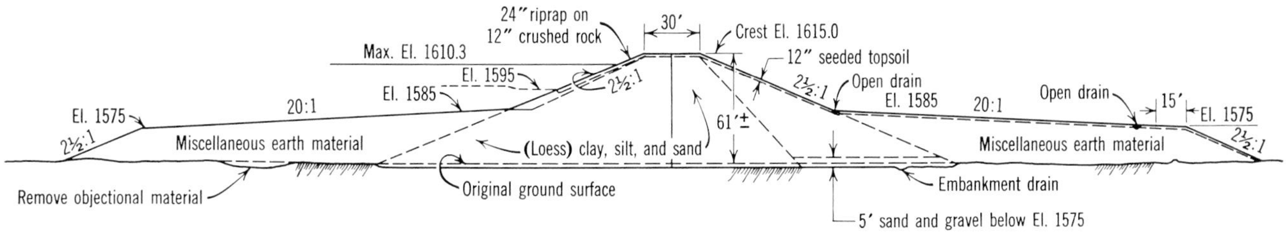

Fig. 2.11:1—Cross section of Lovewell Dam (after Walker, Ref. 147).

cracks were noticed on the embankment surface and no particular concern was felt about drying.

After the dam was finished, an exploratory hole was drilled through the embankment near the downstream edge of the crest for the purpose of testing the strength of the soft foundation clay. The engineers were surprised to find that a considerable amount of drilling water was lost through the hole into the impervious embankment at a level about 5 ft. above the foundation. Water was poured in at the rate of 100 gal./min. and completely disappeared. Later, 35 additional holes were drilled through the embankment on the center line of the crest at various points, and water was lost in 25 of them in the lower 3 to 5 ft. of the embankment. The loss varied from a minor quantity of up to 160 gal./min.; one hole took 7,700 gal. in 24 hr. without filling.

Later a test pit was put down at the location of one of the exploratory holes, and the material exposed in the walls of the pit was carefully examined. At all points the embankment appeared uniformly well compacted, without cracks and with good bond between the layers. From these observations, and from the results of laboratory drying and wetting tests on undisturbed samples from the pit, the engineers concluded that the water had been lost through a system of drying cracks which had developed in the lower part of the embankment during the three weeks when construction was suspended. They decided that these cracks had closed when the water pouring into the drill holes caused the embankment to swell.

This experience is particularly significant since the embankment was constructed of a soil type which has been used for many dams. Similar cracking has undoubtedly developed in other dams, and this probably accounts for some of the cases of mysterious loss of drilling water when exploratory holes have been put down in impervious sections of earth dams. The properties of the embankment material at the Lovewell Dam were approximately: liquid limit 40%; plasticity index 20%; Maximum Standard Proctor density 102 lb./ft.^{3}; 30% by weight finer than 0.005 mm.; 95% by weight finer than the No. 200 sieve.

Because of this episode, specifications for the USBR's Twin Buttes Dam, which was to be constructed of similar material, required that the contractor protect completed portions of the embankment against drying out by sprinkling or covering with loose earth.

three

Measurements of Movements and Pore Water Pressures

As part of the design process the engineer must estimate the movements, strains, and pore water pressures which may develop in his dam at various times in its life. He makes these estimates largely on the basis of laboratory tests and theoretical analyses, but since the theories are still in a state of development he can only rely on them to the extent that they have been demonstrated to check actual measurements on completed structures.[1] Therefore it is essential that he have a good general understanding of the movements and pore pressures which have been measured in existing dams, the factors which influenced them, and the accuracy with which they could have been estimated in advance by theory.

This chapter contains a review of all the current information which the authors have been able to collect. The material presented will be outdated and replaced in the next few years as additional measurements are accumulated.

Before about 1935 the body of knowledge about movements, strains, and pore water pressures contained little that was reliable and much that was misleading. Since then a considerable number of measurements have been made, and the conclusions based on them have had a great influence on current designs. While measurements have been taken by engineers in all parts of the world, the most important have been recorded by a few groups who have built large numbers of dams. The measurements of the engineers of the USBR, who have been installing instruments in their earth dams since the early 1930's, have been particularly valuable because they have been accumulated in one place, carefully studied, and published for the use of the profession.

[1] For details of measuring instruments, see Sec. 8.1.

3.1 EMBANKMENT MOVEMENT DURING CONSTRUCTION

3.1a Embankment Compression—USBR Measurements

Using USBR measurements for more than 20 large dams in the western United States, J. P. Gould in 1953 and 1954 published the results of a comprehensive study of embankment compression and pore water pressures (Refs. 157, 158, and 300). As part of the study he analyzed the embankment measurements during construction and concluded that the compression in the central portions of the dam, where the measuring devices were installed, took place essentially in a vertical direction, as though the material had complete lateral restraint. Gould compiled average values of measured compression for each embankment at various times during the construction and plotted them as a function of the weight of an overlying column of embankment material. After subtracting the pore pressures measured in nearby piezometers, he obtained curves relating the vertical compression and the vertical effective stress.

As seen in Fig. 3.1:1, these compression curves approximate straight lines when plotted on log-log plots and can be defined by any two points, such as the compression at vertical pressures of 10 and 100 psi. Under 100 psi the measured vertical compression for all the USBR dams ranged approximately between 1.0 and 4.0%.

Table 3.1:1 summarizes the results of these measurements. This summary can be used in estimating embankment compression which may take place during construction.[1] In the last few years similar compression-measuring devices have been installed in a few other dams, and the measurements obtained have extended the range of the USBR data slightly. At the present time, however, the USBR measurements constitute the principal source of information available to the profession concerning the compression of rolled earth dams during construction.

FACTORS INFLUENCING COMPRESSION

Table 3.1:1 shows that the compressibility of the embankments depends primarily on the compaction water content and the soil properties, and that the absolute value of the dry density (or void ratio) of the embankment material has little or no influence. For example, the Vallecito Dam embankment, which had an average dry density of over 120 lb/ft.,^{3} was more than twice as compressible as the embankment

[1] In using this material it must be kept in mind that all measurements were made on dams constructed with USBR practice; i.e., compacted in 6-in. layers with 12 passes of a heavy sheepsfoot roller and close moisture-density control.

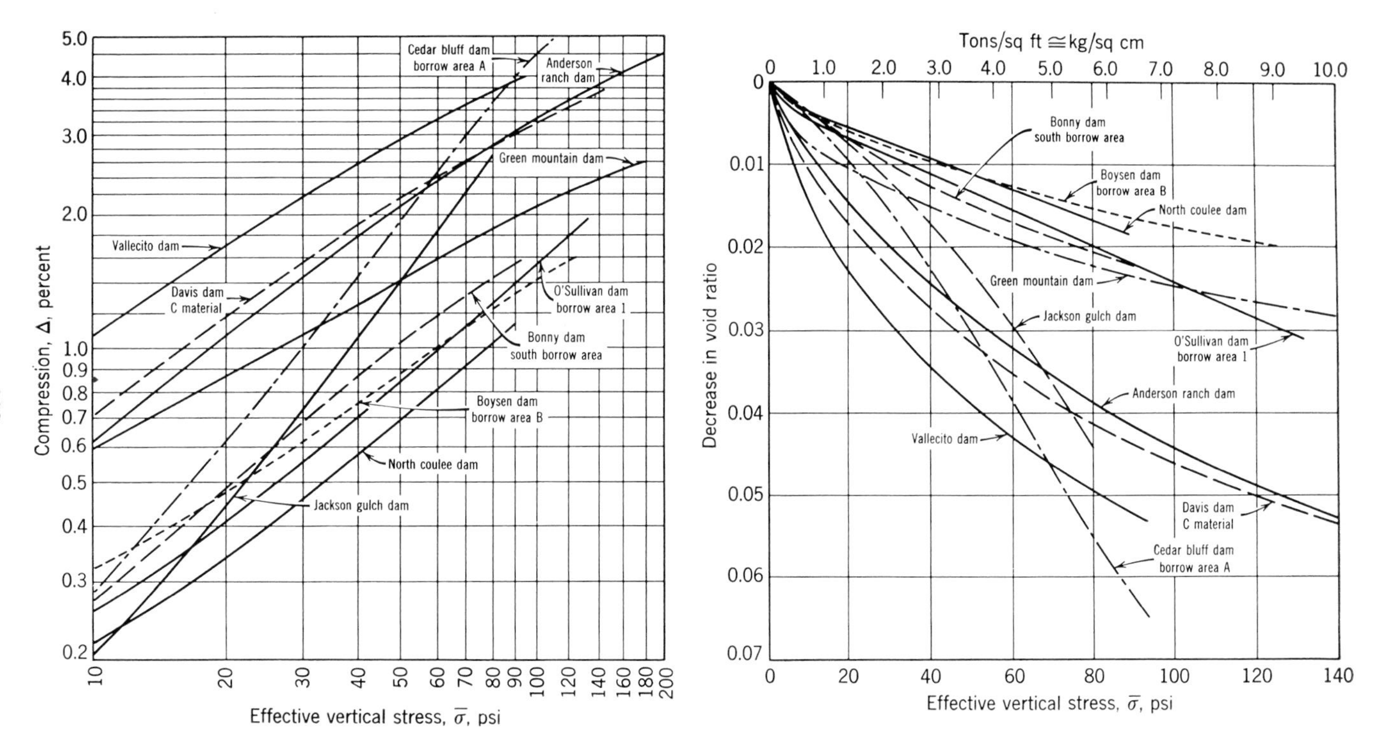

Fig. 3.1:1 Typical measured embankment compression curves (after Gould, Ref. 158). (a) Logarithmic plot. (b) Arithmetic plot.

Table 3.1:1 Summary of Measured Compression of Impervious Sections of USBR Earth Dams During Construction (After Gould)

SOIL TYPE	CLASSIFICATION SYMBOL	DAM NAME (BORROW AREA)	AVERAGE MEASURED EMBANKMENT COMPRESSION (VERTICAL) at 10 p.s.i. (%)[2]	at 100 p.s.i. (%)[2]	AVERAGE SOIL PROPERTIES Dry Density (Fill) −#4 mtl. (lb./ft.^{3})[3]	Water Content (Fill) −#4 mtl. (%)[3]	Opt. Water Content (Lab) (%)[1]	Rock Content (>#4 Sieve) (%)	Liquid Limit (%)	Plasticity Index (%)
Silty gravel	GM-SM	Davis (A & B)	0.2	1.0	122.0	9.8	11.5	45	23	2
and	GM-GP	Boysen (B-1)	0.3	0.9	121.6	10.3	10.4	50	18	2
coarse	GM-SM	Boysen (B)	0.3	1.4	122.5	10.8	11.4	39	20	5
silty	GM-SM	Caballo	0.3	1.1	117.9	11.6	12.8	37	—	—
sand	GM-SM	Cachuma	0.3	1.1	112.7	13.7	14.3	37	25	6
Fine	SM	Davis (B)	0.5	1.8	118.1	11.9	12.2	14	21	1
silty	SM	Long Lake (1)	0.3	1.6	107.2	14.6	17.1	0	20	1
sand	SM	Bonny (S.A.)	0.3	1.6	113.4	12.7	13.9	0	21	2
and	SM	O'Sullivan (1 & 7)	0.3	1.6	107.6	15.5	17.1	0	22	NP
silt	ML	Enders	0.4	2.1	111.6	13.2	14.5	0	20	1
of low	ML	Bonny (N.A.)	0.4	1.9	106.9	15.4	16.2	0	21	2
plasticity	ML	No. Coulee	0.2	1.3	101.6	15.7	18.2	0	22	2
Clayey	SC-GC	Cachuma	0.3	2.0	110.9	14.2	15.3	36	—	—
gravel	SC-SM	Heart Butte (NE)	0.4	2.6	112.0	15.1	16.3	4	26	6
and	SC-SM	Boysen (K)	0.4	1.9	121.8	11.3	11.8	27	21	5

(*Table continued on p. 204*)

Table 3.1:1 ***Summary of Measured Compression of Impervious Sections of USBR Earth Dams During Construction (After Gould) (Continued)***

Soil Type	Classification Symbol	Dam Name (Borrow Area)	Average Measured Embankment Compression (Vertical) at 10 p.s.i. (%)[2]	Average Measured Embankment Compression (Vertical) at 100 p.s.i. (%)[2]	Average Soil Properties: Dry Density (Fill) −#4 mtl. (lb./ft.^{3})[3]	Water Content (Fill) −#4 mtl. (%)[3]	Opt. Water Content (Lab) (%)[1]	Rock Content (>#4 Sieve) (%)	Liquid Limit (%)	Plasticity Index (%)
coarse	SC	Anderson Ranch	0.6	3.3	121.2	12.5	13.0	10	28	12
clayey	SC	Granby	0.8	2.4	126.8	9.6	10.0	27	22	8
sand of	SC	Green Mtn.	0.6	2.1	132.2	8.4	8.9	35	21	8
low	SC	Shadehill	0.4	2.0	122.4	11.4	12.4	30	25	10
plasticity	CL-GC	Deer Creek	0.4	2.6	110.6	15.1	15.5	30	26	10
	CL-ML	Fresno	0.7	3.6	119.1	12.2	12.4	3	20	5
Clay	CL-ML	Long Lake (LC)	0.2	2.9	100.0	18.5	20.5	0	24	4
of low	CL-SC	Vallecito	1.1	4.2	123.2	12.2	11.5	16	—	—
to	CL	Cedar Bluff (C)	0.2	2.8	112.7	12.9	13.9	0	31	14
medium	CL	Horsetooth	0.2	3.0	112.5	11.6	13.9	2	26	12
plasticity	CL	Medicine Ck. (A)	0.3	3.5	104.5	16.8	17.6	0	30	8
	CL	Heart Butte (S.W.)	0.5	3.8	115.3	13.5	14.5	2	24	10
	CL	Jackson Gulch (1)	0.2	3.7	100.7	18.9	21.4	5	40	21
	CL	Carter Lake	0.3	4.0	113.5	13.4	14.9	6	27	11

[1] Standard Proctor compaction test.
[2] Effective stress (pore pressures subtracted).
[3] Measured by control tests during construction.

of North Coulee Dam, even though the latter had a dry density of about 102 lb./ft.^{3} [1]

In the range of stress below a value of approximately 100 p.s.i., the compaction water content had the most important influence on compressibility. The compression curves of embankments constructed with relatively low average water contents show low initial strain and constant or only moderately increasing compressibility under higher pressures. These curves are often slightly concave in the upward direction on the log-log plot, Fig. 3.1:1*a*. The curves for embankments constructed with water content at or above Standard Proctor Optimum are characterized by high initial strains and have a slight concavity in the downward direction (see, for example, the curve for Vallecito Dam, Fig. 3.1:1*a*).

Under pressures of 100 p.s.i. or higher, however, compressibility was more or less independent of the compaction water content and was governed by the gradation and plasticity of the embankment soil. The plasticity of the fines was relatively more important than the coarseness and gradation. As seen in Table 3.1:2, the embankments can be grouped roughly according to their soil properties in order of increasing compressibility:

Table 3.1:2 Relationship between Embankment Compression and Soil Type as Determined from Measurements on USBR Dams

EMBANKMENT SOIL TYPE	APPROXIMATE RANGE OF MEASURED COMPRESSION (%) at 10 p.s.i.	at 100 p.s.i.
1. Silty gravel and coarse silty sand (GM and SM)	0.2–0.3	0.9–1.4
2. Fine silty sand and silt of low plasticity (SM-ML)	0.2–0.5	1.3–2.1
3. Clayey sands and gravels (GC–SC)	0.3–0.8	1.9–3.3
4. Clay of low to medium plasticity (CL and CL-ML)	0.2–1.1	2.8–4.2

[1] An extreme example from outside the USBR data is given by the performance of the Sasumua Dam in Kenya. The embankment was constructed of halloysitic clay with an average compacted dry density of about 70 lb./ft.^{3} and the measured embankment compression curve (Ref. 304) falls nearly in the center of the range of curves obtained from measurements on the USBR dams—all of which had embankment densities of more than 100 lb./ft.^{3} (Fig. 3.3:5).

Although the type and intensity of the compaction effort undoubtedly also have an important influence on compressibility, the measurements in this study gave little information on it because all of the USBR dams were constructed with essentially the same procedures.

ESTIMATING EMBANKMENT COMPRESSION FROM LABORATORY TESTS

In USBR practice, laboratory consolidation tests are performed on the fraction of an embankment soil finer than the No. 4 sieve. For the dams included in the study, the average of the laboratory test results usually roughly approximated the measured field compression, although there was a great spread—frequently greater than the whole range of compressibility measured in all the dams—in the results for any given material. These results offer some promise that in the future it may be possible to predict the average embankment compressibility in advance of construction from laboratory tests. Except in a few cases where the borrow soils are extremely uniform, at the present time it is probable that the embankment compressibility can be estimated in the design stage more reliably by interpolating in Table 3.1:1.

3.1b Compression of Rockfill Embankment Sections

Because of the difficulty of installing instruments, very few measurements have been made of the compression of rockfill dams during construction.[1] The compressibility of rockfill embankment sections undoubtedly varies over a considerably greater range than that of modern rolled-earth embankments and is influenced by:

1. The method of construction—height of lift, compaction, amount of water, and method of sluicing used (Sec. 11.5).
2. The quantity and characteristics of the soil fines and small rock.
3. The character of the rock—strength, shape, and size.

The relative influence of these factors and the range of rockfill embankment compressibility are not known. We know only that rockfills of clean, hard rock constructed in thin, rolled and wetted layers are likely to be less compressible than the least compressible rolled earth embankment, and that the most compressible rockfills are many times more compressible than any well-constructed rolled-earth fill. In the

[1] The authors know of only two published records: at Quioch Dam in Scotland (Ref. 67) and at the Bou-Hanifia Dam in Algeria (Ref. 451).

next few years much more information will be available concerning the compressibility of rockfill embankments since a considerable number of high rockfill dams are currently being constructed, and many of these are provided with instruments for measuring compression during and after construction.

3.1c Foundation Settlement (During and After Construction)

The embankment of a high earth dam is one of the heaviest of man-made structures. Its weight, together with that of the reservoir water overlying the upstream slope, causes settlements in the foundation ranging from a few inches at sites with hard rock[1] to many feet at dams underlain by thick deposits of compressible soils. Depending on the foundation material, nearly all or only a fraction of the total foundation settlement may have occurred before the end of construction.

Settlement of foundations of fine-grained soils can be estimated from laboratory consolidation tests by means of the Terzaghi consolidation theory, which is commonly used for the prediction of settlements of buildings and other structures on clay. In a number of successful dams constructed on foundations of soft soils, settlement of more than 5 ft. has been measured. The USBR Fresno Dam in Montana, which has a maximum height of only 75 ft. but which is underlain by soft silts and clays, settled a total of about 8 ft.; half of this during construction (Refs. 147, 182). One of the most extreme cases of both total and differential foundation settlement is that at the Mission Dam completed in 1960 on the Bridge River in British Columbia. The dam has a maximum height of 150 ft. and is underlain by a thick layer of very compressible clay. It was estimated that a maximum settlement of approximately 15 ft. would occur, about half of it during construction. The abutments at the Mission Dam consist of steep cliffs of hard rock. Early settlement measurements approximated the predicted values reasonably well (the results of the final measurements have not yet been released). In another exceptional case, the measured settlement beneath the 200-ft. Fort Peck Dam constructed on the Missouri River reached 16 ft. in the center of the valley. This was attributed primarily to compression of a foundation clay layer. The rate of settlement

[1] Although the settlement of rock foundations is usually considered to be negligible and is not frequently measured, surprisingly high values (more than 12 in.) have occasionally been measured in what would generally be considered hard rock under the loads of dams of moderate size.

leveled off in 1950, some 10 years after completion and 13 years after diversion (Ref. 569).

Settlements of foundations consisting of sands and gravels usually occur wholly during construction. The magnitude may be only a few inches or several feet depending on the relative density and the gradation of the material. For example, at the 400-ft. Serre-Ponçon Dam in France, which is underlain by more than 300 ft. of dense granular alluvium, the total settlement of the upper surface of the foundation during construction was less than 18 in.[1] At the other extreme, the settlement of the 160-ft. Fort Randall Dam on the Missouri River, which was underlain in the center of the valley by about 170 ft. of alluvium consisting primarily of loose fine sand and silt, was about 6.5 ft.

In addition to the settlement caused by the weight of the embankment and the pressure of the water acting on the upstream slope of the dam, the weight of the water retained behind the dam, which for large reservoirs is much larger than that of the embankment, causes an area-wide or regional settlement centered on the middle of the reservoir. Few reliable measurements of this areal settlement have been made. Since the whole reservoir area, including the abutments and foundation, probably settle about the same amount, it is not of practical concern to the dam designer except in very special circumstances.

At Boulder Dam the maximum areal settlement from the weight of the water in Lake Mead was 7 in. and was located at a point near the geographic center of the reservoir (Ref. 338). In measuring this settlement, engineers ran precise levels for a distance of 50 miles to establish bench marks outside the area influenced by the weight of the reservoir. A similar study made during the design of Garrison Dam on the Missouri River led to the estimate that the area-wide subsidence would be between 1 and 2 ft. Bench marks were located 15 to 18 miles on either side of the reservoir in the hope of getting outside the area of settlement (Ref. 179).

3.1d Foundation Spreading

In the plane which is transverse to the longitudinal axis of the dam, the foundation settlement usually occurs in a dish-like pattern, with the maximum value developing in the central and heaviest-loaded portion of the embankment and the minimum values under the toes. As-

[1] Under the 150-ft. Vermillion Dam in California, underlain by 200 ft. of sand and gravel, the foundation settlement was 3 in. (Ref. 196).

sociated with this settlement is an extension of the foundation, since the shear stresses imposed on it cause the distance between the upstream and downstream toes of the embankment to increase. Knowledge of this foundation spreading is important both because it affects the design of outlet conduits passing through the base of the dam and because it can have a damaging influence on rigid foundation cutoff walls.

The amount of spreading cannot be predicted reliably on the basis of laboratory tests. The only guides available to the designer are field measurements, and few of these are available since instruments are not often installed for this purpose.

One of the few compilations of measurements was made in 1960 by Gould and Rutledge as part of a study for the American Concrete Pressure Pipe Association (Ref. 551). In this they found the stretch of the foundation of a number of small dams (40 to 80 ft. high) by measuring the gap which developed between adjacent individual pipes comprising the outlet conduit. The principal results of the study are given in Table 3.1:3 and can be summarized as follows:

1. The maximum vertical foundation settlement varied from 0.3 to 3.1 ft.
2. The average stretching of the outlet conduit varied between 0.3 and 0.9% of the original length.
3. In some dams the stretching was distributed uniformly over the entire length of the base of the dam. In others the strain was higher under the center of the dam and more or less proportional to the amount of settlement which occurred.

Similar measurements for a rockfill dam in Algeria given by Terzaghi (Ref. 126, p. 498) show that the foundation stretch was dependent on the steepness of the side slopes of the embankment, as would have been expected, and reached a maximum of about 0.5%, Fig. 3.1:2.

3.2 POSTCONSTRUCTION EMBANKMENT MOVEMENT

3.2a Embankment Compression and Swelling

In the USBR earth dams included in the study by Gould (Sec. 3.1*a*), very little postconstruction embankment compression occurred. The relationship between measured embankment compression, rate of con-

Table 3.1:3 Summary of Dam Base Stretching as Determined from and Rutledge, Ref. 551)

NAME OF STRUCTURE	DAM HEIGHT (FT.)	DATE STRUC-TURE COM-PLETED	SIZE AND TYPE OF RC OUTLET PIPE	TOTAL LENGTH OF RC OUTLET PIPE (FT.)	LENGTH OF PIPE SECTION (FT.)	MAXIMUM SETTLEMENT OF PIPE (FT.)	MAXIMUM ROTATION ANGLE AT JOINTS (RAD.)
Cobb Creek—site No. 1	65	Jan. 1959	36 in. I.D. Bell and spigot	468	12	3.14	0.0192
Dead Indian Wild-horse—site No. 2	47	9-3-59	30 in. I.D. Bell and spigot	276	12	1.40	0.0416
Dead Indian-Wild-horse—site No. 4	55	9-11-59	30 in. I.D. Bell and spigot	312	12	1.32	0.0183
Sandstone Creek—site No. 17	50	11-17-52	28 in. I.D. Centrifugal	372	12	1.61 1.41 (at axis of dam)	0.0208
Sulphur Creek—site No. 2	50	5-14-59	28 in. I.D. Centrifugal	274	12	0.28	0.0117
Middle Colorado Clear Creek—site No. 6	43	10-23-58	55 in. I.D. Centrifugal	262	12	0.36	0.0092
East Lateral Trin-ity—site No. 2	36	12-24-54	28 in. I.D. Centrifugal	190	12	0.27	0.0083
West Lateral Trin-ity—site No. 6	36	12-2-48	30 in. × 30 in. RC box	170	40	1.53	0.0177
Linn Subwatershed Dam No. 3A	37	8-30-55	30 in. I.D. Bell and spigot	167	7.5	2.34	0.0840

Note: Joint opening and horizontal strain are taken at midheight of pipe.

struction, and reservoir water level for Deer Creek Dam is representative. As seen in Fig. 3.2:1, the compression in the dam embankment increased only slightly in the first few years after construction.

Even in the USBR dams where high construction pore pressures had developed, the amount of compression associated with their dissipation was very small in relation to the compression during construction. In some cases a small secondary rate of compression (compressive "creep") occurred. Cumulative postconstruction compression of the USBR dams from all causes averaged less than 0.2% in the first 3 years after completion and less than 0.4% in periods up to 14 years.

Even though some of the dams were constructed with average water contents as low as 3% below Standard Proctor Optimum, in no case was any "compression on saturation" measured as the result of the

Measurements inside Concrete Outlet Pipes (After Moran, Proctor, Mueser

Total Extension of Pipe (in.)	Joint Opening at Axis of Pipe			Unit Horizontal Strain at Axis of Pipe			Distribution of Joint Openings along Pipe
	Average (in.)	*90 % Smaller than (in.)*	*Maximum (in.)*	*Average*	*90 % Smaller Than*	*Maximum*	
21.80	0.57	1.19	1.66	0.0039	0.0083	0.0115	Max. at max. settlement
13.78	0.63	1.07	1.69	0.0043	0.0074	0.0117	Max. at max. settlement, decreasing to toe
10.22	0.41	0.75	1.44	0.0028	0.0052	0.0100	Max. at max. settlement, decreasing to toe
23.31	0.78	0.97	1.06	0.0054	0.0067	0.0074	Essentially uniform for entire length
10.03	0.47	0.69	0.72	0.0032	0.0048	0.0050	Essentially uniform, large at downstream toe
27.12	1.23	1.50	1.75	0.0086	0.0104	0.0122	Decreasing slightly from max. settlement to toe
16.19	1.08	1.41	2.12	0.0075	0.0098	0.0147	Decreasing slightly from max. settlement to toe
7.59	—	—	2.72	0.0037	—	0.0057	Approximately equal for 40 ft. lengths
18.23	0.76	2.00	3.01	0.0091	0.0222	0.0335	Erratically spaced, large at expansion joints and downstream toe

penetration of seepage water from the reservoir into the embankment. Neither was there any indication of embankment swelling at any time,[1] because none of the USBR dams was constructed of soils having a large percentage of fines with high plasticity.

The authors do not know of any measurements which have indicated that an embankment has compressed (or expanded) as the result of reservoir drawdown. On the other hand, most instruments have been installed near the central parts of dams, where the influence of reservoir drawdowns would be less than in the upstream slopes.

[1] Swelling of well-constructed earth dams of highly plastic clay has been reliably measured. At two dams in Israel built of clay with liquid limit of about 80, the upper 20 to 25 ft. of both embankments swelled when the reservoir filled, and the crests heaved several inches (Ref. 64).

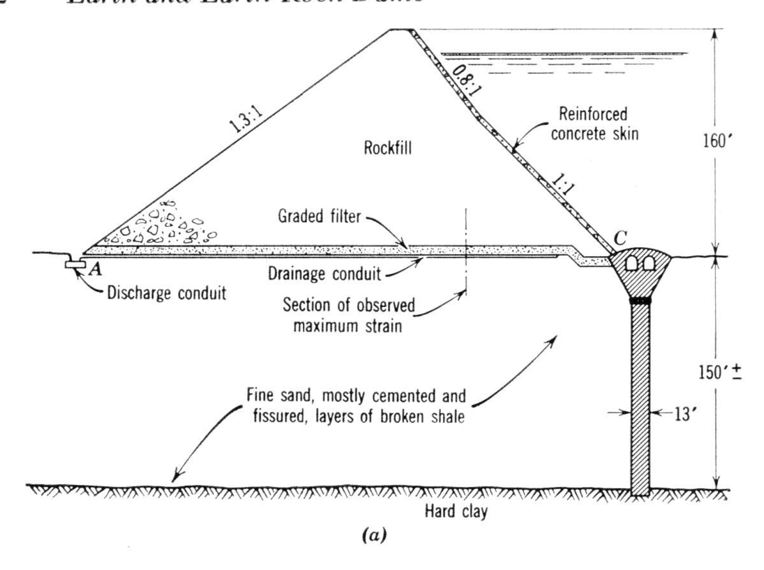

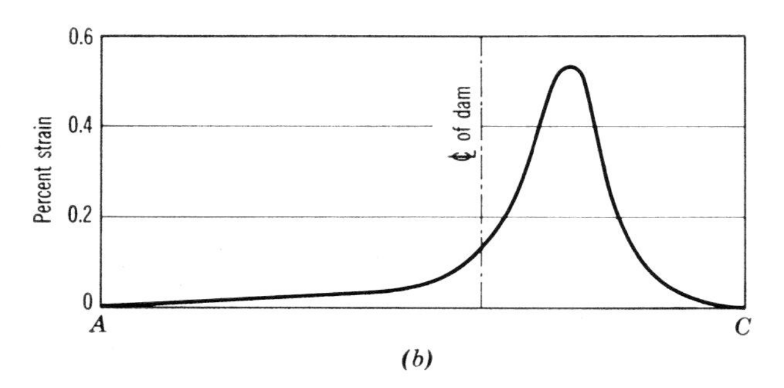

Fig. 3.1:2 (a) Section through rockfill dam on sand, friable sandstone, and shale. (b) Plot showing distribution of horizontal strain due to tendency of fill to spread as determined by measurements of separation of joints in drainage conduits (after Terzaghi and Peck, Ref. 126).

COMPRESSION OF ROCKFILL EMBANKMENT SECTIONS

No good summary of measurements is available to guide an estimate of the maximum compression which may occur in rockfill embankments as a function of rock type, content of fines, and method of construction. It is known that the large part of the total settlement of rockfills occurs during construction.

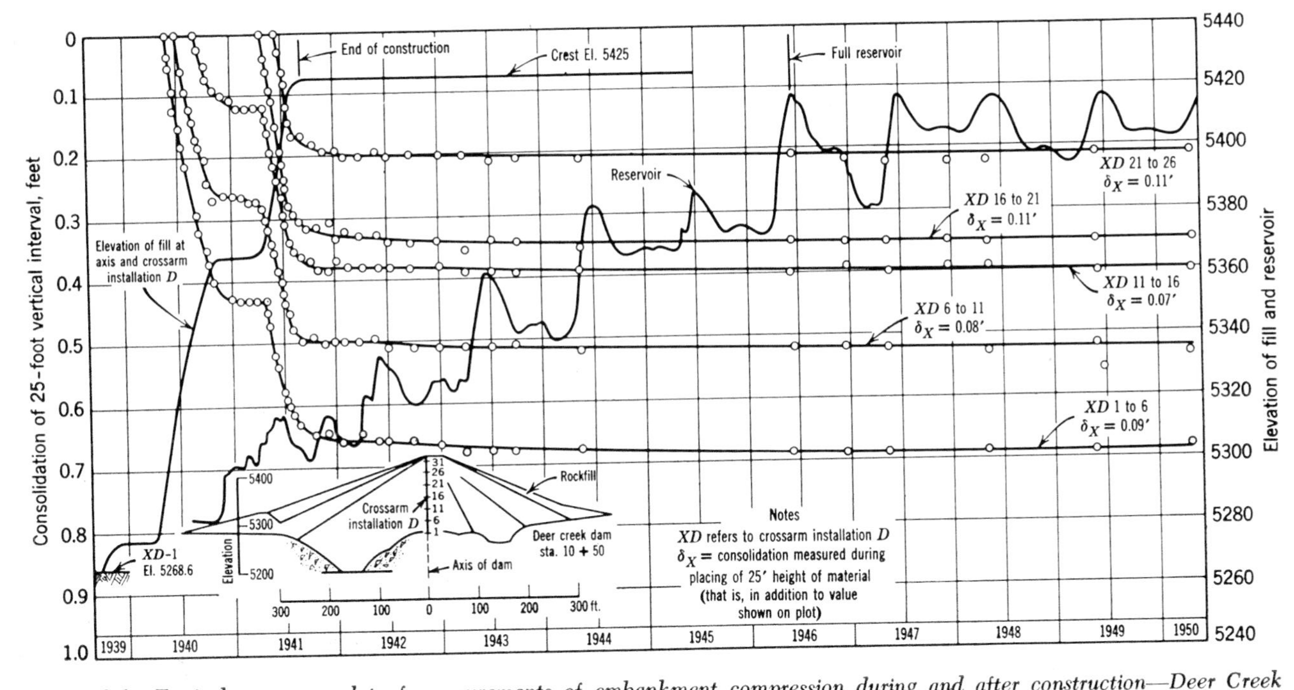

Fig. 3.2:1 Typical summary plot of measurements of embankment compression during and after construction—Deer Creek Dam, Utah (after Gould, Ref. 158).

The postconstruction settlement results from a gradual readjustment of the rock structure. The area at the point of contact between individual rocks is small and the stresses are high. Directly after construction, the larger individual rocks may be separated by smaller rocks or fines in the form of soil or quarry dust. Readjustment of the internal structure occurs when the rocks crush at the points of contact or slip with respect to each other, throwing their loads to adjacent rocks and points of contact.

If the rockfill has been built without sluicing or wetting of any kind, additional postconstruction settlement will follow when the section is wetted by the rain or by seepage. Water causes the additional settlement by softening the fines separating the larger rocks and by reducing the compressive strength of the rock itself at the small, highly stressed contact points (Sec. 11.5*a*).[1]

It can be assumed that the postconstruction settlement of a well-constructed, wetted rockfill placed in layers will be of the same order of magnitude as that of well-constructed rolled earth fills, i.e., between 0.1 and 0.4% of the height. On a number of rockfill dams built in thin, wetted layers up to about 150 ft., the settlement has been too small to measure. There are no recorded measurements of very high rockfills constructed in thin wetted layers.

The postconstruction compression of a rockfill section constructed by dumping in high sluiced lifts is greater than that of a section placed in thin wetted layers, and the settlement continues over a considerable number of years, as shown by the typical measurements of Dix River and Nantahala Dams, Fig. 3.2:2 (Ref. 329). The magnitude is extremely variable, depending on the type of rock and details of construction, and can be estimated in advance only very roughly. Fig. 3.2:3 gives a summary of the maximum settlement of the upstream slope of well sluiced rockfill dams with concrete faces and can be used as a guide to an estimate.

3.2b Crest Movements

Because the installation of apparatus for measuring the compression or swelling of the interior of an earth dam embankment is expensive

[1] An extreme example of the influence of rain water is given by the performance of the 260 ft. Cogswell Dam in California (Ref. 330). The main part of the dam was constructed from dirty rockfill dumped in 25-ft. high lifts which were not sluiced because of a water shortage. In 1934, when the embankment was approximately 80% completed, a 15-in. rainstorm caused the rockfill to settle about 10 ft. and to bulge the lower parts of the slope.

and time-consuming, such measurements have been made only in a very small percentage of earth dams. On the other hand, reliable measurements of crest movements after construction are simple to obtain and are carried out on most large dams.

HORIZONTAL CREST MOVEMENTS (UPSTREAM-DOWNSTREAM DIRECTION)

For an ordinary well-constructed earth dam with a centrally placed impervious core or a homogeneous section, the horizontal movement of the crest, which occurs when the reservoir fills, is less than an inch or two and usually too small to be measured. The pressure of the water against the upstream portion of the dam exerts two counterbalancing influences: (1) It tends to push the dam downstream; and (2) it loads the foundation under the upstream slope and thereby tends to rotate the crest upstream.[1] Ordinarily these actions counteract each other so that the measured crest movement is zero or a very small deflection which can be either upstream or downstream. In the following special cases of earth dams and foundation conditions, however, appreciable horizontal crest movements always develop and can be expected:

1. Central core dams with thick upstream and downstream shells of rock dumped and sluiced in high lifts.
2. Dams with upstream sloping cores and large downstream sections of rock dumped and sluiced in high lifts.
3. Dams of any type in which the reservoir water load or wetting action has a large influence on the foundation compression.

In recent years a number of high dams have been built with central impervious cores, upstream and downstream shells of dumped quarried rock, and a cross section which is more or less symmetrical about the center line of the crest, Sec. 1.2*c*. Because these dams are usually constructed with relatively steep side slopes, and because the dumped rock sections are less rigid than equivalent zones of rolled fill, appreciable downstream deflection of the crest usually develops when the reservoir is filled for the first time. This is true even where the dam is founded on hard rock. Often the crest moves downstream when the reservoir is filled and then, when the reservoir is lowered again, re-

[1] There can be no question about this tendency for the crest of some dams to move upstream when the reservoirs fill for the first time. Occasionally, small upstream movements have even been measured on dams with upstream impervious membranes of concrete or asphalt—when the embankment was well compacted and relatively incompressible.

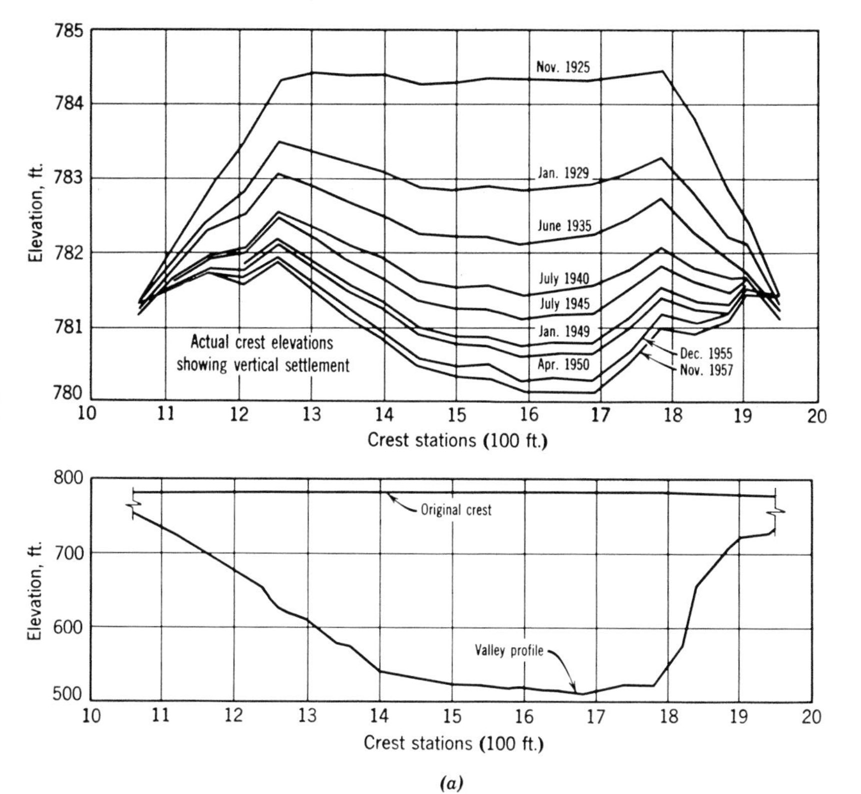

785
784
783
782
781
780
Elevation, ft.
Nov. 1925
Jan. 1929
June 1935
July 1940
July 1945
Jan. 1949
Apr. 1950
Dec. 1955
Nov. 1957
Actual crest elevations
showing vertical settlement
10
11
12
13
14
15
16
17
18
19
20
Crest stations (100 ft.)
800
700
600
500
Original crest
Valley profile

(a)

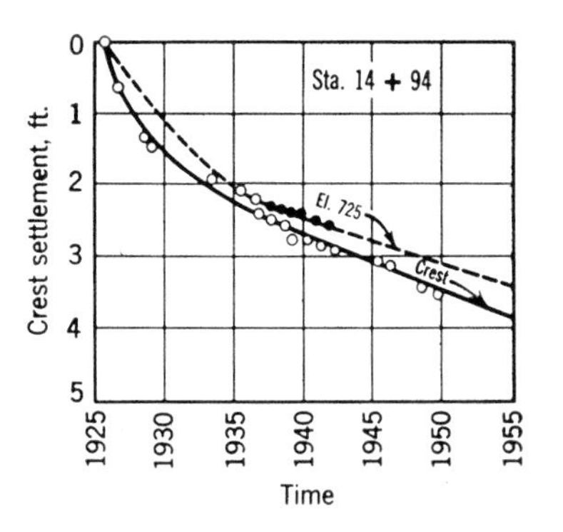

0
1
2
3
4
5
Crest settlement, ft.
Sta. 14 + 94
El. 725
Crest
1925
1930
1935
1940
1945
1950
1955
Time

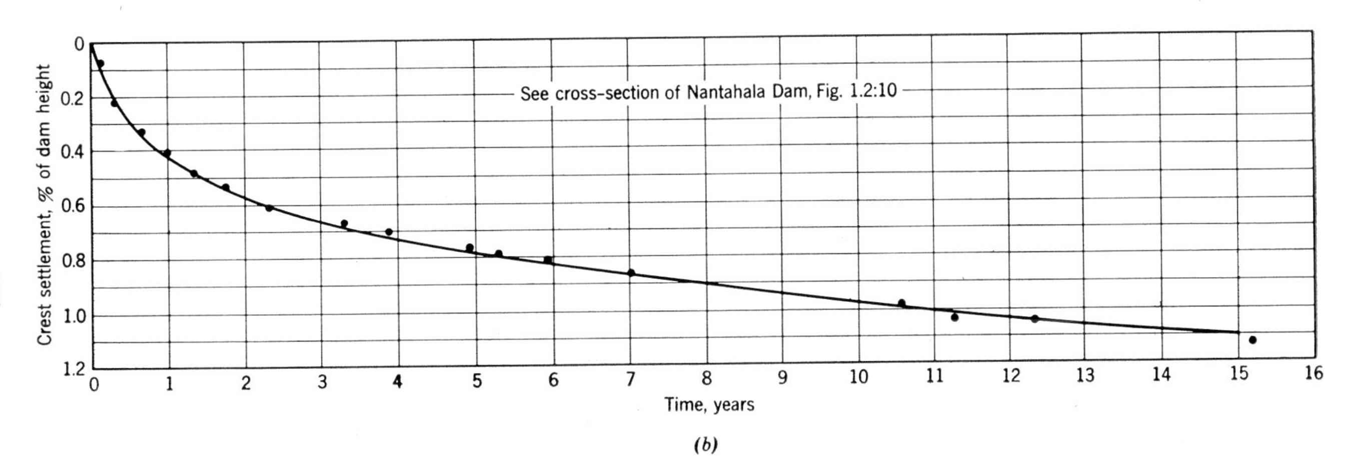

(b)

Fig. 3.2:2 Typical measured long-time crest settlements on rock fill dams. (a) Dix River Dam (after Schmidt, Ref. 329). (b) Nantahala Dam (after Growdon, Ref. 336).

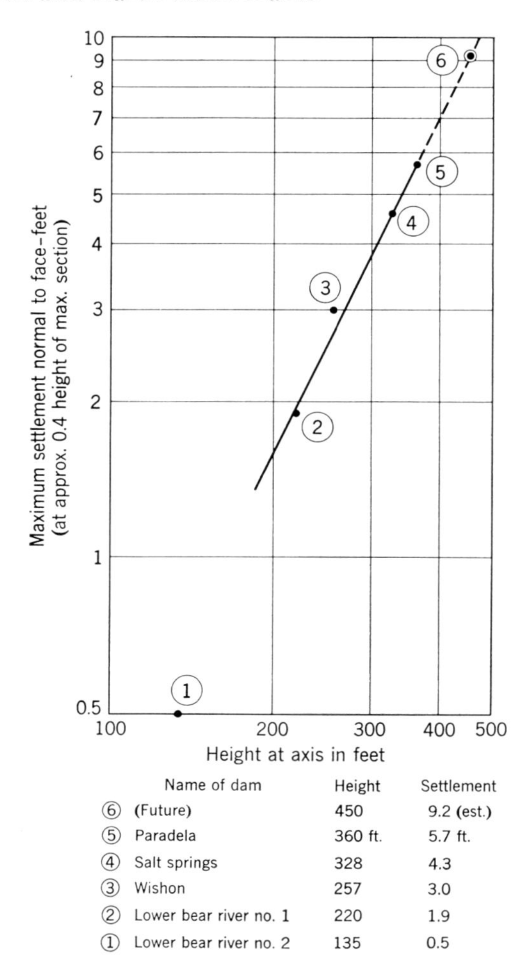

	Name of dam	Height	Settlement
⑥	(Future)	450	9.2 (est.)
⑤	Paradela	360 ft.	5.7 ft.
④	Salt springs	328	4.3
③	Wishon	257	3.0
②	Lower bear river no. 1	220	1.9
①	Lower bear river no. 2	135	0.5

Fig. 3.2:3 Maximum settlement normal to the face of concrete face rockfill dams after second reservoir filling (after Steele and Cooke, Ref. 337).

bounds at least part of the way as though it were an elastic body. The movements measured at the 300-ft. Cherry Valley Dam (Fig. 3.2:4), which was completed in California in 1955, are typical. As shown in Fig. 3.2:5, the dam rests on hard rock, has a large central zone of impervious rolled fill, and has outer shells of quarried rock dumped and sluiced in 30-ft. layers. Before the reservoir was filled for the first time, survey monuments were set along the crest, and measurements of the vertical and horizontal movement were made at frequent intervals during the first 2 years of operation. For the entire period both components reflected the level of the water in the reservoir with a time lag of about one month (Fig. 3.2:4). When the reservoir reached a maximum elevation in July of 1957, the crest had moved downstream about 6 in.; and when the water level was subsequently lowered 110 ft. in the fall of 1957, the crest moved back upstream approximately 2 in.

Most high dams of the type discussed above have shown a downstream movement of several inches and a sensitivity to the position of the reservoir which lasts for a number of years. However, the South Holston Dam of the Tennessee Valley Authority (Fig. 1.2:32), which has a cross section very similar to that of Cherry Valley Dam, moved upstream instead of downstream. The crest shifted 3 in. upstream when the reservoir was filled for the first time and subsequently moved back and forth about 0.5 in. as the water level fluctuated 50 to 80 ft. each season (Refs. 21, 50).

The crests of all dams with sloping upstream cores and large downstream zones of dumped, quarried rock have moved downstream an amount commonly between 50 and 75% of the vertical settlement of

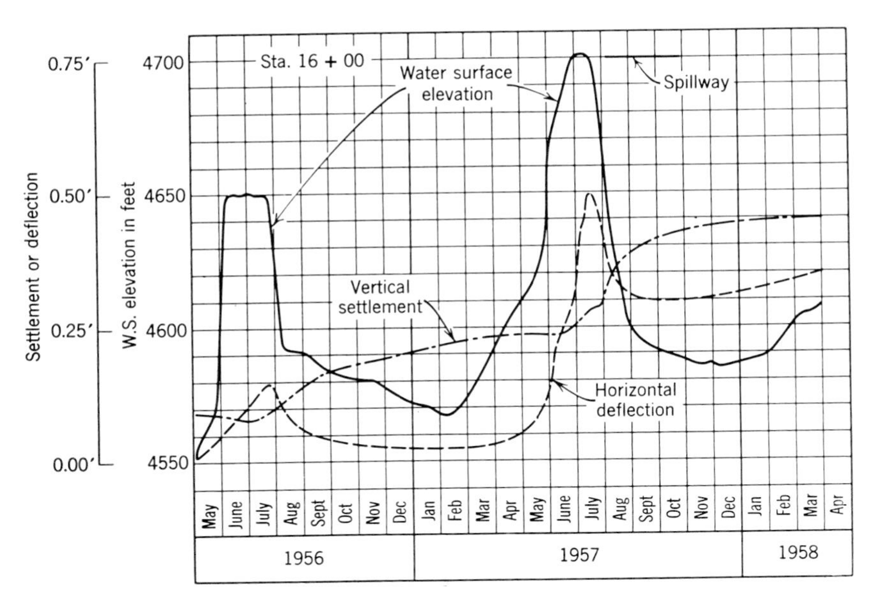

Fig. 3.2:4 Typical measured crest movements during first two years of reservoir operation—Cherry Valley Dam (after Lloyd et al., Ref. 26).

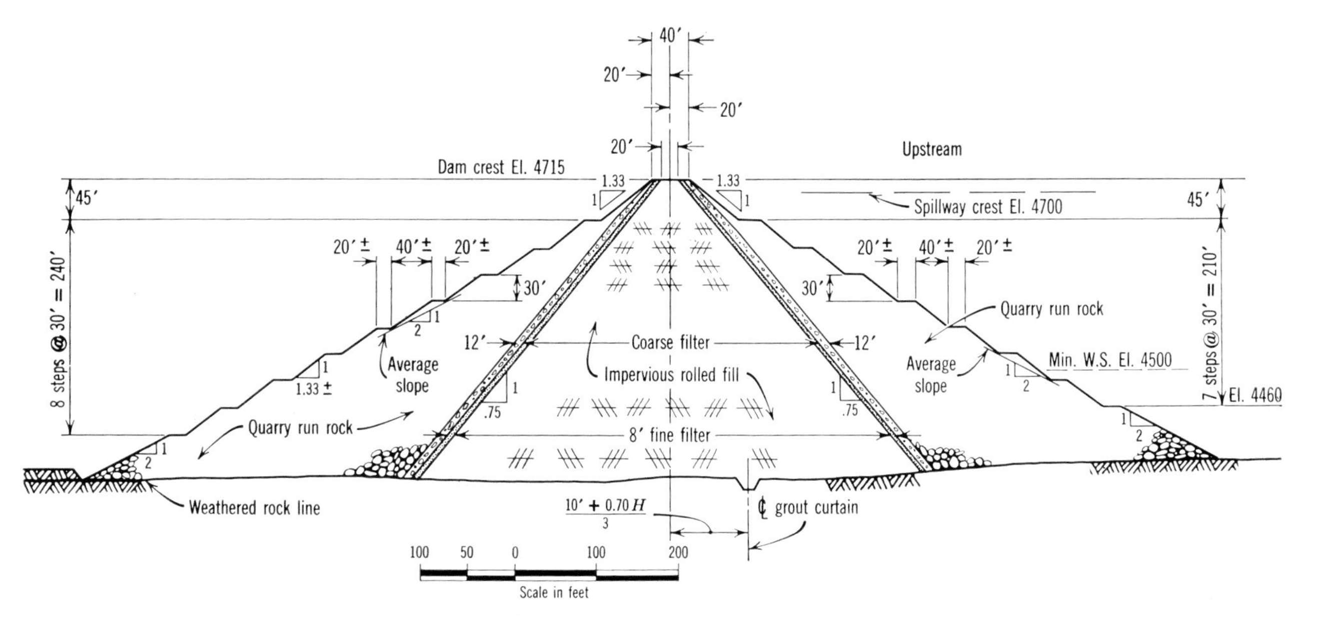

Fig. 3.2:5 Cross section of Cherry Valley Dam (after Lloyd et al., Ref. 26).

the crest. For these there is usually no measurable rebound of the crest when the reservoir is lowered.

At sites where the foundation is relatively incompressible, the only dams which have regularly undergone horizontal crest deflections of more than an inch or two are those with large zones of dumped rock. This is due to the fact that rolled-earth sections are more rigid and resistant to deformation under the reservoir pressures than sections of dumped rock. A few dams comprised wholly of well-rolled earth, however, have had large horizontal crest movements because the seeping water has caused settlement of the foundation material. The experience at Rector Creek Dam, described in the next section, is an example.

LONGITUDINAL CREST MOVEMENTS

While it has been the custom on large dams to measure postconstruction crest settlement and the horizontal movement in the upstream-downstream direction, in only a few cases have measurements been made of the horizontal movement in the longitudinal direction. The lack of such measurements has left a serious gap in our knowledge of the behavior of earth dams. Although it is known that there is a definite tendency for the portion of the embankment located near the abutments to move longitudinally toward the deeper parts of the valley, the range of the magnitude of horizontal movements in the direction parallel to the longitudinal axis of the dam can only be speculated on.

The Rector Creek Dam in California (Refs. 95, 263), which had relatively large postconstruction crest movements, is one of the few dams on which longitudinal measurements have been reliably reported. It is a conventional zoned earth dam with a maximum height of about 150 ft. and a foundation cutoff extending down about 50 ft. to bedrock, Fig. 3.2:6. The main part of the embankment was constructed of a residual soil which resulted from the insitu weathering of a fine-grained igneous bedrock.[1]

[1] The coarser soil particles (gravels) were composed of the parent rock in varying stages of disintegration. Below about ⅛ in. in diameter, they were frequently soft enough to be crushed with the fingernail, and the relatively high postconstruction settlements were probably caused in some degree by these particles breaking down under the action of the sheepsfoot roller. The material became finer with each pass of the roller, and it was difficult to avoid compacting too dry even though careful water content-density control was exercised. (See Sec. 11.3*e* for the difficulties of controlling the construction water content of residual soils.)

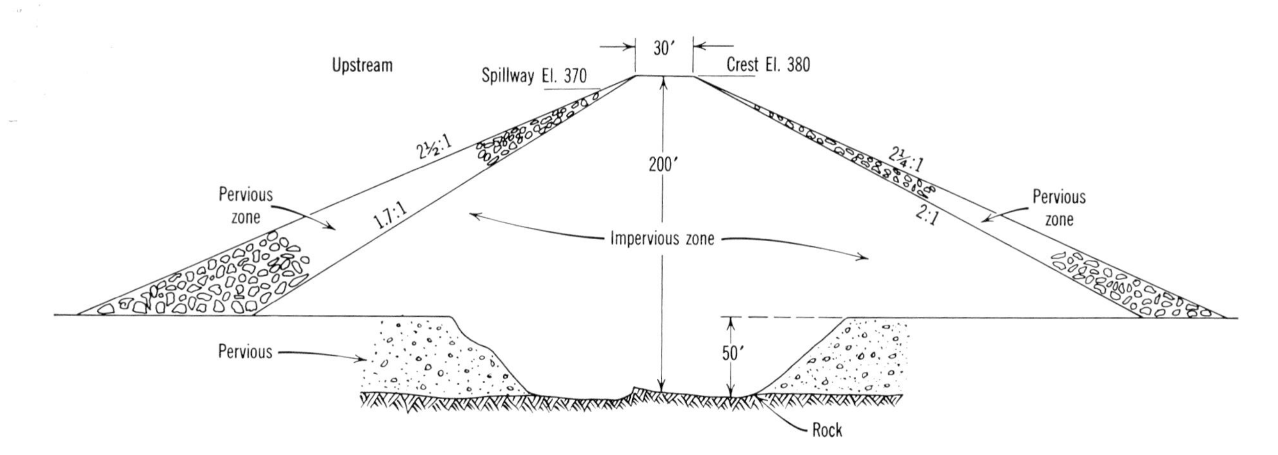

Fig. 3.2:6 Cross section of Rector Creek Dam. (Ref. 95).

The measurements, which were made between monuments established 100 ft. apart on the crest, are particularly instructive. As seen in Fig. 3.2:7, the crest settled about 2.8 ft. in the first 10 years after construction. Associated with the settlement was a longitudinal compression of the crest in the central part of the valley and an elongation of it over the rock abutments. The horizontal distance between Stations 14 and 15 at the center of the valley decreased from 100.00 to 99.41 ft., and the length between Stations 18 and 19 stretched from 100.00 to 100.73 ft.[1]

While the total crest movement at Rector Creek Dam was greater than expected, the general pattern of the movement should be similar on a reduced scale at other dams in valleys of similar shape. It is interesting to note that the magnitude of the elongation over the abutments was directly related to the abutment steepness: a stretch of 0.73 ft./100 occurred on the left abutment, which had an average slope of about 1.5 to 1; while on the right, which had an average slope of 2.2 to 1, the stretch was only 0.5 ft./100.

While measurements of this type have not been made on many earth dams, similar movements have been noted on almost all of the high rockfill dams with reinforced concrete upstream slabs. In these the post-construction movement of the embankment from the abutments toward the center of the valley causes vertical joints in the slab to open at the abutments and to close at the center of the valley.

3.3 PORE WATER PRESSURES DURING CONSTRUCTION

It has been known for more than 20 years that positive, and sometimes alarmingly high, pore water pressures develop in the impervious sections of some earth dams during construction before there is any water in the reservoir to cause seepage pressures. Pressures equal to 50% of the total weight of the overlying embankment are common in the central part of the core, and pressures exceeding 80% have been

[1] In February 1947, a transverse crack with a width of about ¾ in. opened on the crest at Station 18 + 65. It ran from one edge of the crest to the other, making an angle of about 40° with the longitudinal axis of the dam. Because the outer embankment zones consisted of cohesionless material it was not possible to trace the progress of the crack down the slopes, and the engineers had a test pit sunk 24 ft. deep on the crest in order to examine it. The crack was observed to have a slope of about 9° from vertical toward the center of the valley, and at the bottom of the pit it still had a width of about ½ in. with no end in sight. At the end of several months it did not measurably widen, and it was finally grouted with cement grout and gave no further trouble.

Measured Distances Between Stations on Crest								Date
100.00′	100.00′	100.00′	100.00′	100.00′	100.00′	100.00′	100.00′	1/47
100.19							100.25	3/8/49
100.21	100.00	99.76	99.82	99.93	99.88	99.68	100.27	4/29/49
100.25	100.16	99.70	99.73	99.92	99.82	99.50	100.40	4/14/50
100.27	100.23	99.69	99.70	99.91	99.82	99.56	100.46	9/15/50
100.29	100.28	99.70	99.70	99.91	99.82	99.55	100.51	2/6/51
100.34	100.40	99.69	99.65	99.91	99.82	99.57	100.58	10/31/51
100.30	100.39	99.69	99.59	99.90	99.79	99.54	100.60	3/25/52
100.33	100.45	99.68	99.55	99.87	99.75	99.54	100.62	11/8/52
100.30	100.43	99.68	99.49	99.90	99.76	99.60	100.70	4/7/53
100.26	100.50	99.77	99.41	99.73	99.63	99.62	100.73	8/15/56

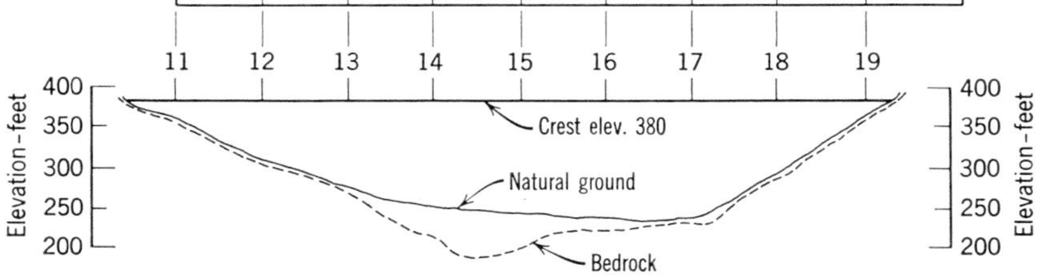

(*a*)

Fig. 3.2:7

measured. The pressures which may develop at a given dam can be predicted from laboratory tests and/or from theory (Sec. 7.5); however, the reliability of the predictions is still somewhat problematical, and the designer must look to measurements on other similar dams as the most reliable support for his estimate of the pore pressures which will develop.

3.3a USBR Measurements

The USBR has been installing piezometers in major dams and measuring pore pressures during construction since the middle 1930's. A detailed analysis of the measurements from 26 of these dams constructed between 1936 and 1952 which was published by Gould (Ref. 300) constitutes one of the largest and most valuable concentrations of experience available to the profession. The dams included in Gould's study cover the whole range of soil types encountered by the USBR and a fairly wide range of average construction water contents (see Table 3.1:1).[1]

[1] As discussed in Sec. 3.1*a*, when using these data it must be kept in mind that all dams were constructed according to USBR practice.

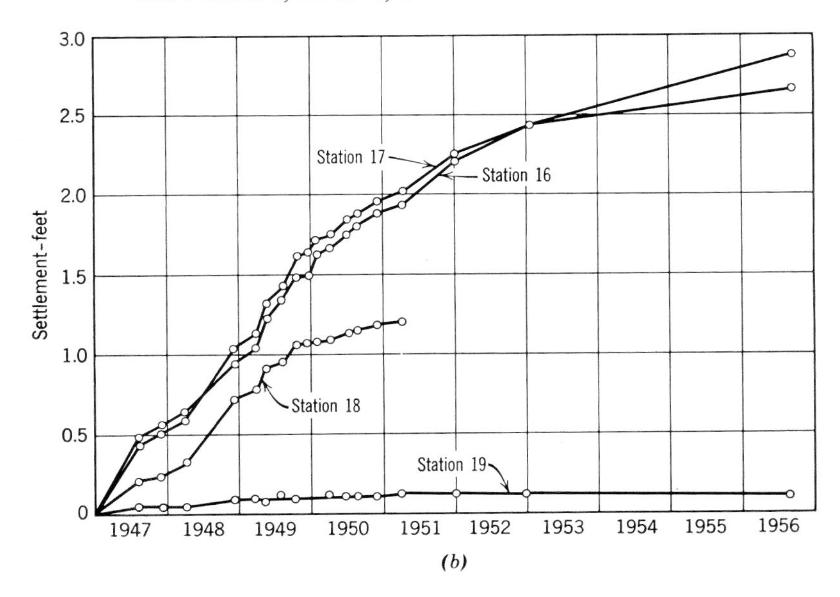

(b)

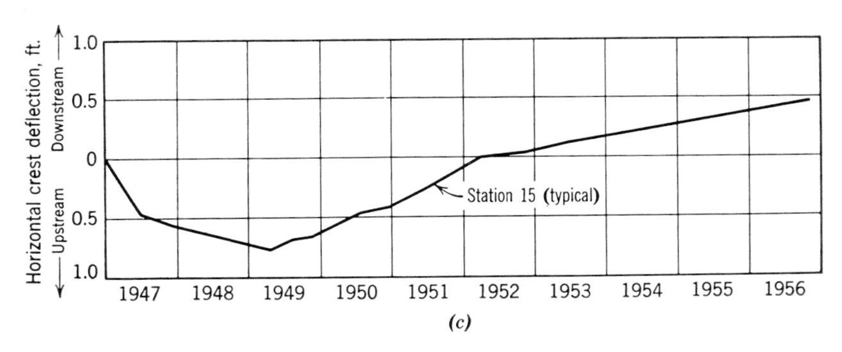

(c)

Fig. 3.2:7 Measured crest movements in first 10 years of reservoir operation—Rector Creek Dam (Ref. 95). (a) Longitudinal crest movement. (b) Crest settlement. (c) Horizontal crest movement in upstream-downstream direction.

Measured construction pore pressures in a typical USBR dam are given in Fig. 3.3:1. Contours of equal pressure generally have an oval configuration conforming to the shape of the central impervious zone of the dam. The highest pressures develop in the central portion of the impervious core, where the overburden pressures are highest and the effect of drainage to the adjacent pervious zones is least.

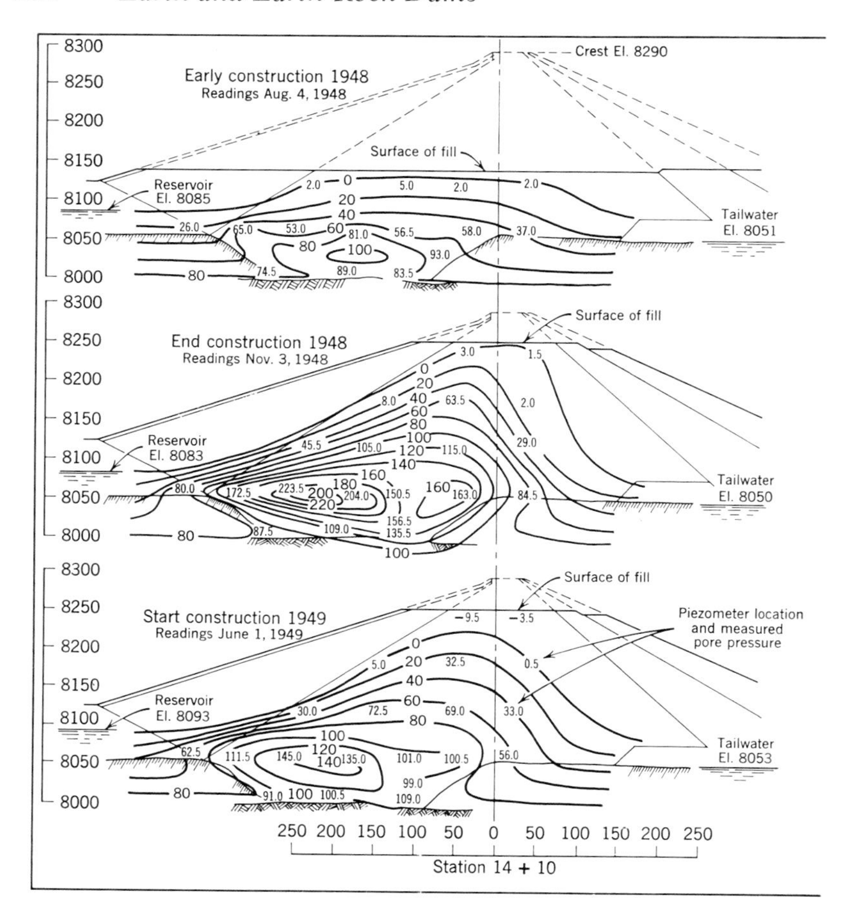

MEASURED VERSUS PREDICTED PORE PRESSURES

As a major part of the study, Gould compared the measured pore pressures with those which would have been predicted on the basis of the USBR Theory (Sec. 7.5).[1] He found that the two agreed roughly, and that the main difference between the predicted values and the pore pressures measured at a given dam could be explained reasonably from the construction history. For example, the measured pressure was higher than predicted when concentrations of wet and compressible soil had been placed in the embankments. Drainage relief and surface

[1] See also Refs. 22, 318.

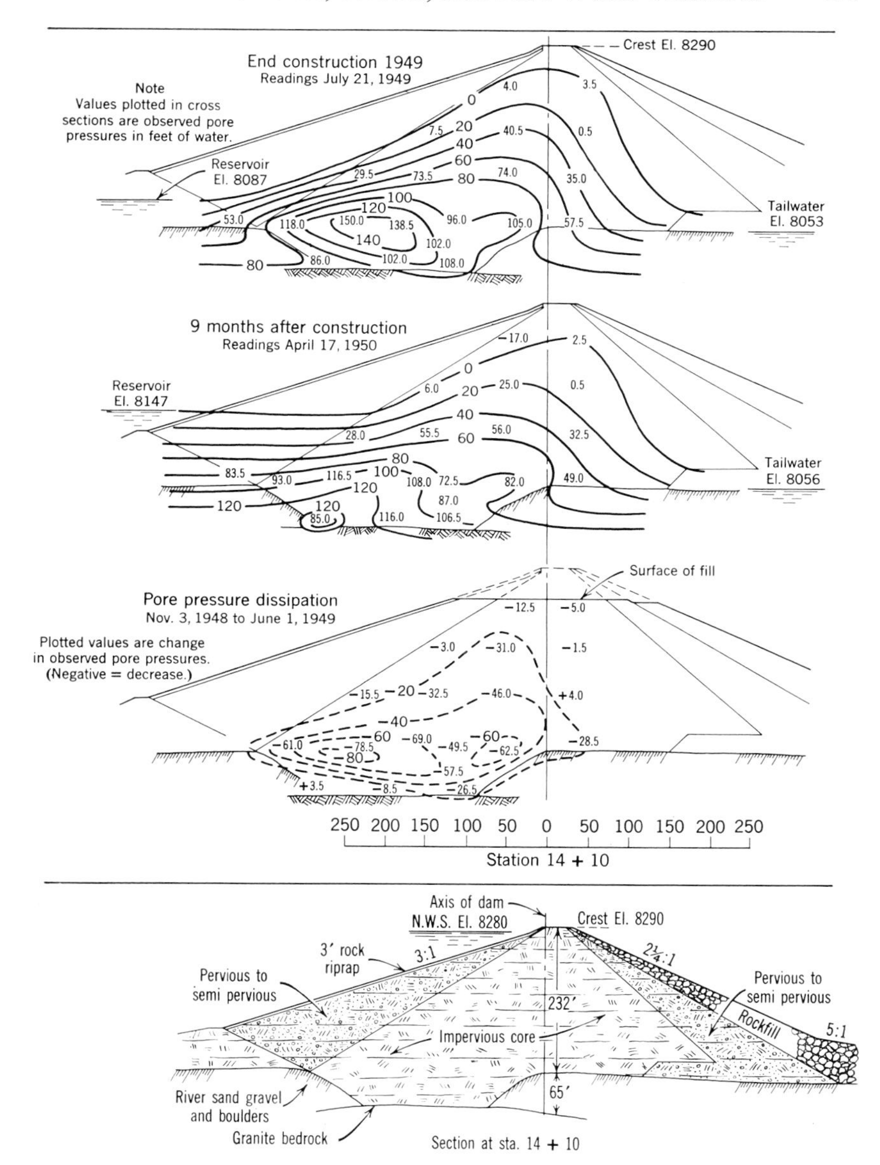

Fig. 3.3:1 Typical USBR measurements of construction pore pressures at various times during construction (after Gould, Ref. 300).

tension in the pore water caused the pore pressures in the impervious zone to be lower than predicted, and differential settlements led to different pressures than predicted because they caused transfers of total stress within the embankments.

Though the study indicated that the measured and calculated pore pressures were roughly equal and, hence, that the theory is probably satisfactory for the types of soils analyzed, it does not by any means follow automatically that the measured pore pressures could have been estimated reliably in advance of construction. In the theoretical prediction (Sec. 7.5), small differences in the assumed construction water content and density or in the average soil type make large differences in the computed pore water pressures. In order to obtain an accurate prediction of the construction pore pressures in the design stage, the engineer would have to be sure that he was testing the average soil from the borrow pit and that the moisture-density control during construction would result in an embankment in which the average water content and density approximated closely the values assumed on the basis of the preliminary laboratory tests. In addition, he would have to estimate the compressibility of the embankment either from previous measurements (Sec. 3.1*a*) or from laboratory consolidation tests. Since all these elements required for the theoretical prediction are difficult to estimate reliably, except perhaps under ideal conditions, the theory should only be considered as a general guide to the order of magnitude of the pore pressures which might develop.

INFLUENCE OF CONSTRUCTION WATER CONTENT

As was expected, the primary factor governing the development of pore water pressure during construction was found to be the average water content at which the embankments were constructed. With only two exceptions, in every dam in which the average water content exceeded 0.6% below Standard Proctor Optimum, substantial pore pressures developed. In embankments compacted a few percent drier, no significant pressures developed.

In recent years it has been the practice of the USBR to prevent high construction pore pressures by maintaining the average construction water content dry of optimum. Since the introduction of this procedure, high pressures have occurred in only a few dams where, because of wet borrow materials or rainy weather, it was not possible to place the embankment material at the desired low water content.

INFLUENCE OF DRAINAGE AND SURFACE TENSION

As seen from the measurements in the typical USBR dam, Fig. 3.3:1, drainage was always effective in keeping the construction pore pressures low on the periphery of the impervious zones adjacent to the more pervious embankment material. The maximum pore pressures often developed at about mid-height in the impervious core, indicating that a substantial seepage gradient existed vertically downward toward the relatively more pervious foundation. In other dams not illustrated, the foundation permeability was equal to or lower than that of the impervious core, and the highest construction pore pressures occurred at or just above the foundation.

In the interiors of the thick cores, very little dissipation of the pore pressures took place as the result of drainage during construction. In only 3 of the 26 dams studied was there any indication of large pressure reduction. From this fact Gould concluded that, for materials similar to those in the USBR dams, Table 3.1:1, it is not excessively conservative to neglect the influence of drainage on the construction pore pressures predicted in the central portions of thick core zones.

In dams where the core material has been more pervious than for the USBR dams studied by Gould, dissipation of pore pressures due to drainage during construction has occurred even in the center of thick cores. A rough relationship, based on USBR experience and measurements in other dams, between pore pressure dissipation and embankment permeability is given in the Summary of this section.

The observed construction pore pressures studied by Gould in USBR embankments of very fine soils (having more than 20% by weight smaller than 0.005 mm.) were usually 5 to 10 p.s.i. lower than the pressures computed from the USBR theory. Gould attributed this discrepancy to the action within the soil pores of surface tension, which the theory did not take into consideration. In a number of large dams constructed with low water contents, the influence of surface tension was strong enough to produce negative construction pore measurements at all piezometers in the impervious section, Fig. 3.3:2.

INFLUENCE OF SOIL TYPE

While the water content is the primary factor governing construction pore pressures, there is also a strong correlation with the embankment soil type. Both theory and the measurements on the USBR dams

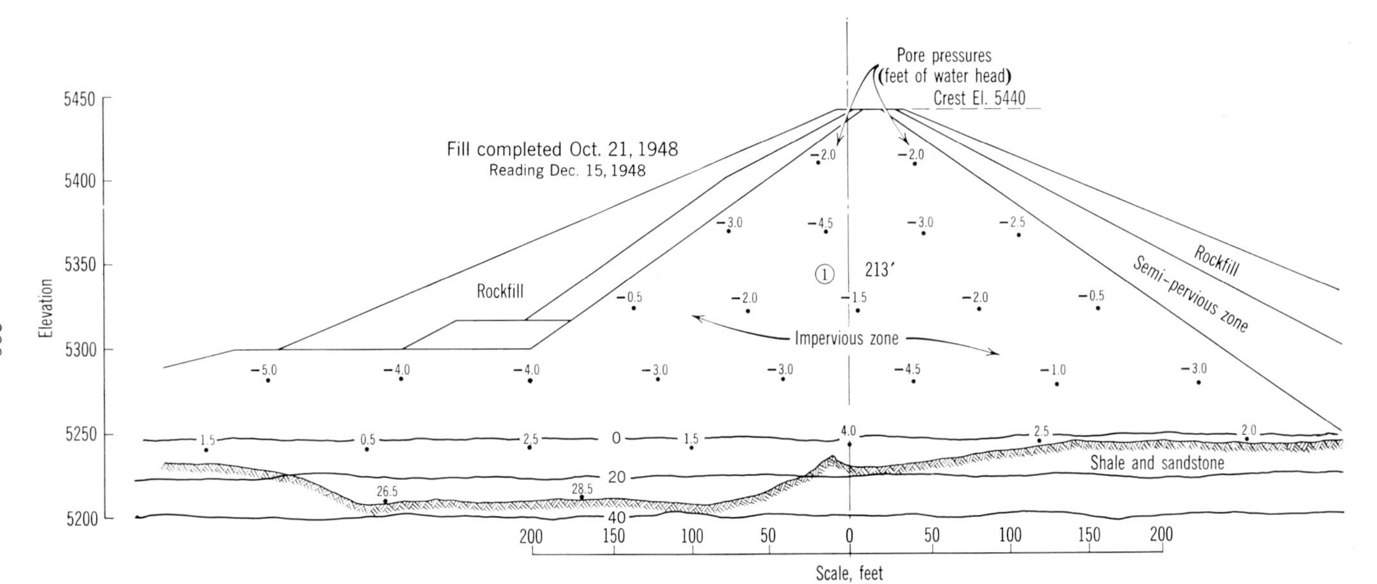

Fig. 3.3:2 Negative (below atmospheric) pore pressures at end of construction—Dixon Canyon Dam, Colorado (after Gould, Ref. 300).

indicate that the highest pressures are generated in embankments constructed of well-graded gravel mixtures with clayey fines, in which the original air content is relatively low and the compressibility relatively high (Table 3.1:2).[1] Conversely, theory and measurements also both agree that embankments of uniform silts and fine silty sands with little or no plasticity are the least susceptible to the development of high pore pressures because of their high initial air content and relatively low compressibility.

No measurements are available from the USBR on dams which have impervious sections of very fine-grained clayey soil with high or even medium-high plasticity, since none were constructed of this material. Other measurements (see Sasumua Dam, Sec. 3.3*b*) suggest the possibility that because of high capillary forces and the relatively large air contents, the construction pore pressures in embankments of very fine-grained clayey soils may always be low.

3.3b Measurements in Other Dams

While the USBR's measurements of construction pore pressures are probably the largest collection of experience available, valuable measurements have been made on many other dams. Results from a few of the more instructive are given in the following sections.

SWIFT CREEK DAM

Swift Creek Dam in Washington (Refs. 77, 159, 160) provides the designer with a reliable landmark in the form of soil so pervious that no construction pore pressures could exist. This is one of the highest earth dams ever constructed, Fig. 3.3:3. It was built very rapidly during 1957–1958 in an area of high rainfall (normally more than 100 in./yr.), and in spite of the dense, well-graded embankment material and the high average construction water content, no pore pressures developed.

Since the construction water content was to be very high, the designers anticipated high pore pressures and planned to control them

[1] More concern over possible sliding during construction has therefore been directed toward embankments constructed of this type of soil than of any other. This is something of a paradox, because all our experience with slides (Sec. 2.4) indicates that dense, granular embankments are the least likely to fail by sliding during any period in the life of a dam. The events at Otter Brook Dam (Sec. 3.3*b*), where a slight embankment strain and bulging caused appreciable dissipation in the construction pore pressures, may offer one explanation.

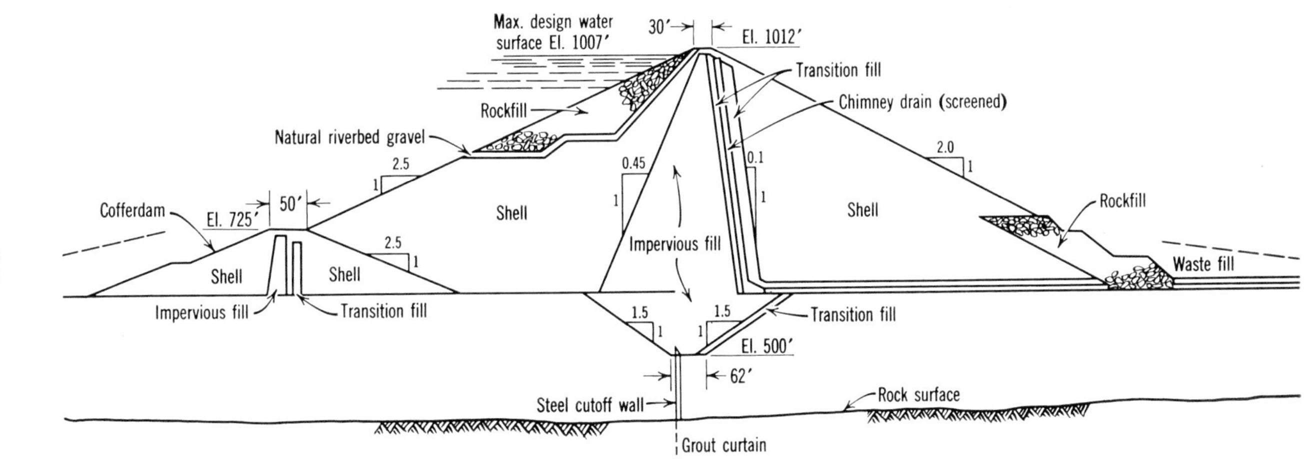

Fig. 3.3:3 Cross section of Swift Creek Dam, Washington (after de Luccia, Ref. 77).

by providing horizontal drains in the slopes. Early in construction, however, the embankment material was used in a very wet condition for the cofferdam, and the piezometers installed there registered only 2 ft. of pore pressure following the addition of 10 ft. of fill in a 24-hr. period. Because of the low pressure and because even this rapidly dissipated, the horizontal drains were not used. During construction the piezometers placed in the lower portion of the central core recorded maximum pressures of only 4 to 6 p.s.i.

The embankment material was obtained from a reworked mud flow deposit. It is a well-graded, gravelly sand containing generally from 5 to 20% nonplastic fines (finer than the No. 200 sieve). The average gradation of the core material and the relationship between permeability and void ratio of compacted specimens is shown in Fig. 3.3:4.

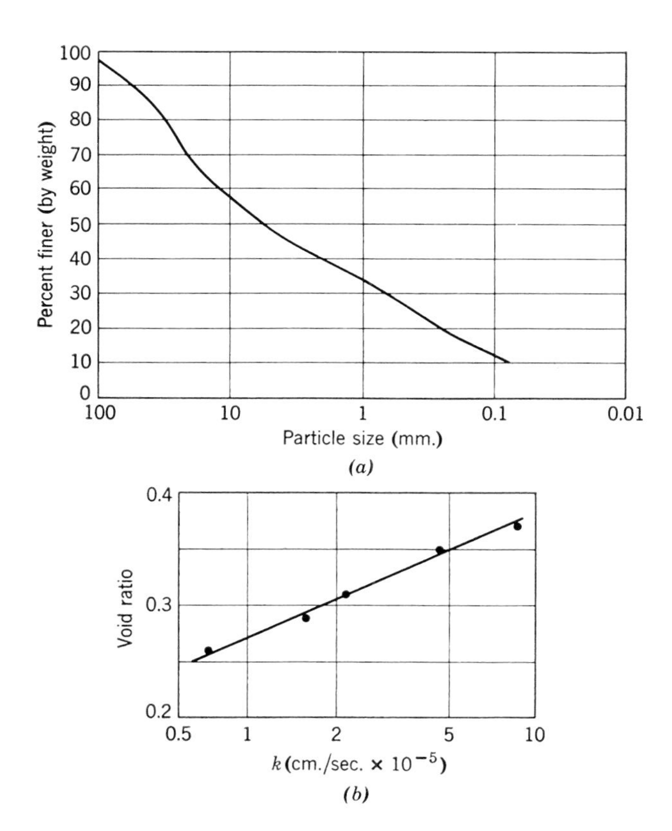

Fig. 3.3:4 Swift Creek Dam—properties of impervious core (after de Luccia, Ref. 77). (a) Typical gradation. (b) Range of coefficient of permeability.

SASUMUA DAM

The records for Sasumua Dam (Refs. 68, 304, 305) are significant because it is one of the few earth dams constructed from very fine clay with high liquid limit and high water content at which reliable measurements of construction pore pressures were made. Built near Nairobi, Kenya during 1954–1955, it is essentially a homogeneous structure with large upstream and downstream rock toes and a maximum height of approximately 115 ft., Fig. 1.2:6.

The typical embankment material is the product of the weathering of volcanic rock. A mineralogical analysis showed that it contained approximately 60% by weight of the mineral halloysite.[1] The average soil properties were:

Liquid limit, 87%.
Plasticity index, 33%.
Coefficient of permeability (3×10^{-7} cm./sec.)
Proctor maximum dry density, 70 lb./ft.^{3}
Optimum water content, 50%.
Specific gravity, 2.83.

The natural water content of the material in the borrow pit was between 60 and 75% and the annual rainfall was about 55 in./yr. During construction the water content was laboriously lowered by drying in the borrow pits. The embankment was constructed in thin layers with heavy rollers and with careful moisture-density control, so that the average compacted density was approximately 71 lb./ft.^{3} and the average compaction water content approximately 49.5%, or very close to Standard Proctor Optimum.

Piezometers and compression measuring apparatus of the USBR type were installed in the embankment, and even though the density was very low and the water content high, the measured compression

[1] A valuable study of the influence of the halloysitic minerals on the performance of the compacted material was made by Terzaghi (Ref. 304). His results showed that because of the inert water held in the voids of the halloysite crystals, the material performs in many ways as though it were a typical inorganic clay of much lower plasticity. Although not one of the most common clays, halloysite occurs with enough frequency in nature that a number of dams have been constructed of it. When its presence is suspected, tests to identify it should be made because it is a superior embankment material to the more common clay with the same liquid limit. For example, at the Djatiluhur Dam in Java in 1960, tests on alternate materials for the impervious core indicated that a halloysitic clay with liquid limit of 110% had higher strength than another clay consisting of the more common minerals which had a liquid limit of 60%.

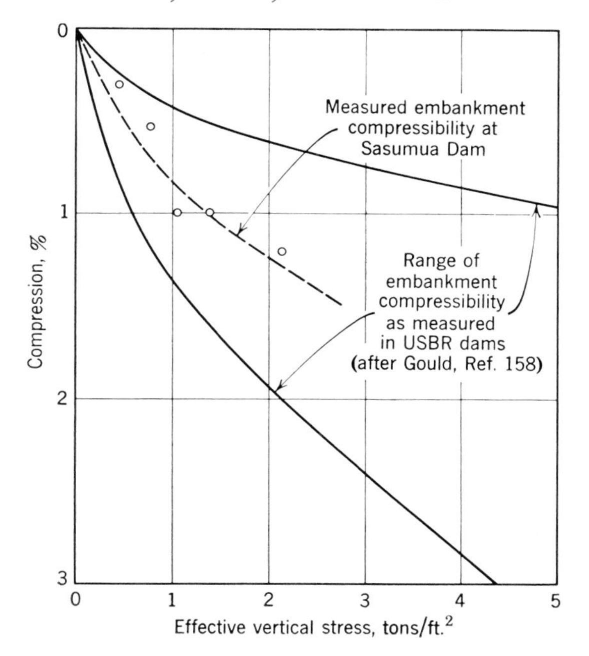

Fig. 3.3:5 Plot showing the relationship of the compressibility of Sasumua Dam to the range of compressibility of USBR dams (after Terzaghi, Ref. 304).

was approximately the average of the values for all the USBR dams, Fig. 3.3:5. In the early part of construction, when the embankment construction surface rose a few feet over the locations of the piezometers, several of them registered apparent pore pressures of about 5 ft. of water. As construction continued, no further pressures developed, and at completion they were practically zero in all cases.

OTTER BROOK DAM

Otter Brook Dam in New Hampshire (Ref. 489) furnishes the only documented case of a dense, impervious embankment of well-graded clayey sand in which high pore water pressures caused movement.[1] It is a homogeneous earth dam constructed in 1957 to 120 ft. with an

[1] Also see Ref. 543 for a recently reported similar experience (1963) in the Troneras Dam in Colombia where high construction pore pressures were associated with slope bulging.

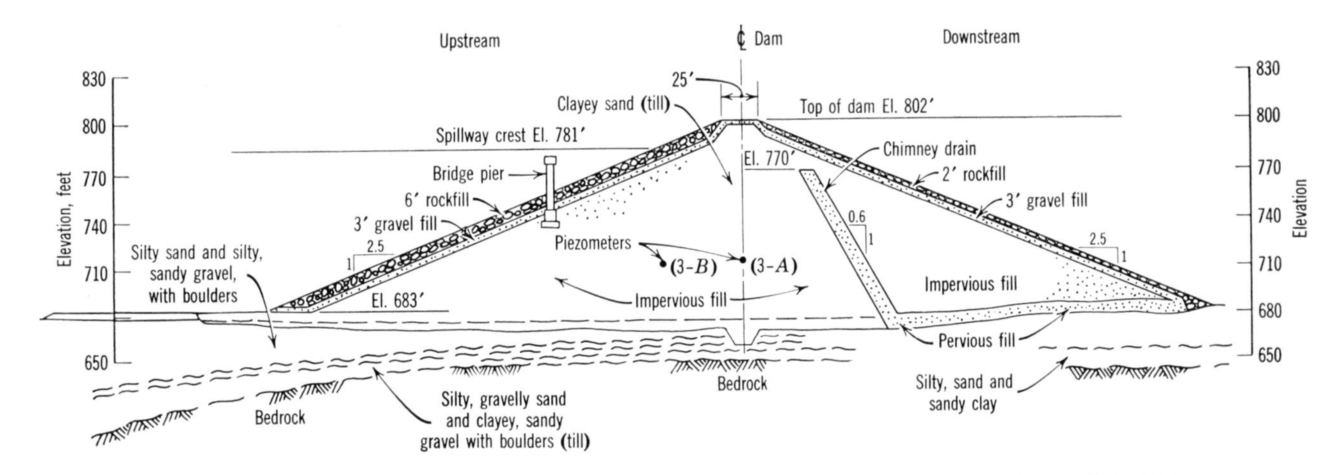

Fig. 3.3:6 Typical cross section—Otter Brook Dam, New Hampshire (after Linell and Shea, Ref. 489).

inclined downstream drain, Fig. 3.3:6. The average properties of the soil (New England Glacial Till) and the compacted embankment were:

	SIZE	PERCENT FINER THAN
GRADATION	No. 4 Sieve	80 to 90
	No. 200 Sieve	45
	0.005 mm.	30

Liquid limit	21 to 29%
Plasticity index	7 to 14%
Standard Proctor Optimum water content	11.3%
Average construction water content (minus No. 4)	10.7%
Average embankment dry density (minus No. 4)	125.7 lb./ft.^{3}
Coefficient of permeability	0.1 ft./yr.*

* This is the estimated permeability of the compacted embankment. Laboratory tests indicated that the coefficient of permeability of the minus No. 4 fraction ranged roughly between 2 ft./yr. (at 117 lb./ft.^{3}) and 0.006 ft./yr. (at 127 lb./ft.^{3}).

As shown in Fig. 3.3:6, the design includes a concrete bridge pier founded on the upstream slope of the embankment at elevation 739. The foundation and the lower part of the pier were constructed on August 12, 1957, at which time the embankment construction surface was several feet higher than the pier foundation. On August 16 it was noticed that the partially completed bridge pier had moved upstream and downward. Reference stakes were subsequently installed at various points on the upstream slope, and measurements of the horizontal and vertical components of the movement were made at regular intervals during the rest of the construction period.

As Fig. 3.3:7 demonstrates, points on the surface of the slope at about midheight moved continuously upstream and downward as additional fill was placed. This bulging was closely related to the rate of increase of embankment height. When embankment placing was temporarily halted, the movement stopped within a day or two and then started again after construction was recommenced. Some settlement of the embankment construction surface occurred in conjunction with the bulging, but the bridge pier remained perfectly plumb during all the movement.

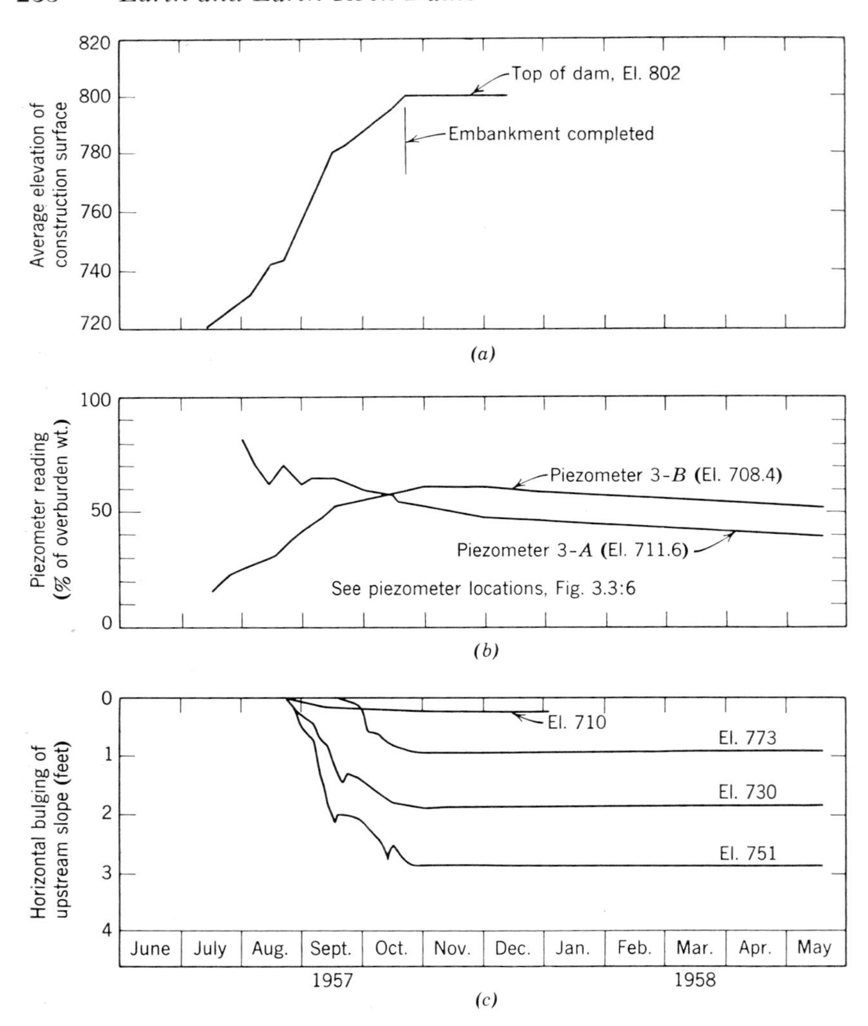

Fig. 3.3:7 Measurements of construction pore pressures and slope bulging on Otter Brook Dam. (a) Approximate rate of embankment construction. (b) Typical pore pressure measurements. (c) Horizontal bulging of upstream slope. (Courtesy U.S. Corps of Engineers, New England Division; also Ref. 489)

At the end of construction (October 1959) the maximum movement of the slope had a horizontal component of about 3 ft. and vertical component (downward) of about 6 in. When the embankment was finished, the movement stopped immediately and completely. Since some movement had evidently occurred before any measurements were made, the total bulging was somewhat more than 3 ft., its horizontal magnitude varying gradually from zero at the toe to a maximum at about elevation 750. The downstream slope also bulged slightly in the same general manner, but the maximum horizontal movement was less than 6 in.

During construction in the latter part of June, six closed-system piezometers were installed. As shown in Fig. 3.3:6, two were placed in each of three transverse embankment sections, one on the longitudinal dam axis and one 75 ft. upstream. The first readings were obtained July 15, before the slope bulging was noticed.

Typical readings plotted in terms of percent of overburden pressure are given in Fig. 3.3:7. The piezometer at the axis (3-*A*) recorded high initial pore pressure (more than 80% of the total weight of the overburden), while that registered by the upstream piezometer (3-*B*) was initially very low. From the first reading until the end of construction, the pressure at 3-*A* decreased and that at 3-*B* increased until at the completion of the embankment both piezometers recorded approximately the same pressure. After the embankment was finished in October, piezometers at both locations indicated a gradual pressure dissipation due to drainage (about 1 to 2% of the overburden pressure per month).

In September of 1957, when the rate of bulging appeared to be accelerating, construction was suspended a week for a thorough reappraisal of the slope stability. Stability analyses using the measured pore pressures and the "consolidated-drained" shear strength (from slow direct shear tests) indicated that the upstream slope was unstable; however, analyses using the "quick" shear strength (from unconsolidated, undrained triaxial compression tests) on specimens cut from the fill indicated that the slope was safe. From these studies it was concluded that the embankment was stable in spite of the bulging, and construction was resumed. Successful completion of the dam testifies to the validity of the conclusions.[1]

These construction measurements on Otter Brook Dam lead to some interesting, if rather speculative, conclusions. The embankment material was very impervious and the rate of pore pressure dissipation

[1] See Sec. 7.5 for discussion of the differences in safety factors computed using effective and total stress analyses.

following the end of construction was low. It seems reasonable to assume that the pore pressures were not influenced appreciably by drainage during construction and that changes in the pressures as measured in the piezometers must have been due primarily to changes in the embankment pore volume caused by changes in the total stresses.

It also seems probable that the pore pressure decrease measured in piezometer 3-*A* between July and October was due to the gradual development of large shear strains in the central portion of the embankment. This is true since the pore pressure at all piezometers (in percent of overburden weight) would have increased gradually with the added weight of the overlying embankment if there had been no large shear stresses imposed, and since the shear strain of a dense, granular embankment creates a tendency for expansion of the grain structure (increase in the void ratio) with consequent reduction in the pore water pressures.

As the construction progressed and the measured slope bulging increased, undoubtedly an increasingly larger portion of the embankment developed high shear strains. This changing area of the cross-section of the dam which had been subjected to high shear strains must be responsible, at least in part, for the differences in the pore pressures measured in piezometers 3-*A* and 3-*B*.[1] It is likely that the embankment zone surrounding 3-*B* had already been subjected to high shear strains before the first readings were taken, so that the densely compacted soil was already dilated. It would be logical in such circumstances that much lower initial pore pressures would be measured at 3-*B* than at 3-*A*. As the dam grew higher, the zone subjected to the shear strains must have extended inward and begun to affect the pressure measured at 3-*A*, until by the end of construction the sections of embankment surrounding both piezometers were included in the zone of high strain and had approximately the same pore pressures.

The triaxial compression tests on compacted specimens of the embankment material showed one somewhat unusual result in that they commonly developed 20% axial strain without reaching a peak stress. The designers speculated that this property of the compacted material had some influence on the performance of the dam.

The bulge which existed on the upstream face at the end of construction could not be easily detected from a visual inspection. Except for the existence of the bridge pier, it is possible that the bulging would have escaped notice. This conjecture leads to the speculation that similar movements may have occurred unobserved on other dams.

[1] Since the toe of the dam did not move, it is apparent that the shear strains were confined to the embankment and that there was no foundation movement.

INFLUENCE OF PERIODS OF WORK STOPPAGE AND DRAINAGE ON PORE PRESSURE REDEVELOPMENT

When fairly high pore pressures develop within the core of a dam during the early part of construction and then are later dissipated to some degree by drainage during a period of work stoppage, they subsequently increase at a rate which is appreciably smaller when construction commences again. Hence, delays in construction may have a doubly beneficial influence on the stability of earth dams during construction.

An example is given in Fig. 3.3:8, which shows pore pressure readings in the central part of the core of the Serre-Ponçon Dam during construction. As seen in this plot, the pore pressure increased more or less linearly in the latter part of 1958 as the embankment rose above the level of the piezometer. It reached a maximum pressure head which was about 10 meters above the surface of the fill by the time work stopped at the end of November. During the winter work stoppage the pressure dropped at a uniform rate due to drainage a maximum of

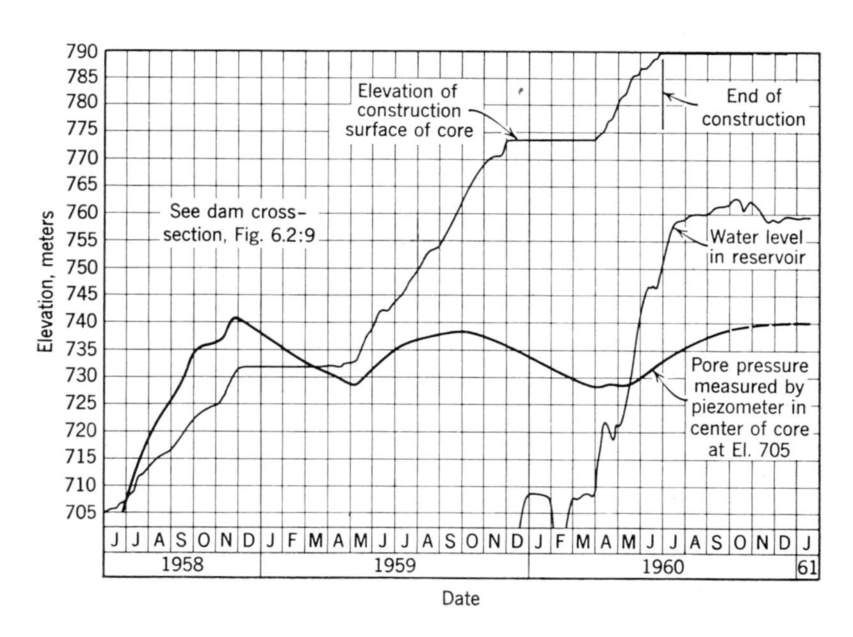

Fig. 3.3:8 Construction pore pressures at Serre-Ponçon Dam (after Guelton et al., Ref. 615).

about 15 meters of water (about 3000 psf). When work started again in May 1959, the pore pressure again commenced to increase under the added load, but the rate of increase was much lower than in the previous season. Finally, in the fall of 1959 the rate of pore pressure increase due to added embankment load was less than the influence of drainage and the pore pressure began to drop again even though fill was still being added. (The later increase in the pore pressure starting in May 1960 was undoubtedly because of the fact that the reservoir was being filled).

The reason for the decrease in the rate of pore pressure development after periods of work stoppage is that the embankment becomes denser and less compressible as the result of the consolidation which occurs. When work starts again, the soil structure is stiffer than before, and a smaller percentage of the added weight of the new embankment is thrown to the pore fluid. Some other examples and a theoretical treatment of the subject are given by Bishop in Ref. 108 (also see Ref. 339).

3.3c Summary

1. The water content at which the embankment is constructed has the largest influence on the magnitude of the pore pressures which develop. A high dam constructed of almost any type of soil with an average water content near Standard Proctor Optimum water content will develop construction pore water pressures.
2. The magnitude of construction pore pressures also depends on the embankment soil. Embankments of well-graded, clayey sands and sand-gravel-clay mixtures have the highest construction pore pressures, and those of uniform silts and fine, silty sands are the least susceptible to the development of pore pressure. Also, while less information is available to support the conclusion, there is evidence that embankments of moderate height of very fine-grained clayey soils, compacted near optimum water content, may not develop appreciable construction pore pressures.
3. Study of the maximum construction pore pressures which have occurred in USBR earth dams has shown that the measured values check reasonably well with values computed on the basis of the one-dimensional compression theory for predicting construction pore pressures (Sec. 7.5). Pore pressures which have

developed in embankments with the highest contents of clayey fines, however, appeared to be lower than predicted values.

4. Contours of equal measured construction pore pressures usually have an oval configuration generally conforming to the shape of the impervious core. For thick cores of very impervious material, the pore pressures at the center usually are not reduced appreciably by drainage during the time of typical construction periods. At the outside edges of the impervious core, construction pore pressures are always reduced by drainage of pore water to the adjacent more pervious zones.
5. The relationship between the coefficient of permeability of the impervious core and the development and dissipation of construction pore pressures is not well known. On the basis of the limited information available, the authors' rough estimate of the relationship between the coefficient of permeability of relatively thick impervious cores and the dissipation of pore pressures in the central part of the core is as follows:

COEFFICIENT OF PERMEABILITY	DISSIPATION OF PORE PRESSURES DURING CONSTRUCTION
Less than 0.5 ft./yr. (0.5×10^{-6} cm./sec.)	No dissipation of pore pressure
0.5 to 5 ft./yr. (0.5 to 5.0×10^{-6} cm./sec.)	Some dissipation
5 to 50 ft./yr (5 to 50×10^{-6} cm./sec.)	Appreciable dissipation
More than 50 ft./yr. (50×10^{-6} cm./sec.)	Complete dissipation

6. For dams with thin central cores (width at any level equal to 40 percent or less of the water height), appreciable dissipation of construction pore pressure always occurs in the central part of the core. While pressure equal to 80 percent or more of weight of the overlying material may develop initially if the core is compacted wet, they rapidly reduce and seldom exceed 40 percent at any point one year after the material has been placed.
7. Any appreciable dissipation of pore pressures due to drainage will decrease the rate at which construction pore pressures subsequently develop under additional load. For thin core dams where drainage gradients are high, this phenomenon can have an important influence on the maximum construction pore pressures which develop.

3.4 PORE PRESSURE MEASUREMENTS AND OTHER SEEPAGE OBSERVATIONS DURING RESERVOIR OPERATION

3.4a Full Reservoir Condition

There are a great number of measurements available of "full reservoir" pore pressures. Review of these indicates that the magnitudes and distributions of the pressures vary widely among dams of different characteristics. In dams with relatively narrow cores or cores constructed from materials which are not excessively impervious (coefficient of permeability greater than 50 ft./yr.), the measured pore water distribution through the core closely approximates the theoretical distribution of pressures for gravity flow (Chap. 5). However, measurements in dams with thick cores constructed of very fine-grained impervious materials (coefficient of permeability less than 0.1 ft./yr.) frequently show pore pressures quite different from those which would be predicted on the basis of the gravity flow theory. These differences result primarily from the influences of desiccation, capillarity, and rainfall on the seepage pattern.

For a very impervious embankment in which a particle of water flows through the core at the rate of inches per year under the gravity head, the influence of water seeping into the crest and slopes of the dam during the rainy season, and of water being sucked out of the dam during the dry season by evaporation, vastly outweighs the influence of gravity on the pore pressure distribution. In embankments of very fine soils capillary forces are much greater than the forces due to gravity, and as a result almost any distribution of pore water pressures can be obtained.

THE UPPER SEEPAGE SURFACE

In earth dams with coarse-grained or relatively pervious cores, well-defined upper seepage surfaces develop. Interesting observations of the phreatic surface were made at the Costilla Dam in New Mexico, a homogeneous embankment approximately 125 ft. in height constructed of a well-graded, gravelly, clayey sand with a probable coefficient of permeability between 10 and 100 ft./yr. A test pit was excavated in the downstream slope at approximately mid-height to inspect the seepage condition. In the walls of the pit a well-defined phreatic surface was exposed. "Above this surface, the walls of the shaft, while moist, did not indicate active seepage; below it, free water could be readily seen as it appeared on the walls, and dripped into the shaft" (Ref. 95). The line was so distinct that it was possible to make measurements of its slope to the nearest hundredth of a foot.

Such seepage surfaces have also been observed intersecting the downstream slopes of homogeneous dams of coarse soils. The downstream slope is dry above the line of intersection and wet below, and on a dry day the demarcation frequently can be seen from a distance because of the color difference between wet and dry sections.

It can be concluded that for dams with cores of coarse soil and a coefficient of permeability greater than 10 to 100 ft./yr., a well-defined upper seepage surface will develop if the reservoir is kept full for a sufficient length of time. For rolled embankments of fine-grained clay or silt, however, there is no distinct surface. In the walls of test pits sunk in the cores of dams constructed of very fine soils, the embankment materials have the same appearance from crest to foundation even if the reservoirs have been filled for long periods.

AIR IN PORES OF COMPACTED EMBANKMENT

It has long been a matter of speculation among earth dam engineers whether a core ever becomes saturated, i.e., whether the water seeping through the core from the reservoir to the downstream portion carries the air bubbles and dissolved air with it. Few reliable measurements on this subject are available. It is probable that in embankments constructed of fine-grained soils there is little change in the air content with time.

At Belle Fourche Dam (Ref. 95), a homogeneous dam constructed in 1909 of inorganic clay with medium to high plasticity, undisturbed samples of the embankment clay taken near the upstream slope after the reservoir had been in operation for 25 years still retained an air content equal to 8% of the total volume, which is probably about the same as it had when constructed. Since this is an irrigation reservoir in which the water level is lowered each year, the embankment is not subjected to continuous inflow of seepage. At the other extreme, more pervious embankments retaining reservoirs which are not emptied regularly probably have all or nearly all of the air eventually washed out of the pores, though little quantitative information is available on the subject.

In the absence of better information, the authors believe it is reasonable to assume that the initial air contained in embankments with coefficients of permeability less than 0.1 ft./yr. will not be influenced by seeping water, and that embankments with coefficients of permeability of more than 100 ft./yr. will probably have all of the initial air washed out of the voids within a fairly short period.

EMBANKMENT ANISOTROPY

Since a compacted earth dam embankment is placed and rolled in horizontal layers, it is stratified to some degree regardless of the care taken in the construction control. Because of the stratification, it becomes more pervious in the horizontal direction than in the vertical. The ratio of the coefficients of permeability in the two directions (anisotropy) has an influence on the pore pressure distribution which will develop, and it is necessary to estimate this ratio in order to predict the seepage pattern by means of theory (Chap. 5).

Little reliable quantitative information is available which will permit a realistic estimate of the permeability ratio in advance of construction. The degree of anisotropy depends on a number of factors, the influence of which cannot be evaluated quantitatively, such as the borrow pit stratification, the method of excavation in the borrow pit, the tendency of the construction material to segregate while being handled, and the method of placing and compacting the embankment material.

The permeability ratio can only be predicted from observations on similar dams. The degree of anisotropy in a dam which has piezometers installed can be estimated roughly by comparing the measured pore pressure distribution with the theoretical distribution drawn for various ratios of horizontal to vertical permeability (Chap. 5). The few such studies which have been made indicate that the ratio between horizontal and vertical permeability usually does not exceed 4 for embankments constructed of uniform deposits of fine-grained soils. The coarser the soil and the more erratic the deposit, the higher the probable anisotropy. At the Guntersville Dam in Northern Alabama, for example, a study showed that the ratio of horizontal to vertical permeability must have been about 50 for the embankment to have produced the pore pressure distribution which developed (Ref. 319).

It is reasonable to assume that the permeability ratio will not exceed 4 in dams built with good moisture-density control in such a way that the embankment appears visually to be uniform and homogeneous. In circumstances where the soils in the borrow area are coarse and variable and are not blended well by the excavation equipment, it is inevitable that layers of varying permeability will be placed in the dam and that a higher ratio of horizontal to vertical permeability will develop. In such circumstances the ratio can only be estimated from the appearance of the embankment stratification as it is exposed in test pits during construction. In Sec. 7.6, suggested assumptions for use in stability analyses are given. There is little doubt that embankments with less anisotropy are obtained when compacted with a sheepsfoot roller than with a rubber-tired roller because of the blending action of the sheepsfeet (Sec. 11.3*c*).

3.4b Pore Pressures Following Reservoir Drawdown

From the standpoint of the stability of the upstream slope, the most critical period is during a "rapid" drawdown of the water level after the reservoir has been full for some time (Sec. 2.4). Much analytical effort goes into the theoretical prediction of the drawdown pore pressures (Sec. 7.7), but few field measurements exist to compare with the results of the theoretical computations. What measurements have been made are not adequate in scope or number to allow reliable generalizations concerning the relationship between the drawdown pore pressures and the embankment soil type or construction method.

The measurements taken during the first 5 years of the operation of central core earth-rock Kajakai Dam in Afghanistan (Ref. 321) are typical. As seen in Fig. 3.4:1, the pressures recorded by all the piezometers dropped with the reservoir level, showing little or no time lag. This relationship between the water level and the piezometer readings is characteristic of a majority of the measurements compiled, even from piezometers located near the center or at the far side of a thick impervious core.[1] No good measurements of drawdown pore pressures are available on dams with impervious sections consisting of very fine-grained, clayey or silty soils.

Even though the field measurements which exist are not adequate to support any reliable generalizations about the magnitude of drawdown pore pressures, some facts are known:

1. In most of the dams where measurements were made, the pore pressure distribution roughly approximated the pressures which would be predicted from a simple gravity flow net drawn using

[1] See, for example, the correlation between measured drawdown pressures and reservoir levels as summarized for five Tennessee Valley Authority Dams (Ref. 319).

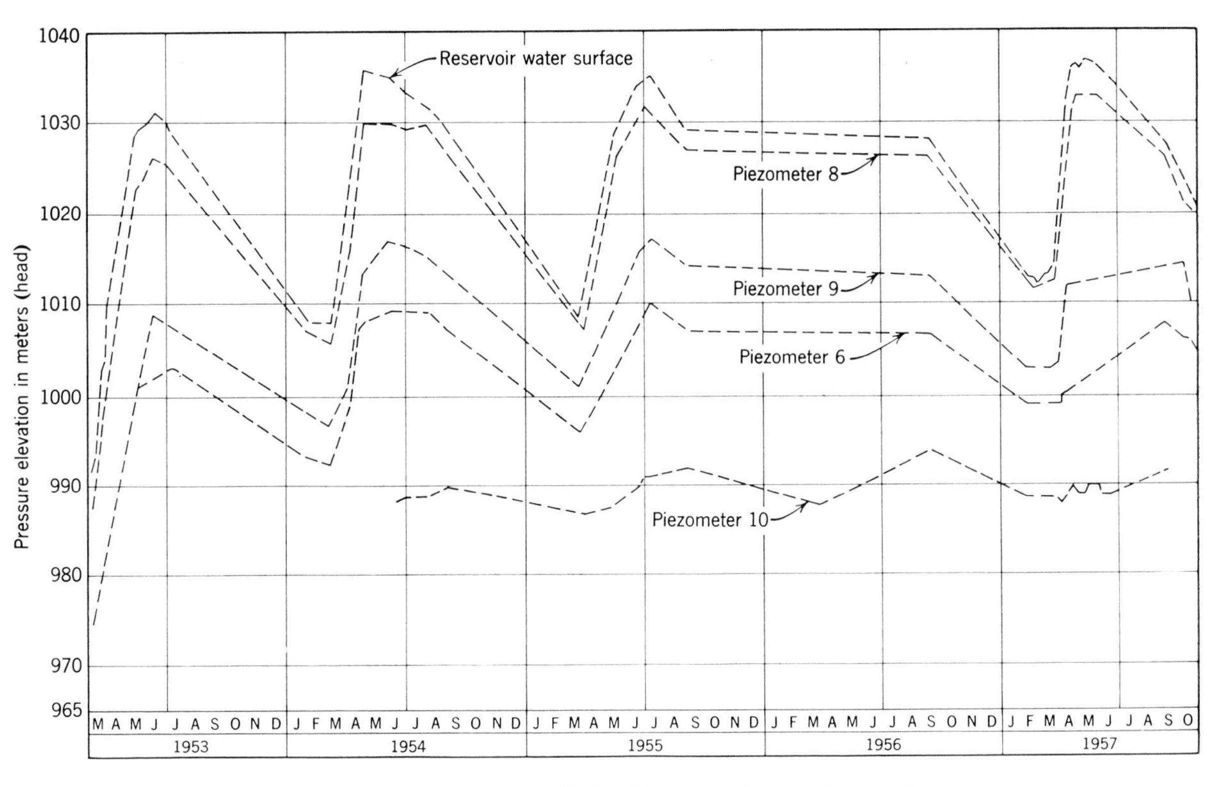

Pore pressure measurements during first years of reservoir operation

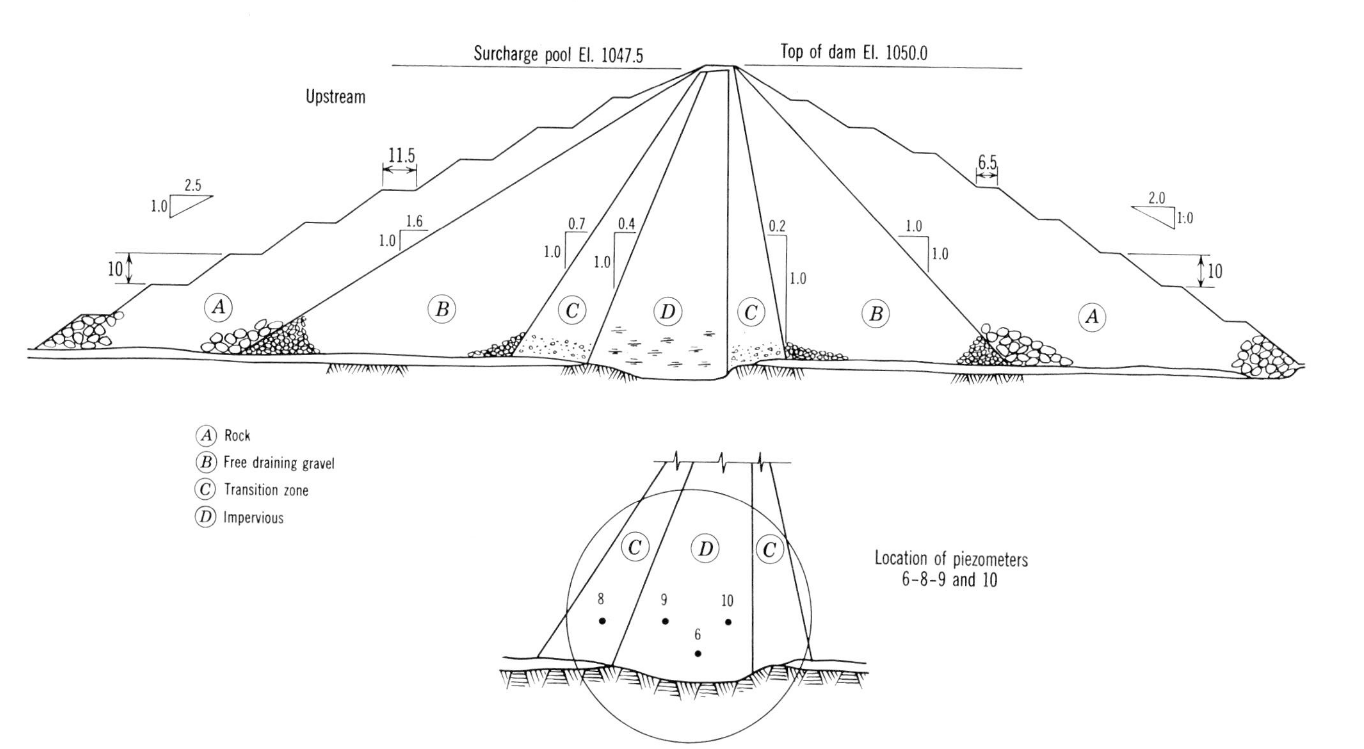

Fig. 3.4:1 Cross section of Kajakai Dam showing the locations of piezometers and measurements of pore pressures following reservoir drawdown (after Sudman, Ref. 321).

the drawdown boundary conditions and assuming the embankment is incompressible and no capillarity exists.[1]

2. In one set of careful measurements on a small dam in Scotland, the actual pore pressures were considerably higher than the values predicted on the basis of gravity flow theory (Ref. 540).
3. It has been speculated that in the case of incompressible embankments of fine soil with high capillary rise, the drawdown pressures might be negative and, hence, might increase the stability of the slope during reservoir lowering. In no case, to the authors' knowledge, have negative drawdown pore pressures been reported to confirm this hypothesis.

[1] See the summary of drawdown pore pressures measured in Alcova Dam following a reservoir drawdown of 126 ft. in 49 days (Ref. 317).

four

Explorations for Foundations and Embankment Construction Materials[1]

4.1 INFLUENCE OF TOPOGRAPHY AND SUBSOIL CONDITIONS ON SITE SELECTION

After the reservoir site has been selected, which is a process outside the scope of this book, several alignments or specific locations for the dam are often possible and must be considered. The selection among these is made by preparing preliminary designs for each and weighing the various construction problems and costs. The main factors are:

1. Embankment volume and construction difficulties.
2. Embankment construction materials available.
3. Location and details of spillway and other appurtenant structures.
4. Foundation treatment.

The best site will be the one at which the project has the lowest cost and can be constructed within the time available. The basis for rough but reliable cost estimates for the dam and appurtenant structures must be provided by preliminary surficial and subsoil explorations at all the possible dam locations.

The same site generally will not have ideal topographic and subsoil conditions for the embankment, the foundation and the spillway. The element of the structure which is estimated to have the largest cost will probably have the greatest influence on the design. For example, if the spillway constitutes the major portion of the project's expense, the best site will probably be the one best adapted to the spillway con-

[1] In this chapter, no attempt is made to cover exhaustively the subject of subsurface exploration since it is a major topic by itself and the principles used for dams are, in large part, common with those used for other civil engineering projects. The treatment here is brief and is intended only to pinpoint some of the major special problems involved in explorations for earth dams.

struction even though the embankment at this location has to be much larger than at others.

The cheapest site for the concrete dam is usually in the narrowest section of the valley, where the concrete yardage can be kept at a minimum. For an earth dam, however, the embankment volume is not always the most important item governing cost, although earth dam sites are too often selected as if it were. Economy in an earth dam embankment depends to an important degree on using large, rapid moving equipment at its maximum efficiency, and this means that it must be possible to build good access roads at convenient locations. With steep abutments, relatively expensive hand labor is needed for much of the stripping, whereas bulldozers can do most of this work on gradual slopes. Thus a large embankment at a site which has gentle sloping abutments and easy access may cost less than a smaller embankment in a narrow confined portion of the canyon with precipitous abutments.

Special problems arise at a site at which the geologic conditions are such that the hillsides forming the abutments and reservoir basin are just barely in equilibrium before a dam is built or reservoir filled. At some such sites any little excavation during construction may trigger large landslides of a progressive nature, and the problems of coping with these slides during construction may be of sufficient economic importance to dominate not only the selection of the dam location in the valley but the design of the dam itself. In recent years, on a few dams where this contingency was not adequately anticipated, the quantity of earth moved to clean up the slides was appreciable with respect to the volume of the dam. In a few cases landslides into the reservoir caused simply by the buoyancy associated with raising the water level have appreciably reduced the water storage capacity and have even created barriers in the form of natural dams across narrow sections of the reservoir. Abutment slides which result in blocking the openings of the outlet conduits have created the most severe problems at some dams.

4.2 FOUNDATION SUBSURFACE EXPLORATION

The purpose of the exploration is to determine the dimensions, location, and physical characteristics (strength, permeability, and compressibility) of the various earth and rock strata which comprise the dam foundation and abutments. The extent and cost of the foundation exploration required for a dam of a given size varies greatly from site to site depending on the subsurface conditions and cannot be estimated reliably in advance.[1]

As a general rule, the more erratic and the poorer the qualities of the foundation material and the greater the probable cost of foundation construction, the more thorough the exploration should be. Its extent will also depend to some degree on the evolving design of the dam. For example, if the designer knows beforehand that very little water loss from underseepage can be tolerated, he may decide on the basis of the first few borings that it is necessary to provide a rolled earth cutoff through the overburden soil down to bedrock. A detailed study of permeability in the natural soil layers overlying bedrock will not then be necessary. But if some underseepage is tolerable, and the cost of constructing a complete seepage barrier is high, it may be advisable to spend a large amount of effort in field exploration to obtain a reliable estimate of the effective permeability of the various subsoil strata.

"Off-channel" reservoirs, which are constructed in dry valleys away from the main source of water and then filled by pipe lines or canals, pose special problems. Dams retaining these reservoirs so often have developed foundation trouble that their subsurface conditions should always be explored with great care. Since no stream flows through the foundation of an off-channel dam, there is often no ground water table to indicate whether the valley will hold water. The statistical frequency of excessive leakage and settlement is so high for off-channel dams, that their foundations should be always considered suspect and explored with greater care than for similar dams in normal stream valleys.

Just as the extent of the subsurface study will depend on individual conditions, so the details of the procedures must be adjusted to the particular site. Determining these, and in fact conducting the exploration as a whole, is one of the earth dam engineer's most important and demanding functions.

The exploration should proceed in stages, the details of each growing out of the one before. The first stage will consist of a general field inspection of the topography and outcrops, and a review of all published information on subsurface conditions in the region.[2] Aerial

[1] For this reason, it is in the best interest of the engineer and the client to have the cost of borings, field and laboratory testing, and geological and soil engineering consultants paid for separately outside the main engineering fee for design and preparation of drawings and specifications.

[2] Geological maps and aerial photographs can be obtained for a large part of the continental United States, Alaska, and the possessions from the U.S. Geological Survey, Washington 25, D.C.

photographs are especially valuable, because by showing details not readily apparent on the ground they facilitate interpretation of the geology and surface soils. They can also be used as maps for orientation, and for plotting observations made in the field.

This preliminary stage leads to a rough guess about the depths and characteristics of the soil strata. Next begin the various steps of field exploration, which should be under the continuous supervision of the engineer so that he can interpret the results and modify the procedures as the work progresses.

Geophysical (seismic) surveys, and to a lesser degree, electric resistance surveys, have been used to assist in the evaluation of the foundation of many dams. The reliability of such surveys has been found to be quite variable and depends to a large extent on the skill of the operators and their familiarity with the local conditions. In some cases, these surveys have been used effectively to provide an economical and reliable means to fill in the details of the subsurface conditions between the exploratory borings. In others, the results of these surveys have proved to be consistently misleading.

During all phases of the explorations, the services of an experienced engineering geologist are valuable, and they are imperative where the subsurface conditions are erratic or complex. Knowledge of the geology not only helps in locating the exploratory holes where they can give the most useful information but also serves as a guide for projecting the data between and beyond the holes.

It is very often the case that the water loss from a reservoir will not be determined by the permeability of the average material comprising the dam foundation and abutments but by the permeability of the most pervious zones—the cracks, the faults, and the lenses of very coarse gravel. These leakage paths often comprise the minor details of the general geology of the site and their existence can best be anticipated by a competent geologist.

The first exploratory borings or pits are put down at wide spacings, the engineer selecting equipment and methods which are best adapted to the anticipated subsoil conditions.[1] Additional field investigations

[1] Test pits should be used wherever feasible in order to allow a visual inspection of the subsoil profile. At some dam sites it has proved beneficial, even though the expense may be great, to install pits below the water level for visual inspection of the details of the stratification, to obtain large undisturbed samples, and to perform "in-place" tests. Freezing (Refs. 57,583), electro-osmotic stabilization (Ref. 173), and compressed air caissons (Refs. 91 and 92) also have been used to put down pits in soil below the water table. However, recent improvements in techniques for obtaining large-diameter samples from bore holes, especially in loose sands, have reduced the number of situations where these expensive measures appear justified.

are based on the results from these holes. During all stages of field exploration, tests are performed on samples sent to the laboratory. The field explorations are continued until the accumulated information is sufficient for the design.

As discussed in more detail in Sec. 11.1, the process of foundation exploration should be considered to continue during the construction of the dam. In all cases additional information about the characteristics of the foundation is obtained during construction.

The thickness of surface soil in the dam foundation which it may be desirable to excavate and waste or reuse in the dam can have an important influence on the cost of the structure and consequently deserves careful consideration. The quantities of stripping are relatively more important for long, low dams than for high dams. Usually, organic top soil containing fibrous roots is automatically excavated, and sometimes layers of loose or soft soils are taken out. Since every cubic yard of material taken out of the foundation must be replaced with a cubic yard of dam embankment, this portion of the dam is more costly than any other, and consequently the engineer must study very carefully the reasons for excavating the material and the amounts to be excavated. Often the depth of desirable stripping will vary erratically over the foundation, and consequently the engineer is justified in exploring the depth of stripping carefully. In cases where the foundation is covered with closely spaced large trees, each of which leaves a crater in the ground when removed, it is common that the depth of stripping is greater than originally anticipated; there is always a tendency to excavate the ridges between the holes rather than to fill them with compacted soil. For all these reasons it can be taken as an axiom that the cost of the most careful field exploration with frequent shallow test pits which might be proposed by the most conservative engineer will be small compared to the value of results in minimizing over-excavation.

MEASURING FOUNDATION STRENGTH

Rock Foundations. Generally rock foundations are stronger than either the natural soils overlying the bedrock or the embankment itself. Consequently, in the large majority of cases where bedrock is close to the surface it is not necessary to consider that a failure surface could pass through the rock or to try to determine the strength of the rock.

One main exception to this is the situation where the foundation consists of relatively soft sedimentary rock classified broadly as shale and sometimes called claystone, siltstone, mudstone, and occasionally, marl. The properties of materials falling into these general classifications range over very wide limits. Depending on their basic constituents and geologic history, the range in density of the materials may vary between 60 and 155 lb./ft.^{3} and the water content may vary from a minimum of 2% to 40% or more. Some are cemented with various cementing agents of different degrees of permanence and some are in the form of rock only because of the high pressures to which they have been subjected in the formation ("compaction" shales). While the great majority have easily adequate strength to support an embankment dam, there have been a sufficient number of failures of dams and other similar structures in recent years to indicate clearly that there are exceptions in which the over-all shearing resistance may be equal to that of a soft clay.

Recognition of these suspect materials, and determination of their strength is an almost impossibly difficult task and at the present state of knowledge on the subject must be put outside the scope of a book such as this. The authors can only call attention to the fact that shale foundations must always be given the most careful and conservative consideration—even for relatively low dams.

Some of the main recognizable characteristics which may reduce the strength of a rock foundation to a dangerously low value are:

1. Continuous seams of clay or other weak material. Perhaps having a thickness of only a small fraction of an inch.
2. Closely spaced crack systems often associated with severe twisting movements and faults.
3. Basic constituents consisting of highly plastic and very fine clay with no sand sizes so that the individual rock fragments have a greasy surface texture and very little frictional resistance when sliding in respect to each other.
4. Horizontal bedding is likely to be more dangerous for two main reasons: (*a*) Weak seams are more likely to be continuous, and (*b*) it is more likely that high pore water pressures will be transmitted horizontally.

Laboratory tests of individual rock specimens are generally of no practical value except for tests on specimens of the clay seams themselves which may provide a conservative lower limit of the strength of the formation. In developed areas of the world where some history of construction is available, experiences with the typical shale under heavy structures and in excavations may give the best clue as to whether the material has a low over-all strength. Since shale forma-

Fig. 4.2:2 Installation used to test the shearing resistance of the surface of shale bedrock at the Djatiluhur Dam, Indonesia.

tions generally extend over a large geographic area, valuable experience of this type may be available at a considerable distance from the dam site. Natural slopes of hill sides and particularly the river valley walls can also be used in some areas as a guide to the strength of the material. (For discussion of this technique, see Ref. 536.)

In some cases, large-scale field shear tests may be a reasonable approach to the problem.[1] At the Djatiluhur Dam in Indonesia (Refs. 614, 218), a high rockfill dam founded on compaction shale, the shearing resistance of the interface between the shale and the rockfill embankment was tested by 15 field shear tests using details shown in Figs. 4.2:1 and 4.2:2. The tests were performed under different loads and at different lengths of time, some with the shale surface protected by asphalt.

Sedimentary rock has generally been precompressed sometime during its geologic history under loads which are greater than those imposed by a dam embankment, and consequently no appreciable com-

[1] For example, see the descriptions of the elaborate field shear tests performed on chalk deposits with bentonite seams under major dams of the Army Engineers in the Missouri Valley Basin (Refs. 229, 234). Other descriptions of field shear tests on rock formations are described in Refs. 92, 367, 610, 611, 612, 613, 659, 663, and 684. Also see Ref. 600 for one of the few interesting experiments to study the shearing resistance of shale sliding on shale in the laboratory.

pression of the shale itself will usually occur during construction. However, if the formation is badly fissured, considerable settlement can be expected because of the closing of the crack system under the weight of the dam. If the cracks are fine and filled with water, pressures in the water in the cracks can develop with a magnitude equal to an appreciable portion of the weight of the dam. Since the pore pressures which develop in this fashion are related to the compressibility and permeability of the foundation mass, as determined by the crack system, there is no way that the magnitudes of the pore pressure can be estimated in advance from laboratory tests. Under the worst condition, these high water pressures can spread through fine horizontal cracks and exist also under the more lightly loaded toes of the dam causing an unstable condition. Consequently, measurements of the water pressures in the rock formation during construction of the dam should be always considered for suspect shale formations.

In any rock in which the existence of continuous thin seams of clay or other weak material is suspected, core borings alone are not adequate for exploration. Test pits and shafts must be used in order to allow a minute visual inspection of the various strata. Also, subsequent investigations of some recent failures indicates that clay seams may not be seen even in the walls of a test pit. Since the walls of the pit are necessarily rough and somewhat weathered, it is not easy to see or evaluate properly a continuous clay seam of 1/4 in. in thickness or less. The best available method of exploration is the use of calyx borings which give a large diameter hole with smooth walls in which the clay seams are easier to see plus a large core which can be pulled apart for a close examination of the nature of the clay seam. While calyx holes are quite expensive, in some cases they may offer the only hope of reliable evaluation of the nature of weak seams and therefore are indispensible.

Soil Foundations. When the soil overburden consists of natural strata of clean, pervious sand and gravel, the shear strength can be approximated by means of a rough statistical relationship between the strength and the relative densities and grain size characteristics, Table 4.2:1, or determined from laboratory tests on samples recompacted to densities simulating the field conditions.[1] Strength in fine-grained, impervious soils is estimated from laboratory tests on undisturbed samples taken from bore holes or test pits, or from vane shear tests (Sec. 8.4). The reliability of laboratory tests on natural fine-grained soils, and of the stability analysis itself, frequently depends to a large extent on the

[1] See Sec. 8.3 for methods of measuring the relative density and a review of current knowledge about the "liquefaction" potential of loose sand strata.

Table 4.2:1 Approximate Statistical Relationship between Relative Density and Angle of Internal Friction for Pervious Sands and Gravels

FINENESS AND GRADATION	DENSITY		
	Dense R.D.[a] > 70%	*Medium* 70% > R.D. > 50%	*Loose* R.D < 50%
Fine to coarse sand, uniformly graded (i.e., single-sized)	34–38°	32–34°	28–30°
Coarse sand, well graded from coarse to fine and poorly graded sand-gravel mixtures	37–45°	33–36°	30–33°
Well-graded sand-gravel mixtures	40–45°	36–41°	33–36°

[a] R.D. = relative density (see Sec. 11.4*a*).

method of sampling and the degree of disturbance of the samples. The softer and more sensitive the material, the more effort and cost are justified in securing undisturbed samples of good quality. If the soil is clay with slickensided fracture planes or has a jointed structure, the laboratory analysis should be supplemented by a careful field examination in the test pits or trenches in order to investigate the nature of the cracks.

MEASURING FOUNDATION PERMEABILITY

Layers of silt and clay in a dam foundation are so impervious that the underseepage through them is usually negligible. The effective over-all permeability is determined primarily by the continuity, regularity, thickness, and grain size characteristics of interbedded layers or lenses of coarser materials. The influence of the layers of coarse material on the effective average permeability of natural deposits of fine-grained soil can be studied only in field tests, and laboratory permeability tests on undisturbed specimens of silts and clays are seldom performed since they are of little assistance.

Where the subsoils consist of relatively uniform layers of sand or gravel, on the other hand, it is possible to obtain a rough indication of the foundation permeability from laboratory tests. Although undisturbed samples are preferable, remolded ones can be used for relatively clean materials which are sampled in such a way that the fines are not lost. At best, however, the tests give only a general clue to the effective permeability. There have been few opportunities to check estimates of seepage quantities based on laboratory results against measured quantities of underground flow, but in most of these the actual flow has been less than the estimated value.

For a reliable appraisal of the permeability of sand and gravel deposits, as well as of silt and clay deposits with interbedded coarser soils, field permeability tests must be made. Field tests are expensive and can be misleading when poorly executed or improperly interpreted. In order to obtain the maximum possible information from field permeability tests, their details should be planned only after the foundation exploration is essentially completed, and the final picture of the subsoil stratification is available for study.

References 458 and 518 describe procedures for field permeability tests and methods of interpreting the results. Above the ground water table field tests are performed by pumping water into bore holes (pump-in tests) or by measuring the downward seepage under the bottoms of test pits. Below the water table, ground water is pumped from a central well, and the drawdown is measured in observation wells located at various distances from the pump well. Well pumping tests below the ground water level are more reliable than the pump-in tests made above it.

When properly interpreted, well pumping tests can furnish information on the vertical permeability of the more impervious horizontal strata as well as on the average coefficient of permeability in the horizontal direction. The horizontal permeability is studied in the main pervious layer, where the quantity and pressure gradient of the radial horizontal flow toward the well are measured.[1] The shape of the drawdown curve at a distance from the well reflects the vertical permeability of the less pervious horizontal layers which bound the main pervious acquifer. In addition to open observation wells, piezometers which measure pore pressures at various elevations within the zone of flow toward the pump well can help substantially in interpretation of the seepage pattern.[2]

[1] By using flow meters and other special devices within the well, it is possible to estimate the quantity of ground water entering through the aquifer at different elevations (Ref. 226).

[2] In Germany, attempts have been made to measure the velocity of seepage by tracing radioactive isotopes (Refs. 80, 97).

At important projects in which the permeability of the sand and gravel foundation has a strong influence on the design, field tests from bore holes and wells have been supplemented by pumping tests from large, machine-excavated pits. For example, at the Wanapum Dam on the Columbia River, where the foundation had more than 80 ft. of "open-work" gravels (Fig. 6.2:12), field pumping tests in wells showed that the coefficient of permeability ranged between 1,000,000 and 5,000,000 ft./yr. Subsequently, on each side of the river a large pit was excavated with a dragline to a maximum depth of about 30 ft., Fig.

Fig. 4.2:3 Aerial view of excavations for field permeability tests at site for Wanapum Dam on the Columbia River. (Courtesy Harza Engineering Co.)

Fig. 4.2:4 Closeup of excavation for large-scale pumping (permeability) test, Wanapum Dam. (Courtesy Harza Engineering Co.)

4.2:3. Pumps with a total capacity of 10,000 gpm lowered the water level in the pits by approximately 5 ft., Fig. 4.2:4.

If the foundation soil is erratically deposited, a condition frequent in glaciated regions, it may be hopeless to try and form a reliable estimate of the potential underseepage. Sometimes it is not even possible to guess if or where the seepage will discharge in the form of springs downstream. Under these circumstances unless a complete seepage cutoff is provided, the designer must plan for the worst conceivable conditions, and must be prepared to control all leaks which develop when the reservoir is filled for the first time (Sec. 6.4).

In evaluating rock abutments or foundations, drill holes are sunk to a distance greater than the maximum probable depth of a grout curtain. As discussed in Sec. 10.2, pressure tests in the holes, together with studies of the crack and joint system as observed visually in exploratory adits,[1] allow an estimate of the permeability and the necessity for grouting. In the case of faults explorations must be very thorough so that the designer will have a sound basis for decision on how to treat them.

Cavernous limestone foundations and reservoir bottoms constitute one of the most difficult problems, both from the standpoint of exploration and treatment. At a number of dams in limestone areas the

[1] Cracks observed in rock walls of adits and shafts excavated by rock tunnelling methods may be exaggerated by both stress relief and the shocks caused by blasting.

leakage has been so great that it has been finally impossible to fill the reservoir which then had to be abandoned (Ref. 634, pp. 333 and 581). In other cases, very expensive grouting has been required to reduce the leakage to tolerable quantities (Refs. 580, 639). Much more extensive explorations are required for limestone formations than for other types of rock. Large diameter calyx holes which permit visual inspection of the nature of the rock in-situ may be the only practical means of evaluation of the foundation. Many of the large dams of the Tennessee Valley Authority were constructed on limestone formations which presented formidable problems of foundation treatment. The excellent publications describing the explorations and the foundation treatment at the TVA projects should be studied by all engineers planning a dam on cavernous limestone (Refs. 578, 579, 580, 583, 645, 646, 647).

In the last few years, bore hole cameras have become available for studying the crack systems and the permeability of rock foundations under dams. These allow almost as good a view of the width, frequency and direction of the cracks as can be obtained by visual inspection of the walls of a large diameter shaft. Consequently, they will undoubtedly be used widely in the future.[1] Two types are available: (1) a television camera which is lowered gradually into the hole and which rotates at each elevation to give a view of all faces of the walls (Ref. 680); and (2) a camera which takes a photograph of the whole 360° perspective at one time (Refs. 241, 452). Both have certain advantages.

4.3 STUDIES OF EMBANKMENT CONSTRUCTION MATERIAL

FIELD EXPLORATION

Almost all soils can be used successfully to construct rolled earth dams.[2] Properties of embankments of different soil types vary consider-

[1] In parts of the Ruhr Valley in Germany where the basic rock is a thinly bedded, badly fractured schist, and where it is impossible to get any good idea of the nature of the crack system from core borings, the television camera is used almost exclusively for dam foundation studies to the extent that most exploratory holes are drilled with plug bits without taking cores.

[2] Theoretically, earth dam embankments can be designed in such a way that soils of any type can be used. Practically, highly organic, fibrous soils (peat) are not chosen because of low shear strength and high compressibility. Because of the construction difficulties, inorganic clays of very high plasticity are not desirable except

ably, however, and the design and cost of an earth dam depends to a large extent on the materials near the site.

The exploration for prospective embankment materials begins at the same time as the foundation investigation and follows the same pattern. The possible borrow areas are selected first through field reconnaissance and from studies of aerial photographs and geologic maps. The exploration proceeds in stages, moving outward from the dam site, until much more than enough borrow soil has been found.

The following factors must be studied in considering each borrow area:

1. *The nature and thickness of the organic topsoil* which may have to be wasted. A difference of a few inches in the stripping thickness required for large borrow areas results in an appreciable difference in total cost and the problem justifies considerable study. If the borrow pit is excavated with a power shovel on a vertical face of 10 ft. or more, a few inches of topsoil mixed into the material has a negligible influence on the final properties of the compacted embankment. On the other hand, when scrapers are used for excavation much less blending action can be obtained in the borrow pit to dilute the mixture and good practice requires that a larger depth of topsoil be wasted. The dark color of the surface soil should not be the only criterion used to determine the thickness of the stripping. The soil is often discolored to a considerably greater depth than that to which the properties are appreciably influenced by the organic content.
2. *The relative elevation of the borrow area* with respect to the dam and to the grades on which haul roads would have to be constructed.
3. *The natural water content,* its probable seasonal fluctuation, and the need for draining or irrigating of the pit.
4. *The quantity of oversized cobbles* which would have to be removed from the soil before compaction.

---

where no other materials are available, or in climatic regions where water content control is not excessively difficult. Research is badly needed to increase our knowledge concerning the relative suitability of various soil types, particularly with respect to piping resistance, cracking potential, and strength. The need for this increased knowledge is reflected by the fact that there is very little in the literature to assist in the differentiation of embankment construction materials on the basis of their important properties and most of this is of a very general nature. The bulk of the important contributions to this subject is included among the following: Refs. 37, 95, 279, 432, 474, 629, 636.

5. *The influence of the excavation on the appearance or* operation of the reservoir. When borrow pits are opened within the reservoir basin, for example, they sometimes expose underlying pervious formations which increase the leakage. These excavations have caused trouble, especially at small dams without complete foundation seepage barriers, and at dams in valleys underlaid with broken or porous rock formations. Borrow pits in the reservoir basin should not be considered at all unless they are advantageous economically, and they should never be used close to the dam or near thin abutments or natural saddles. Excavations for borrow in flat areas downstream from the dam can also be troublesome if they stay full of water and cannot be drained. The ponds created may interfere with the construction of irrigation canals or other facilities. At dams on pervious soils, with no complete foundation seepage barrier, ponds at the downstream toe may hinder inspection of foundation leaks erupting as springs (Sec. 6.4).

Since undisturbed soil samples are not required, borrow pit explorations are usually made either with auger holes, or by means of pits and trenches dug by bulldozers or back hoes.[1] The recently developed large-diameter power augers, which can drill holes with 32- to 42-in. diameters (or larger) to depths of 50 ft. in a few minutes, are very economical and effective except where the material contains boulders.

Careful evaluation of the water content of fine-grained soils is especially important in the situation where the natural water content is above the value at which the material will be placed in the dam. In such a case the designer must carry out enough studies to allow a realistic evaluation of the time that will be required to lower the water content and of the cost of doing it. For some types of fine-grained soils it is impossible to lower the water content appreciably by gravity drainage alone. Such a soil can be dried only by evaporation of the exposed excavation surface, and this is a relatively slow process (Sec. 11.3f). A number of very difficult problems have arisen during construction of major dams in recent years when it became apparent that it was essentially impossible to reduce the water content of the soil at a rate that would have been required to meet the construction schedule. Although often the ineptitude of the contractor may make the situation worse, the engineer cannot escape all responsibility for such difficulties.

[1] Soils from "required excavations" for cutoff trenches, spillways, and other structures are explored and catalogued in the same manner as borrow soils.

The density of the natural soil in the borrow pits is usually not of great importance to the designer; it is of considerable practical importance to the contractor. It is common practice to make only a few tests to obtain a rough estimate of the "shrinkage" between the volume of a given weight of soil in the borrow pit and its volume when compacted in the dam. The value commonly ranges between 5 and 20% but can fall outside these limits for very dense and loose natural soils. Under some circumstances, it has been necessary to determine the shrinkage accurately through the use of a sufficient number of tests to give a reliable statistical average. This situation arises most frequently when there is a shortage of soil located within an economical haul distance from the site and additional material must be obtained from a considerable distance.

When samples are taken from sand and gravel deposits below the water table, the fines must be carefully preserved. A small difference in the quantity of fines has a large influence on the permeability of an embankment made from such deposits, so that a loss in the sample of even a few percent of silt sizes may be critical. At many dams, where gravel has been sampled below the water table with dragline buckets, for example, lost fines have resulted in misleading estimates of permeability and has necessitated major changes in design during construction (Sec. 1.4*b*).[1]

TEST EMBANKMENTS

For dams of small or moderate size and especially of the more common soil types, there is usually no reason to build a test embankment during the design stage. For small dams the cost is usually not justified since the maximum possible saving in the construction cost is not great and for ordinary types of soils we already know, in general at least, how they will behave.

For large dams, on the other hand, a small increase in our understanding of the best method of placing the material and consequent improvement in the specifications can have a large total influence on the cost of the structure; therefore, for large dams and for certain special purposes, the construction of test embankments during the design stage can sometimes be of considerable value.

[1] Also see Sec. 11.4*b* for a description of unexpected construction difficulties which have arisen as the result of inadequate evaluation of the fine content of sand and gravel materials excavated below the water table.

Such test embankments can, of course, be made in the early part of construction; however, if there is some major problem to be studied which may have an appreciable influence on the construction method specified, these tests should be made during the design stage under a separate contract supervised by the engineer.

Some of the main problems which can be profitably studied with test embankments are:

1. The best method and difficulty of excavation and placing the materials with respect to water content, mixing and blending, problems of oversized rocks, segregation, scarification of layers, and compaction.
2. The shrinkage problem.
3. The shear strength and the permeability of the material in the test fill as contrasted to materials compacted in the laboratory.

In certain cases, test fills may be absolutely necessary in order to determine if certain materials can be used to construct embankment sections as planned. Some examples are: (1) to determine if a coarse, well-graded soil with only a small percentage of fines can be used to construct an impervious core without excessive segregation; (2) to determine if a proposed material can be placed during rainy weather; (3) to determine the best way to mix two different materials; (4) to determine if the water content of a wet material can be practically lowered during construction. In most cases, however, the lessons learned are less vital but can nevertheless have a large influence on the cost of the structure. In a few circumstances valuable lessons may be learned which are not anticipated in advance. Tests on rockfill to be placed in layers are always of interest since we know much less about the compaction of rock than of soil.

In recent years large quantities of weathered and decomposed rock have been used in many major dams (Sec. 11.5d). In some cases these materials have been excavated with rippers and scrapers and then have been broken down on the construction surface with rollers into the form of an impervious soil. In others it has been necessary to resort to a crushing plant to break down the harder fragments. Making the decision in the design stage regarding the handling of a given material is a difficult problem at best and can be studied only by the use of test embankments. In the common situation in which the rock in the proposed borrow pit becomes less weathered and sounder with depth, deep excavations with necessarily large volume may be required in the exploratory stage to be sure that the material used in the test embankment is representative of the average material that would come out of the proposed excavation.

SOIL CLASSIFICATION

In order to catalogue the soils available for construction of the embankment, and to select representative samples for testing, all samples from prospective borrow pits should be carefully inspected by the embankment designers. A detailed written description should be made of the pertinent properties of all samples and they should be classified according to the Unified Soil Classification System.[1] This system has been adopted as a standard by the U.S. Corps of Engineers and the USBR, and is used by most consulting engineers. The widespread adoption of the method makes possible a basic communication which was not possible only a few years ago. Because of this and its undoubted superiority to other methods of classification for the purpose, it is incumbent on all earth dam engineers to use the system, and to develop a minimum standard of personal proficiency in working with it.

The great advantage of the Unified Soil Classification System is that it permits the experienced engineer to classify soils rapidly in the field or laboratory by looking at them and handling them. One way to develop this technique is to practice it under a skilled soil engineer; another is to compare the appearance and feel of a wide range of soils with the laboratory tests performed on them. A number of excellent publications are available which provide the beginner with detailed instructions for learning the system (Refs. 274, 382).

As shown in Table 4.3:1, the system divides soils into the two main classes—coarse-grained and fine-grained. Coarse-grained soils fall into eight simple subdivisions, based on coarseness, gradation, and percentage and plasticity of fines; fine-grained soils are separated into seven groups, all of them based on plasticity except for peat and organic soils, which are identified by odor, color and spongy feel. Laboratory tests are only made to give the engineer a check on his visual judgment of the soil properties or to define borderline cases. Table 4.3:1 shows the numerical criteria used with these tests for exact classification of soil samples, and Table 4.3:2 illustrates in graphical form

[1] The method was originally developed by A. Casagrande, primarily for use with airfield pavements during World War II (Ref. 381). In a slightly modified form which improved its applicability for earth dam construction, it was renamed the Unified Soil Classification System.

Table 4.3:3 Approximate Correlation Between Embankment Properties and Soil Classification Groups

	EMBANKMENT PROPERTIES[1]				
GROUP SYMBOL	RELATIVE PERMEABILITY	PROBABLE RANGE OF K (FT./YR.)	RELATIVE PIPING RESISTANCE[2]	RELATIVE SHEAR STRENGTH	RELATIVE WORKABILITY (EASE OF MOISTURE-DENSITY CONTROL)
GW	Pervious	1,000 to 100,000	High	Very high	Very good
GP	Pervious to very pervious	5,000 to 10,000,000	High to Medium	High	Very good
GM	Semipervious	0.1 to 100	High to medium	High	Very good
GC	Impervious	0.01 to 10	Very high	High	Very good
SW	Pervious	500 to 50,000	High to medium	Very high	Very good
SP	Pervious to semipervious	50 to 500,000	Low to very low	High	Good to fair
SM	Semipervious to impervious	0.1 to 500	Medium to low	High	Good to fair
SC	Impervious	0.01 to 50	High	High to medium	Good to fair
ML	Impervious	0.01 to 50	Low to very low	Medium to low	Fair to very poor
CL	Impervious	0.01 to 1.0	High	Medium	Good to fair
OL	Impervious	0.01 to 10	Medium	Low	Fair to poor
MH	Very impervious	0.001 to 0.1	Medium to high	Low	Poor to very poor
CH	Very impervious	0.0001 to 0.01	Very high	Low to medium	Very poor

[1] When placed as well-constructed rolled-earth embankment with moisture-density control.

[2] See Sec. 2.2.

the procedures which should be followed. This graph also can be a convenient guide when classifying visually.

EMBANKMENT PROPERTIES AND SOIL CLASSIFICATION GROUPS

Because the Unified Soil Classification System arranges soils by their principal engineering properties, a rough correlation exists between each classification group and the properties of embankments from the corresponding soils. The relative permeability, piping resistance, shear strength, and workability of embankments of the system's principal soil categories are summarized in Table 4.3:3. Properties vary substantially within the categories, of course, and such a table is no substitute for the detailed laboratory tests. But in the early stages of planning, when different borrow areas and design sections are being studied, these generalized predictions of embankment properties can be taken as useful qualitative guides.

five

Theoretical Seepage Analysis

The amount of water seeping through and under an earth dam, together with the distribution of the water pressure, can be estimated by using the theory of flow through porous media.

This theory is one of the most valuable analytical tools available to the engineer.

The computed amount of seepage is useful in estimating the loss of water from the reservoir. The estimated distribution of pressure in the pore water is used primarily in the analysis of stability against shear failure (Chap. 7) and also occasionally to study the hydraulic gradient at the point of seepage discharge which gives a rough idea of the piping potential.

5.1 LIMITATIONS OF THE THEORY

The reliability and accuracy of the results of the calculations depend on the degree to which the seepage conditions in the dam resemble those postulated by the theory. When evaluating the results of theoretical seepage analyses, the designer should keep in mind the theory's limiting basic assumptions that:

1. The rolled embankment and natural soil foundation are incompressible seepage media; the size of the pore spaces through which the water seeps remains the same, regardless of the water pressure.
2. The seeping water flows under a hydraulic gradient which is due only to gravity head loss.
3. There is no change in the degree of saturation in the zone of soil through which the water seeps; the quantity flowing into any element of volume is equal to the quantity which flows out in the same length of time.
4. The hydraulic boundary conditions can be found.

These assumptions are best satisfied by seepage through an embankment of sand or gravel where the quantity of seepage is relatively large, the voids have no air in them, and the influence of capillary action is small.

In a dam constructed of fine-grained soil, on the other hand, the capillary forces commonly have more influence than the gravity head on the pore water pressure, and the actual seepage pattern will be vastly different from the computed one (Sec. 3.4). No theory which includes the effect of capillary action on fine-grained soils is available for practical use by the earth dam designer, and therefore seepage analyses for embankments of fine-grained silts or clays can serve only as rough guides to judgment.

5.2 HISTORICAL DEVELOPMENT OF SEEPAGE THEORY

The basis for rational seepage analysis was developed about 100 years ago by Darcy (Ref. 494), whose experiments led to the conclusion (Darcy's law) that the velocity with which water seeps under a pressure gradient through the void spaces of a fine, porous medium, such as a soil, is directly related to the first power of the hydraulic gradient[1] (Ref. 494). Subsequently in the 1880's Forchheimer demonstrated that the distribution of water pressure and velocity within a seepage medium was governed by the Laplace differential equation (Ref. 210).

Early in the 1900's a powerful graphical method was developed independently by Forchheimer in Germany and Richardson in England to obtain approximate solutions of the Laplace equation. This method was not widely used for earth dams until after Casagrande published his comprehensive paper on the subject in 1937 (Ref. 211). Since then the solution of the Laplace equation by the graphical procedure or with electric models has become standard procedure for seepage analysis.

5.3 DARCY'S LAW AND THE LAPLACE EQUATION

Darcy's law for the seepage of water through soil can be written:

$$V = ki = k\frac{dh}{dl} \qquad \text{(Eq. 5.3:1)}$$

[1] Many experiments have demonstrated that Darcy's law is valid for the range of soils which are used for the embankment or foundation of an earth dam. See Ref. 495, Chap. 2, for a comprehensive treatment of the hydraulic implications and the range of validity of Darcy's law.

or

$$Q = kiA \qquad \text{(Eq. 5.3:2)}$$

where V = discharge velocity of seeping water
k = coefficient of permeability
i = hydraulic gradient
h = the pressure head
l = length of the seepage path
A = cross-sectional area of the soil through which the water is seeping
Q = quantity of seepage per unit of time

The general hydrodynamic equation for steady-state seepage (Laplace equation) can be written:[1]

$$\frac{\partial^2 h}{\partial x^2} + \frac{\partial^2 h}{\partial y^2} + \frac{\partial^2 h}{\partial z^2} = 0 \qquad \text{(Eq. 5.3:3)}$$

For the analysis of seepage through and under earth dams, flow is considered to be two-dimensional and the equation is used in the form:

$$\frac{\partial^2 h}{\partial x^2} + \frac{\partial^2 h}{\partial y^2} = 0 \qquad \text{(Eq. 5.3:4)}$$

The solution of Eq. 5.3:4 gives the variation of the pressure head and direction of flow at all points within the area of soil through which the water is seeping. The results are commonly plotted in the form of contours of equal head (potential), called "equipotential lines." These are illustrated in Fig. 5.3:1, which represents the solution of the Laplace equation for the simple case of seepage through a homogeneous soil formation under a line of partially penetrating steel sheet piling. The pile wall is assumed to be impervious and the ground surface horizontal.

As is shown by the drawing, the pressure head in the water when it enters the soil on the left side of the sheet piling (along line *AB*) is equal to h_1. As the water discharges from the soil on the right side of the pile wall (along line *BC*), its pressure head is h_2. Therefore in seeping under the wall, the water has undergone a pressure head loss equal to $h_1 - h_2$. The horizontal lines *AB* and *BC* are equipotential

[1] The applicability of the Laplace equation to the analysis of seepage through soil can easily be demonstrated by combining Darcy's law with the assumption that the quantity of water which flows into any element of volume of the soil must equal the quantity which flows out in the same length of time; see Ref. 211 or Ref. 212, p. 235.

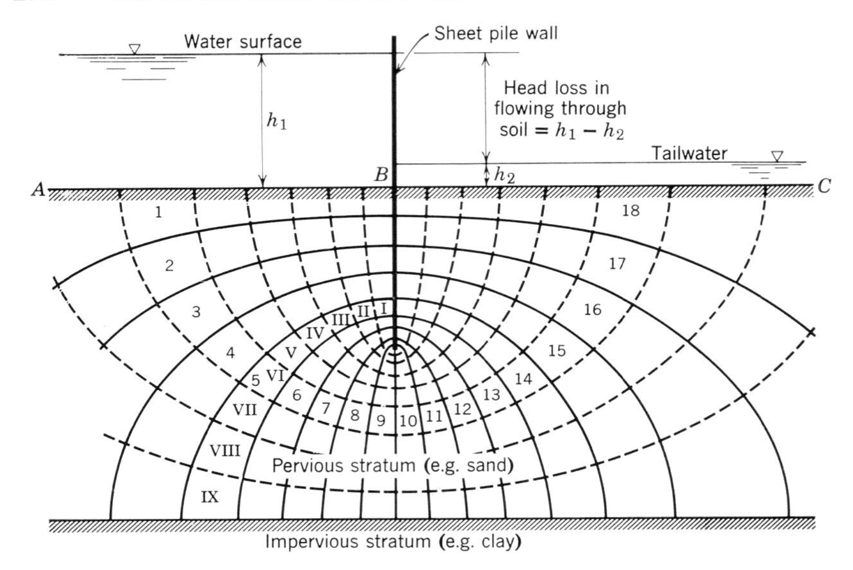

Fig. 5.3:1 Simple flow net for seepage under a single line sheet pile wall (after Casagrande, Ref. 211).

lines marking the upstream and downstream boundaries of flow through the soil. The curved solid lines are intermediate contours of equal head (equipotential lines) within the zone of seepage.

The direction of seepage is always perpendicular to the equipotential lines. The paths along which the individual particles of water seep through the soil (dotted lines in Fig. 5.3:1) are called "flow lines."[1] The horizontal impervious stratum at the bottom is a flow line which limits the lower boundary of seepage. The other limiting flow line is the boundary between the soil and the sheet pile wall.

For every seepage problem there is a unique solution of the Laplace equation which can be plotted as equipotential and flow lines (in combination, the "seepage pattern") as in Fig. 5.3:1. Though the Laplace equation is relatively simple, the boundary conditions of a typical earth dam are too intricate to obtain a mathematical solution.[2] In all but the most complex situations, however, solutions of reasonable ac-

[1] The family of flow lines also satisfies the Laplace equation in which velocity potential is substituted for pressure head.

[2] Reference 624 gives an excellent recent summary of exact mathematical solutions which have been obtained for various practical situations.

curacy can be found by the graphical method or by using electric models.

5.4 THE GRAPHICAL METHOD

Based on Forchheimer and Richardson's discoveries that a special solution of the Laplace equation can be obtained in which all the spaces formed by the intersecting equipotential and flow lines are approximately equidimensional (approximately squares), the graphical method is the easiest and most widely used technique for solving the Laplace equation. The solution is found by freehand trial-and-error sketches. The flow lines are chosen in such a way that the quantities of seepage between each pair (the flow channels) are equal, and the equipotential lines so that the difference in pressure head is the same between each pair. Thus if two sets of orthogonal curves can be drawn satisfying the boundary conditions and forming squares, they constitute the solution.

The solution produced by the graphical method is called a "flow net." Construction of flow nets requires some manual skill, and the earth dam designer should study their characteristics and practice drawing them until he develops at least moderate competence. For most dams, solutions accurate enough for design purposes can be obtained by a competent operator in less than 30 minutes, although dams with complex zoning or stratified foundations will require more time.[1] Since the specific techniques for constructing flow nets are explained in great detail in Casagrande's comprehensive paper on the subject (Ref. 211), and the procedures are explained in most of the recent textbooks on soil mechanics, they are not covered here.

After the flow net is drawn the rate of seepage can be computed from the geometry of the seepage pattern with the equation:

$$Q = kh \frac{N_f}{N_d} \qquad \text{(Eq. 5.4:1)}$$

where k = coefficient of permeability (ft./yr.)
h = total pressure head loss (ft.)
N_f = number of flow channels
N_d = number of equal potential drops
Q = rate of seepage (ft.^{3}/yr.)

[1] See Sec. 6.4 for the analysis and discussion of methods to control seepage under dams on erratically stratified foundations.

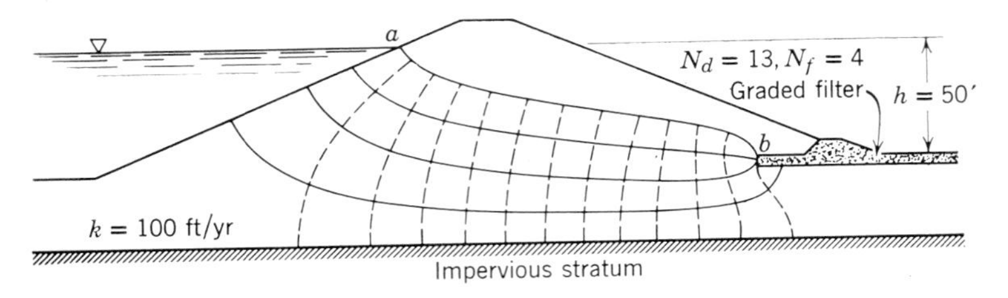

Fig. 5.4:1 Cross section of a dam showing a simple flow net through a homogeneous embankment and foundation (after Terzaghi, Ref. 212).

Fig. 5.4:1 shows a flow net solution of seepage through a homogeneous earth dam and natural soil foundation which has the same coefficient of permeability as the dam embankment. The total quantity of seepage is computed as follows:

$$Q = kh \frac{N_f}{N_d} = (100)(50)\left(\frac{4}{13}\right) = 1{,}540 \text{ ft.}^3/\text{yr./ft. of dam} \quad \text{(Eq. 5.4:2)}$$

Flow nets can also be drawn for seepage through portions of zoned dams or foundations with different coefficients of permeability. More effort and skill are required, however, in direct proportion to the number of zones and the differences in the permeability between the zones.

At boundaries between zones of material with different permeability, both the flow lines and the equipotential lines are deflected (Fig. 5.4:2). More area is required for the same quantity of flow when the

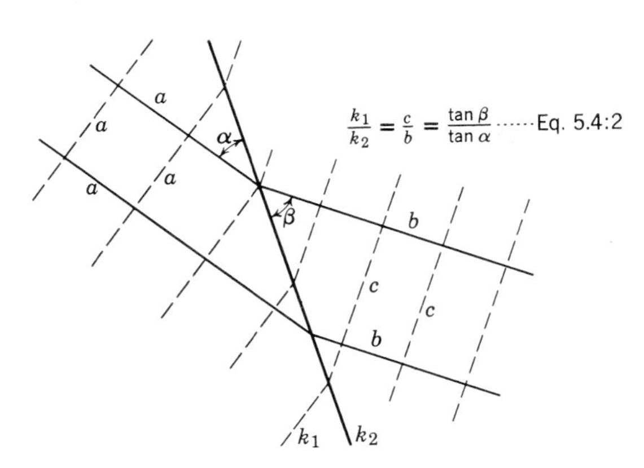

Fig. 5.4:2 Deflection of flow net at boundary of soils of different permeability (after Casagrande, Ref. 211).

seepage moves from a zone of higher permeability to one of lower, so that the flow lines in this case deflect to make the flow channels wider. In the opposite situation where the water seeps to a more pervious zone, the flow channels are narrowed.

Figures 5.4:3*a* and *b* show flow nets for the situation in which the dam and foundation have different permeabilities. Figure 5.4:3*c* is an example of a relatively complex flow net, which could be made only after the engineer has developed considerable proficiency in making these drawings.

The graphical method can also be used to draw flow nets for seepage through embankment zones or foundations in which the coefficient of permeability is not the same in all directions (anisotropic soils).[1] Here the equation governing seepage is:

$$k_x \frac{\partial^2 h}{\partial x^2} + k_y \frac{\partial^2 h}{\partial y^2} = 0 \qquad \text{(Eq. 5.4:3)}$$

The solution will no longer give families of equipotential and flow lines which always intersect at right angles, but the problem can be converted to an ordinary Laplacian solution by transforming the geometric scale. If the coefficient of permeability in the horizontal direction (k_{hor}) is the maximum value and the value in the vertical direction (k_{ver}) is the minimum, it is easy to show that by shrinking the horizontal dimension by the factor:

$$\sqrt{\frac{k_{ver}}{k_{hor}}}$$

the problem is reduced to an ordinary flow net (see proof in Ref. 213, p. 171).

After the solution is found on the transformed section, the scale is again expanded to the true section, Fig. 5.4:4 (Ref. 214). The quantity of seepage is then computed by using Eq. 5.4:1 and a weighted coefficient of permeability given by:

$$k = \sqrt{k_{ver} k_{hor}}$$

The greater the anisotropy assumed, the higher are the pore pressures in the downstream portion of the embankment, Fig. 5.4:5.[2]

[1] In almost all practical cases the coefficient of permeability is highest in the horizontal direction and lowest in the vertical. For natural foundation soils, the anisotropy is caused by the horizontal bedding of the deposit. Anisotropy in rolled-earth embankments results from the fact that the material is placed and compacted in horizontal layers (Sec. 3.4).

[2] See Sec. 7.6 for recommended assumption concerning the degree of embankment anisotropy.

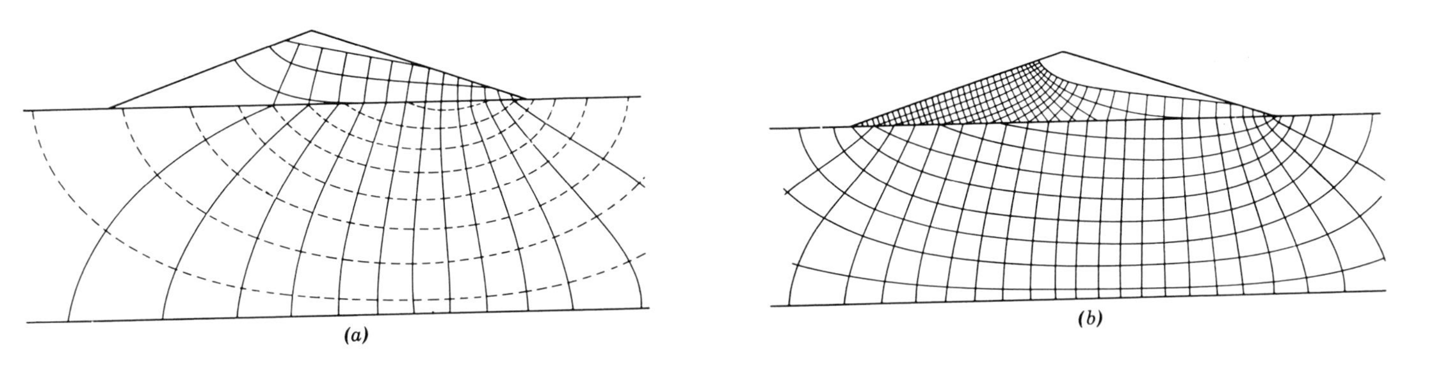

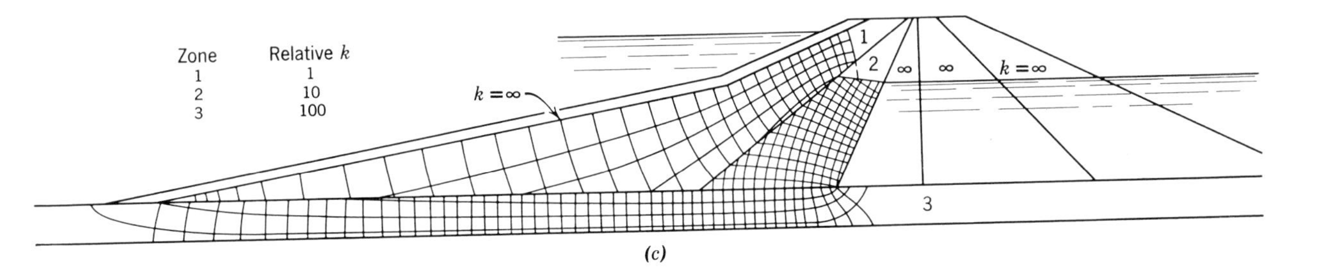

Fig. 5.4:3 Typical flow nets through dams on foundations of different permeability (after Cedergren, Ref. 214). (a) Permeability of embankment 10 times the permeability of foundation. (b) Permeability of embankment 0.1 times the permeability of foundation.

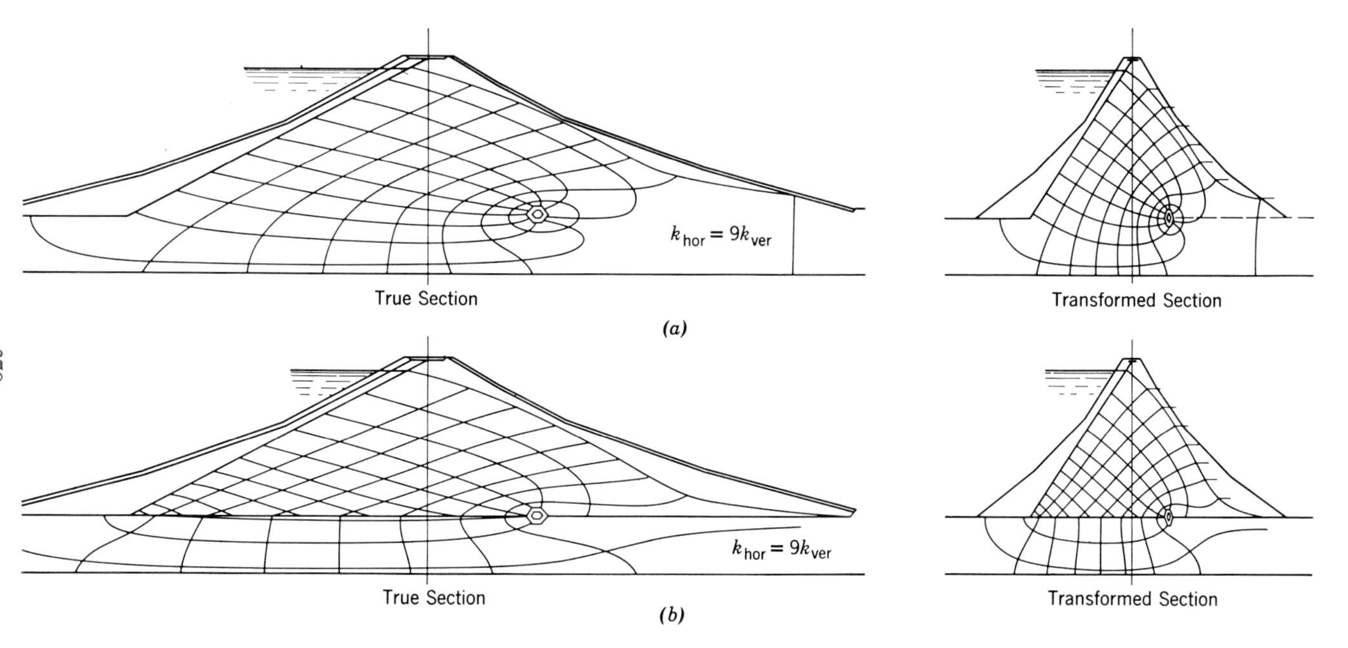

Fig. 5.4:4 *Typical flow nets through anisotropic embankment and foundations (after Cedergren, Ref. 214). (a) Foundation permeability equal to embankment permeability. (b) Foundation permeability 30 times the embankment permeability.*

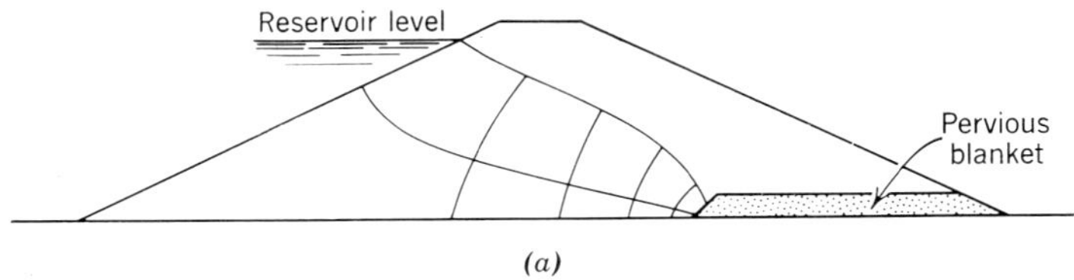

(a)

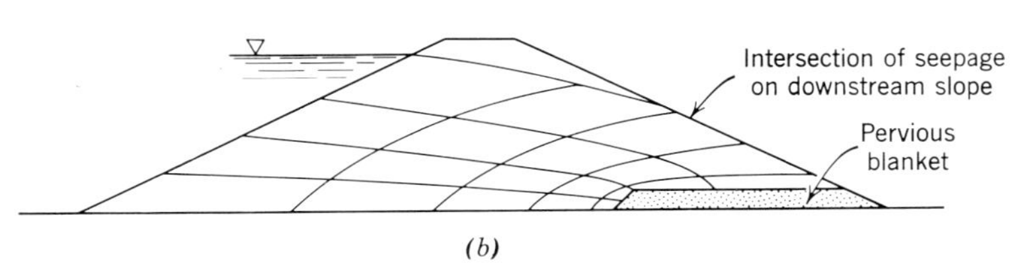

(b)

Fig. 5.4:5 Flow nets showing the influence of anisotropy on the pore pressures in the downstream portion of an homogeneous dam (after Cedergren, Ref. 214). (a) $k_{hor} = k_{ver}$. (b) $k_{hor} = 16k_{ver}$.

5.5 ELECTRICAL ANALOGY MODELS

When the Laplace equation is used to measure the flow of electricity through a conducting medium, the corresponding quantities are as follows:

DARCY'S LAW FOR WATER SEEPAGE	OHM'S LAW FOR ELECTRIC FLOW
$Q = \dfrac{KAh}{l}$	$I = \dfrac{CA'E}{l'}$
Q = quantity of seepage	I = rate of flow of electricity
K = coefficient of permeability	C = electric conductivity coefficient
A = cross-sectional area	A' = cross-sectional area
h = pressure head	E = electric potential (voltage)
l = length of seepage path	l' = length of path of electric current

Electrical analogy models have not been used frequently for earth dam studies because the graphical method is easier and does not require equipment. However, the electrical analogy can provide solutions to some complex problems which are difficult to solve graphically.

Solutions to seepage flow problems in two dimensions or three dimensions can be obtained with electrical models which have the same geometric shape as the soil through which the water flows. The seepage medium is replaced by an electric conductor and the boundary conditions are established by applying electric potentials at the source and exit of flow. The drop in potential across the model simulates the drop in water head in the prototype and is measured with a voltmeter. The model acts as two arms of a wheatstone bridge circuit and a calibrated slide wire potentiometer as the other two arms. To determine a line contour of equal potential, the potentiometer is adjusted to a percentage of the total voltage drop and the probe of a galvanometer is used to find the corresponding balance points on the model. Changes in the coefficient of permeability of soil zones in the seepage analogue are simulated by changes in the electric conductivity coefficient in the model.

No fund of experience is available to indicate the best types of models or testing techniques. A number of conducting materials have been tried successfully, including fluid electrolyte, solid metal sheets, sprayed graphite,[1] and a rectangular gridwork of wires. For the problem of seepage through zones of different permeability, the zones in the model must be constructed with different electric conductivity in the same ratios. Models with liquid electrolyte can be fitted with different depths of solution or different salt concentrations in the various zones (Refs. 59, 493, 497), and models with solid conductors can be modified by varying the thickness of the conductor or the type of material used. Recent developments in conductive rubbers and plastics lend promise of using these materials in three-dimensional models.[2]

Flow through anisotropic soils can be studied with electrical models by using the same transformed geometric sections as in the graphical method. But with models where a gridwork of closely spaced wires serves as the conductor, the anisotropy can be built by varying the wire spacings or weights in the different directions. This procedure has not been widely used, but because it can deal with anisotropy in an untransformed state, it shows promise of flexibility for difficult problems.

As in the graphical procedure, the upper seepage line of an earth dam (phreatic line) is determined by trial and error in electric

[1] Models with sprayed graphite as the conducting medium have been used successfully by the Missouri River Division of the U.S. Army Engineers for the study of seepage problems in which zones with ratios of coefficients of permeability up to 10,000 to 1 have been simulated (Ref. 558; see also Ref. 623).

[2] See Refs. 392, 635, 638 for some interesting applications of flow nets to study of seepage through dam abutments (three-dimensional problem).

models. The location of the upper seepage surface is assumed in the first trial and then moved as necessary in order to find the location where the measured electric potential equals the value corresponding to hydraulic gauge pressure of zero. The pressure head in the water at the phreatic line is equal to its elevation; i.e., gauge pressure equal to zero.

5.6. RELAXATION SOLUTIONS

While exact mathematical solutions of the Laplace equation cannot be obtained for problems of seepage through earth dams, approximate mathematical solutions are possible with the calculus of finite differences and the relaxation technique developed in England by R. V. Southwell and his associates about 20 years ago (Ref. 498). The procedures are laborious and time consuming, but they are simple and can be used to solve problems of any complexity.[1] The method has not been used for the practical analysis of seepage although it has been applied on an experimental basis to a number of specific problems of flow through soils (Refs. 499, 500, 501, 502, 650).

5.7 TRANSIENT FLOW PROBLEMS (NONSTEADY STATE SEEPAGE)

5.7a Time Lag in "steady-state" seepage

The preceding discussions of seepage problems were based on "steady-state" flow, which exists when the reservoir level has been at the same elevation (the maximum water elevation) long enough for the highest pore water pressures to develop everywhere in the zone of seepage. The pore pressures computed from the analysis of steady-state seepage are used to evaluate the stability of the dam's downstream portion during "full reservoir" (Sec. 7.6).

In reality, the reservoir behind the majority of earth dams is never continuously full, and there is a time lag before changes in the reservoir elevation are reflected by changes in the pore pressures of the embankment or foundation. The pressures indicated by the steady-state seepage analysis are maximum values which develop only in pervious materials or at sites where the reservoir level does not change. Since

[1] See Ref. 215, p. 271, for a brief description of the mechanics of the method when used for the solution of seepage problems.

the time lag, which depends primarily on the degree of saturation and permeability of the soil, cannot be calculated for any given dam, the practice is to neglect it and assume that the reservoir will remain full long enough for the steady-state pore pressures to develop.

5.7b Rate of change of "construction" pore pressures and "drawdown" pore pressures

The two main problems of nonsteady-state seepage which interest the designer are the rate of change in pore pressures during construction and the magnitude and rate of dissipation of residual pore pressures in the upstream portion of the embankment after reservoir drawdown. Neither problem is amenable to simple analytical treatment.

In forecasting the rate of dissipation of construction pore pressures it cannot be assumed that the seepage medium is incompressible, that the soil voids are completely filled with water, or that the Laplace equation is applicable. To estimate the rate of change in the design stage, it would be necessary to use the theory of consolidation, but no mathematical solutions of the theory exist for boundary conditions simulating the geometric form of an earth dam (Ref. 212, p. 327). In addition, it is not known to what extent the theory applies to pore pressure dissipation in a semisaturated rolled-earth-dam embankment, so that this approach has not often been used by designers.[1]

The prediction of "drawdown" pore pressures in the upstream zones of dams made with fine-grained soil is another of the major problems for which we have no reliable theoretical solutions (see Sec. 7.7). The most common method of estimating these pressures is to assume that the Laplace equation is applicable and that the reservoir level will be lowered rapidly enough to prevent substantial drainage inside the embankment. A flow net can then be constructed for the transitory seepage condition which would exist directly after "instantaneous drawdown." Assuming an incompressible embankment and no capillary forces, this is the most conservative prediction possible.

The hydraulic boundary on the upstream slope is considered to be a free surface, along which the pressure head is equal to its elevation at every point (that is, the gauge pressure is equal to zero). The flow net then is drawn in the same way as a steady-state flow net.

[1] See Ref. 86 for a brief description of interesting attempts by the engineers at the Swedish Power Board to solve this problem using the coefficient of consolidation obtained from laboratory tests and an electrical model consisting of an electric resistance network with 400 interconnected capacitors.

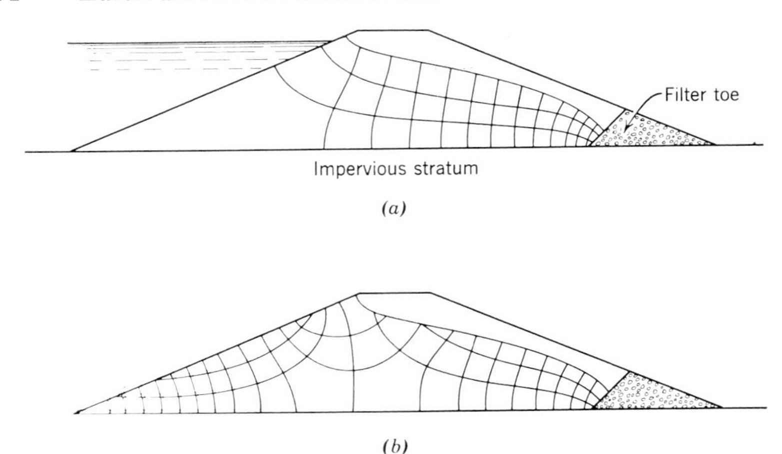

Fig. 5.7:1 Theoretical pore pressures existing in the upstream slope of homogeneous earth dam on an impervious foundation after instantaneous drawdown (after Terzaghi, Ref. 212). (a) Full reservoir state. (b) Drawdown state.

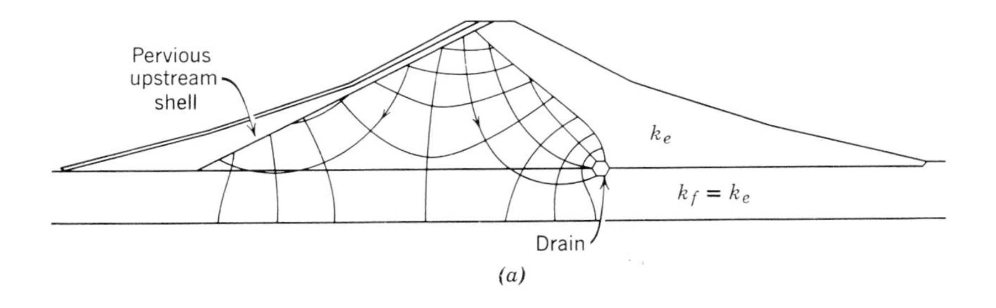

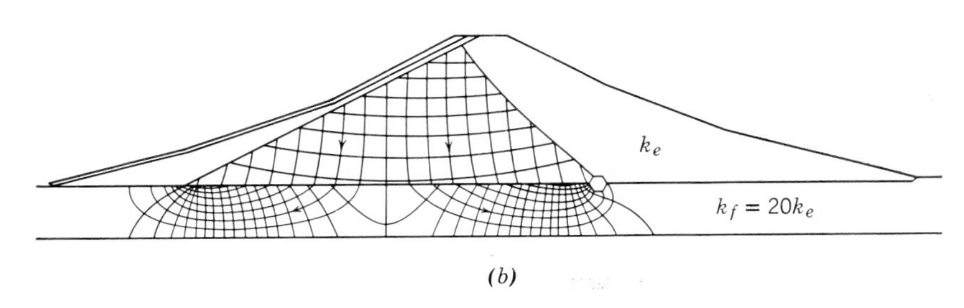

Fig. 5.7:2 Influence of foundation permeability on theoretical drawdown pore pressures (after Cedergren, Ref. 214). (a) Foundation permeability equal to embankment permeability. (b) Foundation permeability 20 times as great as embankment permeability.

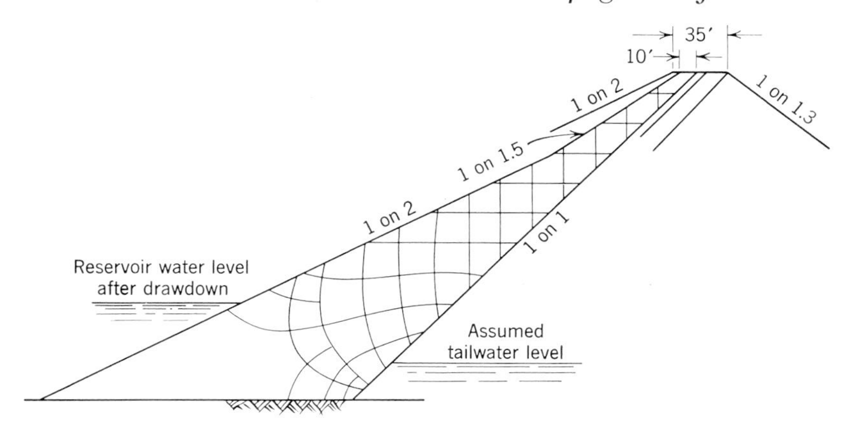

Fig. 5.7:3 Theoretical drawdown pore pressures in thin, sloping cores are essentially zero. (Courtesy D. M. Greer)

In the case of homogeneous dams on impervious foundations, however, the construction of flow nets for drawdown is commonly unnecessary. As Fig. 5.7:1 demonstrates, the equipotential lines within the upstream slope of such a dam are nearly vertical after rapid drawdown (i.e., the pressure is slightly less than hydrostatic), and it is the usual conservative practice to assume that they are actually vertical. In zoned dams or for homogeneous dams where the foundation is appreciably more pervious than the embankment, the pressures are much less than hydrostatic. The influence of a pervious foundation on the theoretical drawdown pore pressures in a typical dam is shown in Fig. 5.7:2.

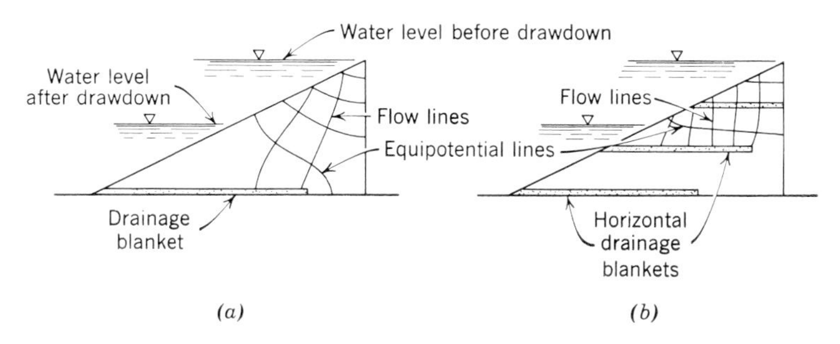

Fig. 5.7:4 Influence of single and multiple horizontal drainage blankets in the upstream slope on the theoretical drawdown pore pressures (after Reinius, Ref. 574).

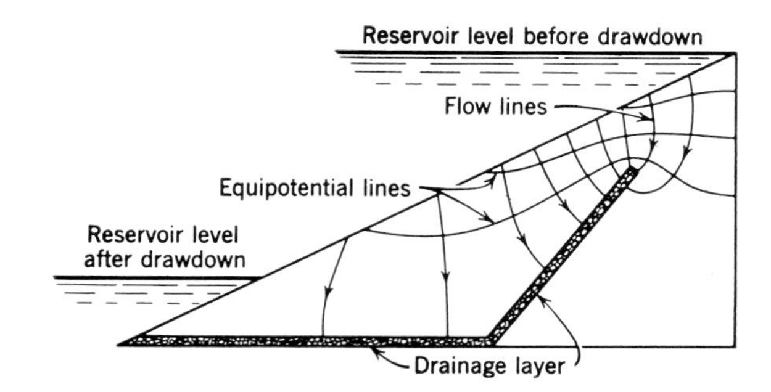

Fig. 5.7:5 Influence of inclined upstream drains in the upstream slope on the theoretical drawdown pressures (after Reinius, Ref. 574).

A central "chimney" drain in an otherwise homogeneous dam has an appreciable influence in reducing the drawdown pore pressures. If the chimney drain is sloped upstream (see, for example, the cross section of North Hartland Dam, Fig. 1.2:6), the effect is the same as if the dam had a thin upstream sloping core, Fig. 5.7:3, and the theoretical drawdown pressures are very low. Theoretically, the drawdown pore pressures in the upstream slope of an homogeneous earth dam can be reduced greatly by the use of internal drains which make the pore water flow away from the slope instead of toward it. Figure 5.7:4 shows the influence of a horizontal drainage blanket under the upstream slope and of intermediate horizontal drainage blankets at various elevations. By using an inclined drainage blanket of the type shown in Fig. 5.7:5, the theoretical pressure in the pore water above the drain is less than atmospheric pressure.

six

Earth Dams on Pervious Soil Foundations

6.1 METHODS OF FOUNDATION TREATMENT

In many parts of the world nearly all of the sites with rock foundations have been developed, and it is becoming necessary with increasing frequency to build dams which are founded on thick deposits of pervious soil. At a site with a pervious foundation three basic methods of foundation treatment are possible:

1. Eliminating the seepage or reducing it to a negligible amount by constructing a complete vertical foundation seepage barrier (cutoff).
2. Reducing the seepage, either with a partial vertical cutoff or with an upstream impervious blanket, and providing for the control of the water that does seep through.
3. Taking no steps to reduce the seepage, and providing for its control.

Before about 1940 almost all high earth dams built on pervious soils were treated by the first method, primarily because of concern that a safe dam could not be constructed without a complete seepage barrier. The majority of dams are still designed in this manner. However, in the last 20 years, confidence has grown in the safety of seepage control methods, and an increasing number of dams have been founded directly on pervious soils at sites where complete cutoffs would have been very expensive.[1] There is no practical limit to the height to which a dam can be constructed without a complete seepage cutoff, except in the cost of facilities to accomodate the underseepage. Completely

[1] Probably the first major dam constructed on a pervious foundation without a cutoff was the Franklin Falls Dam completed by the U.S. Corps of Engineers in New Hampshire in 1942 (Ref. 17), Fig. 1.2:24.

safe dams have been built without foundation cutoffs at sites where the average coefficient of permeability of the foundation material has exceeded 100,000 ft./yr.

At sites where the construction of a foundation seepage barrier is expensive, difficult, or time consuming, the decision to either dispense with it or to provide a partial seepage barrier should rest on cost studies showing whether the value of the water lost by seepage is less than the amortized cost of the cutoff. In the extreme case, where the estimated maximum seepage loss is less than the required water discharge past the dam, the value of the water lost by seepage is zero. Frequently the water loss may have no value at reservoirs used only for flood control. As a consequence the U.S. Army Corps of Engineers, which has designed and constructed the majority of major flood control reservoirs in the United States, has been largely responsible for developing modern methods of controlling underseepage (Ref. 314).

At sites where the cost of a complete seepage cutoff is not great, however, it should always be provided even if the loss of water due to underseepage would have no importance. This is true because dams with no seepage losses require less inspection and maintenance during their lifetime than dams with considerable, albeit controlled and safe, leakage through the foundation.

6.2 PREVENTING UNDERSEEPAGE WITH COMPLETE VERTICAL BARRIERS

Foundation cutoffs have been constructed primarily from one of three materials: (1) Compacted impervious earth in excavated trenches; (2) steel sheet piling; and (3) concrete walls constructed by one of several methods. Since about 1920, rolled-earth cutoffs have been used more frequently than the other two methods. In recent years, because of the improvement in earth-moving equipment, they have become increasingly more favorable economically, and are now installed under the great majority of earth dams. Steel sheet piling is still preferred under certain circumstances, but concrete walls constructed by traditional methods generally are uneconomical in the United States today. Two new methods which show promise for the future are alluvial grouting and the slurry trench.[1]

[1] Reviewers of the book in manuscript have pointed out with justification that more attention and space have been given in this chapter to these new methods of foundation treatment than their use to date justifies and that, as a consequence,

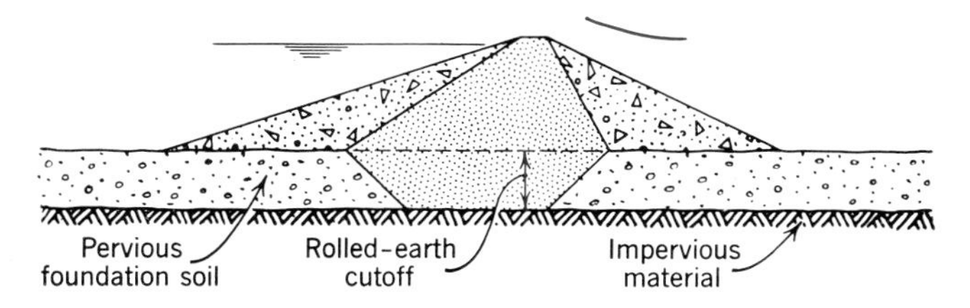

Fig. 6.2:1 Typical foundation cutoff trench.

6.2a Rolled-Earth Cutoffs

For a rolled-earth cutoff (Fig. 6.2:1) the same impervious soil and construction methods are employed in constructing the trench backfill as in the impervious core of the dam. The work is carried on in a de-watered excavation where it can be inspected and controlled at all times. As a result this type of cutoff provides a tight and effective barrier to seepage. Rolled-earth cutoffs are most commonly constructed to depths up to 75 ft. The cost increases rapidly below this depth, and although trenches have been sunk to more than 150 ft. for high dams, the construction tends to be expensive and troublesome.

The principal difficulty in the construction of a rolled-earth cutoff is frequently the dewatering of the excavation and holding down the water level until the trench is backfilled. A number of experiences indicate that it is difficult to make a reliable estimate of the quantity of pumping which will be required, even where the most careful field pumping tests are performed as part of the explorations. The estimate is usually higher than the quantity actually pumped, primarily because it is difficult to measure the over-all permeability of the foundation material (Chap. 4). In some cases also, silting occurs in the still ponds which form behind the cofferdam, and the bottom of the reservoir becomes tighter than anticipated.

Dewatering is carried out with well points installed on the slopes of the trench as the excavation is being made or with deep-well pumps

---

the reader may receive an overoptimistic impression of the applicability of these methods. In fact, all these methods are very expensive and less positive than an open cutoff trench and, consequently, should be adopted only after thorough study of all possible alternatives at any given site. On the other hand, there is also no question that these methods will be used more widely in the future than in the past. For this reason and, also, because there is no existing summary of the subject in the literature, the authors have felt justified in devoting a perhaps disproportionate amount of attention to these new methods of foundation treatment.

located upstream and downstream from the excavation (Ref. 455).[1] In a few cases, sump pumping at the bottom of the excavation has proved adequate for controlling the seepage inflow. Except in very coarse soil or relatively impervious material, however, some well points have been necessary on almost all jobs where the foundation trench extended for any considerable depth below the ground water table, Fig. 6.2:2.

Regardless of the type of dewatering equipment employed, when the cutoff trench is extended down to bedrock it is usually necessary to drain the water running into the bottom of the excavation by catching it in narrow trenches excavated in the bedrock at the upstream and downstream edges of the cutoff trench, Fig. 6.2:3. In order to keep the bottom of the excavation dry while placing rolled earth in the lower elevations of the trench, these small continuous trench-sumps are filled with gravel so they can continue to function even after they are cov-

[1] At a few dams, contractors have driven tunnels in the rock formation below the alluvium and dewatered by draining the water downward into the tunnels (Ref. 643). At the Anderson Ranch Dam this procedure was attempted but abandoned (Ref. 31).

Fig. 6.2:2 General view of typical cutoff trench dewatered with well points and sump pumping. (Courtesy USBR)

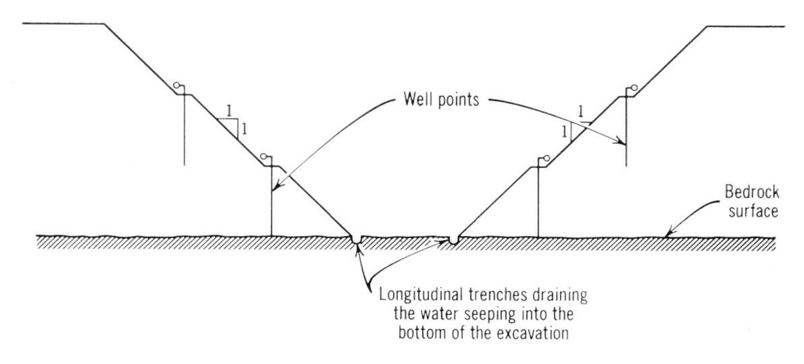

Fig. 6.2:3 Common method of dewatering cutoff trench excavations.

ered with the earth. A riser pipe is then brought up from the sump, either vertically through the fill or on the slope of the trench, and the ground water level in the adjacent natural foundation soil is kept 10 ft. or more below the surface of the rolled-earth core during construction. After the backfilling is completed, the riser pipe and gravel-filled drains may or may not be grouted with clay or cement. When this is to be done, grout and bleeder pipes are placed in the fill, Fig. 6.2:4.

6.2b Steel Sheet Piling

Steel sheet piling was among the several important types of foundation seepage barrier used in pervious soils 30 years ago. Because it is relatively expensive, and because experience has shown that leakage through the interlocks between the individual piling is considerable, it is chosen much less frequently today. However, in some circumstances where the alternative cutoff methods are expensive or time consuming, steel sheet piling can provide the best underseepage barrier. [Lane and Wohlt in a recent summary of experience with underseepage control for the large U.S. Corps of Engineers dams on the Missouri River (Ref. 569) give an excellent picture of current knowledge concerning the use of sheet pile cutoff walls.]

The leakage (and pressure head loss) of water flowing through a steel sheet pile wall depends primarily on the average width of the openings in the interlocking connections. Both theory and laboratory tests indicate that, if no stresses exist in the interlocks, the average flow channel between individual piles (with the typical sections used in the United States) is such that the seepage resistance of a sheet pile wall driven into a given soil will be approximately equal to the seepage resistance of 30 to 40 ft. of the soil (Ref. 116). Few measurements

(a)

(b)

Fig. 6.2:4 Views showing dewatering and backfill of typical cutoff trench excavation, Cachuma Dam (1951). (a) Cleaning up rock surface. Note drainage trenches in bedrock at upstream and downstream edges of excavation. (b) Backfilling trench. Note riser pipes on slopes. (c) Close-up of riser pipes. Note 2-in. grout and bleeder pipes for grouting gravel drains. (Courtesy USBR)

(*c*)

of the pressure head loss across sheet pile cutoff installations under dams are available. However, these few indicate that the resistance to seepage actually offered by the sheet piling has usually been equivalent to that offered by a length of the soil at the site varying between 400 and 2,000 ft. (Refs. 116, 164, 174, 569). The fact that the measured resistance is higher than indicated by theory and laboratory tests is due to: (1) interlock stresses (either tension or compression close to the interlock openings), and (2) to plugging of the interlock openings by corrosion and/or soil fines.[1]

It can be concluded that the seepage resistance offered by a sheet pile wall may be considerable under a low dam where the path of seepage is no more than several hundred feet. The sheet pile wall also is more effective in a homogeneous than in a stratified foundation. In a typical stratified alluvium, in which the permeability in the hori-

[1] At some dams, a definite increase in the effectiveness of sheet pile cutoffs with time has been recorded, probably due primarily to corrosion of the interlocks. At the Fort Peck Dam, which had a sheet pile wall varying in depth from 100 to 160 ft. in the central part of the valley, the initial head loss measured across the wall was only about 15 ft. or about 12% of the total effective head. In the following 17 years, this increased to about 30% of the total head. A similar increase in the apparent tightness of the sheet piling under the Garrison Dam occurred in the first 5 years of reservoir operation (Ref. 569).

zontal direction is many times that in the vertical, a large part of the total pressure head loss is due to seepage in the vertical direction as the water enters the foundation upstream from the dam and leaves it downstream. In such cases a relatively smaller part of the pressure head loss occurs as the water travels horizontally under the dam, so that the influence of the sheet pile wall is relatively much lower than for a dam with a homogeneous foundation.

Another disadvantage of steel sheet piling is the constant possibility that the wall may be damaged by cobbles or boulders during the driving. Often, with no indication of trouble at the surface, the interlocks tear open or the piling curls. There are no reliable guides to the relationship between the coarseness or density of a soil and the type of steel sheet piling which should be used in it. Driving and pulling tests should be made at sites where there is any doubt, and sheet piling should always be jetted through coarse material. In some cases heavy piling has been driven successfully through coarse gravel and cobbles.

For the most satisfactory use of sheet pile walls, the underlying bedrock should be sufficiently weathered and soft at the surface to allow the pile to penetrate a few inches. During this distance the driving resistance increases rapidly. Very hard rock with an irregular upper surface may cause the piles to deflect and break. Figure 6.2:5 shows

Fig. 6.2:5 Failure of sheet piling driven in test sections through Columbia River gravels to hard bedrock. (Courtesy Harza Engineering Company)

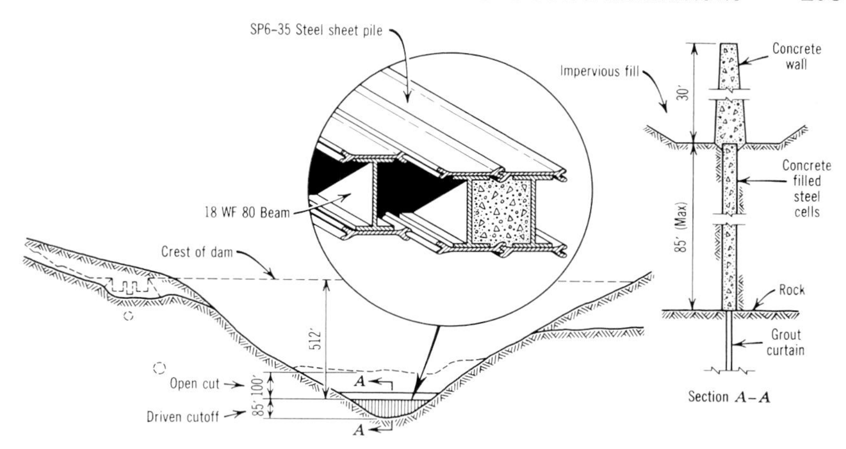

Fig. 6.2:6 Special sheet pile wall used at Swift Creek Dam, Washington (after de Luccia, Ref. 77).

views of damaged piling which were driven and pulled in a test section made during the design of the Wanapum Dam (Sec. 6.2*e*), for which sheet piling was considered as a method of foundation seepage control.

A steel sheet pile wall with special design to reduce leakage in the interlocks and to prevent breaking or curling was used in 1957 in the lower 85 ft. of the foundation material at Swift Creek Dam in Washington (Refs. 77, 160). As seen in Fig. 6.2:6, it consisted of a double row of straight section sheet piling welded to 18-in. wide-flange steel beams. After the driving, the spaces between the beams were excavated and the cells were filled with concrete.

6.2c Concrete Walls

Vertical concrete walls of various types have been installed under a few major earth dams as foundation seepage barriers. When founded on rock over their whole length, concrete walls are perfectly satisfactory and provide almost complete imperviousness. However, they have not been used in the United States to any large degree for a number of years. The high cost of hand labor and concrete usually makes the rolled-earth cutoff in a machine-excavated trench more economical in this country.

One of the last major dams in the United States at which a cutoff of this type was used is the USBR's Tieton Dam completed in 1925, Fig. 9.5:2 (Refs. 147, 377, 378). There, a 5-ft. wide concrete wall was carried down more than 130 ft. Difficulty with dewatering and shoring during construction discouraged further use of similar designs. In the 1930's another type of very deep concrete cutoff wall which was installed as a compressed air caisson (Refs. 167, 460) was constructed under a few major dams in the eastern United States. It is doubtful that such a design would be economically feasible under any circumstances now.

Concrete cutoff walls are still common in England. (Ref. 44). The excavation is made in the form of a sheeted and braced trench usually of constant width (about 6 ft.). The trench is backfilled with unreinforced concrete, and the water which enters the excavation through the walls and bottom is pumped from sumps. On several English dams these walls have been put down from 150 to 200 ft. (Refs. 163, 166, 461).

A new form of concrete cutoff wall which shows promise in special cases (where its brittleness would not result in cracking and where a deep barrier is not needed), is a continuous row of overlapping concrete piers installed with special drilling rigs or other equipment. The method is adapted to the use of machinery, and a continuous barrier can be built relatively economically. A number of different systems for constructing the wall without dewatering the ground have been devised, the most common being the Intrusion-Prepakt Co. method in the United States and the "Veder" process of the construction firm ICOS in Milan, Italy. In the Intrusion-Prepakt method, the piers are made by mixing Portland cement with the natural foundation soil, or by pumping a sand-cement grout into the hole from the bottom up as the drill bit is removed, Fig. 6.2:7. In the Veder process, the holes are held open with a bentonite mud slurry, and the concrete is placed with a tremie from the bottom of the hole.

The Intrusion-Prepakt process has been used to build continuous underground impervious walls for many types of civil engineering structures (see, for example, Ref. 471); however, up to the present it has been employed for permanent seepage barriers under only two small diversion dams of the USBR (Ref. 565). With the Veder method cutoffs have been constructed under a considerable number of low earth dams in Europe and South America (Refs. 371, 470, 472, 473, 488).[1] Figure 6.2:8 shows in closeup an uncovered portion of a wall constructed by the Veder process under a small dam in Italy.

[1] Probably the most ambitious project in which an ICOS pile cutoff has been used (completed and tested satisfactorily in 1962) is for the upstream cofferdam at the Manicouagan 5 Dam in Quebec. Here a 2-ft. thick wall is being put down to a maximum depth of about 250 ft. through coarse alluvium containing boulders up to 3 ft. in diameter. The wall has an area of about 28,000 ft.^{2}

Fig. 6.2:7 Construction of mixed-in-place, soil-cement pier as part of a continuous cutoff wall under a small diversion dam. (Courtesy USBR)

6.2d Grouting Alluvial Deposits[1]

While cement grouts have been used for many years to reduce leakage through fissured rocks under dams (Sec. 10.2), until recently all attempts to grout alluvial sand and gravel deposits have been notably unsuccessful. Cement grout could not be injected uniformly except in

[1] See Footnote 1, p. 288.

Fig. 6.2:8 View of portion of concrete cutoff wall under an earth dam constructed by the Veder (ICOS) process exposed in an excavation at the rock abutment. (Courtesy ICOS)

coarse materials. Chemical grouts, which could be injected into sand deposits, were usually too expensive, and the few attempts made to grout alluvium with asphalt were largely unrewarding (Ref. 417).

The primary difficulties in grouting sands and gravels arise from the necessity of keeping the hole open with casing, the impossibility of using packers, and the lack of an injection technique which would assure a uniform penetration of the grout. However, at present, largely as the result of developments in Europe, satisfactory procedures for grouting alluvium are available and are being more widely used each year.

More than 20 years ago, French engineers started experimenting with clay and admixtures for use in grouting sands, and they carried out some successful projects in the years before 1950 (Refs. 9, 195, 230, 475–478, 622). About this time the design was begun for the 400-ft. Serre-Ponçon Dam, which was to be located at a site in the French Alps underlain by more than 300 ft. of erratic river alluvium consisting of sands and gravels with lenses and layers of silt and clay. The typical sand and gravel layers were graded as follows (Ref. 615):

SIZE (MM.)	PERCENT FINER
200	100
100	80–95
5	18–40
0.08	0–5

The estimated overall coefficient of permeability of the foundation soils was 50,000 to 100,000 ft./yr. Extensive tests with clay-cement grout were performed, and a grouting method was developed by which the coefficient of permeability was reduced to approximately 50 to 100 ft./yr. (Refs. 219, 479, 615). The foundation was grouted for the full depth over a width of approximately 50 ft., Fig. 6.2:9.

The Serre-Ponçon Dam was completed and the reservoir filled in 1960, and measurements of pore water pressure in the foundation indictate that the grouting was very successful. When the reservoir level is approximately 100 meters above the tailwater level, the pore pressure measured by piezometers located in the foundation alluvium just downstream from the grouted zone is about 2 meters above the tailwater level.

Following the work at the Serre-Ponçon Dam, the same French contracting firm, Soletanche, was engaged to grout alluvial deposits which extended to a maximum depth of 523 ft. under the Mission Dam on the Bridge River in British Columbia 1957–1960 (Ref. 677). There, in what was essentially the same procedure as that at the Serre-Ponçon Dam (the "Soletanche" method), five lines of holes, spaced 10 ft. apart, were drilled and grouted, i.e., the distance between the outer rows was 40 ft. The holes (5 to 10 in. in diameter) were drilled through the alluvium to bedrock by using drilling mud and tricone rock bits, and 2½-in. diameter grout pipes were set to bedrock in the drilled holes. The pipes had circumferential rows of holes at vertical intervals of 12 in., and these holes were covered with 4-in. long, tightly fitting rubber sleeves, Figs. 6.2:10, 6.2:11. The 2½-in. grout pipes were then sealed in the ground by filling the annular space outside each with a

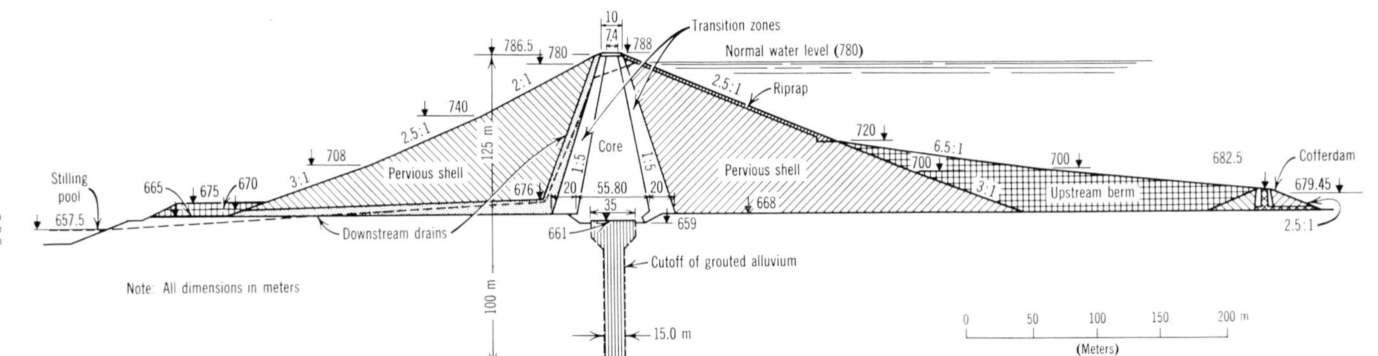

Fig. 6.2:9 Cross section of the Serre-Ponçon Dam showing the grouted alluvium cutoff (after Guelton et al., Ref. 615).

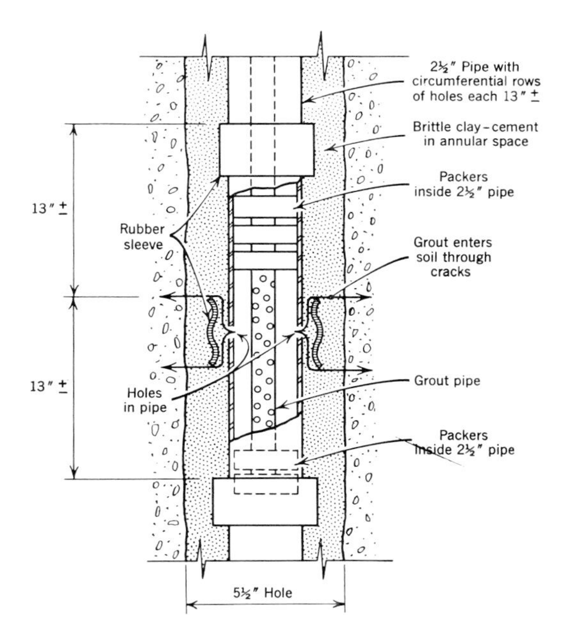

Fig. 6.2:10 Details of alluvial grouting procedures—Soletanche process.

brittle clay-cement grout (designed to have an unconfined compressive strength at three months of about 15 tons/ft.^{2}).

Grouting of the alluvium was then carried out from the bottom up through each of the rows of circumferential holes in the pipes. Grout was pumped into each hole under pressure through a smaller interior grout pipe, which had two rubber packers spaced 1 ft. apart. The grout pressure was confined to the short length of the 2½-in. pipe opposite each of the individual rows of circumferential holes and rubber sleeves. In this way the grout was forced out of the small holes in the pipe so that the rubber sleeve expanded slightly, cracking the thin cylinder of clay-cement grout surrounding the pipe. The fluid grout then entered the soil through these cracks.

The outer rows of bore holes were grouted with a mixture of about 20% clay, 13% Portland cement, and 66% water, and the inner rows were grouted with a mixture of 25% clay, 5% cement, and 70% water. The center row was grouted with clay grout stabilized with chemicals in the ratio 25% clay, 70% water, and 5% of silicate and phosphate (all percentages by weight). A predetermined quantity of grout was pumped at each elevation (50% of soil volume or 175% of soil voids),

Fig. 6.2:11 Closeup of 2½-in. grout pipe with 4-in.-long rubber sleeves. The sleeves are held in place as shown by a wire band which is tack-welded to the pipe.

while the grout pressures were allowed to build up as high as necessary to get the desired quantity of grout into the surrounding soil. Sometimes the pressures rose to 1,000 p.s.i. or higher.

One of the contributing factors to the success of the Soletanche method is the fact that the discharge point of the grout in the ground is reliably controlled. In the last few years it has been used for grouting sands and gravels below a number of earth dams in France and Germany. See, for example, the experience described at Sylvenstein Dam in Germany, where pervious alluvial soils were grouted to a depth of about 300 ft. in 1956–1958 (Ref. 555). Present plans for the High Aswan Dam on The Nile River contemplate grouting the pervious alluvium under the axis of the dam to a depth of about 400 ft. (Ref. 592). The only grouting which has been done by the Soletanche method in the United States up to the present time (1961) is the minor work on the lower river gravels at the Wanapum Dam below the main slurry trench (Sec. 6.2*e*).

At the Notre Dame de Commiers Dam near Grenoble, France, a grout curtain of about 80,000 ft.^{2} in area with a maximum depth of about 160 ft. was completed in 1961 in a 12-month construction period through a sand-gravel alluvial deposit using the Soletanche process.

The work was carried out to meet a specification which required that the final hydraulic gradient in the foundation alluvium downstream from the grout curtain should not exceed the average slope of the river bed at the site (about 6 ft. in 1,000 ft.). A maximum of five lines of holes was used in the upper part of the curtain. Grouts used were clay-cement deflocculated bentonite, and chemicals. The total cost for the work was about 7,000,000 NF or about $17.00/ft.^{2} of grout curtain. During the grouting process, the ground surface along the grouted area was heaved 8 to 12 ins.

In 1958–1959 the deep pervious gravels forming an abutment of the Rocky Reach Dam on the Columbia River were successfully grouted with a method developed by W. F. Swiger (Ref. 453).[1] The soils were the typical erratically bedded Columbia River deposits, which include lenses of very pervious, coarse openwork gravels with a maximum coefficient of permeability estimated at 5,000,000 ft./yr. After field tests, the material was grouted on three lines of holes spaced 10 ft. on centers using clay-cement and bentonite-cement grout mixtures. The

[1] Similar grouting was carried out for the alluvium under the Noxon Rapids Dam on the Clark Fork River in Montana in 1957 (Refs. 296, 297).

Fig. 6.2:12 Typical openwork gravels through which slurry trench cutoff was made at Wanapum Dam (6-in. by 9-in. notebook gives scale).

grouting was carried out from the surface of the gravel layer down in 5-ft. stages with nearly the same practice as is used for grouting rock in stages (Sec. 10.2). Some additional final grouting was done with chemicals (American Cyanamid Co., AM-9 solution) and the results were reported to be very satisfactory.

6.2e Slurry Trench Foundation Cutoff[1] (Refs. 602, 603, 705)

The "Slurry Trench" procedure had been employed to construct ground water seepage barriers for various purposes but, except for some dikes below the McNary Dam (Ref. 463), had never been used under a permanent dam structure until the Wanapum Dam was built in 1958. The foundation materials at the site consisted of an erratic deposit of sands, gravels, and cobbles with some openwork gravel, Fig. 6.2:12, extending to a depth of more than 80 ft. The coefficient of permeability of the various layers ranged roughly between 100,000 and 3,000,000 ft./yr., and the average permeability was estimated roughly at 1,000,000 ft./yr. Studies indicated that the costs of installing a satisfactory sheet pile cutoff or of opening and dewatering a trench

[1] See Footnote 1 on p. 288.

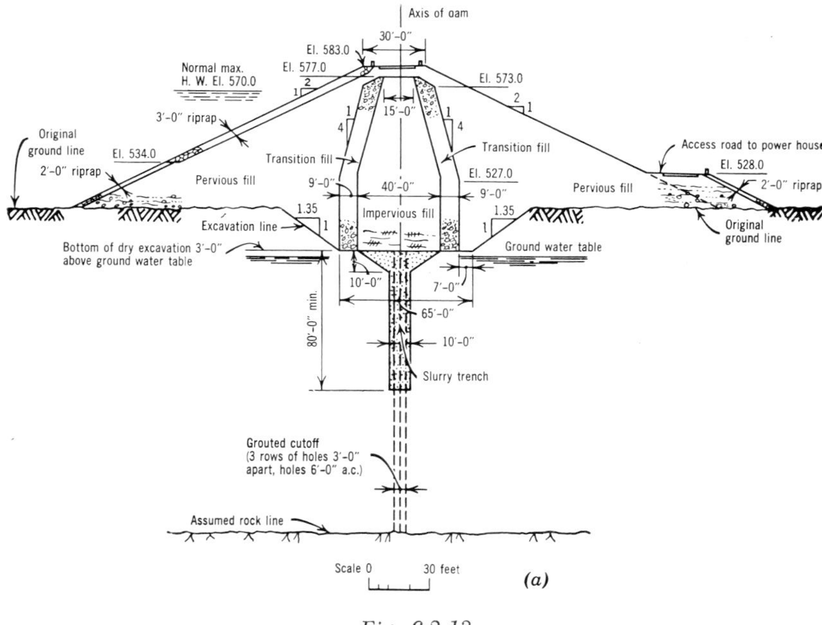

Fig. 6.2:13

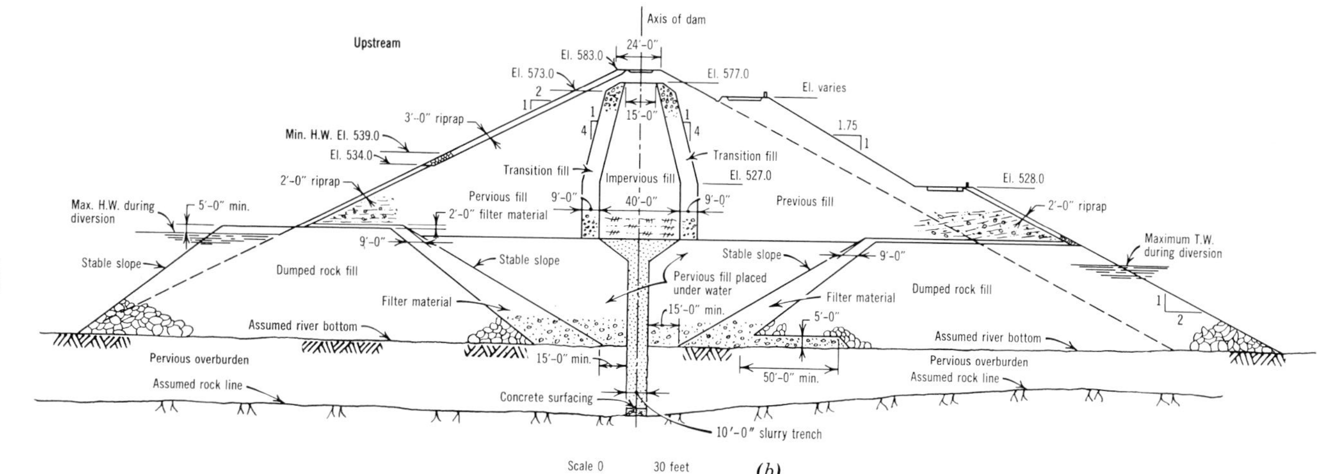

Fig. 6.2:13 Cross section of Wanapum Dam showing the slurry trench cutoff. (a) Abutment section. (b) River channel section. (Courtesy Harza Engineering Co.)

for a rolled earth cutoff through the pervious foundation soils would have been very high.

After successful large-scale field tests (Ref. 454), a 10-ft. wide foundation seepage barrier was constructed to a maximum depth of 80 ft. by the slurry trench procedure, Fig. 6.2:13. In some portions of the dam length where the thickness of the foundation soil exceeded 80 ft., the soil below the slurry trench was grouted with a soil-cement mixture by the Soletanche method (Sec. 6.2*d*).

The excavation, made with a dragline, was kept continuously full of bentonite mud slurry, which prevented the walls from caving so that no sheeting or shoring had to be used (Fig. 6.2:14). After the trench was excavated to bedrock with the dragline, the bottom was cleaned with clamshell buckets and air lifts. The excavated material (a mixture of sand and gravel with bentonite slurry) was stockpiled in windrows adjacent to the trench, and blended by means of draglines and bulldozers with a quantity of 15 to 20% of natural silt brought in from a borrow pit. This mixture was then dumped from one end of the trench, displacing the bentonite slurry until the backfilling was completed.

Construction was carried out with a minimum of trouble. Although heavy equipment worked very close to the edge of the trench, no cave-ins of the trench walls occurred, even though the trench was kept open for lengths of as much as 1,000 ft. before backfilling commenced. The total cost (roughly \$7/ft.^{2} of cutoff wall) was much less than the estimated cost of any comparable seepage barrier. Based on the results of a few air pressure tests made in bore holes drilled in the slurry cutoff, the material was estimated to have a coefficient of permeability of less than 0.1 ft./yr. The construction method was so successful that the contractor used similar slurry trenches to make cutoffs beneath his 7,000 ft. of cofferdam.

The slurry trench mixture remained in a very soft condition for many months after construction. When it was necessary to excavate through the foundation for the construction of the spillway and powerhouse (which were founded on bedrock), the material was still so soft that it was necessary to drive short rows of steel sheet piling to prevent it from flowing into the excavation. As exposed in the walls of the excavation several months after construction of the cutoff, the material had a consistency somewhat like that of stiff butter, Fig. 6.2:15.

The slurry trench method has many advantages and, now that it has a precedent on a major project, it will undoubtedly be used more frequently in the future. At sites where the depth of cutoff is not so

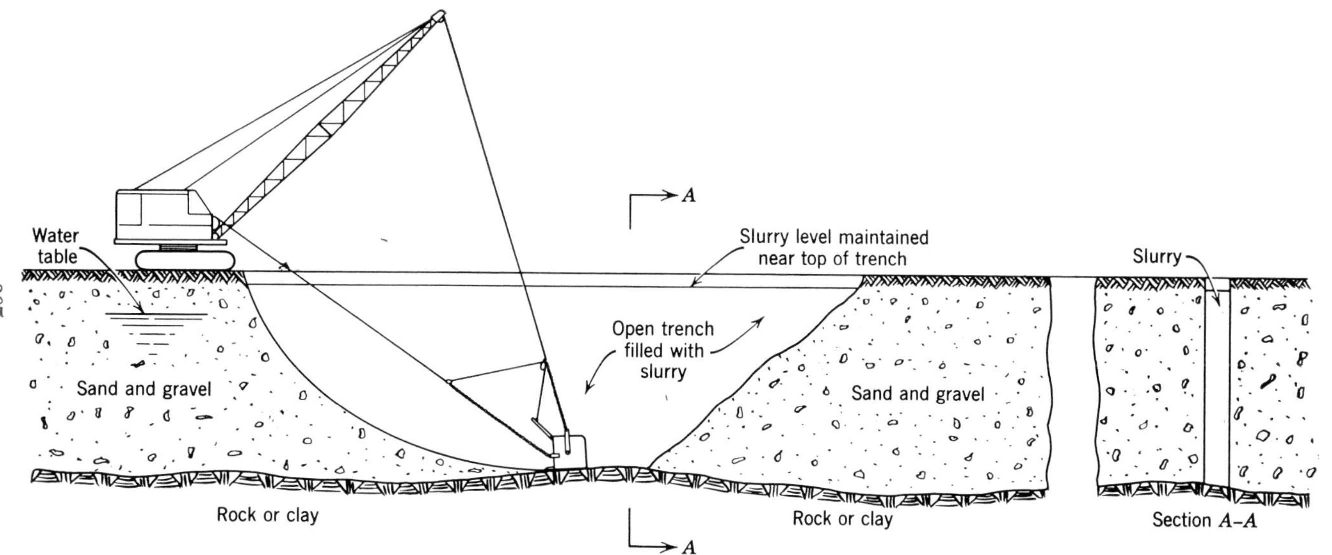

Fig. 2.6:14 Construction of slurry trench (Courtesy Cronese Products, Inc.)

Fig. 6.2:15 Views of the appearance of slurry trench cutoff material as exposed in an excavation three months after completion—Wanapum Dam.

great, the excavation can be made with trenching machines or backhoes, and trench widths of 3 to 4 ft. (or even less) can be employed and unit costs of \$2.00/ft.^{2} or less, can be obtained.

6.3 REDUCING UNDERSEEPAGE

If the construction of a complete seepage barrier for a dam founded on pervious soil is not practicable or economical, the designer has the choice either of making a partial cutoff or no cutoff, and then providing for the control of the seepage which develops. Underseepage can be controlled by methods discussed in Sec. 6.4. It can be reduced in quantity by one or both of the following principal measures:

1. Constructing a partial vertical cutoff usually extending down to an intermediate soil stratum of lower permeability, Fig. 6.3:1*a*.
2. Increasing the width of the base of the impervious section of the dam by constructing a horizontal blanket of impervious soil connected to the dam core and extending upstream, Fig. 6.3:1*b*.

Each of these methods reduces the quantity of underseepage by lengthening the seepage path. In doing so, each also reduces the downstream pore water pressures, and thus increases the stability. Another method sometimes used to increase the stability is to provide a downstream berm with or without a horizontal drainage blanket, Fig. 6.3:1*c*.

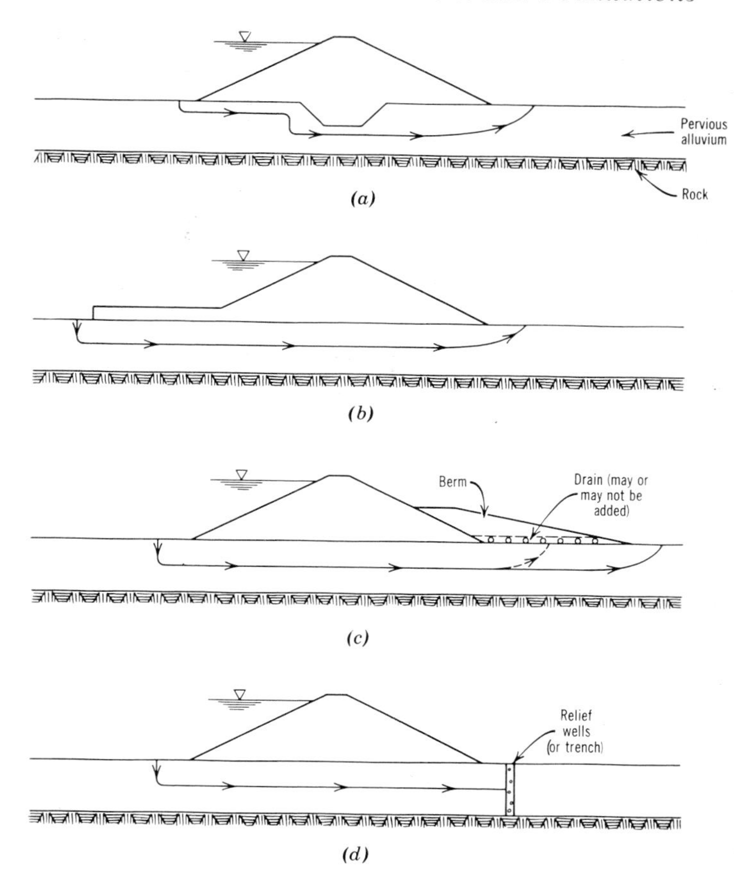

Fig. 6.3:1 Methods of underseepage control for dams on pervious foundations without complete seepage cutoffs. (a) Partial vertical cutoff. (b) Upstream impervious blanket. (c) Downstream berm. (d) Relief wells.

If no drain is provided, the berm will act to increase the seepage path in the same way as the upstream impervious blanket.

6.3a Partial Vertical Cutoffs

At sites where the average over-all coefficient of permeability of the foundation soil is practically the same in both directions and does not decrease with depth, a partial seepage cutoff has little influence on the underseepage quantities or pressures. Theory and model tests (Ref.

120) indicate that it is necessary for a cutoff to penetrate a homogeneous pervious soil foundation at least 95% of the full depth before there is any appreciable underseepage reduction. For this reason, only a complete cutoff should be considered at a site with a homogeneous foundation.

On the other hand, a vertical cutoff extending partially through the pervious soil may be of considerable value in reducing the underseepage in certain circumstances. Partial seepage barriers are valuable at sites where the average permeability of the foundation soil decreases with depth below the surface or where there is a single continuous impervious layer into which the cutoff can be connected.

In the circumstance where a complete or even a partial vertical cutoff is not warranted, the designer should always specify a nominal minimum cutoff trench with a depth of 6 to 10 ft. The purpose is to provide a continuous excavation through the upper few feet of the soil for inspection. Frequently a much better understanding of the subsoil profile may be obtained from examining the walls of this trench, and sometimes conditions are exposed which indicate that deeper exploration should be carried out. The trench is subsequently backfilled with rolled impervious core materials. This gives a secondary advantage by cutting off localized leakage which may tend to develop through the surface soil layer in drying cracks, animal burrows, or root holes. The cost of such a trench is relatively very small and is always well justified even for low dams.

6.3b Horizontal Upstream Impervious Blankets

The horizontal upstream impervious blanket, which increases the horizontal length of the average path of underseepage, is more effective in controlling seepage through a homogeneous soil foundation than the partial vertical cutoff.[1] If the blanket is very impervious compared to the natural foundation, so that relatively little seepage through the blanket occurs, the reduction in the seepage quantities and pressures at the downstream toe is directly related to the length of the blanket. If the blanket is only slightly less pervious than the foundation material, there is a maximum length of blanket beyond which no appreciable additional value is obtained by increasing the length.

Good practice for a blanket which is relied upon to control the underseepage requires that it be constructed of impervious soil in the

[1] See Ref. 271 for a review of the performance of upstream blankets installed at three major dams of the U.S. Corps of Engineers.

same manner and with the same care as the impervious core of the dam. At some dams nominal upstream blankets have been constructed by dumping impervious soil (sometimes waste stripping or other material) in a random manner and compacting only by the travel of the hauling equipment. At many sites a natural surface blanket of impervious material already exists, and it is necessary only to fill the holes or gaps to make a continuous seal. In such cases at least the upper surface of the natural blanket should be scarified, brought to a good water content, and compacted.

The necessary thickness and length of a blanket depend on the permeability of the blanketing material, the stratification and thickness of the pervious foundation, and the reservoir depth. Thicknesses varying from 2 to 10 ft. are most frequently used. If the blanket is not very tight with respect to the natural underlying soil, its effectiveness can be increased by making it thicker in the portion of its length which is directly upstream from the core of the dam.

Although the effectiveness of the blanket can be analyzed by using flow nets obtained from graphical solutions or electric models, these methods are tedious and cumbersome when many solutions are desired for varying sizes of blankets and estimates of permeability. A convenient mathematical solution has been developed by P. T. Bennett for the condition where the foundation consists of a single horizontal pervious layer with a more impervious surface blanket (Ref. 90). The errors resulting from the simplifying assumptions made in the development of Bennett's theory are small compared with the accuracy with which the coefficients of permeability of the various layers can be determined. The equations for underseepage quantities and pressures using Bennett's theory are given in Fig. 6.3:2.

6.4 CONTROLLING UNDERSEEPAGE[1]

Underseepage threatens the safety of a dam in two ways:

1. The pressure in the water through the foundation at some point below the downstream toe may approach or equal the weight of the overlying soil. Such high water pressures reduce the stability of the slope against sliding and in the extreme situation could theoretically heave the upper layers of the foundation.
2. The discharging seepage water may lead to piping failure.

[1] Suggested general reading on underseepage control and analysis—Refs. 98, 101, 314, 424, 464–469.

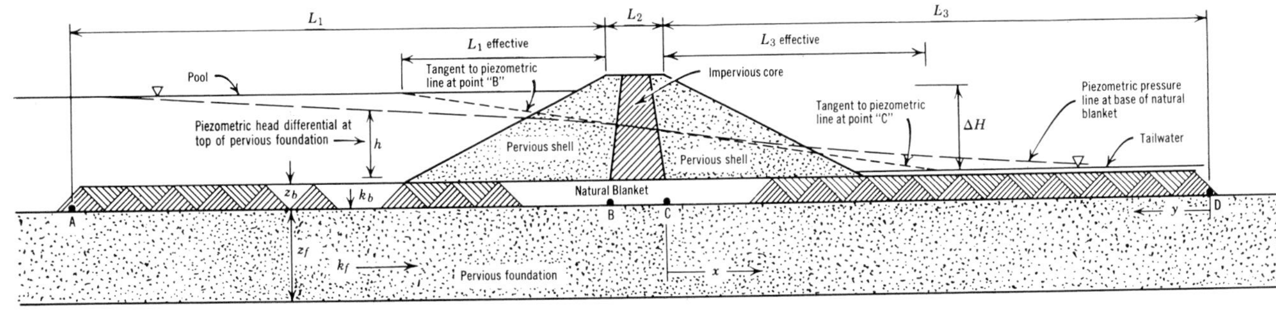

Because of leakage, the effectiveness of the blankets upstream and downstream from the dam core is reduced. The effectiveness of such blankets is expressed in terms of effective length:—

$$L_{1\text{eff.}} = \frac{\tanh (aL_1)}{a} \; ; \; L_{3\text{eff.}} = \frac{\tanh (aL_3)}{a}$$

$$\text{Where } a = \sqrt{\frac{k_b}{k_f z_f z_b}}$$

Where k_b = the vertical coefficient of permeability of the blanket.

k_f = the horizontal coefficient of permeability of the pervious foundation. The approximate value of foundation seepage per unit length of structure is:—

$$Q = \frac{k_f \Delta H z_f}{(L_{1\text{eff.}} + L_2 + L_{3\text{eff.}})} = k_f z_f i \, , \qquad \text{Where } i = \frac{\Delta H}{(L_{1\text{eff.}} + L_2 + L_{3\text{eff.}})}$$

The piezometric head from D to C is $h_x = \dfrac{i \sinh (ay)}{a \cosh (aL_3)}$;

If $L_3 \longrightarrow$ infinity, then $h_x = iae^{-ax}$

If landside outlet is blocked, then $L_{3\text{eff.}} = \dfrac{1}{a \tanh (aL_3)}$;

$$h_x = \frac{i \cosh (\alpha y)}{\alpha \sinh (\alpha L_3)}$$

Fig. 6.3:2 Effectiveness of finite upstream and downstream impervious blankets (after U.S. Army Engineers, Ref. 388).

6.4a Underseepage Analysis

HIGH PORE PRESSURES

The uplift pressure in the foundation near the downstream toe of the dam depends primarily on the configuration and permeability of the various subsoil layers, and it may be high even if the quantity of seepage is not great. The highest pore pressures develop in the upper elevations of the foundation in the circumstance where there is a relatively impervious horizontal surface layer which prevents the free discharge of the seepage water. If this impervious surface layer is not sufficiently thick, the seeping water usually breaks through a crack or other hole, and may form a sand boil (Sec. 2.2*b*). On the other hand, if it is thick, the underseepage may never be seen at the surface even though dangerously high pressures can be measured with piezometers.

There are no reported failures of earth dams in which the foundation soil below the dam heaved, although there has been much concern at many dams when very high pressures were measured.[1] Undoubtedly, high downstream seepage pressures have contributed to the instability of most of the dams which have had downstream slope slides (Sec. 2.4). The influence of the pore water pressures on the stability of the downstream portion of the dam is subject to analysis (Sec. 7.6). Where high pore pressures are anticipated, the maximum values which may develop can be estimated in advance and the dam section proportioned to have an adequate factor of safety against sliding under the worst conceivable conditions. The actual pressures which develop after the dam is built and the reservoir fills can then be measured to assure that the stability condition is not more severe than anticipated.

PIPING

The piping potential, or factor of safety against internal erosion, in a given dam cannot be analyzed, but piping can be completely and reliably prevented by controlling the underseepage in such a way that (1) the exit velocities are not high, and (2) the water discharges

[1] Since the seeping water is flowing under a gradient toward the downstream toe, the pressure head in the foundation below the dam is always above the tailwater elevation. Therefore, if the surface of the ground downstream from the dam is relatively flat, the pressure head measured in a piezometer will nearly always be above the elevation of the ground surface.

through adequate thicknesses of progressively coarser soils which meet the gradation requirements for filters (Sec. 1.3*b*).

Early efforts to analyze the factor of safety against piping under dams on pervious foundations were made through the use of a concept called the "line of creep," which was defined as the shortest path that a particle of water would have to travel in seeping under the dam. The ratio between the line of creep and the pressure head loss was termed the "creep ratio," which is the inverse of the average hydraulic gradient. Later, when it was realized that the foundation stratification had a major influence on the piping potential, a "weighted creep ratio" was defined for which the length of the line of creep was computed as the sum of the vertical components of the shortest seepage path plus one-third of the length of the horizontal seepage path.

Using this weighted creep ratio as a criterion of piping potential, E. W. Lane made a study of 280 dam foundations, which included 150 failures (Ref. 480). As a result of this study, safe values of weighted creep ratio were established as a function of the foundation soil type. They ranged approximately from a minimum of 3 for foundations of gravels and boulders to a maximum of 8 for very fine sands. This correlation between foundation soil type and piping potential has some value as a guide to judgment for the engineer designing an earth dam on a pervious soil foundation, but it should not be used as a design criterion. Regardless of the average hydraulic gradient (or creep ratio), any earth dam foundation can be made safe against piping by the installation of properly proportioned graded filters.

SEEPAGE FLOW PATTERN

If the subsoil strata comprising the foundation are relatively uniform, the seepage flow pattern under the dam can be estimated readily with graphical flow nets (Chap. 5). This procedure rapidly becomes difficult and tedious for non-uniform subsoil conditions; however, by making simplifying assumptions, the designer can arrive at a fair estimate of the seepage pattern. The three-dimensional seepage which occurs through the pervious abutments has also been studied with reasonable success by means of flow nets. In this case electric models have been very helpful because of their rapid construction of the complex flow pattern.

At sites where the subsoil profile is very erratic, experience indicates that a simplified seepage analysis, based on a roughly estimated average permeability of the foundation soil, is likely to be as reliable as the most detailed study which attempts to evaluate the seepage which may develop through the individual lenses or strata. In extremely erratic soils, which frequently occur in glaciated regions, the most thorough possible subsurface investigation, with a great number of exploratory borings, does not give sufficient information to allow a reliable estimate of the seepage pressure distribution, the quantity of flow, or even of the locations, if any, where leakage is most likely to emerge downstream from the dam. Under such circumstances the value of a competent geological study of the subsoil conditions cannot be overemphasized. After accumulating all the information possible, the designer must anticipate the most unfavorable conditions which might develop. Piezometers must be installed to check the actual pressures which develop after construction, and provisions must be made to control whatever leakage occurs.[1]

6.4b Regulation of Leaks

Any uncontrolled seepage erupting in the form of springs in the natural ground downstream from the core of the dam is potentially dangerous from the standpoint of piping. The greater the quantity and velocity of the discharging underseepage, the greater the piping potential. The danger is worse if the natural surface soils are fine, cohesionless sands or silts. It decreases as the soil becomes coarser, because well-graded coarse soils tend to form natural graded filters at the discharge points. Therefore leaks discharging through coarse soils may or may not need to be protected by specially constructed filters. On the other hand, leaks through fine-grained cohesionless soils always need filters to prevent progressive subsurface erosion.

With a dam which has a considerable quantity of underflow, it is necessary to locate and observe the point of seepage discharge, if any, on the ground surface, and to install filters if they are needed to prevent erosion. Underseepage which discharges uniformly over a large, more or less flat area downstream from the dam can be covered with thin layers of progressively coarser filter material. Where the seepage occurs in the form of a few concentrated leaks in one area, it can be controlled by channeling all the leaks into an excavated drainage

[1] Excellent descriptions of the step-by-step procedures used for the design of two dams on very erratic, pervious foundations are given in Refs. 93 and 456. Also see Ref. 463 for a description of the design measures to control seepage through the very pervious abutments at the Chief Joseph and McNary Dams on the Columbia River.

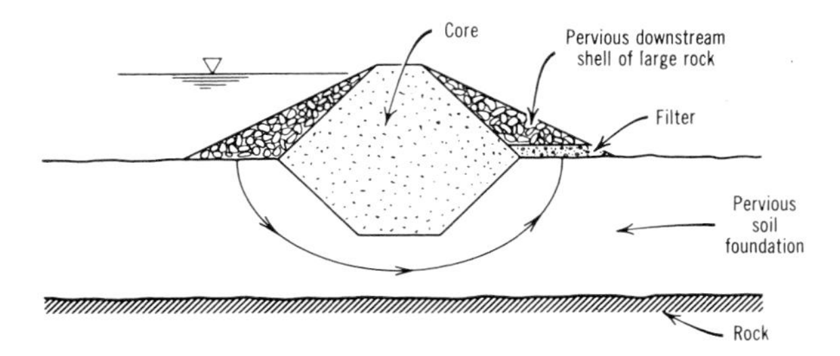

Fig. 6.4:1 Need for filter under downstream rockfill shells on soil foundation.

trench which is backfilled with graded filter material. Drainage tunnels or relief wells (Sec. 6.4*c*) can be installed to control the discharge of the underseepage before it reaches the ground surface. This action decreases the pore pressures deep in the abutments or foundations and thus increases the stability of the downstream portion of the dam.

A dam cannot be considered safe unless all downstream leaks are discovered. For this reason, it is dangerous to allow large ponds of water to stand in the vicinity of the downstream toe of a dam which has appreciable underseepage. Such ponds may form in the borrow pits or as the backwater from the outlet or spillway discharge. If large areas downstream from the dam are covered with water, piping of concentrated leaks may progress unobserved and lead to failure.

Unobserved piping can also occur under downstream embankment sections composed of large rock, as shown in Fig. 6.4:1. For this reason, a horizontal filter is needed between the foundation soil and the downstream rockfill section.

6.4c Relief Wells

Relief wells to control the pressure of seeping water under the foundation of an earth dam (Fig. 6.3:1*d*) were first used by the U.S. Army Corps of Engineers. They were installed below several of the major Army dams in the late 1930's as an emergency measure when unexpectedly high pressures developed during reservoir operation. One of the first times they were included in the design stage was in 1940 at the Arkabutla Dam in Mississippi. Because they are relatively inexpensive and enormously effective in controlling seepage pressures, they have been used with increasing frequency since then. At present, with few exceptions, most engineers consider relief wells at any site where there is a possibility that high pressures could develop.

The primary disadvantages of relief wells are:

1. They require inspection and maintenance and may have to be replaced during the lifetime of the dam.
2. They decrease the average seepage path and increase the quantity of underseepage.

In the immediate vicinity of each well, the pressure head in the seepage water is reduced to a value nearly equal to the elevation of the top of the well (hydrostatic pressure). The wells should be spaced close enough so the water pressure between them is not excessive. In a comprehensive paper on this subject, Middlebrooks and Jervis (Ref. 462) have developed theoretical methods for determining well spacings. The theory, however, is approximate and the results of such computations can only be considered a rough guide.[1] The most practical approach is to install the wells at reasonable spacings and to measure the pore water pressures which develop between them. Wells are commonly spaced between 50 and 100 ft. on centers. Intermediate wells can be added later if necessary to keep the pressure below any desired value. One of the main advantages of a relief well system is this flexibility which allows it to be expanded to meet the need at nominal extra cost.

Relief wells should extend through the full depth of the pervious foundation if possible. This is especially important for erratically bedded soil formations, where the seepage flow pattern cannot be reliably estimated in advance. Extensive laboratory tests on models, performed in the U.S. Army Engineers' laboratory at Vicksburg, Mississippi (Ref. 464), indicate that even in a very homogeneous pervious foundation, the effectiveness of wells which penetrate less than 50% of the thickness of the layer is greatly reduced. Wells which penetrate less than 25% have far less efficiency than fully penetrating wells and if the layer is more pervious in the lower elevations, a partially penetrating well may have almost no effect on the underseepage (Refs. 56, 114).

A relief well should have an interior perforated pipe (well screen) with a minimum inside diameter of 6 in., or larger if heavy flows are

[1] See Ref. 562 for an excellent description of the relief well installation at the U.S. Corps of Engineers' Fort Randall Dam on the Missouri River, including an example of the application of the analytical theory for computing well spacings and discharges.

anticipated. Gravel-filled holes are much less effective than open pipes (Ref. 462). Many kinds of pipe have been used successfully, although wooden pipes have probably been most widely installed because they last longer than other types when completely saturated, Fig. 6.4:2.[1] When it is anticipated that the reservoir may become completely empty and the wells may not remain always full of water, the wooden well screen can be backfilled with concrete in the upper elevations so that the well will remain serviceable even if the wood deteriorates. Asphalt dipped, galvanized metal well screens have been used at a number of large dams in recent years, Fig. 6.4:3. At the USBR's Enders Dam in Nebraska, a 24-in. diameter metal well screen was protected cathodically with buried magnesium anodes (Ref. 42). Perforated well screens of plastic pipe and concrete pipe have also been used on a few major dams, Fig. 6.4:4.

The annular space surrounding the well screen is backfilled with gravel which is graded to meet the filter requirements of the natural foundation soil. At the surface, the annular space is backfilled with impervious soil or concrete in order to prevent an upward flow of water outside the pipe.

Seepage water from the wells is usually discharged on the ground surface at the toe of the dam through a horizontal overflow pipe, Fig. 6.4:5. It should then be collected in a lined drainage ditch. At some dams, the seepage water has been discharged into a horizontal header pipe which is buried below the ground surface, Fig. 6.4:4. This system has the psychological advantage of keeping the leakage hidden in addition to reducing the discharge pressure by a few feet. Whenever the header pipe is buried (at the toe of the dam or inside the dam), the relief well casing should be extended to the surface for inspection and maintenance.

The holes for relief wells should be drilled by a method which will not seal the pervious soil strata exposed in the wall of the hole with fines. Relief wells should be developed in the same way as wells drilled for water supply, and after the hole is completed and cased, it should be pumped out and "surged" to remove all muddy water and "to develop the filter." The surging can be carried out with a heavy rubber piston a little smaller than the inside diameter of the pipe. After cleaning and surging, a pumping test should be made to determine the rate of inflow for a given drawdown and the rate of sand

[1] Corrosion of the initially installed steel relief well pipes at Fort Peck Dam on the Missouri River made it necessary to replace the system with wood pipe after about six years of operation (Ref. 569).

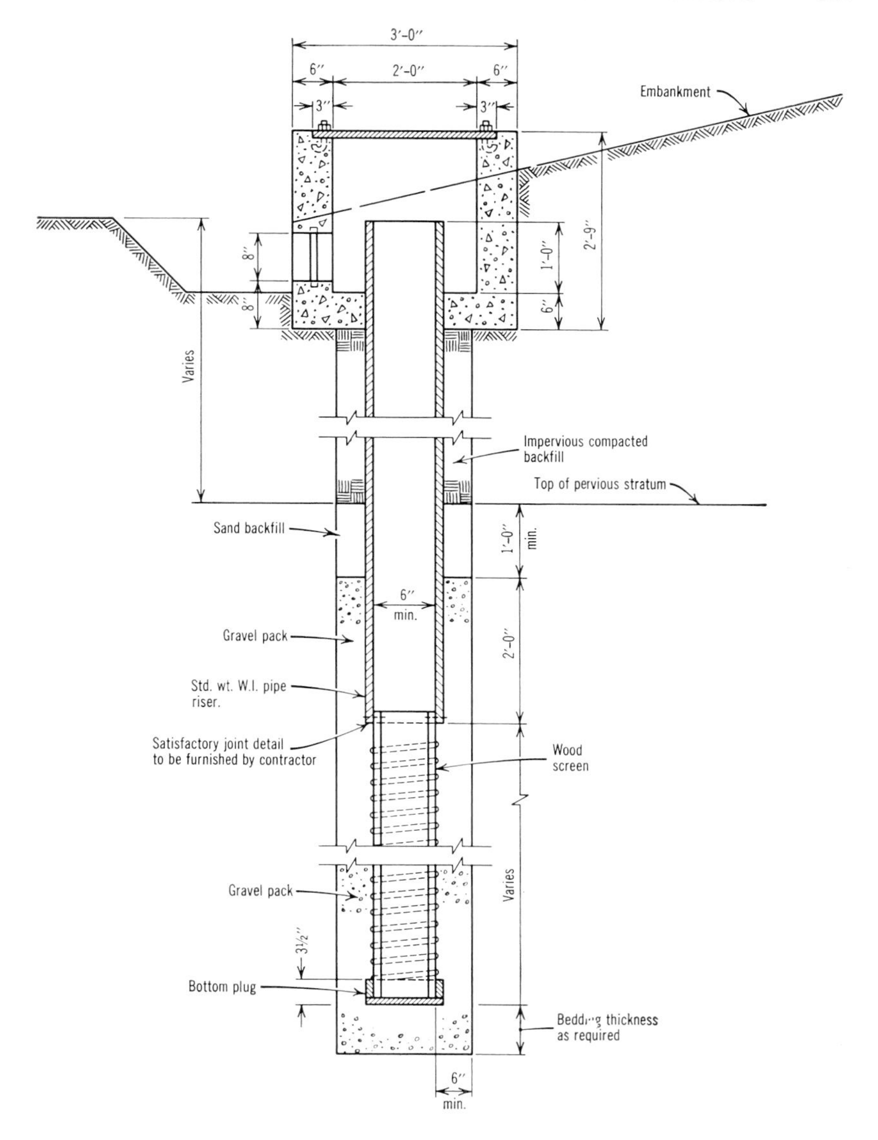

Fig. 6.4:2 Gravel-packed relief well with screen (after U.S. Army Engineers, Ref. 388).

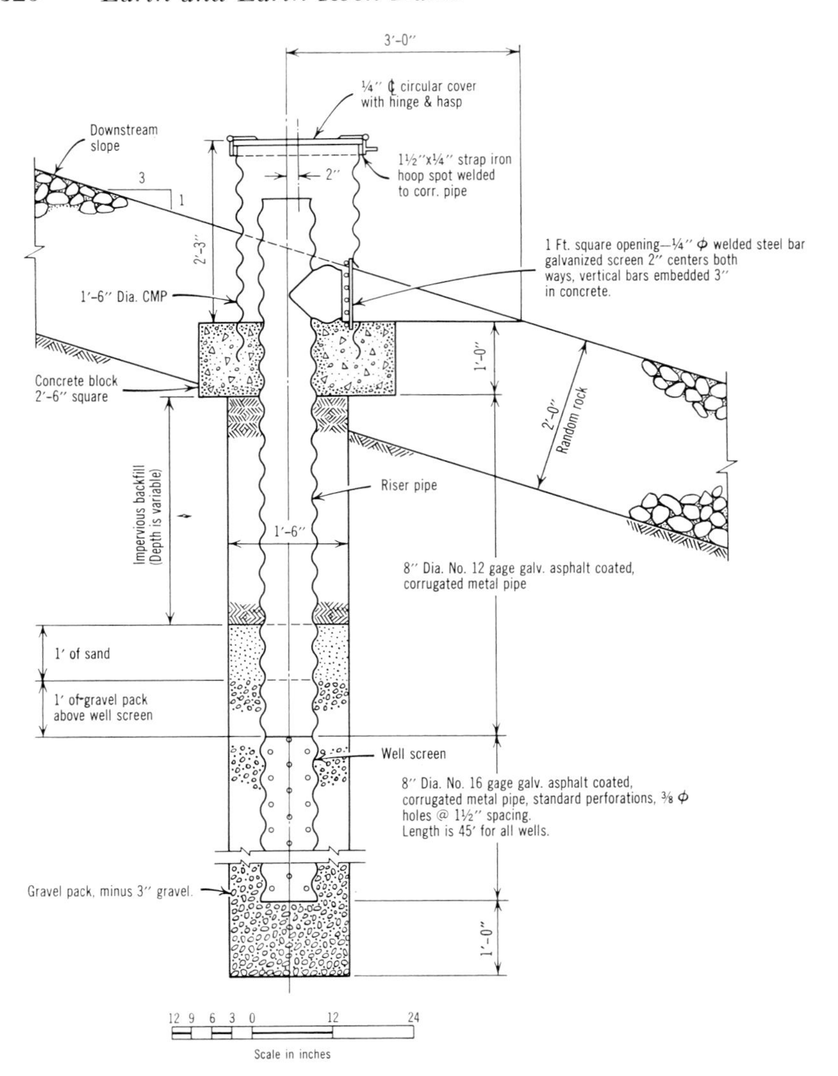

Fig. 6.4:3 Relief well detail—Success Dam. (Courtesy Sacramento District, U.S. Army Corps of Engineers)

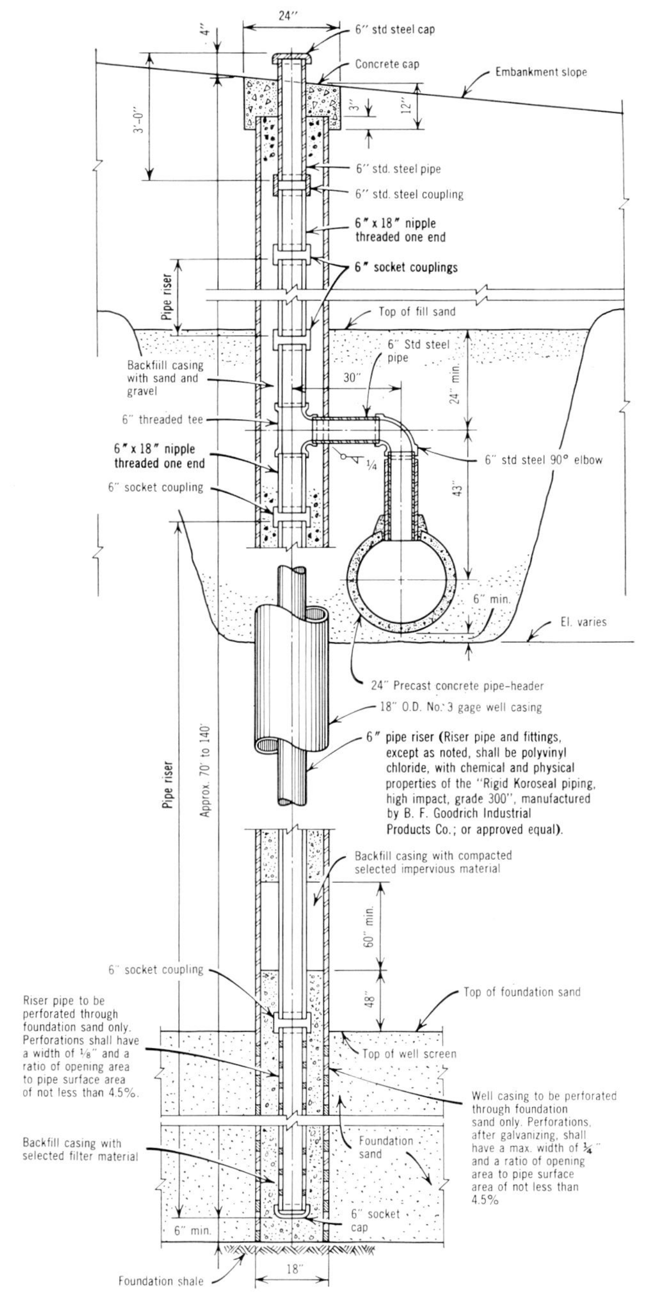

Fig. 6.4:4 Relief well using plastic pipe—Karnafuli Dam, East Pakistan, 1959. (Courtesy International Engineering Co.)

Fig. 6.4:5 Typical overflow discharge pipe at top of relief well.

infiltration. The inflow measurement can be used to compare with subsequent tests on the well to determine if it is becoming less effective with time.

Relief well systems may be supplemented at the abutments with horizontal drainage tunnels. Sheeted and braced tunnels have been provided at several major dams in recent years to tap seepage in earth and rock abutments; such tunnels may have lengths of many hundred feet. In rock abutments the holes are usually left open, and they may or may not be provided with a concrete lining which has weep holes in it. Tunnels in earth abutments may be concrete lined or may be backfilled with pervious filter material.

seven

Stability Analyses

7.1 HISTORICAL DEVELOPMENT

Prior to 1935 few experienced engineers placed much reliance on theoretical embankment stability analyses. Before this time, earth dam side slopes were selected wholly on the basis of past experience, and local rules evolved reflecting local experience, or the opinions of the principal designers in the area. In the period between 1905 and 1915, for example, during a boom in the construction of irrigation works in Colorado local engineers built a number of major earth dams which had steep upstream slopes (1.5:1) and flatter downstream slopes (2:1 to 3:1). But during the same years in California many large reservoirs were constructed with the use of earth dams in which the upstream slope was usually made flatter than the downstream slope.

In France in the late nineteenth and early twentieth centuries, the upstream slope was generally made 1.5:1 and the downstream slope flatter (Ref. 277):

> This belief in the 1.5:1 dogma was so firmly enrooted in the minds of certain French engineers that, having been called into consultation during 1934, on an earth dam 16 meters high, having all the characteristics of a slide, the contractor said in answer to our observations on the stability of the dam . . . . "A 1.5:1 dam always holds" . . . .

In his distinguished book *Earth Dam Projects* Joel D. Justin suggested that slopes be selected on the basis of past experience and judgment, and he summarized his remarks on the subject with the following recommendation (1932):

> Some engineers contend that the upstream slope should be flatter; others have claimed the opposite. A thorough study leads to the conclusion that neither is correct, as sometimes the upstream slope should be flatter and vice versa.

In general, the slope of the upstream face or that part of it to be under water, should not be steeper than one-half that at which the same material would stand out of water . . . (Ref. 278, p. 158).

In an influential series of articles covering the design and construction of earth dams, R. R. Proctor (Ref. 209, p. 372) suggested in 1933 that slopes should vary from 2:1 to 4:1, depending on the foundation conditions. However, he proposed no specific means of analysis or slope selection.

At the First World Congress of Large Dams in 1933, the consensus of opinion among eleven authorities representing Europe, Russia, and Japan was that the designer should pattern the cross section of a new dam after successful dams with similar dimensions (Ref. 279). Only five of these authorities believed that stability analyses had any value for the earth dam designer.

One of the first suggestions found in engineering literature that earth slopes could be analyzed on the basis of the results of laboratory tests of soil strength was made about 1850 by A. Collin, a French engineer (Ref. 294). At this early date Collin proposed a method of stability analysis very similar to that which we use now. Another early pioneer in recommending the use of a theoretical analysis was an eminent California consulting engineer, B. Bassell. In a small book, *Earth Dams* (Ref. 205), Bassell said that a dam embankment should have a base width that provides enough frictional resistance to prevent the dam from sliding downstream on its foundation under the reservoir pressure. He suggested a coefficient of friction between the dam and foundation of 1.0 and a factor of safety of 10.

The modern sliding circle method of analysis was first applied to the analysis of a quay wall failure in Sweden in 1916; hence the common name, "Swedish circle analysis" (Refs. 280, 291).[1] Over the next decade the applicability of the method to earth dam problems was hotly discussed, especially by European engineers, and a comprehensive discussion was published for the first time in English by Terzaghi in 1929 (Ref. 275).

In the years following 1930 a number of investigators checked the sliding circle stability analysis by computing the factor of safety in earth slopes which had suffered shear failures. But because so little was known about soil shear strength at that time, these studies produced inconclusive results. One of the more valuable studies, published by Terzaghi in 1933, was an analysis of a number of natural slope and embankment slides (Ref. 281). In it Terzaghi showed that slides in

[1] See other early treatment of the subject, Refs. 286–289.

some clay materials occurred at safety factors greater than unity when they were computed from the shear strengths measured in the laboratory at that time.

By the mid 1930's many earth dam engineers were interested in the sliding circle method, but they could find little factual information to support its reliability. On the contrary, most of the few slide investigations made indicated, like Terzaghi's, that the method could not have predicted the slides. One of the most interesting and valuable of these studies was the analysis by A. Mayer in 1936 of four upstream slides in clay dams in France (Ref. 277). Mayer also concluded after making direct shear tests on undisturbed and remolded samples of the embankment material, that the slopes would have been presumed stable from the sliding circle stability analysis and the soil strengths measured.

At the Second World Congress on Large Dams in 1936, 15 papers were presented which discussed earth dam stability analyses. But with the exception of Mayer's paper described above, they were devoted to variations in the methods of computations. In the last 20 years, our understanding of soil shear strength and methods of laboratory testing have been vastly improved. In addition, improved methods of computations for the stability analyses have been developed. Using these improved procedures, some recent comparisons of the results of computations of a number of natural slope slides have shown reasonable agreement. As a result of this progress, and also because of the great need which exists for an analytical means of estimating the margin of safety of earth dam embankments against shear failure, stability analyses have become firmly entrenched design procedures. It must be kept in mind, however, that the computation procedures are based on assumptions which can only be viewed realistically as gross simplifications. Stability analyses have not given a good check on all the natural slides which have been studied, and there have been almost no shear failures of rolled earth dam embankments in recent years which could be used to verify the theories. Consequently, some caution is required in the interpretation of the meaning of the results obtained.

The authors firmly believe that, at the present state of our ability to handle the problem, stability analyses should be considered primarily of value as a tool to compare the relative stability of various possible designs at a given site. For example, if during the design stage of a given dam two alternate designs are analyzed by using the same method of calculation, it is probable that the one with the highest computed safety factor is the one with the greatest actual reserve of safety against the shear failure—though this is far from being certain when radically different alternative designs are being studied. However, when one employs the absolute numerical value of the safety factor to justify the acceptibility of a given design, one is placing reliance on several important assumptions whose validity and limitations are still not well known.[1] For any dam which is built in any unprecedented manner in some major respect (particularly for dams of greater height or steeper slopes than used previously and for dams on foundation with suspect strength), regardless of the results of stability calculations, we must place primary reliance on measurements of movement and pore pressures in the completed structure to give final confirmation that the dam is performing as assumed.

7.2 METHODS OF ANALYSIS

Of the numerous methods of slope stability analysis which have been devised over the last 40 years, two general categories are in current use for earth dam embankments and foundations:

1. *Method A (sliding surface methods):* This general category includes procedures in which continuous surfaces of potential shear failure are assumed within the embankment and foundation. Calculations are made of the shear stresses along the potential failure surface which would be necessary for equilibrium, and of the shear strength forces which can be developed on the same surface.
2. *Method B (unit stress method):* This general category includes procedures in which the unit stress is estimated using elastic or plastic theories. The margin of safety is evaluated by comparing the computed unit shear stresses in the area analyzed with the shear strengths.

Both methods require simplifying assumptions whose validity cannot be readily evaluated and it is not clearly established which method may be intrinsically superior. In practice, most engineers prefer and work exclusively with some variation of Method *A*. Some use both methods. Few, if any, earth dams in recent years have been analyzed using only the procedures of Method *B*. The theories for Method *B* are discussed further in Sec. 7.8, but, except as otherwise noted, the term "stability analysis" is used in this book for the procedures of Method *A*.

[1] See Sec. 7.6 for a further discussion of the minimum tolerable safety factor.

The following general procedures are used in a stability analysis:

1. The problem is usually considered to be two-dimensional.[1]
2. A continuous surface of potential shear failure passing through the dam foundation and/or embankment is assumed. The shape and location of the surface are chosen arbitrarily on the basis of judgment as being a possible failure surface, Fig. 7.2:1. The trace of the shape of the failure surface on the dam cross section may be a circle, a series of straight lines, or any arbitrary curve. The portion of the dam embankment and foundation lying above the assumed failure surface is called "the trial sliding mass."
3. The shear stresses on the potential sliding surface which would be necessary to prevent movement are computed.
4. The shear strength forces available to resist movement are computed.

[1] See Sec. 7.4*e* for treatment of the stability problem in three dimensions.

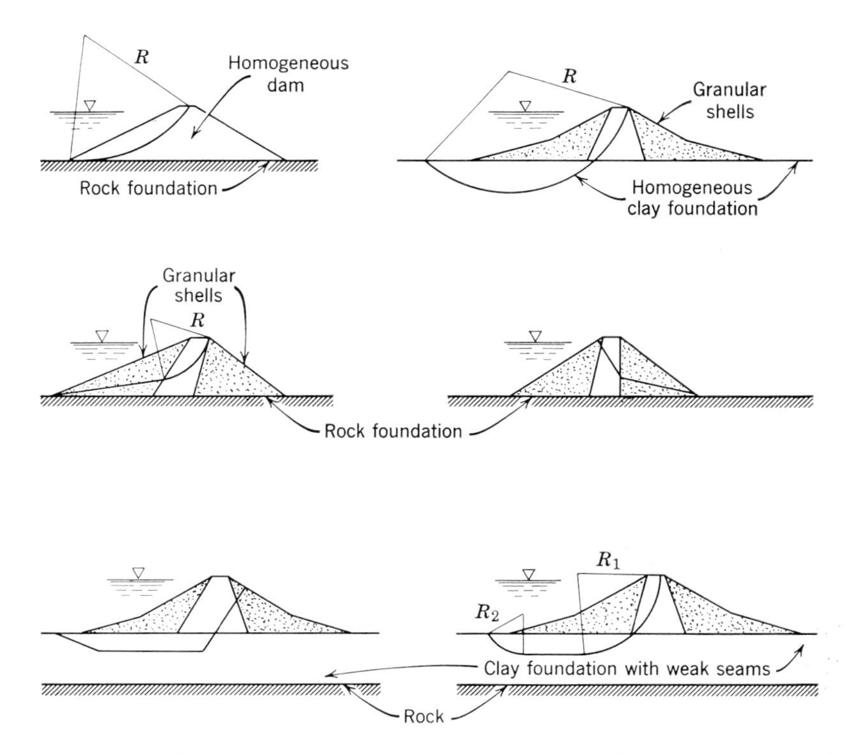

Fig. 7.2:1 Typical trial failure surfaces for stability analyses.

5. The margin of safety is obtained by a comparison of the shear stresses needed for equilibrium with shear strength available. Several definitions of the "safety factor" are in common use. These vary in the method of comparing the strength forces available to the forces needed for equilibrium. The forces involved and the methods of computation are described in Sec. 7.4.
6. The procedures are repeated for other potential failure surfaces until a "critical surface" is found which gives a minimum factor of safety.

Both theory and experience with failures show clearly (Sec. 2.2) that there are three periods in the life of a dam which may be critical from the standpoint of shear failure and which must be analyzed:

1. *"During construction" condition:* The factor of safety for either the upstream or downstream slope may be lowest during construction. This is particularly true of dams on soft foundations.
2. *"Full reservoir" condition:* When the reservoir has been full long enough for seepage water to percolate all the way through the embankment, the pressure in the pore water in the downstream portion reaches its highest values. Under this condition the downstream slope may have its lowest factor of safety against sliding.
3. *"Rapid drawdown" condition:* After the reservoir has been operating for some time and seepage water has penetrated the embankment, the upstream slope may fail by sliding after the reservoir is lowered.

At each of these stages the mechanics of failure are different, and slightly different analytic techniques are used. Details of the stability analyses for the critical periods are given in Secs. 7.5, 7.6, and 7.7.

7.3 SHEAR STRENGTH

Earth dam stability analyses require an estimate of the strength of the soil comprising the embankment and foundation. The embankment strength is estimated by performing laboratory tests on soil specimens which are compacted to densities and water contents simulating the conditions anticipated in the dam. The strength of foundation soils is determined by laboratory tests performed on samples obtained by bore holes or test pits.

At the present time the determination of shear strength is probably the most controversial and rapidly evolving single element of modern soil engineering. While great strides have been made in this field in the last 20 years, our knowledge of the subject is still far from adequate. Much research is being carried out in various parts of the world to clarify and extend our understanding of the fundamental factors influencing soil shear strength. As a result of this concentration in research effort, improved equipment and procedures are continually being developed and there is no general agreement at the present time on the best methods of testing and interpreting the test results.[1]

Laboratory determination of the shear strength of earth dam embankments requires more care and professional scrutiny than the routine testing of other construction materials. The work requires skilled technicians, relatively elaborate equipment, and professional supervision. Laboratory test results depend to a large degree on the test procedures and equipment used. The planning of the test program and the interpretation of the results are the most important components of the stability analysis, and require a thorough grounding in the specialty of modern soil mechanics.

A complete review of present knowledge of soil strength would require a volume of its own and, if written, such a volume would soon be out of date. Consequently, no attempt has been made here to present an exhaustive treatment of the subject. Since the problem of determining the strength of natural foundation soils is common to most other civil engineering problems, and is covered in detail in numerous professional papers and books, the discussions included here are limited generally to the strength of earth and rockfill embankment materials. Even with this limitation, no pretense is made that the subject is covered completely. It is assumed in the following sections that the reader has some familiarity with soil mechanics terms and concepts.

7.3a "Total" and "Effective" Stress Methods of Analysis

The total compressive stress acting at any point on a given surface within a compacted earth embankment consists of stress between the soil grains, called "effective" stress, and pressure in the pore fluid, called "pore" pressure. The combined effective stress and pore pressure acting on any surface is called the "total" stress.

Since the pore fluid cannot develop shear resistance, all shear stresses are resisted by the intergranular or effective stresses. The shear stress at failure on any surface within an embankment (shear strength) is

[1] See, for example, Refs. 62, 110, 524, 528, 529, 575.

directly related to the intergranular normal stress on that surface, and the relationship can be stated with sufficient precision for practical purposes by the equation

$$s = c' + (\sigma - u) \tan \phi'$$ [1]

or

$$s = c' + \sigma' \tan \phi' \qquad \text{(Eq. 7.3:1)}$$

where s = shear stress on the surface at failure
c' = cohesion intercept } in terms of effective stress
ϕ' = angle of shear resistance } in terms of effective stress
σ = total normal stress acting on the failure surface
u = pore water pressure acting on the failure surface
σ' = effective normal stress acting on the failure surface

Two basically different approaches to the stability problem are in common use by the earth dam designer: (1) the effective stress method and (2) the total stress method. In the effective stress method, the pore pressures which exist on the potential failure surface within the dam and foundation are estimated and the shear strength is given by Eq. 7.3:1. In the total stress method, laboratory tests are performed in a manner which is designed to simulate the conditions in the embankment, and the shear strength is determined in terms of total stresses without pore pressure measurements: it is assumed that the pore pressures which develop in the sample during the laboratory test will be equal to those which would develop in the embankment at failure.

Both methods of approach have been used from the earliest application of stability analyses to earth dam design. However, before the early 1950's there was little reliable information available from piezometer installations in dams in operation with which to check the theories; consequently, the problem of estimating the pore pressures which might exist in a dam at various times of its life, for use in the effective stress analysis, was not well established and the total stress analysis was widely considered to be the most realistic approach. More recently, considerable performance history has become available from measurements in actual dams (Secs. 3.3 and 3.4) and, concomittantly,

[1] The equation is strictly applicable only to materials in which the pore spaces are completely filled with either water or air. Where the voids contain both air and water, the pore pressures in the air and water are somewhat different and the equation is not correct. However, since pore pressures in an earth dam are not likely to be sufficiently high to be important unless the embankment is nearly completely saturated, the error can be neglected in earth dam analyses.

laboratory testing has developed to the point where much more is known about the strength of the soil in terms of effective stress, as well as the relationship between embankment strains and the resulting pore pressures.[1]

At the present time, some engineers use the effective stress method only (Ref. 522). Others continue to use the total stress method of analysis because they believe it is easier and more reliable. A complete discussion of the relative merits of the two methods of approach is beyond the scope of this book. The authors believe that the methods are equally reliable when applied with the same degree of understanding and judgment.

In the latest manual of practice of the Army Engineers which is used by all their district offices and, hence, is being applied to studies for a large number of dams, the total stress method is recommended almost exclusively because of the difficulty of estimating pore pressures as

[1] Measurements of pore water pressure in triaxial compression tests have been standard practice for only a very short time. In 1940, only a few research laboratories were experimenting with the necessary equipment, and it seemed that the laboratory problems were so great, especially for very fine-grained impervious soils, that pore pressure measurements would never be part of routine engineering practice. The first considerable application in the United States came soon after World War II when the engineers of the engineering laboratories of the USBR at Denver began to use and improve the measuring equipment. Although few private firms followed suit before 1955, today almost all soil engineering laboratories now make the test. The tests are still intricate and unless they are performed with great care and the experimental errors evaluated with accuracy, the results can be very misleading. One of the principal errors in the measurements, and one difficult to evaluate, results from the fact that the main zone of shear failure is in the center of the sample. The rigid loading caps and rubber membrane prevent the specimen from straining as much at the end as at the center, so that in rapidly performed tests the pore pressure measured at the ends may be quite different from that in the zone of failure. Early tests used an internal probe or needle for measuring the pore pressure near the center of the sample. However, difficulties developed with the testing equipment, and the solid probe had an indeterminable influence on the pore pressures inside, leading most investigators to conclude that it is preferable to make the measurements at the end of the sample. The only practical means so far devised to minimize this error is to run the test slowly so that the pore pressures have an opportunity to equalize throughout the sample. No specific length of time can be prescribed for this although researches indicate that the time required is much greater than has been generally suspected (see, for example, Ref. 525). Other difficulties in the testing procedure which are not yet completely solved arise from the action of air in the sample and from the thixotropic influence of clayey soils. The best recent summary of knowledge concerning the possible errors in pore pressure measurements is given in the papers and discussions of the ASCE Research Conference on the Shear Strength of Cohesive Soils, Boulder, Colorado (June, 1960).

explained in more detail in Sec. 7.5 and 7.6 (Ref. 619). On the other hand, the engineers of USBR carry out nearly all analyses in terms of effective stresses.

The principal advantage of the total stress method of analysis is its greater simplicity. The advantage of the effective stress method is the fact that the analysis is carried out with a somewhat more fundamental definition of the shear strength and it is possible to compare the pore pressures assumed in the design with those which develop in the dam and foundation as measured by piezometers. For major projects, the authors use the effective stress method as the primary analytical procedure and the total stress method as a check for comparison purposes.[1]

7.3b Types of Laboratory Tests for Compacted, Impervious Soils

For determining the strength of compacted, impervious soils, three types of tests are commonly used which vary in the method of consolidating the sample before it is failed in shear:

1. *Undrained test.* No drainage or dissipation of pore pressure is allowed during any part of the test (sometimes called "quick" or "unconsolidated-undrained" test).
2. *Consolidated-undrained test.* The sample is first allowed to consolidate (with full pore pressure dissipation) under a given consolidation pressure and then is failed in shear with no drainage allowed (sometimes called "consolidated-quick" test).
3. *Drained test.* Drainage and complete dissipation of pore pressures are allowed at all stages in the test (sometimes called "slow" test).

Shear strength testing for earth dam embankment material is carried out with triaxial compression or direct shear testing apparatus. Triaxial tests are used almost exclusively for undrained and consolidated-undrained tests, since the drainage cannot be controlled readily in a direct shear test (Ref. 311). The drained test can be performed with either triaxial or direct shear apparatus; the direct shear test is frequently adopted because the use of a thin sample allows more rapid

[1] As discussed in more detail in Sec. 7.5, the two methods often give radically different computed safety factors since the pore pressures assumed for the effective stress analysis are those which are not affected by large embankment shear strains, while, the pore pressures which enter the total stress analysis are those which occur at failure.

pore pressure dissipation and permits the test to be performed in less time.

7.3c Undrained Tests

Undrained tests are performed primarily to determine the relationship between the shear strength and normal pressure in terms of total stresses for use in the analysis of the stability of the dam during and immediately after construction. For this period in the life of the dam, the water content of the embankment will not be influenced as yet by seepage from the reservoir and the influence of rainfall, drying and pore pressure dissipation due to internal drainage is neglected. The samples are compacted and tested at a water content and density which are chosen to simulate the conditions that will be obtained during construction. A series of samples compacted to equal water contents and densities are tested in the triaxial apparatus with varying confining pressures, and the total principal stresses at failure are plotted in the form of stress circles on a Mohr diagram as shown in Fig. 7.3:1. The upper envelope of these circles ("failure envelope") defines the relationship between the shear strength (s) and the total normal stress (σ) which is used in the total stress analysis.

The failure envelope is typically concave downward, as shown in Fig. 7.3:1, with a slope which is steepest in the lower range of normal stress and which decreases gradually with increasing stress. This cur-

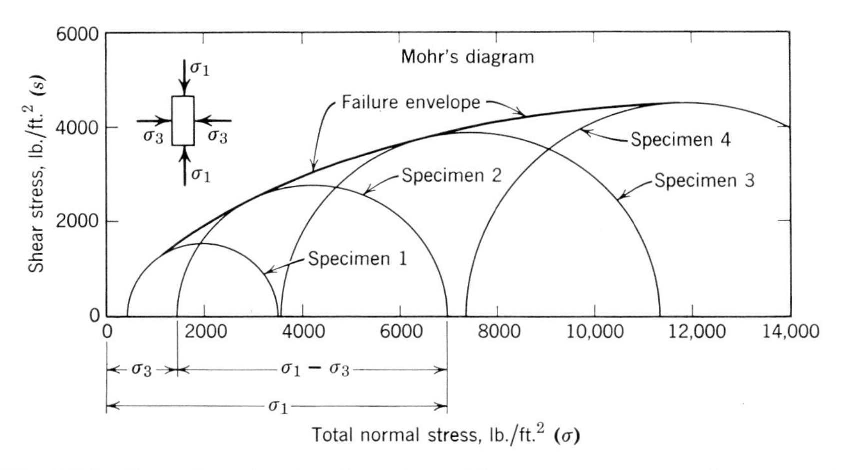

Fig. 7.3:1 Typical results of undrained triaxial tests on semisaturated compacted soil (all samples compacted to the same original water content and density).

vature in the failure envelope is due to the fact that in the lower stress range there is more air in the pore voids, and for these samples a larger percentage of the total normal stress on the failure surface is carried by the grain structure and consequently is effective in increasing the shear resistance. As the stresses are increased, the volume of free air in the voids is gradually compressed and forced into solution in the pore water. Thus an increasingly greater proportion of subsequent increments of total normal stress is carried by the pore water pressure. When the confining pressure is high enough to force all the air into solution, and the sample is completely saturated, the failure envelope flattens to an essentially horizontal line since, at this stage, all additional increments of confining pressure are carried wholly by the pore water, and do not result in any further increase in the shear strength.

For a given soil, the relationship between shear strength and total normal stress obtained in the undrained test varies greatly as a function of the water content of the specimens. Only a few percent difference in the water content dry or wet of Standard Proctor Optimum water content has a very large influence on the shear strength, Fig. 7.3:2. Because the total stress failure envelopes for undrained tests are curved lines, the strength relationship cannot be expressed conveniently in terms of a cohesion and an angle of friction except as an approximation for a given range of normal stress.

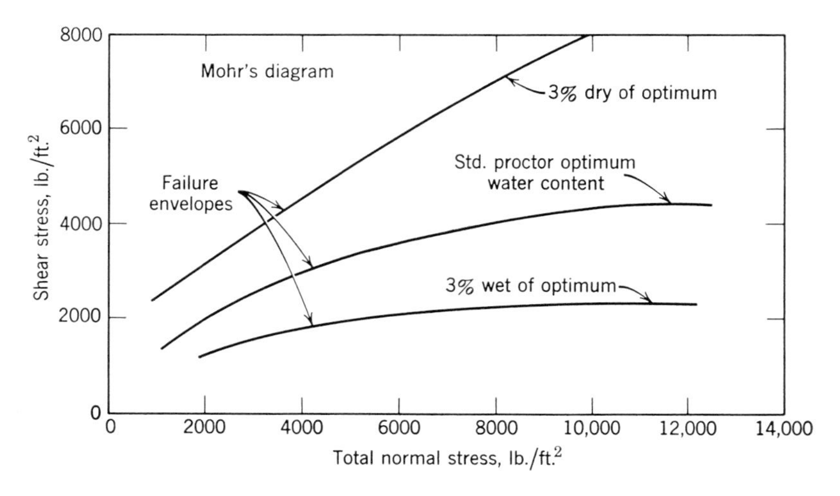

Fig. 7.3:2 Typical influence of water content on undrained shear strength semisaturated impervious compacted soil.

7.3d Consolidated-Undrained Tests

Consolidated-undrained tests are performed for two primary purposes:

1. With pore pressure measurements on saturated samples to determine the strength parameters in terms of effective stress (c' and ϕ').
2. Without pore pressure measurements on saturated or partially saturated samples to determine the relationship between shear strength and total normal stress for use in the total stress analyses.

Because of difficulties in measuring pore pressures in semi-saturated samples, specimens for tests performed to determine c' and ϕ' usually are saturated as well as possible before shearing. Two methods are used for saturating specimens: (1) As in permeability tests, by percolating water through the specimen under a pressure gradient in order to wash the air out of the voids; and (2) by applying sufficient "back pressure" to the pore fluid to force the air into solution.[1] Often, both methods are used on clayey specimens.

If the purpose of the test is to obtain the strength for use in the total stress method of analysis, the sample theoretically should be tested at a water content equal to that which may exist in the dam during the period being analyzed. However, since it is impossible to anticipate reliably what the final water content of a given embankment may be, and since the most conservative results (lowest shear strengths) are obtained with completely saturated specimens, it is common practice to perform the tests on saturated specimens. Another possible approach is to perform tests on two sets of specimens, one set completely saturated and another set partially saturated by some reasonable procedure, and to select the design shear strength as an arbitrary intermediate value.

[1] The former method will not replace all the air in the voids in most of the impervious materials used for the core of earth dams so that the back pressure method must be used when full saturation is desired. The magnitude of the back pressure necessary to cause complete saturation can be estimated in advance from tests on unsaturated specimens (Refs. 525 and 527) or computed using Boyle's and Henry's laws for the compression and solution of air in water, Sec. 7.5. The degree of saturation obtained can be checked before the sample is failed by adding an increment of confining pressure (σ_3) and then measuring the pore pressure induced under undrained conditions.

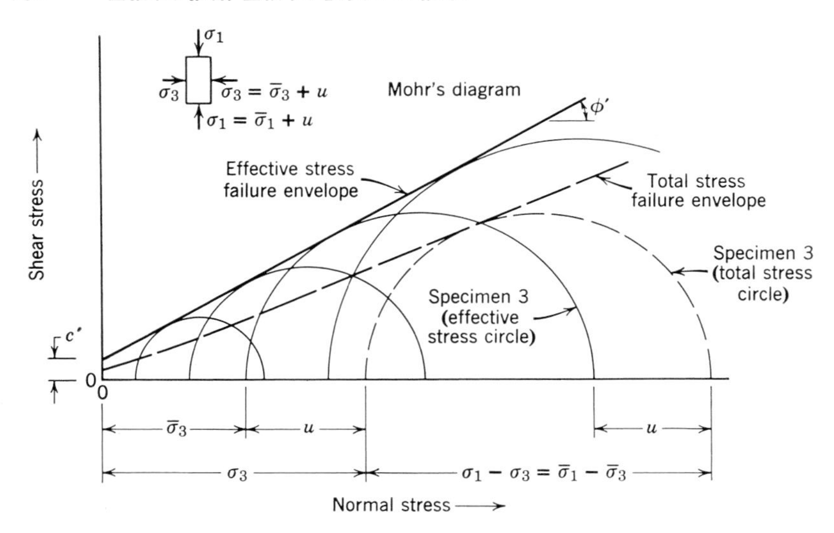

Fig. 7.3:3 Typical results of consolidated-undrained tests on saturated specimens of compacted impervious soil with pore pressure measurements.

Figure 7.3:3 shows typical results of consolidated-undrained tests on a saturated specimen of impervious soil in which the effective principal stresses at failure are plotted in the form of stress circles on a Mohr's diagram. For practical purposes, the rupture envelope for these circles usually can be approximated by a straight line which defines the effective stress strength parameters c' and ϕ' as shown. The results of this type of test on saturated specimens are not influenced to any appreciable degree by the water content at which the specimens are compacted. The test results vary considerably depending primarily on the degree of saturation obtained, the pore water pressure imposed before the sample is sheared, and the mode of the initial consolidation of the samples; i.e., whether they are consolidated under a hydrostatic pressure or in an anisotropic condition; i.e., $\sigma_1/\sigma_3 > 1.0$. The method of compaction used to prepare the specimen may have an important influence; however, no comprehensive researches have been carried out on this point.

7.3e Drained Tests

Drained tests are performed only for the purpose of determining the strength parameters in terms of effective stresses (c' and ϕ'). The values c' and ϕ' obtained with drained tests should be essentially the same as those obtained in consolidated undrained tests on saturated specimens with pore pressure measurements.[1]

Although comparisons have not been made for a great number of soil types and the direct shear test has been in disfavor for several technical reasons, there is considerable evidence to suggest that essentially the same results are obtained with direct shear and triaxial tests.

Drained tests should be performed on samples which are allowed to soak up sufficient water in advance to prevent any possibility that capillary pressures within the pores of the specimens may be acting to influence the results. For the direct shear test, the specimens are submerged at all times during the test in a pot of water. In triaxial compression tests, water is given access to the sample through the porous loading stones and sometimes through filter strips running through the sample. In addition, for triaxial tests performed at low confining pressures, it is usually necessary to saturate the specimen in the triaxial cell or to soak it before testing.

7.3f Strength of Embankment Sections Composed of Pervious Sand, Gravel, and Rock

Sections of embankments composed of clean sand and gravel are so pervious that the pore pressures which develop are caused primarily by gravity seepage flow; pore pressure changes due to changes in the embankment volume (as a result of compression or shear strains) are short-lived and revert to gravity seepage pressure so rapidly that they can be neglected. Consequently, the stability analyses are carried out using the effective stress method of approach.

For coarse pervious materials, the cohesion (c') should be assumed equal to zero since the intergranular attractive forces which are responsible for the cohesion component of the shear strength are negligible. The magnitude of the friction angle ϕ' ranges roughly between 33° and 45° and depends primarily on the relative density, the gradation, the coarseness, and the angularity of the material.[2]

[1] Some investigators have shown that the results are not identical and there is some controversy on this point.

[2] In past years, engineering reference books have listed lower ϕ' values for granular soils when saturated but many experiments show that in clean sands and gravels the value is not reduced either by partial or complete saturation. If the material contains clayey fines, however, there may be a reduction in the angle of internal friction when the material is wetted.

The strength is usually determined in the laboratory using drained tests. Material consisting of clean sand and gravel may be tested dry. Materials with a sufficient content of clayey fines to influence the results are saturated before testing. In general, laboratory strength tests are more reliable and much easier to make than for impervious soils since there is little problem in controlling pore pressures and drainage.

Because of the limitations of the size of laboratory testing equipment, it is not usually possible to test specimens containing large gravels. Research indicates that the maximum diameter of gravel in a triaxial compression test should not exceed 15% to 20% of the specimen diameter (Refs. 112, 224), and that the particle size in direct shear tests cannot exceed 5% of the minimum dimension of the shear surface without distorting the results (Ref. 153). It can be assumed that the results obtained in tests performed on the finer fraction of a given soil are conservative estimates of the strength of the entire soil.

Little reliable information is available on the strength of embankment sections composed of large and angular boulders or of hard quarried rock. There is no way to test such material, except possibly by failing test embankments in the field.[1] It is probable that, when loosely dumped, embankments of rock have an angle of internal friction between 40° and 45°. However, when placed in thin layers and compacted by the hauling equipment or by vibratory rollers (Sec. 11.5) in such a way that the smaller rocks are wedged into the voids between the larger, the strength may be much higher than given simply by $s = \sigma \tan 45°$.

7.4 FORCES INVOLVED AND CALCULATION PROCEDURES

Except where the total stress method of analysis is being used in connection with saturated clays, the strength force which can be developed to resist movement at any point on a potential failure surface is a function of the normal compressive stress acting on the surface at that point. Consequently, in order to make the calculations comprising the stability analyses, it is necessary to determine in some way the distribution of normal compressive stresses on the potential failure surface. The determination of the actual normal stresses which

[1] Large-scale, direct shear tests have been performed in the field at some damsites (Refs. 332 and 395). These tests also must be limited to relatively small rock and to low total normal pressures.

may exist is an indeterminate problem which can only be solved by making simplifying assumptions. A number of methods of calculation are in common use, each based on somewhat different simplifying assumptions.[1]

This section describes the most common methods of analysis currently in use by earth dam designers and the underlying assumptions regarding the distribution of stresses within the embankment. The discussion presupposes that the strength envelopes of the embankment and foundation materials have been already selected. The methods of calculation described can be used either with effective stresses, in which case it is necessary to estimate the distribution of pore water pressure, or with total stresses, in which case the shear strength is assumed to include the influence of pore pressures. The problem of estimating the pore pressures for the critical periods in the life of the dam is covered in Secs. 7.5, 7.6, and 7.7.

For all the analytical procedures in common use, possible failure surfaces through the embankment and foundation are assumed, and then, for each surface, computations are made of the shear forces needed for equilibrium and the strength forces available. A "safety factor" is then computed for each trial failure surface by comparing, by one of several procedures, the sum of the resisting forces available with the sum of the driving forces.

The problem is usually assumed to be two-dimensional.[2] Computations are made for that part of the dam's length where a slide would be most likely to develop; i.e., where the height is greatest or the foundation weakest. The procedure is repeated until the "critical" potential failure surface is found which gives a minimum factor of safety.

For the typical earth dam, potential failure surfaces pass through several zones of embankment and foundation material, each with different strengths and pore water pressures. Because of the variation of these properties along the potential failure surface, the most direct and almost universally applied method of analysis is to divide the trial sliding mass into slices (or wedges) in such a manner that the strength and the pore pressure forces on the bottom of each individual slice (or wedge) can be calculated. In order to make the problem statically determinate, assumptions must be made regarding the directions of the forces between the slices or wedges (side forces). The assumptions

[1] Over the last 25 years, many different methods of calculation were proposed. Some of the more interesting are described in Refs. 63, 132, 233, 284, 285, 293, 295, 523, 533, 537.

[2] See Sec. 7.4*e* for a discussion of three-dimensional stability analyses.

for the directions and magnitudes of the side forces have a large influence on the calculated factor of safety.[1]

The computations are laborious for all the main methods of analysis. The time required for a single failure surface commonly varies in the range between one and six hours depending on the complexity of the problem and the method of computation. As a consequence, programs for electronic computers have been devised recently for some of the methods of computation and at some dams literally hundreds of potential failure surfaces have been studied (see, for example, Refs. 54, 199). Computers have not been widely used, however, for several reasons. First, they have not been on the market long and the practicing civil engineer is only gradually taking up their use for his problems. The more important practical consideration which has deterred many engineers from using computers is that the assumptions underlying the methods of stability analyses themselves are so dubious that the whole procedure is at best supported only very tenuously by experience and reason.

For a given problem, the results of the calculations vary greatly as a function of the methods of computation and assumptions adopted. These differences, and our inability to evaluate the meaning of these differences, are so great as to overwhelm the relatively small benefits in increased precision which can be obtained by computing a great number of potential failure surfaces. For these reasons, the authors have not felt the desirability of using computers for this purpose, preferring to try a relatively small number of failure surfaces passing through critical zones, computed carefully by hand, using different methods of calculation as described in the following sections. In the future when better methods of computation are devised, the authors have no doubt that electronic computers may become extremely valuable or even essential tools for this purpose.

7.4a Standard Method of Slices

By far the most widely used method of analysis is the so-called standard method of slices in which it is assumed that the forces acting

[1] No comprehensive study has been made of the influence of the assumption used for side forces on the results of the various kinds of computation. This needs doing badly since it is clear that the influence is large. Some comments on this subject are given in Refs. 63, 320, 442, and 487. A recently completed study by N. Morgenstern (Ref. 702) in which the relations among the forces involved in the slice method of stability analyses are set up for the use of an electronic computer shows promise of providing the tools which may lead to a better understanding of the problem.

on the sides of a slice have no influence on the maximum shear resistance which can develop on the bottom of the slice. This simplifying assumption was first proposed by Krey (Ref. 299) and was introduced into the English technical literature in 1929 by Terzaghi (Ref. 275). At the present time, the standard method of slices is the principal analytical method used by the Earth Dam Section of the USBR (Ref. 35), it is one of the main methods used by the U.S. Army Corps of Engineers (Ref. 272), and it is used by the great majority of other engineers in the United States and abroad.

The method of calculation is shown with an illustrative example in Fig. 7.4:1 using the effective stress method of analysis. The procedure is as follows:

1. The trial sliding mass is divided into a number of slices (usually, but not necessarily, of equal width). Since the problem is two-dimensional, the section of the dam being analyzed is assumed to be one unit of thickness (1 ft. or 1 meter).
2. For each slice, the following forces are computed:
   (*a*) The total weight (W) which equals the area of the slice multiplied by the appropriate gross unit weight (soil plus water).
   (*b*) The normal force acting on the bottom of the slice, $N = W \cos \alpha$.
   (*c*) The total water force acting on the bottom of the slice which equals the average unit pore water pressure multiplied by the length of the bottom of the slice, $U = ub/\cos \alpha$.
   (*d*) The value $T' = W \sin \alpha$.
   (*e*) The total shearing resistance component due to cohesion which equals the unit cohesion multiplied by the length of the bottom of the slice $C = c'b/\cos \alpha$.
   (*f*) The total shear resistance which can be developed on the bottom of the slice at failure, $S = C + (N - U) \tan \phi'$.
3. The results of these computations are tabulated as shown in Fig. 7.4:1, and the sums of the forces $T' = W \sin \alpha$ and $S = C + (N - U) \tan \phi'$ are obtained.
4. The safety factor is computed by the formula

$$F = \frac{\Sigma S}{\Sigma T'} = \frac{\Sigma[C + (N - U) \tan \phi']}{\Sigma W \sin \alpha} \qquad \text{(Eq. 7.4:1)}$$

An alternate method of making the calculations, which is used by the Army Engineers (Ref. 619), the USBR (Refs. 35, 276) and many others, is to plot the forces T, C, and $N - U$ in the form of continuous curves and to measure their sums graphically with a planimeter as the

Adopted Design Criteria

Embankment zone	Strength		Gross unit weight	
	c'(psf)	tan ϕ'	Above seepage line	Below seepage line
Clay core	500	0.50	120	125
Gravel shell	0	0.80	125	

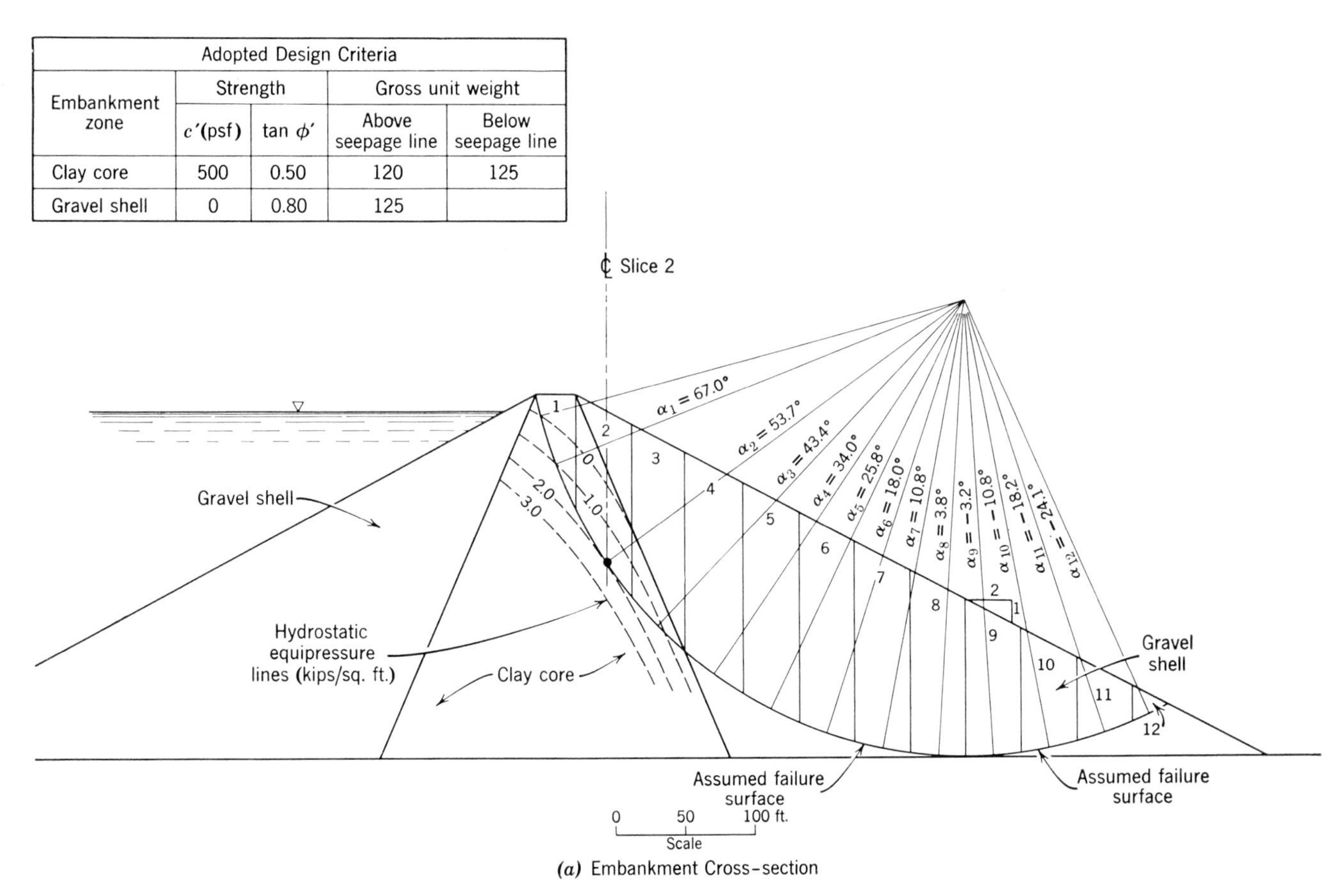

(a) Embankment Cross-section

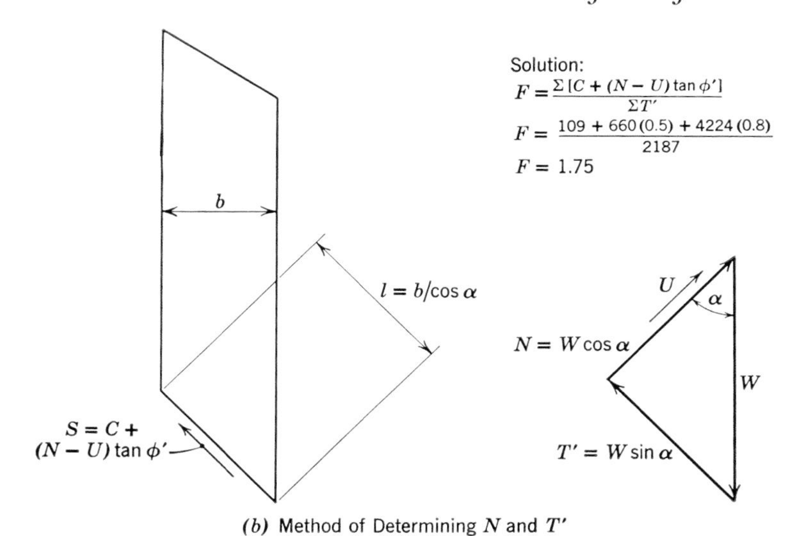

(*b*) Method of Determining N and T'

SUMMARY OF COMPUTATIONS—FORCES IN KIPS (1000 lb.)

Slice	W	$\sin\alpha$	$\cos\alpha$	$T' = W\sin\alpha$	$N = W\cos\alpha$	$C = c'b/\cos\alpha$	$U = ub/\cos\alpha$	$N - U$	
								Clay Core	*Gravel Shell*
1	195	0.920	0.391	179	76	44	60	16	
2	515	0.841	0.540	433	278	37	148	130	
3	783	0.687	0.727	537	569	28	55	514	
4	725	0.559	0.829	406	601				601
5	740	0.435	0.900	322	666				666
6	725	0.309	0.951	324	689				689
7	665	0.187	0.982	124	652				652
8	590	0.066	0.998	39	589				589
9	490	−0.061	0.998	−30	489				439
10	365	−0.187	0.982	−68	358				358
11	210	−0.312	0.950	−65	199				199
12	34	−0.418	0.909	−14	31				31
Sums				2187		109		660	4224

W = total weight of slice (soil plus water).
U = total hydrostatic force on the bottom of the slice
C = total cohesion force acting on the bottom of the slice.
N = total normal force acting on the bottom of the slice.
T = the total driving force exerted by each slice.
F = the safety factor.

Fig. 7.4:1 Standard method of slices.

area under the curves, Fig. 7.4:2. This procedure gives the same results and has the principal advantage that errors in the calculations of the individual quantities which are not seen in the tabular form of computation can be picked up easily since the plotted curves must be smooth and continuous lines. The difference in the precision is negligible.

The principal recommendations for the standard method of slices are its simplicity and relative ease of computation. In addition, the

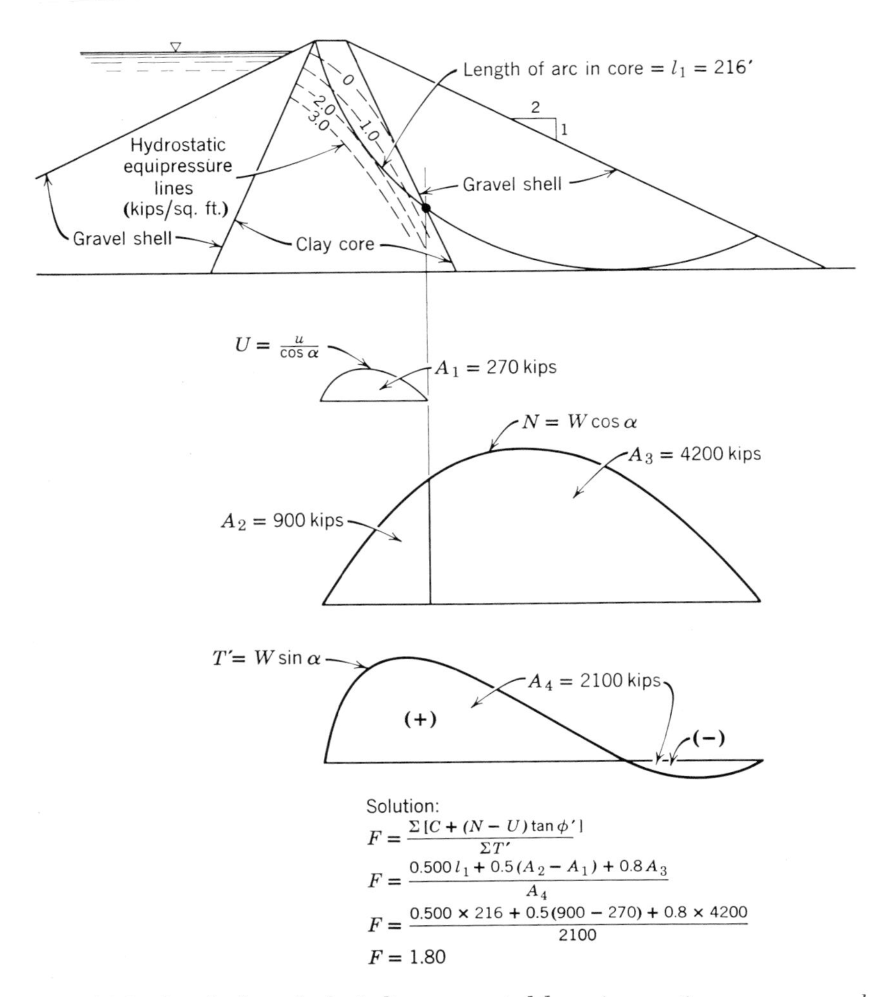

Fig. 7.4:2 Standard method of slices computed by using continuous curves and a planimeter for computing total forces. (**Note:** *See Fig. 7.4:1 for adopted design criteria.*)

method usually gives more conservative results (lower computed safety factors) than computations based on other reasonable assumptions concerning the influence of side forces on the slices. Another recommendation for the use of the method is the fact that it has been used for the great majority of earth dams constructed in the last 20 years, and there is, consequently, a large volume of experience to indicate that an embankment will probably be safe if it has a reasonable computed factor of safety based on the standard method of slices.[1]

There are several ways to view the basic assumptions concerning the side forces on an individual slice in the standard method of slices; however, none of these satisfies the conditions of static equilibrium for the entire sliding mass. The assumption concerning the forces on an individual slice can only be considered as a simplification which allows the computation of the distribution of the normal stresses along the assumed failure surface.

Occasionally, the method has been modified to include consideration of the pore water pressures acting on the sides of the slices even though the influence of the intergranular side forces is neglected. If stability analyses are to be refined to include side forces, the authors prefer to adopt a method of calculation in which all the forces assumed within the trial sliding mass are in static equilibrium, as discussed in Secs. 7.4*b* and *c*.

The standard method of slices has generally been limited to trial sliding masses bounded by a circular failure surface as shown in Figs. 7.4:1 and 7.4:2. Although the method was introduced for use with circular failure surfaces and in practice has been limited almost exclusively to such surfaces, it can be employed for noncircular failure surfaces using exactly the same computational procedures[2] and the results have the same meaning. An example of the standard method of slices applied to a nonuniform failure surface is given in Fig. 7.4:3. The widths of the slices are selected roughly in proportion to the degree of curvature of the assumed failure surface in such a way that the arc subtended by the bottom of the slice can be approximated reasonably by a straight line chord.

[1] It should be pointed out, however, that there are some exceptions to this general statement. There are a few well studied failures which indicate that under certain conditions, the safety factor may not be conservative. See, for example, the studies of the shear failures at the North Ridge Dam and Three Sisters' Dike in Canada by Peterson et al. (Refs. 109, 519).

[2] There is some controversy regarding the applicability of the standard method of slices to noncircular surfaces. In the authors' opinion there is no reason to limit the method to circles: the computations make are not correct the standpoint of statical equilibrium in either case and there cannot be much difference between circles and other surfaces of reasonable shape.

SUMMARY OF FORCES, KIPS

Slice	W	$T' =$ $W \sin \alpha$	$N =$ $W \cos \alpha$	$C =$ $cb/$ $\cos \alpha$	U	$N - U$	
						Clay Core	*Gravel Shell*
1	125	121	28	48	138	0[1]	
2	444	381	222	31	192	30	
3	623	402	479	24	168	311	
4	638	307	558	20	48	510	
5	1520	357	955	0	0		955
6	1610	368	1568	0	0		1568
7	1575	0	1575	0	0		1575
Sums		1936		123		851	4098

[1] This is an example in which because of the mechanics of the calculations and the shape of slice 1, we lose some of the uplift effect. The error is usually negligible; however the planimeter method of computation is probably preferable.

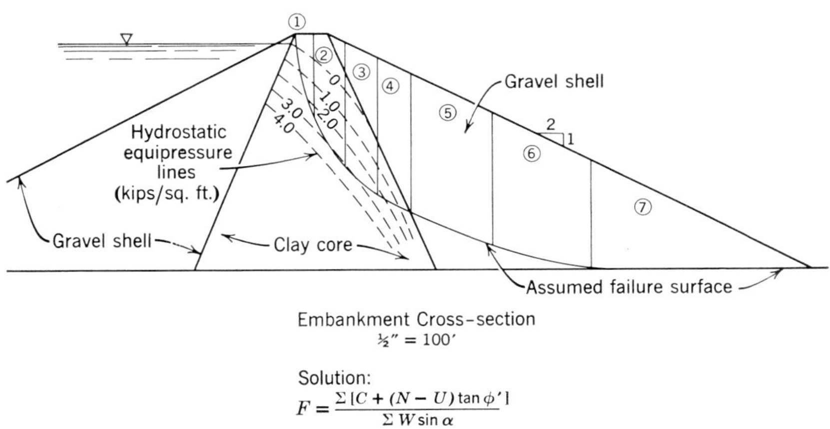

Solution:

$$F = \frac{\Sigma\,[C + (N - U)\tan\phi']}{\Sigma\, W \sin\alpha}$$

$$F = \frac{123 + 851\,(0.5) + 4098\,(0.8)}{1936}$$

$$F = 1.97$$

Fig. 7.4:3 Standard method of slices applied to a noncircular failure surface. (See Fig. 7.4:1 for design criteria.)

7.4b Taylor's "Modified Swedish" Method Including Side Forces Between Slices

This method was developed about ten years ago by Taylor (Ref. 298) as a practical means for including the influence of side forces between the slices. While it is not nearly so widely used as the standard method of slices, it is the only procedure for computation including side forces which has received any general acceptance to date in the United States and is used by many practicing engineers as a standard method of approach. A description of the method is included in a widely used handbook on hydraulic engineering (Ref. 88) and has been adopted as one of the accepted methods of analysis by the U.S. Army Corps of Engineers (Ref. 272, Appendix G).

Since there have been so many modifications and local adaptations of the slice method, it may seem somewhat presumptuous to label any one of them the "modified" procedure. However, the method described in this section is being rapidly accepted among practicing engineers and it needs a name. The name "modified Swedish" method was adopted by the U.S. Army Corps of Engineers in their Engineering Manual for Civil Works Construction and has been continued here.

The method of calculation is relatively simple. The problem is made statically determinate by assuming (1) the directions of the intergranular forces acting on the sides of the slices, and (2) that an equal proportion of the shear strength available is developed on the bottom of all of the slices; i.e., that the shear strength force which is developed on all slices is equal to

$$S_D = C_D + (N_D - U) \tan \phi_D \qquad \text{(Eq. 7.4:2)}$$

where $C_D = C/F_D$, $\tan \phi_D = \tan \phi' F_D$, and F_D is defined as the factor of safety. The force N_D is the normal force acting on the bottom of a slice when the trial sliding mass is at a state of equilibrium (not at failure). An illustrative example is shown in Fig. 7.4:4. The procedure is as follows:

1. An assumed potential failure surface is chosen for investigation. This may be of any arbitrary shape.
2. The trial sliding mass is divided into a number of slices. The slices do not have to be of equal width. The width of each slice is chosen roughly in proportion to the degree of curvature of the failure surface so that the chord and the arc subtended at the bottom of the slice are not greatly different in length and that the portion of the failure surface subtended by each slice passes through material of only one soil type.
3. For each slice, the forces which are known are computed as follows:
   (*a*) The total weight (W) which is equal to the area multiplied by the appropriate gross unit weight (soil plus water).
   (*b*) The total water forces acting on the sides and bottom of the slices (U_L, U_R, and U_B, Fig. 7.4:4*b*) which are equal to the

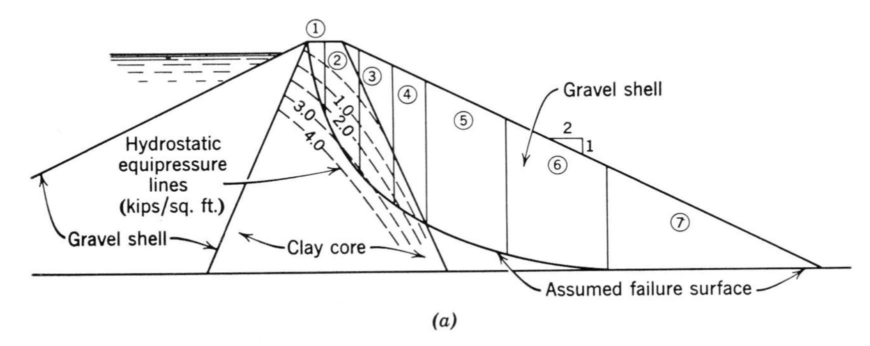

(a)

SUMMARY OF FORCES, KIPS (1000 LB.)

Slice	W	U_B	U_L	U_R	C	C_D		
						$F = 2.1$	$F = 2.3$	$F = 2.5$
1	125	138	0	75	48	23	21	19
2	444	192	75	131	31	15	14	13
3	623	168	131	71	24	11	10	9
4	638	48	71	0	20	10	9	8
5	1520	0	0	0	0	0	0	0
6	1610	0	0	0	0	0	0	0
7	1575	0	0	0	0	0	0	0

W = total weight of slice (soil plus water).
U_L = total hydrostatic pressure force acting on left side of slice.
U_R = total hydrostatic pressure force acting on right side of slice.
U_B = total hydrostatic pressure force acting on bottom of slice.
E = total intergranular force acting on the sides of the slice.
C = total cohesion force acting on the bottom of the slice at failure.
C_D = total cohesion force developed on bottom of the slice $= C \div F_D$.
ϕ_D = developed angle of internal friction ($\tan \phi_D = \tan \phi' \div F_D$).
F_D = safety factor.

Fig. 7.4:4 (*Continued on p. 349*).

average unit pore water pressure multiplied by the area over which the pressure acts.

(*c*) The cohesion force acting on the bottom of the slice at failure (C) which is equal to the unit cohesion multiplied by the area of the base of the slice.

4. For each slice the known weight and pore pressure forces are resolved into a resultant (R), Fig. 7.4:4*c*.

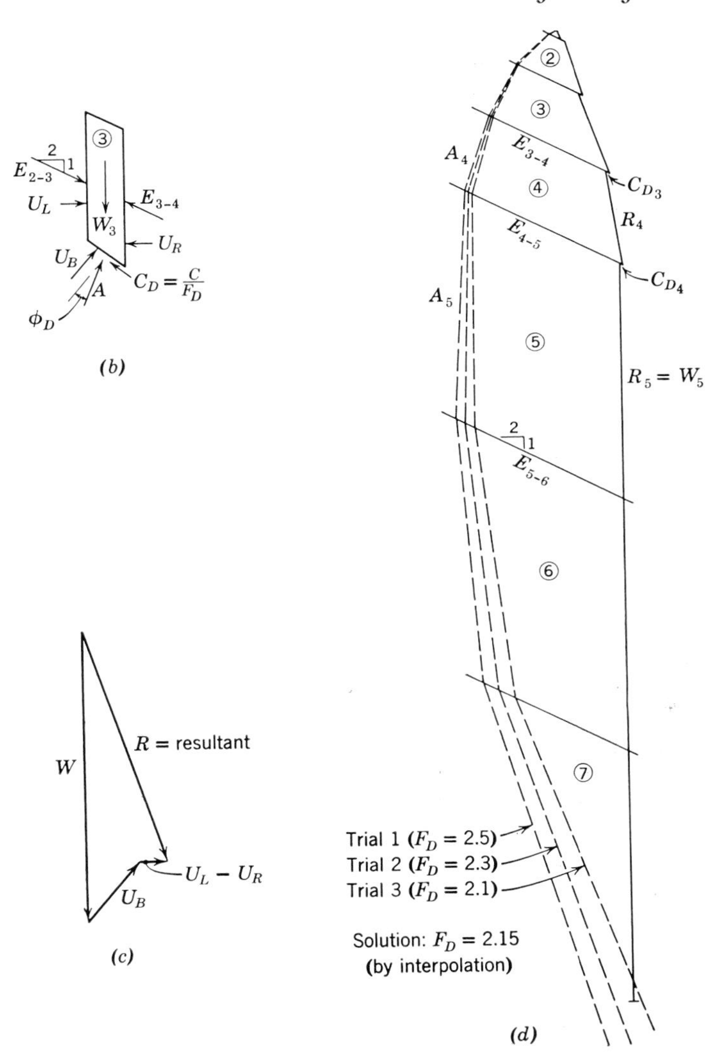

*Fig. 7.4:4 Modified Swedish method. (a) Embankment cross section. (b) Forces acting on an individual slice. (c) Resolution of known forces on individual slice. (d) Composite force polygon. (**Note:** See Fig. 7.4:1 for adopted design criteria.)*

5. The direction of the intergranular forces acting between the slices is assumed, usually as being constant and parallel to the average exterior slope of the embankment.[1]
6. The final solution is made graphically by progressive trials. For the first trial, a safety factor is assumed, values of C_D are computed for each slice ($C_D = C/F_D$), and a composite force polygon for the whole trial sliding mass including all the forces acting on the individual slices is drawn graphically starting from either end, Fig. 7.4:4*d*. If the force polygon does not close, another estimate is made for the safety factor and the procedure is continued until a safety factor is obtained which causes the polygon to close. In the illustrative example, a safety factor of approximately 2.15 would cause the polygon to close and is the solution to the problem.

In this manner, a safety factor is obtained which assures that the conditions of translatory equilibrium are satisfied, i.e., that the summation of the vertical and horizontal forces acting on and within the trial sliding mass are equal to zero. In these calculations, the positions of the lines of actions of the resultant forces acting on the individual slices and the condition of moment equilibrium are not considered. Any attempt to satisfy the moment equations for the individual slices adds another degree of indeterminancy to the problem and requires assumptions regarding the points of application of the forces which are, at best, rough approximations. It can be shown that the forces obtained in the modified Swedish method as described above can also be made to satisfy the moment equations if a reasonable pattern for the points of application for the forces acting on the various slices is assumed; however, the use of the moment equations does not refine the precision or reliability of the calculations and, therefore, it has appeared practical to solve the problem simply for translatory equilibrium as shown in Fig. 7.4:4 and to ignore the moment equations.

While still based on an arbitrary assumption concerning the side force directions, the modified Swedish method is a considerable improvement over the standard method of slices since all the forces involved in the computations are in statical equilibrium. The authors recommend that the method (or the alternative simplified method described in Sec. 7.4*c*) be used as the final analytical procedure for all important dam structures.

[1] The assumed direction has a major influence on the computed results. Some engineers assume the slope of the intergranular side force to vary from slice to slice and be approximately equal to the average slope of the top and the bottom of the slice.

7.4c Simplified Method Including Side Forces Between Slices

In Taylor's modified Swedish method, the final solution must be obtained by progressive trials to close the force polygon as shown in Fig. 7.4:4*d*. With a minor change in the definition of the safety factor, the authors have used a method of calculation in which essentially the same results are obtained by direct computation without the necessity for progressive trials. An illustrative example is given in Fig. 7.4:5. The procedure is as follows:

1. An assumed potential failure surface is chosen. This may be of any arbitrary shape.
2. The trial sliding mass is divided into a number of slices. The slices do not have to be of equal width. The width of each slice is chosen roughly in proportion to the degree of curvature of the failure surface so that the chord and the arc subtended at the bottom of the slice are not greatly different in length and that the failure surface subtended by each slice passes through material of only one soil type.
3. For each slice, the forces which are known are computed as follows:
   (*a*) The total weight (W) which is equal to the area multiplied by the appropriate gross unit weight (soil plus water).
   (*b*) The total water forces acting on the sides and bottom of the slices U_L, U_R, and U_B (Fig. 7.4:5*b*), which equal the average unit pore water pressure multiplied by the area over which the pressure acts.
   (*c*) The cohesion force acting on the bottom of the slice at failure (C) which is equal to unit cohesion multiplied by the area.
4. Assume that the direction of the intergranular side forces is constant throughout the trial sliding mass and equal to the mean exterior slope of the embankment.
5. For each slice, draw to a convenient scale a polygon of forces which would act at failure, Fig. 7.4:5*c*.
6. From the force polygon, determine the components of the resisting and driving forces which would act in the direction of the assumed side force (R_F and T).
7. Compute the factor of safety as $F_1 = \Sigma R_F / \Sigma T$.

Defined in this manner, the safety factor is the ratio of the sums of two sets of hypothetical forces acting in the same direction and, hence, would at first glance appear to be a different approach to the defini-

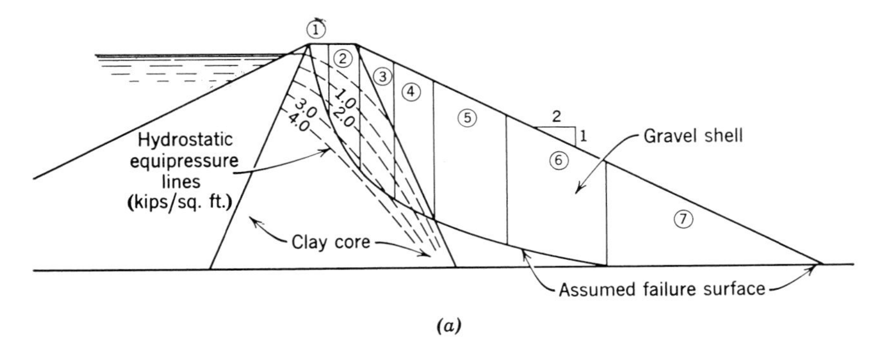

(a)

SUMMARY OF FORCES, KIPS (1,000 LB.)

Slice	W	U_B	U_L	U_R	C	T	R_F
1	125	138	0	75	48	165	91
2	444	192	75	131	31	420	166
3	623	168	131	71	24	457	195
4	638	48	71	0	20	370	258
5	1520	0	0	0	0	539	1202
6	1610	0	0	0	0	378	1500
7	1575	0	0	0	0	0	2377
						2329	5789

W = total weight of slice (soil plus water).
U_L = total hydrostatic pressure force acting on left side of slice.
U_R = total hydrostatic pressure force acting on right side of slice.
U_B = total hydrostatic pressure force acting on bottom of slice.
ΔE_F = net intergranular side force acting on the slice at failure (the vector sum of the forces on the two sides).
S_F = shear force at failure on the bottom of the slice.
T = the driving force in the assumed direction of the side forces (the outside force which would be necessary to hold the slice in equilibrium if there were no shear strength developed on the bottom of the slice).
R_F = the resisting force at failure in the direction of the assumed side forces. (The component of S_F when resolved in the direction of the assumed side forces).
N_F = the total normal force acting on the bottom of the slice at failure.

Solution:
$$F_1 = \frac{\Sigma R_F}{\Sigma T} = \frac{5789}{2329} = 2.48$$

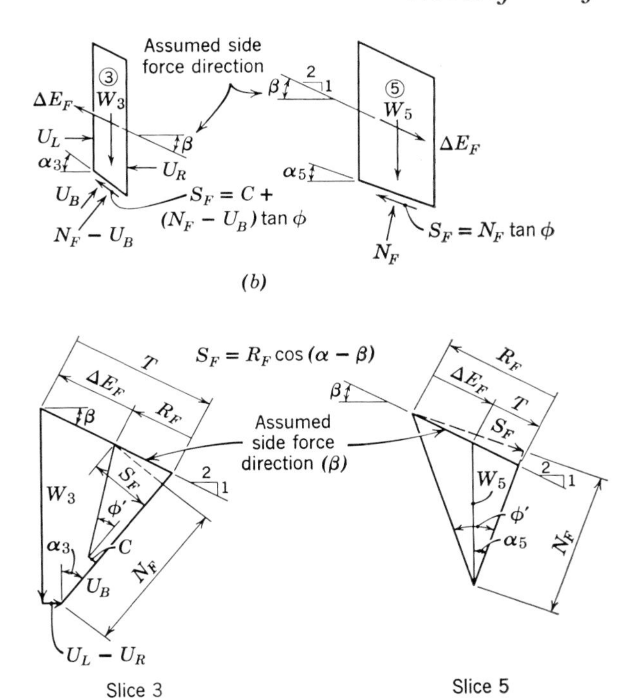

Fig. 7.4:5 Simplified method of slices with side forces. (a) Embankment cross section. (b) Forces acting on individual slices at failure (not shown here to the same scale). (c) Force polygons for individual slices at failure. (**Note:** *See Fig. 7.4:1 for adopted design criteria.*)

tion of the safety factor and not directly comparable to the results obtained in the other methods of analysis. However, it can be readily shown that the safety factor so defined is also exactly equal numerically to the ratio of the sum of the shear strength forces which can be developed on the potential failure surface to the sum of the shear forces necessary for equilibrium; i.e.,

$$F_1 = \frac{\Sigma R_F}{\Sigma T} = \frac{\Sigma \text{ (shear strength forces available on the assumed failure surface)}}{\Sigma \text{ (shear forces needed on the failure surface for equilibrium)}}$$

$$F_1 = \frac{\Sigma[C + (N_F - U)\tan\phi']}{\Sigma S_D} \qquad \text{(Eq. 7.4:3)}$$

The factor of safety defined for this method of computation (F_1) has a slightly different base than that used for the modified Swedish method (F_D). The difference is due to the fact that the normal force included in the numerator of Eq. 7.4:3 is the force which would exist at failure (N_F) whereas a similar equation for modified Swedish safety factor (F_D) contains the normal force which acts at a condition of equilibrium (N_D). Both safety factors are measures of the margin of safety of the slope and neither is preferable in any absolute sense. Both definitions give exactly the same result for an embankment on the verge of failure; i.e., $F_1 = F_D = 1.0$ (since $N_F = N_D$ for all slices at failure). Above failure, the safety factors diverge and the greater the stability of the slope, the greater the difference in the computed values of F_1 and F_D. For ordinary earth dams, in which one factor of safety is of the order of magnitude of 1.5, the difference between the two safety factors is usually small. The difference in the safety factors for the illustrative example of Figs. 7.4:4 and 7.4:5 is typical; i.e., $F_D = 2.15$ and $F_1 = 2.48$.

7.4d Wedge (Sliding Block) Method

In this method, the trial sliding mass is divided into two or three large sections or wedges. The upper wedge is called the driving or active wedge. The lower wedge is called the resisting or passive wedge. In a three-wedge system, the middle segment is sometimes referred to as the sliding block. The potential failure surface is simplified to a series of planes.

The wedge method is used by almost all engineers in circumstances where it appears that the failure surface can be best approximated by a series of planes rather than a smooth continuous curve. It is one of the principal methods of the U.S. Army Corps of Engineers.

The method is most frequently applied under two circumstances as shown in Fig. 7.4:6:

1. Where a thin horizontal stratum of weak soil exists in the upper part of the foundation (Fig. 7.4:6*a*).
2. Where the foundation consists of strong rock which would not be involved in the failure and the dam has a core of fine-grained soil with relatively large shells of dense granular material (Fig. 7.4:6*b*).

In the former case, failure is most likely to occur as a result of movement on the weak horizontal foundation layer; in the latter case, it is

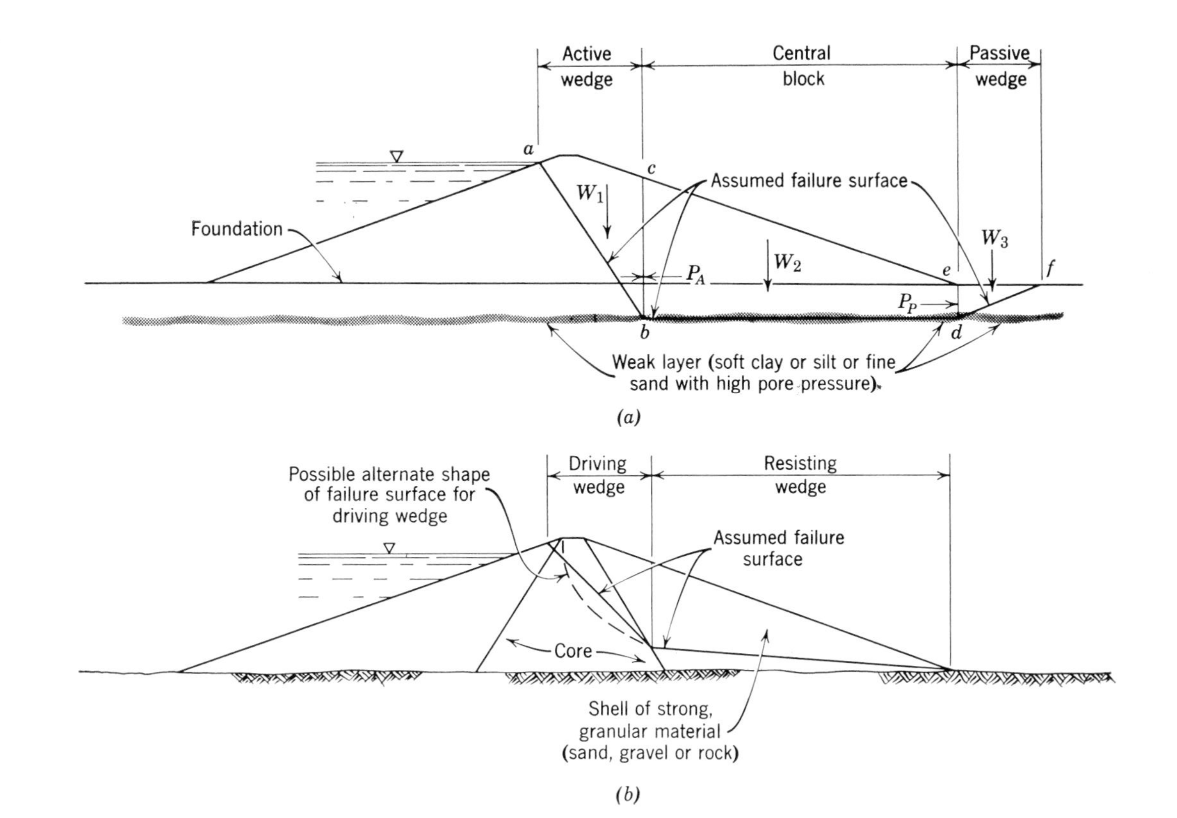

Fig. 7.4:6 Situations in which wedge (sliding block) analyses are applicable. (a) Dam on foundation containing weak horizontal layer. (b) Dam on strong foundation with granular shells.

generally believed that the portion of the failure surface passing through the granular shells is likely to be approximated reasonably well by a plane surface.[1]

The calculations of the safety factor can be made in exactly the same manner as for the methods of slices in which side forces are considered; i.e., by the procedures of the modified Swedish or the simplified methods described in Secs. 7.4*b* and 7.4*c*. The only difference is that there are two or three wedges instead of a larger number of slices. An illustrative example is given in Fig. 7.4:7 in which the safety factor is computed by both methods.

As in all methods of analysis, various possible failure surfaces are

[1] Some interesting recent analytical and model studies indicate that the most critical failure surface in a downstream slope of granular material may actually have an upward convexity (Refs. 534, 625). See also discussions of the subject in Refs. 2, 153, 532.

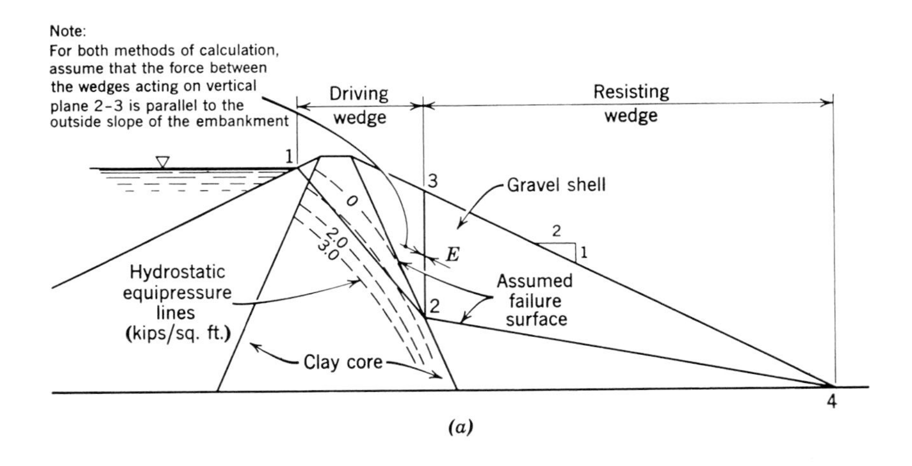

(*a*)

SUMMARY OF FORCES, KIPS

Wedge	*W*	*U*	*C*	*Modified Swedish* C_D			*Simplified Method*[2]	
				$F = 1.9$	$F = 2.1$	$F = 2.3$	R_F	T
Driving	1690	177	98	52	50	48	587	1170
Resisting	4690	0	0	0	0	0	3675	655
						Sums	4262	1825

[2] Obtained from individual force polygons (not shown).

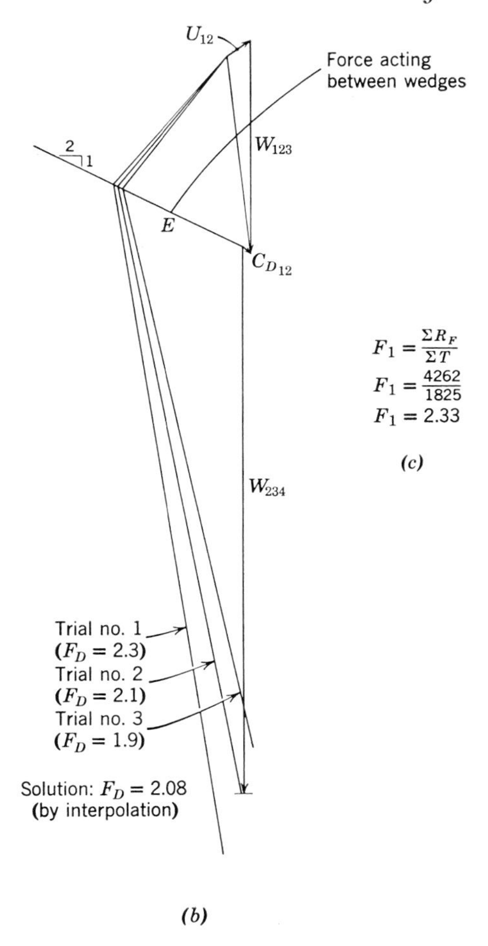

Fig. 7.4:7 Calculation of two-wedge problem with modified Swedish and simplified methods. (a) Embankment cross section. (b) Solution by the modified Swedish method. (c) Simplified method solution. (**Note:** *See Fig. 7.4:1 for adopted design criteria.*)

systematically investigated until the combination having the minimum factor of safety is found. For example, in Fig. 7.4:7, several positions of the vertical plane 2–3 should be investigated and, for each, several inclinations of the failure surfaces should be tried.

Occasionally, a somewhat smaller safety factor will be obtained for such a problem if the failure surface bounding the active wedge is taken as a curve, as shown by the dotted line in Fig. 7.4:6*b*. For the computations using the curved surface, the active wedge is subdivided into several slices and the problem reverts simply to that of the method of slices with side forces.

Another method for calculating the safety factor which is frequently used for the problem shown in Fig. 7.4:6*a*, is as follows:

1. Assume that sufficient horizontal movement has occurred to place the active and passive wedges at failure.
2. Assume the total forces on the vertical planes (*bc* and *de*, Fig. 7.4:6*a*) are horizontal and compute the active force on plane $bc(P_A)$ and the passive force on plane $de(P_P)$.
3. Define the safety factor under these conditions with reference to the stability of the central block; i.e.:
   (*a*) The unbalanced force acting on the central block is the difference between the active force exerted by the driving wedge and the passive force exerted by the resisting wedge:

$$P_1 = P_A - P_P$$

   (*b*) The force available to resist the movement of the central block equals:

$$P_2 = C_{bd} + (W_2 - U_{bd}) \tan \phi'_{bd}$$

   (*c*) Safety factor = $$F_2 = \frac{P_2}{P_1} \qquad \text{(Eq. 7.4:4)}$$

Defined in this manner, the safety factor is not directly comparable to the other safety factors (F_1 and F_D) for which it is assumed that the same proportion of the available shear strength is developed on all segments of the failure surface. (At failure, all the safety factors converge on F = 1.0.)

The definition of the safety factor given in Eq. 7.4:4 may be preferred in the situation where (1) the material comprising the active and passive wedges is stiffer (has less shear strain at failure) than the material along the weak plane in the foundation (*bd*) and (2) the total shearing resistance generated on surface *bd* is an appreciable fraction of the total resistance which can be developed on the entire failure surface (*abdf*).

7.4e Analysis in Three Dimensions

In the above discussions the problem is considered two-dimensional; i.e., the length of the sliding mass is considered so great that the end effect is negligible. In reality each potential sliding mass has a definite length, and the shearing forces acting on the ends of it increase the resistance to movement. It is conservative to ignore the end effects, and for long dams these are so small that they can be neglected justifiably. Where the height of the dam is great compared to the crest length, however, it is desirable to estimate the influence of the end resistance on the computed factor of safety.

This problem is important for many high dams, since they are often constructed in relatively narrow valleys with steep rock abutments where the crest length is considerably shorter than the distance between the upstream and downstream toes. At such a site any deep trial sliding surface must have appreciable curvature to conform to the shape of the rock valley wall. Where the average slopes of the dam abutments are steeper than 1:1 and the valley bottom is narrow, the problem must be considered in three dimensions.

The three-dimensional effect can be computed readily with roughly the same degree of reliability as the results of the ordinary two dimensional analysis. The procedure is as follows:

1. Divide the length of the dam into a series (3 to 5) of segments of equal length with vertical planes perpendicular to the longitudinal axis.
2. Analyze the average cross section of each as though it were a standard two dimensional problem; i.e., find the critical potential shear surface and compute the sums of the resisting and driving forces.[1]
3. Assume that the two-dimensional critical surface for each segment of the dam length is contained in the most critical three-dimensional failure surface.
4. Define the factor of safety as the ratio of the sum of the resisting forces to the sum of the driving forces for all the segments of dam length.

The result of this calculation essentially gives a "weighted" average of the stability of the various sections of the embankment. Often for high dams in narrow rock walled valleys the three-dimensional factor of safety is 25 to 50% greater than the two dimensional factor of safety computed on the basis of the maximum dam height. The difference is accentuated in narrow valleys which are curved in plan over the length of the dam which is not an uncommon occurrence.

[1] Either the standard method of slices (Sec. 7.4*a*), the simplified method including side forces (Sec. 7.4*c*), or the wedge method (Sec. 7.4*d*) can be used. The procedure cannot be applied readily using the modified Swedish method since in this method, the sums of the driving and resisting forces are not obtained.

7.5 STABILITY CONDITION "DURING CONSTRUCTION"

Slides of rolled-earth dams during construction have not occurred as frequently as slides during the operation of the reservoir and have not resulted in failures of the catastrophic type (Sec. 2.4). Nevertheless, the pore pressures which develop in the embankment or foundation during construction may be higher than at any subsequent time (Sec. 3.3), and it is usually advisable to analyze the stability of the embankment for this condition.[1] The construction condition is especially likely to be critical for dams on soft foundations.[2]

TOTAL STRESS ANALYSIS

In the total stress method of analysis, the results of undrained shear tests on unsaturated samples (Sec. 7.3*c*) are used to determine the shear strength. The results of the laboratory tests are greatly influenced by the water content at which the samples are compacted and it is difficult to estimate exactly the average water content which may be obtained in the field during construction. Therefore, several sets of laboratory tests on samples compacted at different water contents should be performed and average as well as conservative estimates of the shear strength should be used in the analyses.

ESTIMATING PORE WATER PRESSURES FOR USE IN EFFECTIVE STRESS ANALYSES

Though we have theory and laboratory test procedures which can be used to predict construction pore pressures, as well as a considerable number of measurements with piezometers in dams under various conditions, we still are not able to predict reliably or precisely in the design stage what these pressures might be in a given dam. The primary reason for this is that the construction pore pressures depend on factors which the designer cannot evaluate well in advance such as the amount of construction which will be done in wet weather, the influence of drying in dry weather, and the effectiveness of moisture and

[1] Section 1.4 contains a discussion of design methods used to control construction pore pressures, so as to avoid the necessity for making the embankment larger because of the stability during construction than it would have to be under reservoir operating conditions.

[2] Section 8.4 treats the special stability problem for an earth dam founded on very soft soil where the strength of the foundation dominates the analysis.

density control during construction.[1] Because of this inability to estimate the pore pressures reliably, and also due to the fact that the during-construction stability usually should not govern the design section (except in the case of the dam on a soft foundation), the construction pore pressures are usually estimated conservatively based on past experience (Sec. 3.3).

An additional guide to the estimate of construction pore pressures can be obtained from laboratory tests in two ways:

1. Direct measurement of pore pressures in sealed laboratory specimens which have been compacted to water contents and densities expected during construction and which are then subjected to increasing stresses with a test procedure designed to simulate the stresses anticipated on the various elements of the embankment by the added weight of the overlying material during construction.
2. Computation of the pore water pressures from theory based on Boyle's and Henry's laws for the compression and solution of the air in the void spaces of the soil (USBR method, Ref. 302).

The range of applicability and the best methods for carrying out these procedures will be better established in future years by soil mechanics research.

The pore pressures estimated by these procedures are those which would develop if no drainage occurred during construction. The amount of drainage (and pore pressure dissipation) which will occur depends primarily on the permeability and dimensions of the impervious zone, the length of the construction period and the magnitude of the maximum pore pressures; it can only be roughly approximated by the designer on the basis of pore pressure measurements in similar dams (Sec. 3.3). Because of the irregular boundary conditions and the existence of the air in the pore voids, it has not generally been considered feasible to treat the problem theoretically (Sec. 5.8).

[1] On a few occasions, pore pressure measurements on cofferdams or in the lower part of the dam have been very valuable as a guide to the estimate on the pore pressures which may develop in the main dam. At the Djatiluhur Dam in Java, the main 155-ft. cofferdam was built during the first construction season in 1961 at the same time that the analysis for the 340-ft. main dam was being carried out. Since the core for the cofferdam was of the same clay as was to be used in the main dam, the pore pressure measurements were of great assistance. At the Swift Creek Dam in Washington in 1957, pore pressure measurements made in the cofferdam had a large influence on the evaluation of the pore pressures which would develop in the main dam (Sec. 3.3*b*). Reference 73 describes the construction of a test embankment during the design stage of a large dam in which pore pressures were measured.

COMPUTATIONS USING BOYLE'S AND HENRY'S LAWS

The method developed and used by the USBR is based on the following primary assumptions:

1. Only vertical embankment strain (compression) takes place during construction; there is no lateral bulging.
2. The relationship between embankment compression and effective stress is known.
3. The pressures in the pore water and the air in the pores are always equal, and directly after compaction of the embankment material, they are equal to atmospheric pressure.[1]
4. The decrease in the embankment volume at any given elevation under the weight of the fill placed above is caused by compression of the air in the voids and solution of the air in the pore water.
5. Boyles' and Henry's laws are valid for this compression and solution.
6. No dissipation of pore water pressures from drainage occurs during construction.

From these assumptions the following equation relating pore pressure and embankment compression was derived (Ref. 302):

$$u = \frac{p_a \Delta}{V_a + 0.02 V_W - \Delta} \qquad \text{(Eq. 7.5:1)}$$

where u = induced pore pressure (p.s.i)
p_a = absolute atmospheric pressure (p.s.i); i.e., 14.7 p.s.i. at sea level
Δ = embankment compression, in percent of original total embankment volume
V_a = volume of free air in the voids of the soil directly after compaction, in percent of original total embankment volume
V_W = volume of pore water, in percent of original total embankment volume

[1] Starting in 1961, the procedure has been changed in USBR practice to account for the fact that the pore water pressure in fine-grained soils directly after compaction is less than atmospheric pressure. Following the research of Hilf (Ref. 507), a special one-dimensional consolidation test has been devised which allows the measurement of the capillary pore pressure and the determination of its influence on the compressibility of the compacted soil (see Ref. 526, pp. 86–89).

The numerical values for the initial air and water volume for use in Eq. 7.5:1 can be estimated from laboratory compaction tests and from previous experience with average embankment densities. The embankment compressibility can be derived from laboratory consolidation tests, or from the average of field measurements for similar embankment materials (Sec. 3.1*a*).

Equation 7.5:1 can also be used to compute the theoretical pore pressure at which all the air is forced into solution; i.e., when the embankment becomes completely saturated. This state, which occurs when the compression of the specimen (Δ) becomes equal to the initial volume of air in the voids (V_a) is described by the equation:

$$u = \frac{P_a V_a}{0.02 V_W} \qquad \text{(Eq. 7.5:2)}$$

The results of computations from Eq. 7.5:1 are plotted in a family of "pore pressure-total stress" curves for various compaction water contents. Figure 7.5:1 gives a typical family plotted by Gould (Ref. 300) for the average embankment properties of Granby Dam in Colorado. Also shown on the plot are typical "during-construction" piezometer readings demonstrating that the general shape of the family of curves for average construction pore pressures as actually measured was about the same as that of the computed family and the magnitudes were similar. The principal exceptions, piezometers No. 60 and 61, were located near the edge of the impervious embankment section and were influenced by drainage.[1]

DIFFERENCES IN SAFETY FACTORS AS COMPUTED BY EFFECTIVE STRESS AND TOTAL STRESS METHODS OF ANALYSIS

When employing the effective stress analysis, as discussed in the preceding section, we commonly use the pore pressures which would be expected to exist in a stable embankment; i.e., we make an estimate based on experience with other dams, or on the basis of laboratory tests in which the specimens are not failed. On the other hand, in the

[1] In the preparation of these curves representing piezometer readings, the pore pressure was plotted against the vertical height of fill directly above the piezometer at various times during construction. The final vertical segments of the curves for piezometers No. 56 and 57 are undoubtedly due to the transfer of total stress by arching from the central portion of the dam to the outer slopes. Similar results have been obtained at many dams.

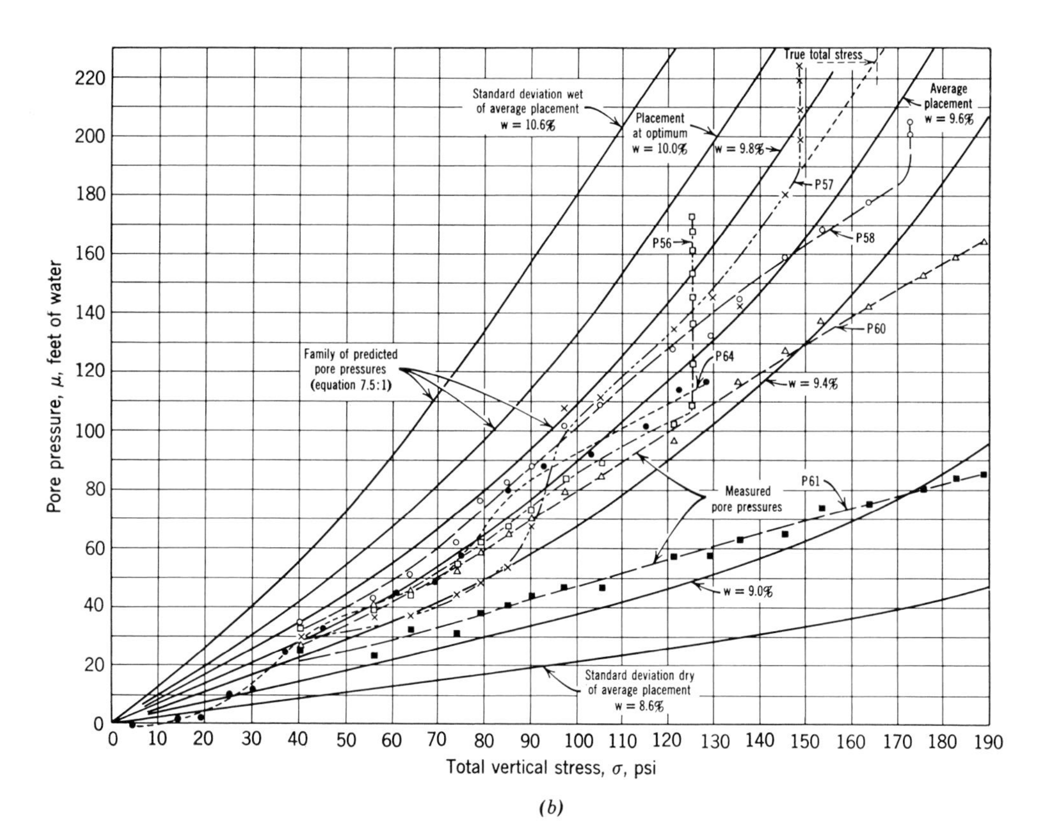

Pore pressure, μ, feet of water
Total vertical stress, σ, psi
True total stress
Standard deviation wet of average placement w = 10.6%
Placement at optimum w = 10.0%
w = 9.8%
Average placement w = 9.6%
w = 9.4%
w = 9.0%
Standard deviation dry of average placement w = 8.6%
Family of predicted pore pressures (equation 7.5:1)
Measured pore pressures
P56
P57
P58
P60
P61
P64

(b)

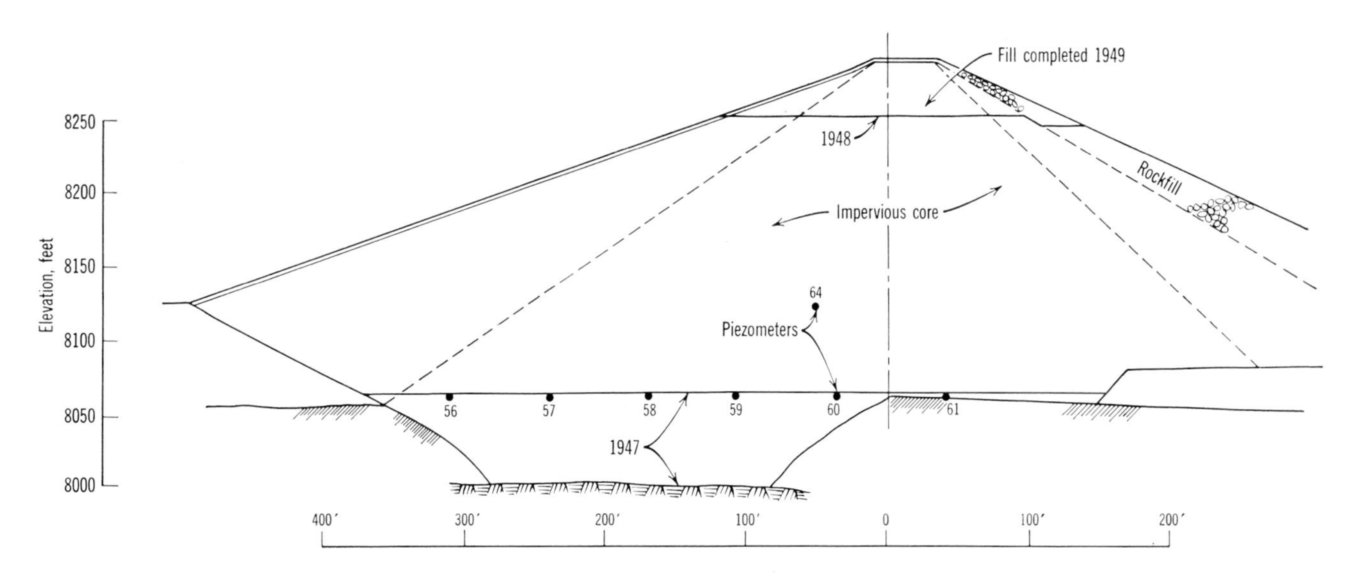

Fig. 7.5:1 Relationship between predicted and measured pore pressures at Granby Dam (after Gibbs et al., Ref. 526). (a) Locations of piezometers. (b) Family of predicted pressures compared with pressures measured with specific piezometers. (Note relatively great difference in computed pore pressure for small changes in construction water content.)

total stress method of analysis, the pore pressures are those which have developed in the soil at failure, since the shear strengths are used in terms of total stresses at failure. Since different pore water pressures are introduced into the analyses by the two methods, different safety factors result.

The differences in safety factors obtained are dependent to a large degree on the soil type. The difference is greatest for dense, well-graded mixtures of gravel, sand and clay which dilate when sheared with a resulting reduction in the pore pressure. The difference is least for very fine-grained clayey soils in which shear strains may cause higher pore pressures. The engineer must consider the fundamental difference between the two methods of approach when evaluating the results of the computations.

MINIMUM TOLERABLE SAFETY FACTOR

Since shear movement or an actual slide during construction will not cause loss of life or great property damage under the worst conceivable circumstances, we are justified in accepting lower safety factors than would be considered reasonable when there is water in the reservoir. Because of this and the fact that the numerical result of the stability analysis greatly depends on the type of analysis used, it is not feasible to establish here minimum tolerable safety factors for the construction condition. In selecting the minimum value to be used in a given situation, the engineer should attempt to evaluate both the degree of conservatism with which he has chosen his assumptions concerning soil strength and pore water pressures, and the influence of the method of analysis which he has used (Sec. 7.4). The problem is discussed in more detail in Sec. 1.4*a*.

7.6 "FULL RESERVOIR" STABILITY CONDITION

Since a shear slide when the reservoir is full can lead to a disastrous failure (Sec. 2.4), the stability analysis should be treated more conservatively than during construction and reservoir drawdown conditions. For the full reservoir state, only the downstream portion of the dam need be analyzed since an upstream slope slide during full reservoir is theoretically conceivable only if the strength of the foundation were to be reduced very greatly by wetting. Because of this, and since there are no records of upstream slope slides in earth dams when the reservoir was full, we can assume that upstream slides occur only during construction or following reservoir drawdown.

The pore pressures which exist within an embankment at any given time are generated as the result of two actions which can be considered independent for practical purposes: (1) gravity seepage flow and (2) changes in pore volume due to changes in the total stresses. The full reservoir stability condition is nearly always analyzed using the effective stress method of analysis and the pore pressures acting are assumed to be those governed by gravity flow through the embankment (Chap. 5). For most well-compacted embankment materials, this approach is conservative since any shear strains which may be imposed on the embankment after the construction is completed and the reservoir is full, are likely to cause the soil to dilate (expand in volume) and to reduce the pore water pressures temporarily.

While the use of the effective stress method with gravity seepage pore pressures is almost universally accepted practice for the full reservoir condition, studies of a few failures have indicated that the procedure may not be conservative when the material consists of very fine-grained and highly plastic soils.[1] For such soils, it is possible that shear strains may cause a temporary increase in the pore pressures over the values computed using the gravity seepage theory. Consequently, for stability analyses where a large portion of the potential failure surface passes through highly plastic clay, which does not contain an appreciable quantity of sand or gravel sizes, it is advisable to analyze the embankment stability also using the total stress method. The shear strengths used should be those determined from consolidated-undrained tests on saturated samples to which pore pressures are applied simulating those which may exist under the gravity flow in the dam and foundation.

The pore pressure distribution in the embankment due to gravity seepage is estimated from graphical flow nets or electric models (Chap. 5). If the downstream slope is relatively pervious, the pore pressures can be increased appreciably by the penetration of rain water, and the gravity flow net should be constructed assuming that the dam crest and downstream slope are sources of supply for seepage water, Fig. 7.6:1.

Anticipating the possibility that horizontal layers of relatively more pervious material than the average may be built into an impervious section, a conservative assumption should be made concerning the ratio of the permeability coefficients in the horizontal and vertical direc-

[1] See, for example, the experiences recorded by Peterson with failures of several clay dams in Canada (Refs. 109, 512, 519).

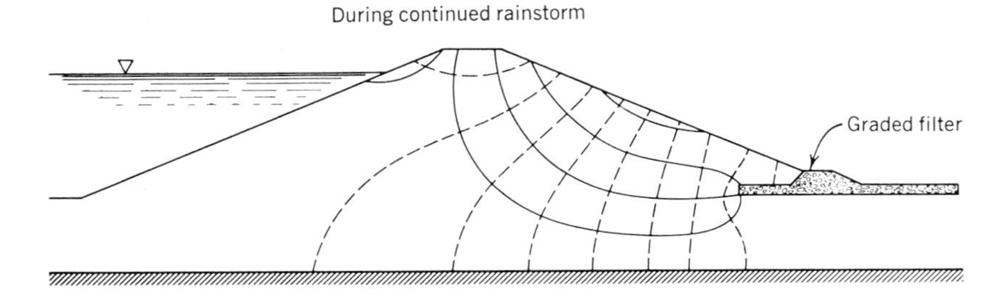

Fig. 7.6:1 Flow net for downstream slope of dam during continued rainstorm (after Casagrande, from Ref. 212).

tions. The degree of anisotropy must be chosen arbitrarily, but the following values are conservative:

DESCRIPTION OF SOIL IN BORROW AREA	k_{hor}/k_{ver}
Very uniform deposit of fine-grained soil (CL and ML)	9
Very uniform deposit of coarse soils with fines (GC and GM)	25
Very erratic soil deposits	100 or higher

These figures are higher than should be expected in earth dams constructed with good practice and control, and consequently the pore pressure estimate for an impervious downstream portion of the core of the dam will be somewhat high. Designs based on analyses using such conservative estimates will not substantially raise the cost of a typical dam, and the estimates will be closer to the actual pore pressures in the few dams where the embankment is definitely stratified because of poor soil blending.

TOLERABLE SAFETY FACTOR FOR DESIGN

As discussed earlier in this chapter, the results of stability analyses depend on many factors which are difficult or impossible to evaluate reliably. Small changes in the assumed soil strength and pore water pressures give appreciable differences in the results. The safety factor computed depends to a large degree on whether the effective stress or total stress methods are used, and on whether the calculations are carried out with the standard method of slices or with other methods in which the influence of side forces are considered.

At the present state of our ability to handle this problem, it is not possible or reasonable to establish any hard and fast criteria for a minimum tolerable safety factor. In the past two decades, and particularly in the last few years, dams with increasing heights have been built, and there is a definite trend towards steeper slopes and lower margins of safety. No shear failures have occurred except for a few relatively low structures on clay foundations (Sec. 2.4).[1] In practice, a great variety of methods of laboratory shear testing, pore pressure assumptions and methods of calculations have been used in stability analyses by different engineers. In spite of these differences, a safety factor of 1.5 has been adopted as a minimum tolerable value for the full reservoir condition on the great majority of projects. The probable end result of this evolutionary process, assuming that no failures occur to alter present thinking, will be that a safety factor of 1.5 will be considered satisfactory when computed with a method of calculation in which side forces are considered (Sec. 7.4) and with pore pressures as estimated from a steady-state flow net.

Although we cannot handle the problem analytically, the designer must keep in mind that the tolerable safety factor should depend on the type of soil through which the potential failure surface passes for most of its length. If we consider the extreme case of an embankment of pervious granular soil or rock, with an upstream impervious membrane and a sound rock foundation, it is apparent that there is no possibility of a shear failure in the downstream slope of the type considered in the stability analysis, even though the computed safety factor may be low. Also we know that we must be more conservative with embankments consisting primarily of fine-grained soils because (1) almost all shear slides have occurred in dams constructed of or founded on very fine-grained and highly plastic soils, and (2) the difficulty in

[1] An exception to this is the failure of the Waco Dam in Texas which occurred just before this book went to press (Refs. 403, 676). In this case, the movement occurred in the foundation at a depth of more than 20 ft. below the surface of a horizontally bedded shale formation. The embankment was only about 90 ft. high at the time of the failure and the side slopes were relatively flat. Consequently, the shear stresses in the shale were low and no conservative engineer using the best modern methods of stability analysis could have anticipated the trouble in advance. This is only one of several failures of civil engineering structures in recent years which have been attributed to weaknesses in rock foundations (see Sec. 4.2 and Ref. 620).

determining the strength and the pore pressure of clay embankments is greater and the possible error larger.

7.7 STABILITY CONDITION FOLLOWING RESERVOIR DRAWDOWN

7.7a Analytical Procedures

The computed safety factor of the upstream portion of the dam is usually lowest following reservoir drawdown. From studies of dam failures, we know the following salient facts about upstream slides which can be used to guide the stability analyses (Sec. 2.4*c*):

1. Except for slides during construction, all upstream slides have occurred following reservoir drawdown.
2. Slides have usually occurred in poorly compacted dams constructed of very fine, highly plastic soil.
3. Most slides occurred within the first few years of reservoir operation. In all cases, slides probably have been caused by reservoir drawdowns which were unprecedented either in the rate or in the extent of the water surface lowering.
4. Slides have typically been caused by drawdowns from maximum reservoir level to about dam mid-height at rates of less than one foot per day.
5. Upstream slides have never caused breaching of the dam. The reservoir level at the time of failure was usually at the dam mid-height or lower.

The total exterior driving forces acting on the trial sliding mass (that is, the weight forces) following reservoir drawdown are not appreciably greater than the forces acting at the end of construction; the only difference is caused by a slight increase in the unit weight of the soil due to a higher degree of saturation or to some minor influence of buoyancy at the toe. Consequently, the reduced safety factor following a reservoir drawdown is due primarily to the existence of higher pore pressures (drawdown pore pressures) acting inside the upstream slope.

Both the effective stress and total stress methods of analysis are in common use. Probably the effective stress method is adopted more widely at the present time. On the other hand, it is more difficult to make a reliable estimate of the pore pressures following reservoir drawdown than for other critical times in the life of the dam; therefore, there is still good reason to use the total stress approach. The methods give different safety factors for the same reasons as discussed in Sec. 7.5.

TOTAL STRESS METHOD

In the total stress method of analysis, the results of consolidated-undrained tests on saturated specimens (Sec. 7.3*d*) usually are used to determine the shear strength. The samples are tested under conditions which simulate the stresses and pore pressures which would exist in the embankment, as well as the designer is able to predict them. There is no widely adopted standard method of testing. Recently, methods of analysis have been employed in which the laboratory samples are consolidated anisotropically before testing in undrained shear, and higher safety factors are obtained by the procedure (see, for example, Ref. 320).

ESTIMATING PORE PRESSURES FOR THE EFFECTIVE STRESS ANALYSIS

At the present time, we have an insufficient number of measurements on dams in operation (Sec. 2.3*c*) to allow the formation of any reliable generalizations about drawdown pore pressures as a function of soil type, embankment strain, and rate of reservoir lowering. For design purposes, we can only estimate drawdown pore pressures on the basis of simplified, albeit very rough and approximate, theory which is still largely unverified by field measurements. For this reason, the pore pressure estimate for the drawdown condition must be considered as somewhat less reliable than for the during construction and full reservoir conditions.

In a dam which is not near failure, the pore volume changes which occur as a result of reservoir drawdown are likely to be very low; consequently, the pore pressures will be dominated by the influence of gravity seepage. If the safety factor is low, large embankment strains may develop and change the pore pressures considerably. If the soil structure expands during shear, the pore pressures will decrease; hence pore pressures due to gravity seepage will constitute the most severe condition which could develop. If the embankment strain associated with reservoir drawdown causes a reduction in the pore volume, the pore pressures at failure may be higher than those due to gravity seepage. The pore pressures in this latter case can only be estimated from measurements in laboratory triaxial tests performed in such a manner as to simulate the embankment stresses which may develop and lead to failure.

Since the reservoir drawdown requires a certain length of time, some drainage and dissipation of drawdown pore pressures will occur; however, for most of the materials used for the impervious sections of earth dams, the quantity of water which will tend to flow out of the pores of the soil is low and it is impossible to evaluate reliably the influence of drainage on the pore pressures. It is conservative and advisable to neglect the influence of drainage unless the material consists of sand and gravel with only a small content of fines. Methods for estimating the influence of drainage are discussed in Sec. 7.7*d*. In Sec. 7.7*b* procedures for estimating pore pressures are given on the basis of the conservative assumption that the drawdown is "instantaneous"; i.e., no drainage occurs during the time period required for lowering the reservoir.

7.7b Pore Pressures Due to Gravity Seepage after Instantaneous Drawdown

For this limiting condition, we assume that changes in the embankment stresses caused by the reservoir drawdown are so small that volume changes can be neglected. The problem then reduces to that of seepage through a porous medium governed by the Laplace equation, and the pore pressure distribution is determined by the construction of graphical flow nets (or with electric models), Chap. 5. Before the reservoir is lowered, the pore pressure distribution is given by the "steady-state" flow net. After drawdown the hydraulic boundary conditions governing the seepage are changed. Instead of being an equipotential line, the upstream slope becomes a free surface along which the pressure head in the pore water is equal to its elevation at every point. Since the boundary conditions are known, the pore pressure distribution can be determined, Sec. 5.8.

If the pores of the soil are completely filled with water, the hypothetically instantaneous change in the hydraulic boundary conditions will be reflected immediately by changes in the pore water pressure inside the dam, since no seepage is required; i.e., the pore pressure distribution will change instantaneously to the drawdown flow net. If the pore water contains air bubbles (or appreciable quantities of dissolved air), the bubbles will expand when reservoir is lowered and the pore pressure is reduced and some flow of water out of the pores will be required to establish the gravity drawdown pressure distribution with a resultant time lag.[1]

In the computations of pore water pressures due to gravity seepage, the influence of capillary forces is neglected since we have no way to evaluate them. Capillary forces tend to prevent the water from flowing out of the pores of fine-grained embankments following reservoir drawdown by forming menisci at the seepage boundary, and also within the dam, with resulting tension in the pore water. Therefore, ignoring capillary action results in a conservative pore pressure estimate. On the other hand, it is quite conceivable, and even probable, that for many soils capillarity will completely dominate the gravity flow pattern, making it impossible for any appreciable positive drawdown pore pressure to exist within the embankment above the reservoir elevation. In future years additional laboratory researches, and measurements of drawdown pore pressures in dams, may demonstrate that our present procedures are excessively conservative except for certain soil types.

7.7c Influence of Reservoir Level

For a homogeneous dam, the most critical drawdown condition exists theoretically after the reservoir has been completely emptied. However, there is only a minor decrease in the factor of safety after the reservoir has been lowered to mid-height of the dam. For a dam with a large upstream zone of pervious soil or rock, calculations often indicate that the safety factor is actually lower when the reservoir is partially full than when the reservoir is empty. This is due to the fact that, when the reservoir is partially full, the shear strength of the upstream pervious zone at the upstream toe, which provides the primary resistance to movement, is dependent on the buoyant weight of the material; whereas when the reservoir is empty, the strength is dependent on the full weight. For this reason it is advisable to calculate the stability of the upstream slope for several assumed reservoir levels, including at and below mid-height of the dam, as well as for complete drawdown.

[1] It has been speculated by some investigators that this time lag might cause the drawdown pore pressures to be considerably higher than those computed on the basis of gravity seepage alone (see, for example, Ref. 168). At the present time little is known about this action and most engineers have ignored its influence. Probably the error on the unsafe side caused by this phenomenon is more than counteracted by the unknown influence of capillary action on the computed pore water pressures.

7.7d Influence of Drainage and Rate of Reservoir Lowering on Drawdown Pore Pressures

The rate of pore pressure dissipation due to drainage following reservoir drawdown is dependent primarily on the permeability, degree of saturation, and capillarity of the material comprising the upstream slope (and on the pore volume change, if any). Our present ability to evaluate these factors even in a rough manner in advance of construction for embankments of fine-grained soils by theoretical procedures is practically nil. Also, we have very few measurements of pore water pressures in operating dams to guide the development of theories. Consequently, for embankments constructed of fine-grained soils, we can only assume that the influence of drainage be neglected.[1]

The engineer is justified in attempting to evaluate the influence of drainage on the dissipation of drawdown pore pressures only in the circumstance where the upstream slope of the embankment consists of relatively pervious sand and gravel, which contains a small percentage of silt or clay sizes (finer than the No. 200 sieve). For material of this kind, capillary action cannot have an overwhelming influence and the permeability is so high that any conceivable embankment volume changes are not likely to cause appreciable pore pressure changes.[2]

TRANSIENT FLOW NETS

The rate of dissipation of drawdown pore pressures can be roughly estimated by drawing a series of flow nets with a gradually lowering upper phreatic surface. Each flow net represents the condition existing at some definite time following reservoir drawdown.[3] Using this procedure, it is assumed that all the water drains out of the pores. Figure 7.7:1 gives a general relationship worked out by Reinius for a simple

[1] As discussed earlier, this assumption may be grossly conservative because capillary action in well-rolled embankments of silt or silty sand may be so great that any slight tendency for drainage will immediately cause the development of menisci at the boundary and create negative pore pressures in the zone of the embankment above the reservoir level. It is to be expected that field piezometer measurements obtained in future years will add greatly to our understanding of this problem.

[2] As discussed in Sec. 1.4*b*, the practical necessity for evaluating the effect of drainage frequently arises during construction, when sand and gravel construction materials which were assumed by the designer to be completely free draining are found to be dirtier and less pervious than anticipated.

[3] See Refs. 168, 324, and 325 for examples of calculations.

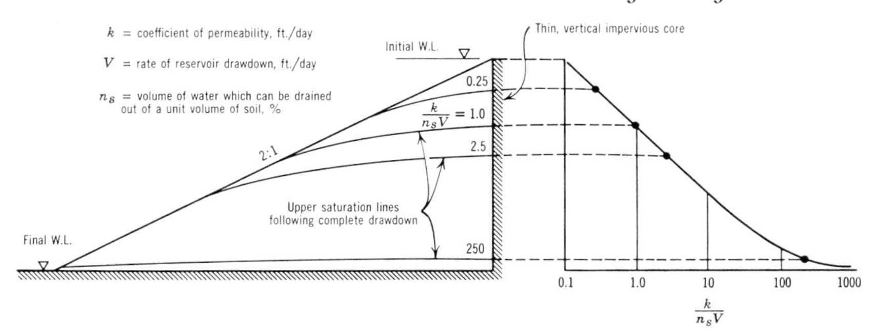

Fig. 7.7:1 General solution for upper saturation lines in upstream slope of dam with thin impervious central core following reservoir drawdown for different soil properties and rates of drawdown (after Reinius, Ref. 574).

dam cross section with an impervious central core using the method of transient flow nets.

CASAGRANDE APPROXIMATE THEORY

A simplified approach to the solution of the problem of the rate of drainage of the upstream shell of an earth dam was developed by Casagrande (Ref. 326). In this solution it is assumed that the upper phreatic surface at various times following instantaneous drawdown may be approximated by inclined straight lines passing through the upstream toe, Fig. 7.7:2.

At the start of the drainage period, immediately following reservoir drawdown, the phreatic line is at the upstream surface of the dam. At any subsequent time t_u, the phreatic line is a straight line with an intermediate slope as shown. The volume of water which will drain out of the slope in an increment of time dt is equal to the volume of the incremental triangular slice multiplied by the amount of water which will flow out of a unit volume of soil (yield $= N_s$) or

$$dq = \frac{dh}{2}(L + h \cot \beta)N_s$$

By using Darcy's law and approximating the area of drainage and the hydraulic gradient, the rate of flow of water out of the toe of the dam is given by

$$\frac{dq}{dt} = kia = k\frac{h}{2}\frac{h}{L + h \cot \beta}$$

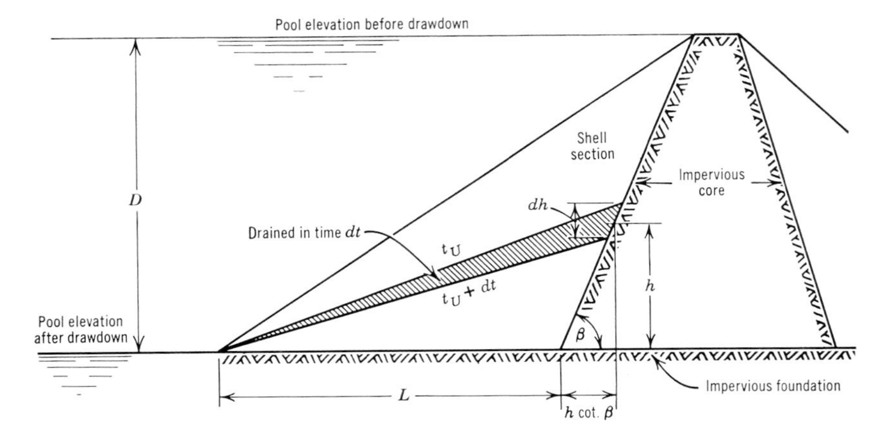

Fig. 7.7:2 Approximate mathematical treatment by A. Casagrande of the drainage of the upstream slope of an earth dam following reservoir drawdown (after Shannon, Ref. 326).

The solution of this equation can be given in the form

$$T = \frac{U}{100 + U} + J \log_e \frac{100}{100 - U} + J^2 \frac{U}{100} \qquad \text{(Eq. 7.7:1)}$$

where U is the percentage of drainage (the quantity of the water which has drained out of the embankment at any time in percent of the total amount of water which can drain out) and J and T are dimensionless quantities defined as follows:

$$J = \frac{D}{L} \cot \beta \qquad T = \frac{t_u k D}{N_s L^2}$$

Where $\beta = 90°$, the solution becomes simply:

$$T = \frac{U}{100 - U} \qquad \text{(Eq. 7.7:2)}$$

Using these equations, the percentage of drainage U can be calculated as a function of time t_u. A trial flow net can be constructed for each value of U if desired which will allow an estimate of the pore pressure distribution at any given time following instantaneous drawdown. Figure 7.7:3 is an example of a flow net representing the seepage condition which exists at 16% drainage.

Tests with viscous fluid models indicated that the equations predicted the actual drainage very closely. Consequently, it is probable that the method is as reliable as the method of transient flow nets and it is much simpler to use.

INFLUENCE OF CAPILLARY ACTION ON RATES OF DRAINAGE

In the two methods described above, the influence of capillary forces is neglected. Both theory (Ref. 212, p. 315) and experiments (Ref. 327) indicate that capillary forces not only reduce the amount of water which will drain out of the upstream slope but they also reduce the time rate of drainage. Consequently, the estimates of the time rates of pore pressure dissipation by both methods will be too high. From the practical standpoint, we are not able to evaluate the influence of capillarity on the rate of drainage; however, it is probable that fine sands and coarser soils with small percentages of fines will drain considerably more slowly than indicated by computations which ignore the influence of capillarity.

For this reason, the designer is not justified in spending much analytical effort in predicting rates of drainage. Rough estimates based on Reinius' summary, Fig. 7.7:1, or from Eqs. 7.7:1 and 7.7:2, are all that are justified in most cases.

7.8 THEORIES OF ELASTICITY AND PLASTICITY

In the 1930's, when the profession was in the first stages of groping about for analytical means to compute the stability of dams, it was felt

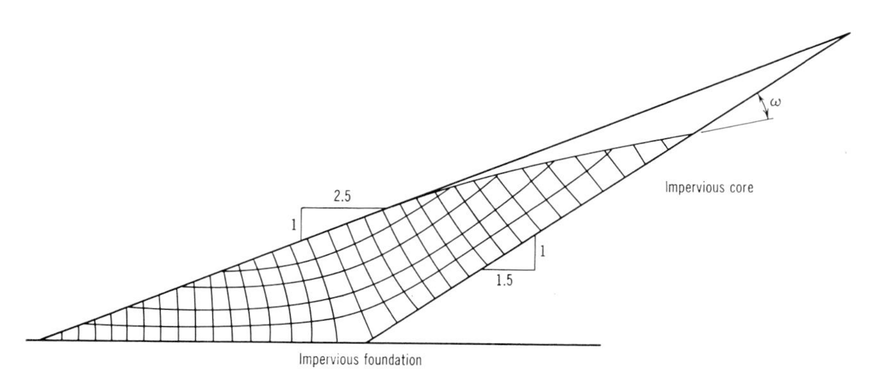

Fig. 7.7:3 Example of drawdown flow net after 16% drainage has occurred (U = 16%) (after Shannon, Ref. 326).

by many that the theories of elasticity and plasticity offered the most promise. The theories were used to compute the stress at various points within the embankment and its foundation and the stress was compared with the strength.

For a number of reasons, not the least of which was the difficulty in making the computations, the methods never received wide acceptance and at the present time are used only by a few engineers as a secondary approach, and occasionally to study special problems. Hence, this section, in which the main contributions to earth dam problems are briefly described, has really only historical interest to the practicing engineer. The only exception to this is the use of the method to approach the problem of computing the magnitude of the stretch of the base of the dam (Sec. 3.1*d*) which is not often done but for which we have no other tool.

On the other hand, if our more common methods of stability analysis are ever to develop beyond the rather unsatisfactory present stage of evolution, we must find some way to compute the internal stresses in an embankment. This need and the development of electronic computers which has so reduced the labor of computation, formerly so formidable even for the simplest boundary conditions, may reawaken interest in this approach to the problem and provide us with some valuable additional tools.

7.8a Theory of Elasticity

EARLY SOLUTION FOR EMBANKMENT STRESSES

Several solutions of the stresses which exist at the base of an elastic vertical wedge under its own weight were made more than 50 years ago (Ref. 212, pp. 406–409). These indicated that the shear stress on a horizontal plane varies from zero at the center of a symmetrical wedge to a maximum at the exterior slope, Fig. 7.8:1. Obviously this condition can never exist in an earth embankment, which cannot support high shear stresses at the outer slope.

JURGENSON'S SOLUTION FOR FOUNDATION STRESSES

The first serious proposal that the theory of elasticity might have practical value for earth dam stability analysis was made by Jurgenson in the early 1930's (Ref. 313). Jurgenson gave a complete solution for

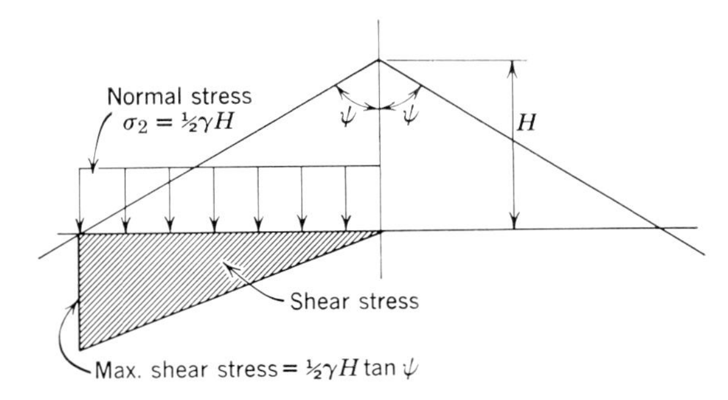

Fig. 7.8:1 Distribution of normal and shearing stress on a horizontal section through an elastic wedge due to the weight of the wedge (after Terzaghi, Ref. 212).

the stresses in an elastic foundation, the earth dam being simulated by a triangular distribution of vertical pressure acting on the surface of a semi-infinite elastic body, Fig. 7.8:2. In the manner commonly used for settlement analyses of buildings and other structures, he computed the stresses by integrating those given by the Boussinesq equation for a point load acting perpendicular to the surface of a semi-infinite elastic medium. The main theoretical deficiency of the Jurgenson solution, which rendered it practically useless for earth dam analysis, was its assumption that a dam transmits only vertical pressure to the foundation and its neglect of the shear stresses at the base of the embankment.

MIDDLEBROOKS' SOLUTION (STRIP LOADING)

In an attempt to reduce Jurgenson's error of disregarding the shear stresses at the base, Middlebrooks devised a method of computation in which the embankment weight was divided into a series of horizontal strips (Ref. 312). For the lower portion of the dam and for the foundation, the stresses due to the weight of each incremental strip were computed and added as if the weight of each strip were applied to the surface of a semi-infinite elastic body. This procedure embodies a major error, since the results do not give zero stresses on the exterior slopes of the dam; but studies by Bennett indicate that such a discrepancy can be ignored providing the slopes of the dam are flatter than approximately 3:1 (Ref. 164). Taking the Middlebrooks strip

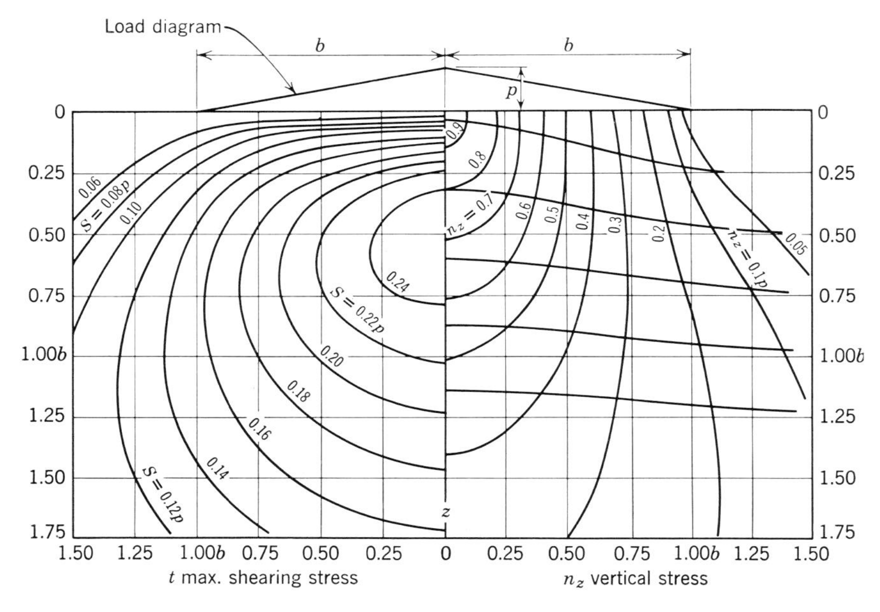

Fig. 7.8:2 Theoretical distribution of maximum shear stress and normal stress in semi-infinite elastic foundation under triangular vertical load. All values are given as a fraction of the maximum stress p (after Jurgenson, Ref. 313).

method and including pore water pressures in it, Patrick has applied it to zoned dams underlain by soft foundations (Ref. 121).

BISHOP'S SOLUTION (RELAXATION TECHNIQUE)

Using relaxation techniques, Bishop (1952) made a general solution of the Airy stress function for the case of a symmetrical, triangular elastic mass (3:1 side slopes) with a horizontal foundation of the same elastic properties (Ref. 58). The solution of this single problem was a laborious project, but the results give a theoretically correct solution for the one case. Figure 7.8:3 shows the solution in terms of the distribution of total major principal stress, maximum shear stress, and shear and normal stresses on horizontal planes at various elevations. In Fig. 7.8:3*a* the distribution of major principal stress is plotted in terms of percent of the overlying embankment weight. Notice that the error associated with assuming the major principal stress to be vertical and equal to the embankment weight, is not large except at the toe of the dam and directly under the crest.

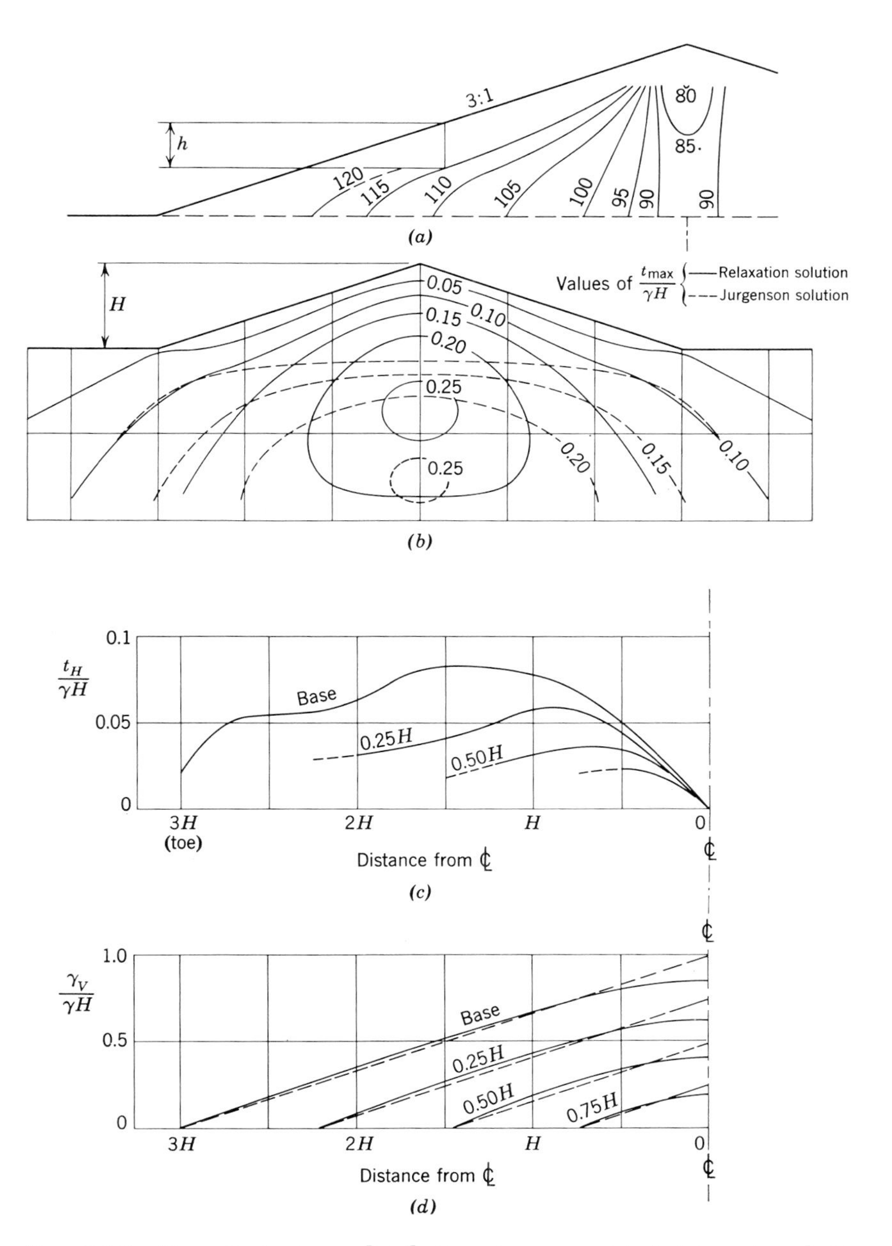

Fig. 7.8:3 Theoretical stress distribution for symmetrical triangular elastic wedge on horizontal foundation (after Bishop, Ref. 58). (a) Major principal stress (in per cent of the weight of the vertical column of soil directly overlying the point, γh). (b) Contours of maximum shear stress. (c) Shear stress on horizontal planes at dam base, one-quarter height and mid-height. (d) Normal stress on horizontal planes at dambase and each quarter-height above.

From Fig. 7.8:3*b*, which compares the maximum shear stresses obtained by the Jurgenson method and the relaxation technique, it is seen that the Jurgenson stress distribution is greatly different in the upper elevations of the foundation as would be expected. Its shear stresses on horizontal planes through the embankment are zero at the center, reach maximum values under the upstream slope, and have appreciable magnitude at the outside slope, as in the case of the theoretical solution more than fifty years ago (Fig. 7.8:1). In an earth dam, however, the shear stresses at the outer slopes must be close to zero.

The relaxation, Middlebrooks, and Jurgenson solutions are compared in Fig. 7.8:4 with respect to the maximum shear stress on the vertical

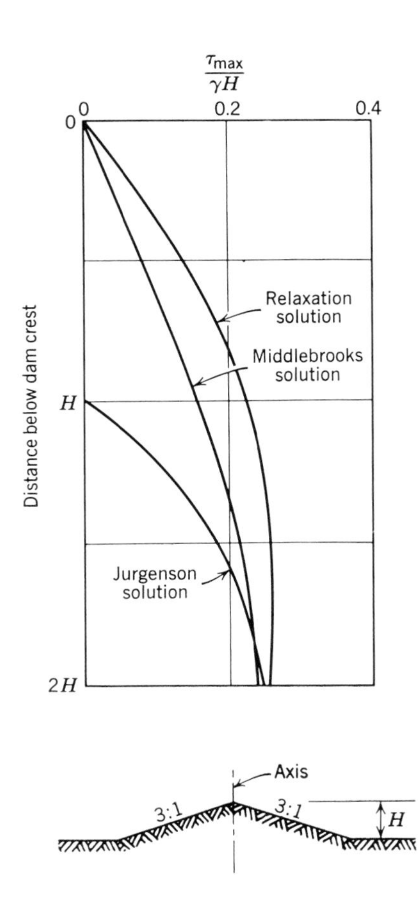

Fig. 7.8:4 Comparison of computations of maximum shear stress on vertical axis (after Bishop, Ref. 58).

dam axis. As would be presumed, the Middlebrooks method gives a better approximation of the foundation shear stresses than the Jurgenson solution. However, the embankment shear stresses computed by the Middlebrooks solution are 40% lower than those of the Relaxation method.

7.8b Theory of Plasticity

Whenever the soil strength is exceeded by the embankment or foundation shear stresses obtained from solutions of the elastic theory, the computations are invalid. The soil cannot carry stresses higher than its strength, so that strain in the areas of excessive shear stress creates plastic zones within which the stress equals the strength. Since the strength can be estimated, it would in theory be possible to define the areas of these plastic zones and recalculate the stress distribution in a series of steps. In practice, however, no methods have been devised to define the zones in the overstressed regions or to determine the distribution of stress in the remaining elastic zones.

To obtain solutions based on plastic theory, either the whole embankment has been assumed on the verge of failure or the ratio of principal stresses have been assumed to be equal at every point. An approximate solution for the normal and shear stresses on a horizontal plane through the base of a sand embankment which is on the verge of failure at every point was obtained by Rendulic (Ref. 212, p. 177). The procedure is not widely used in practice.

A method of Brahtz was proposed for earth dam studies in which an arbitrary ratio of principal stresses was assumed at every point within the embankment (Ref. 308). Glover and Cornwell developed another method in which the embankment and foundation is divided into a series of zones which are in states of plastic and elastic equilibrium. The latter two methods are discribed in the U.S. Bureau of Reclamation's *Manual of Design and Construction* (Ref. 35, Appendices B and C); however, they are not used in practice. Another method deserving mention has been proposed by Trollope in which the stress distribution is analyzed by computing the statical equilibrium of a model built up from single-sized, smooth and rigid spheres (Ref. 520).

Depending on the assumptions used for these various methods, considerably different distributions of stresses are obtained. Until more information is available from measurements in actual dams under various conditions, it is not possible to evaluate the applicability of the various methods which have been proposed.

eight

Special Design Problems and Details

8.1 MEASURING INSTRUMENTS AND PERFORMANCE OBSERVATIONS

8.1a Purposes and Types of Instruments

Instruments for measuring movements and pore pressures have been installed in earth dams with increasing frequency. The results of these measurements on a large number of dams have provided a better understanding of earth dam performance than was available a few years ago (Chap. 3) and, consequently, they have had a great influence on current designs.

The following types of instruments have been used most often:[1]

[1] If satisfactory instruments could be made to measure the total pressure (pore water pressure plus earth pressure) acting on a plane surface within a dam, they would be of great potential value in increasing our ability to analyze earth dams. A number of different ingenious types of pressure cells have been developed for this purpose (Refs. 65, 503, 530, 539.) These allow measurement quite accurately of the pressure acting on the cell; however, because of the difference in the compressibility of the cell and the surrounding soil no technique has been developed as yet to the authors' knowledge which will allow a reliable determination of the actual pressure in the dam. Nevertheless, total pressure cells of various kinds have been installed in some of the larger dams constructed within the last few years including the Serre-Ponçon in France, the Messaure Dam in Sweden, the Gepatsch Dam in Austria and the Oroville Dam in California. In addition to attempting to measure the actual pressures which develop and to testing the instruments, in some of the dams the designers have installed the cells with the hope that any trouble that might develop during construction such as an imminent slide, or a great transfer of load within the dam, would be reflected by a non-linear relationship between the reading of the cell and the height of the embankment over the cell. Some of the pressure cells are very sensitive and surprisingly small differences in pressure can be measured. At the Messaure Dam, for example, where pressure cells of the vibrating wire principle (Maihak) were cast in concrete cubes about one meter on a side, a pressure difference due to the added weight of a 6-inch soil layer could be easily measured when the construction surface was 40 meters above the cell.

1. Piezometers for measuring pore water pressure.
2. Internal instruments for measuring horizontal movements, foundation settlement, and embankment compression.
3. Surface monuments for measuring horizontal movement and settlement.

The main purpose of instruments is to furnish reliable confirmation that the pore pressures and movements which actually develop in a given dam do not exceed appreciably the values assumed by the designer. Also, the records of the measurements can have an important secondary function in providing a sound basis for the analysis of any difficulties that may develop in the performance of the dam. A third and very important purpose for dams of great height or dams which are unprecedented in some major respect is the value which accrues to the profession in general when the results of the measurements are published. Such measurements constitute the primary information which allow dams of greater and greater height to be designed with confidence.[1]

The number of instruments installed differs widely from dam to dam, and many major dams without unusual design features or unusual height have been constructed with practically no instrumentation. Some agencies which build large numbers of dams, instrument most of them as a matter of routine, because the cumulative record of performances is valuable in designing future dams. On the other hand, the engineer for a private owner building a single dam may not feel justified in recommending expensive instrumentation unless it will make the dam less expensive or is necessary for safety. But however much these decisions about instruments may vary where conventional dams are concerned, there is no difference of opinion when it comes to dams of great height or unprecedented design. All of these should be equipped with the best instruments available.

Characteristics of the principal instrument types are presented in the following sections.

8.1b Piezometers

Piezometers (devices for measuring pore water pressure) are used in both embankments and foundations. Foundation piezometers are

[1] A discussion of the vital need for measurements in thin core rockfill dams which are being designed and constructed with rapidly increasing heights in recent years is given by Cooke in Ref. 608.

installed in drill holes; embankment piezometers, though sometimes also located in drill holes, are usually placed in the fill as it is being constructed. Piezometers should be located in such a way that they will provide information on the magnitude and distribution of pore pressure in critical areas. In most dams a main battery of piezometers is set into one or more transverse sections of the dam to give a general picture of the pore water distribution. Additional piezometers are then placed at spots where specific information is needed, such as along abutment contacts or along the contact between the embankment and the concrete structures.

Two basic types of piezometers have been used in earth dams: (1) hydraulic piezometers in which the water pressure is obtained directly by measuring the head in a standpipe or by using a Bourdon Gauge; and (2) electric piezometers in which the pressure deflects a calibrated membrane and the deflection is measured electrically. Poor experiences with electric piezometers have caused most engineers in the United States to prefer hydraulic piezometers at the present time. However, electric piezometers based on the principle of the vibrating wire strain gage recently used widely in Europe have many advantages and may very well find wide application in the future.

Reliable readings from piezometers require a perfect combination of good instruments, good installation, and competent maintenance and reading. Poor or suspect readings are so common as to be almost the rule. Consequently, in any case where the engineer is seriously counting on the validity of the measurements to assure him of the safety of the dam, he should install at least two, and preferably three, different types of piezometers in order to have some assurance of success.

HYDRAULIC PIEZOMETERS

In pervious sands and gravels, pore water pressures at any point can be measured by sinking a simple vertical open pipe or a "well point" to the elevation and measuring the height to which the water rises in it. For fine-grained soils other types of piezometers have been designed, because the water takes too much time to seep from impervious soils into open pipes and to reach equilibrium.

Hydraulic piezometers in earth dams usually consist of a porous stone tip with a pipe or tube leading to a point on the surface of the embankment where the pressure can be measured. Experience has demonstrated that small-diameter plastic tubes are most satisfactory for the purpose, and now these are used generally. Where the piezometer is embedded in the impervious section of the dam, the porous tip is often, but not always, surrounded by a small pocket of sand which acts as a collection reservoir for the pore water and increases the effective area of the porous tip.

On perhaps half of the earth dam projects in recent years, the tubing has been carried out horizontally to a central measuring point near the downstream toe of the dam. On the other half the tubing has been brought up vertically through the embankment inside one or more riser pipes.[1] Each method has certain advantages, which are discussed in a following section (Vertical and Horizontal Piezometer Tubes). But regardless of the location of the measuring point, the tubing must be carefully backfilled with impervious soil to prevent seepage along the line of the tubes which would change the pore water pressure at the piezometer tip.

The success of piezometers depends on how carefully they are installed. For this reason many engineers exclude piezometers and other instruments from the main construction contract and install them with their own personnel.

USBR PIEZOMETER INSTALLATIONS

The USBR has been placing piezometers in embankments and foundations since the early 1930's. As a result of their experiences, Bureau engineers began around 1940 to use closed system piezometers constructed of plastic and porous stone tips with two small diameter plastic tubes for each tip, Fig. 8.1:1. Specifications and detailed directions for the installation are described in Ref. 274.

Originally, single tubes were used for each piezometer tip, but difficulty with air bubbles in the lines led to a double tube design which allows water to be circulated through the system in order to remove the air. In almost all Bureau installations the tubes are led to a central measuring point, the terminal well, at the downstream toe of the dam, Fig. 8.1:2. The terminal well is a reinforced concrete box completely equipped with pressure gauges, pumps, air trap, and a clean water supply.

The horizontal runs of tubing between the piezometers and the terminal well are placed in shallow trenches in the embankment and carefully backfilled with compacted, fine-grained impervious soil, Fig.

[1] At a number of dams in Europe where internal concrete galleries were provided in the foundation of the dam, the piezometer tubes were led into the gallery (See, for example, Ref. 505).

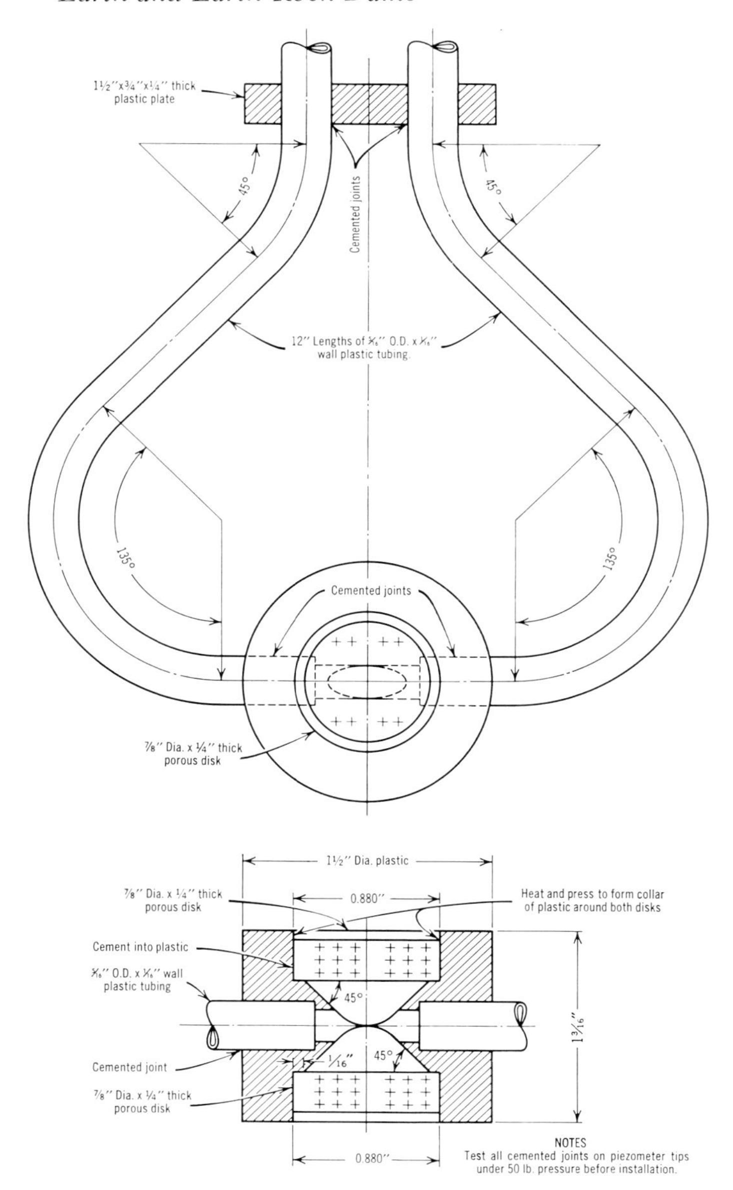

Fig. 8.1:1 Piezometer used in USBR earth dam embankments (Ref. 274).

8.1:3. This closed system piezometer (or slightly modified versions of it) is probably the most widely used at the present time. In addition to their use in USBR dams, they have been adopted by other engineers throughout the United States and abroad. The primary difficulty with them comes from the fact that air in the voids of the partially saturated compacted soil is forced under pressure into the porous tips and the plastic tubes. When the pore water pressure is very high, the air is dissolved in the water and causes no trouble, but under low and moderate pressures it remains in the form of bubbles which cause air locks in the line and erratic readings. Of course, with the double tube system fresh water can be circulated to catch the bubbles in the air trap. But on some dams it has been necessary to make this a continuous

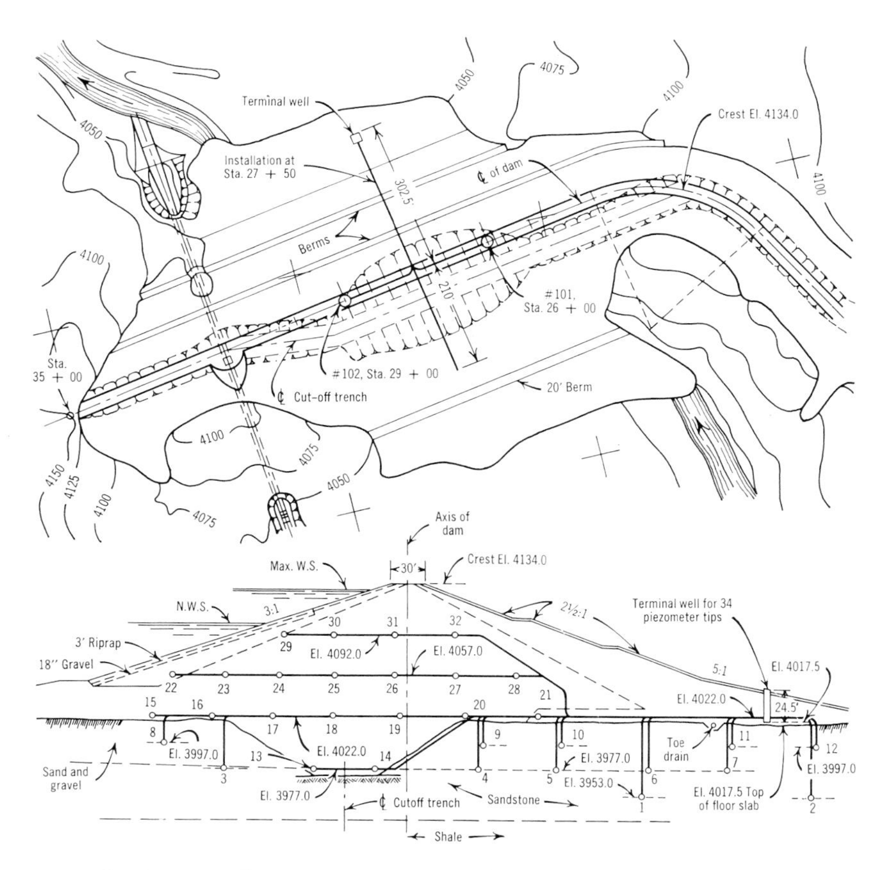

Fig. 8.1:2 Typical USBR piezometer installation—Keyhole Dam (Ref. 274).

Fig. 8.1:3 Placing and backfilling plastic piezometer lines in trench excavated in embankment construction surface. (Courtesy USBR)

process, since more bubbles enter the lines to replace those which are flushed out.[1] Often the troubles caused by the flushing operation have been so great that no reliable readings of pore water pressure could be secured.

Several approaches have been made to the solution of this problem. At the Building Research Station in England, experiments have led to the development of better methods for de-airing the water; these are described in Ref. 506. In addition, the engineers of the USBR have recently begun to use porous stone tips which have much finer pores and higher "air entry" (or "bubbling") pressure; because laboratory researches indicate that air bubbles enter fine-grained stones less easily

[1] In piezometers with vertical tube leads brought up through the fill, it is possible to hear the air hissing as it is continuously forced from the embankment into the piezometer.

than coarse-grained ones (Ref. 507).[1] This innovation seems very promising, although there is not enough experience available now (1961) to indicate the degree of improvement. But since the fineness of the stone should have little, if any, influence on the measured pore water pressure (in a no-flow system), while it probably has a considerable influence on the penetration of air, it seems advisable to use fine stones for future piezometer installations of this type.

[1] Piezometers with high air entry pressures were also used at the Selset Dam in England, 1957–1959 (Ref. 531).

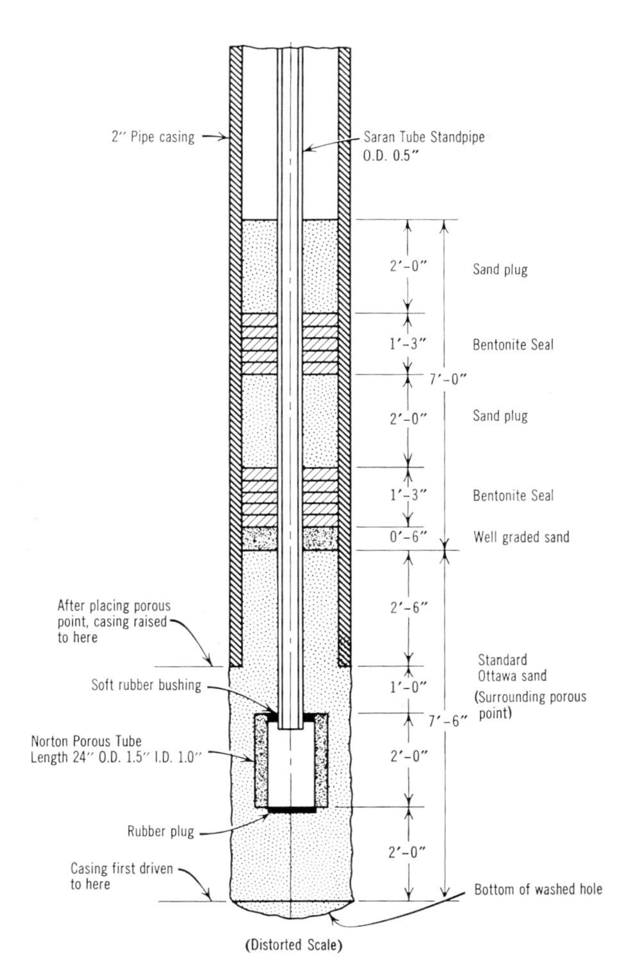

Fig. 8.1:4 Simple piezometer for placing in drill hole (after Casagrande, Ref. 16).

CASAGRANDE PIEZOMETER

During construction of the hydraulic clay fill for the Logan International Airport in Boston, Casagrande developed a successful and widely used piezometer for fine-grained soils (Ref. 16), Fig. 8.1:4. For earth dams this piezometer has been installed most frequently in drill holes in foundations or embankments, but it can also be placed in the embankment during construction. The single plastic tube is extended vertically to the surface, and if the head in the pore water at the piezometer elevation is less than the elevation of the embankment surface, the height of the water in the tube (pore pressure) is determined by a simple electrical sounding device. If the head is above the embankment surface, the pressure is measured by extending the tube as a standpipe above the construction surface or, if necessary, by using a Bourdon gauge. Details for construction, installation, and reading the piezometer are given in Ref. 16.

The main disadvantage of a piezometer of this type is that some flow of water in or out of the piezometer system is necessary in order to accommodate it to changes in pore water pressures. Consequently, there is a time lag before the piezometer reflects the change of pressure, though for the majority of soils from which earth dams are constructed, experience has indicated that this lag is not important unless the pore pressure changes very rapidly.[1]

Using the Casagrande apparatus, the water level in the small tube is measured by dropping into it an electrical sounding device, which short-circuits when it touches the water surface and registers on an ohmmeter. The Army engineers have developed an alternative measuring instrument with two vertical riser tubes connected at the top of the piezometer (Ref. 508). Air under pressure is forced down one of the tubes, and the maximum air pressure which can be developed then equals the water pressure. This Army piezometer has been used successfully at a number of major dams in recent years.

Another piezometer devised by Army engineers and used on a number of major dams consists of simple pockets of sand with a volume of about 1 cubic foot and vertical plastic riser tubes, Fig. 8.1:5. At Hills Creek Dam these functioned very well with negligible time lag, Fig. 8.1:6. They were brought up through the fill inside a corrugated metal pipe, Fig. 8.1:7, and Bourdon gauges were used to measure the pore pressures as shown. When it was necessary to remove a gauge in order

[1] See Ref. 504 for a detailed theoretical treatment of time lag in open piezometers, and Ref. 559 for results of recent laboratory research on the subject.

to add an extension to a tube, water spilled out on the construction surface. When the dam was low and the tubes were short, the pressure built up again rapidly after the gauge was reconnected. As the dam became higher, however, more time was required to regain the lost pressure.

VERTICAL AND HORIZONTAL PIEZOMETER TUBES

One main advantage of vertical piezometer tubes is that less troubles develop with air bubbles.[1] Those which enter the tube rise rapidly to the water surface without influencing the measurements; however, if a Bourdon gauge is necessary to measure the pressure, some difficulty with air bubbles may develop. In addition, piezometers with vertical riser tubes have another advantage in that the tubes are usually shorter. *The greater the distance between the piezometer and the measuring point, the more likely are poor results from all causes, including tube leakages, bubbles, and time lags.* In some large dams where the tubes are carried horizontally to a terminal well at the downstream toe, piezometer lines have been as long as 1,500 to 2,000 ft. Also, where the dam foundation is a relatively soft soil, there is a tendency for the base of the dam to spread (Sec. 3.1*d*) and consequently to stretch horizontal piezometer lines. At some dams, extra lengths of tubing have been placed in the trenches in the form of snakelike undulations in order to allow the base of the dam to spread without placing the tubing in tension.

Horizontal piezometer tubes have two main advantages over vertical tubes:

1. Regardless of the pore pressure, a closed system is obtained in which no flow from the soil surrounding the piezometer is necessary and, hence, theoretically there is no time lag.
2. The lines are buried and gotten out of the way; whereas vertical tubes installed during construction have to be continuously protected and extended above the construction surface.

Cold weather can affect the readings of Bourdon gauges materially. Consequently, terminal wells in cold climates should be heated.

[1] In recent years there have been a number of experiences with foundation piezometers in which gas from some unknown source enters the system causing very high pressure readings in Bourdon gauges. In some of these cases, when the gauge was removed and the pressure measured in a vertical stand pipe, the gas simply bubbled to the surface and the water pressure in the stand pipe was much less than was previously measured with the Bourdon gauge.

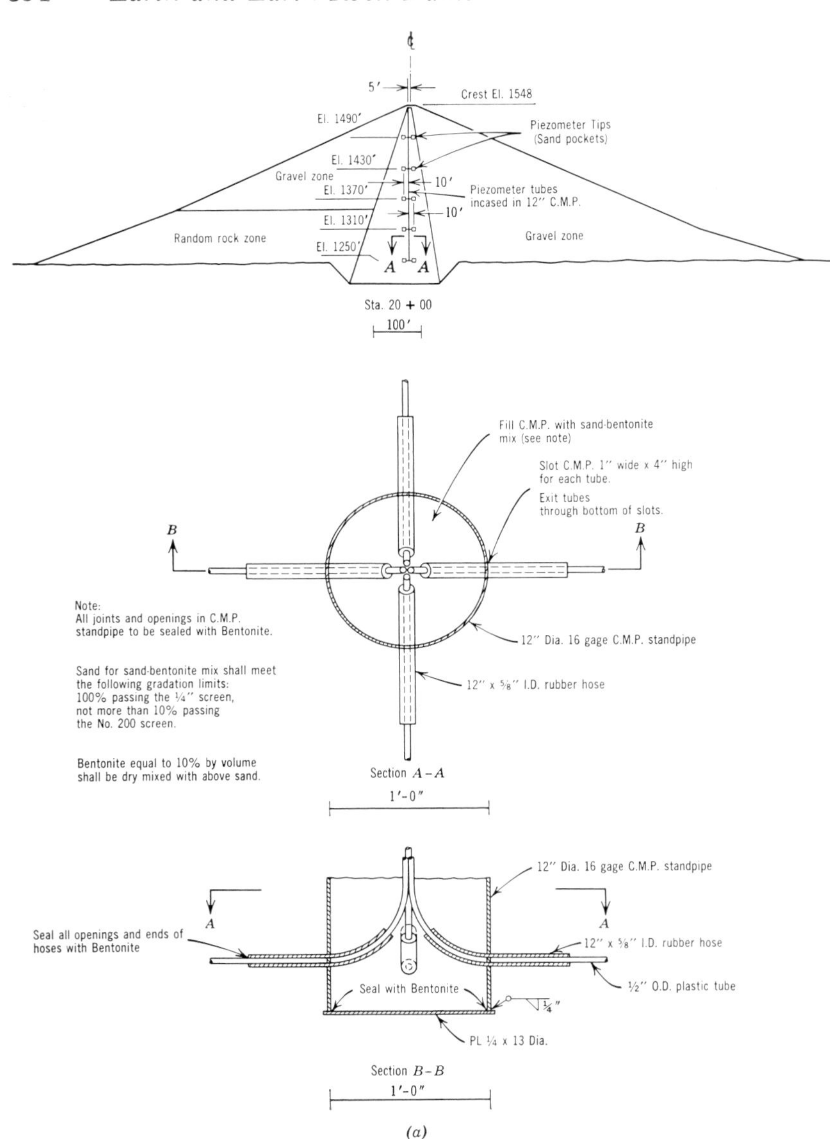

Fig. 8.1:5 Piezometer details used at Hills Creek Dam (1959). (Courtesy Portland District, U.S. Army Corps of Engineers)

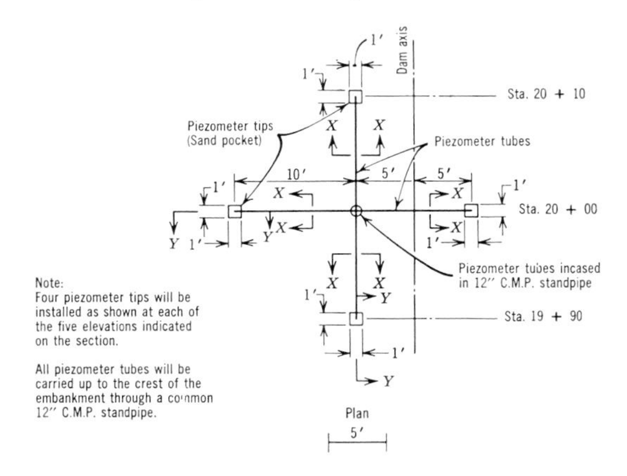

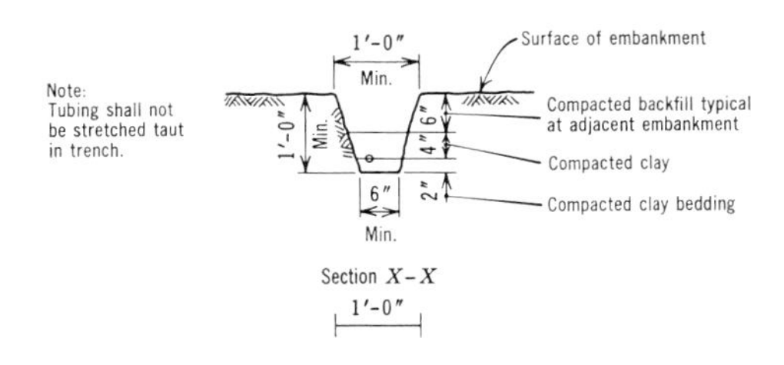

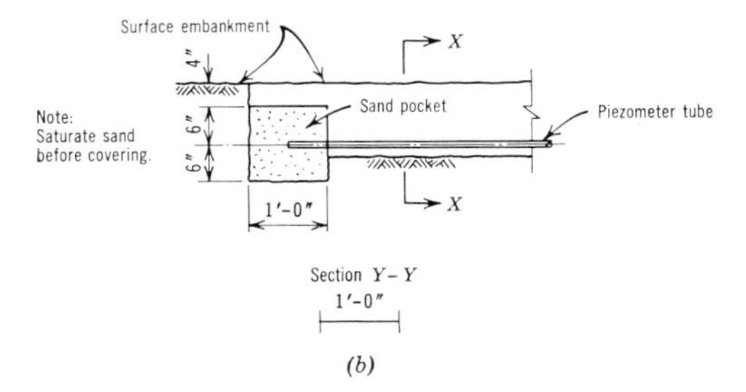

(b)

ELECTRIC PIEZOMETERS

In principle, electric piezometers are much superior to hydraulic piezometers for measuring pore pressures in fine-grained soil since no appreciable flow is required in and out of the piezometer tip in order to register a change in pressure. Also, they eliminate all the troubles associated with leaks, airlocks, and volume change in the piping of the hydraulic piezometer. In fact, in the core of a large dam constructed of highly impervious clay, the authors are quite doubtful that it is possible from the practical standpoint to measure the pore pressures reliably with hydraulic piezometers—especially if the hydraulic lines are long. In addition, it is much easier to install an electric cable and to make electric readings than it is to install hydraulic lines and to measure hydraulic pressures. Also, there is much more flexibility in the installation of an electric cable since it can be brought out at any point in the dam, whereas, with hydraulic piezometers there are definite limitations, especially if the pressures are to be measured in the lower part of the dam or foundation cut-off trench.

The oldest of the widely adopted electric piezometers (Goldbeck cells) employed air pressure to balance the water pressure and to break an electrical contact. These were used in many dams in the United States starting approximately in the mid-1930's. Usually they failed either by rupture of the diaphragm or by internal corrosion as a result of water condensation from the compressed air inside.

Another type of electric piezometer uses SR-4 strain gauges to measure the deflection of the diaphragm (Refs. 459, 503). These piezometers can be made quite precise; however, for a number of reasons (principally drift of the gauges and changes in the electric resistance of the cables), they have not proved reliable for long term

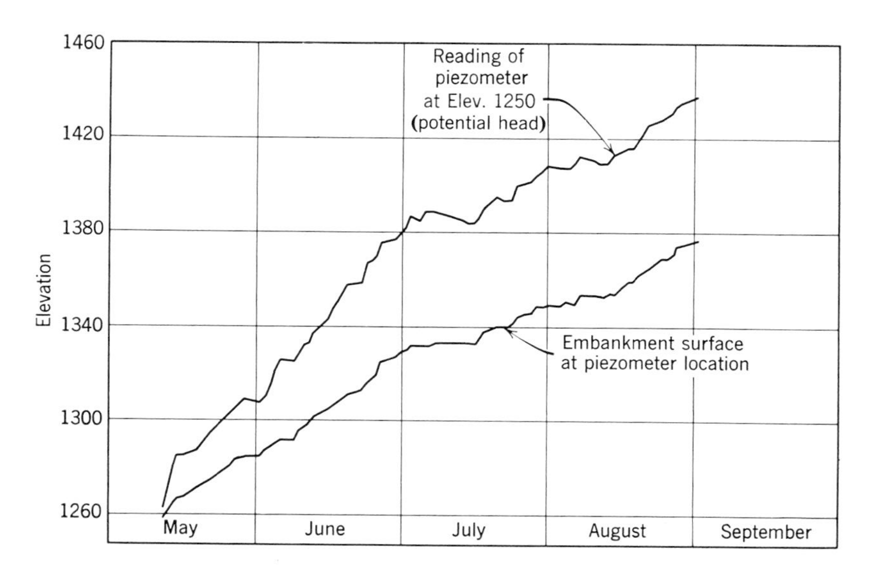

Fig. 8.1:6 Piezometer readings during first part of construction of Hills Creek Dam, 1959. (Courtesy Portland District, U.S. Army Corps of Engineers)

measurements and have not been installed frequently in earth dams. Also, when cables are cut accidently in the field—as they frequently are—and must be spliced, it is difficult to know the influence of the new cable on the calibration curve for the piezometer.

Since about 1955, piezometers using the vibrating wire strain gauge principle have been installed in perhaps a dozen major dams by European engineers and the experiences have been generally quite favorable (Refs. 70, 303, 560, 604, 668, and 679). In these piezometers, changes in pore water pressure acting on a diaphragm or piston cause changes in the tension and frequency of vibration of a fine steel wire (Fig. 8.1:8). In the instrument the wire is set into vibration with a magnet and the frequency is measured allowing a very precise determination of the pore water pressure.[1] While instruments of this type have been used in the United States in the petroleum industry and even for measuring pore pressures in laboratory soil tests, they have not been used here in earth dams to the authors' knowledge.

The primary advantage of the vibrating wire-type piezometer is that the readings are not dependent on the electric resistance of the cable or of a strain gauge and, consequently, corrosion or temperature of the cable and the necessity for splicing it when cut during construction does not influence the results. Because of this factor and the gen-

[1] Most of these instruments have been manufactured by two firms: Telemac (Paris, France) and Maihak (Hamburg, Germany).

Fig. 8.1:7 Bringing plastic piezometer lines up vertically through the fill in corrugated metal pipe, Hills Creek Dam, Oregon (1960).

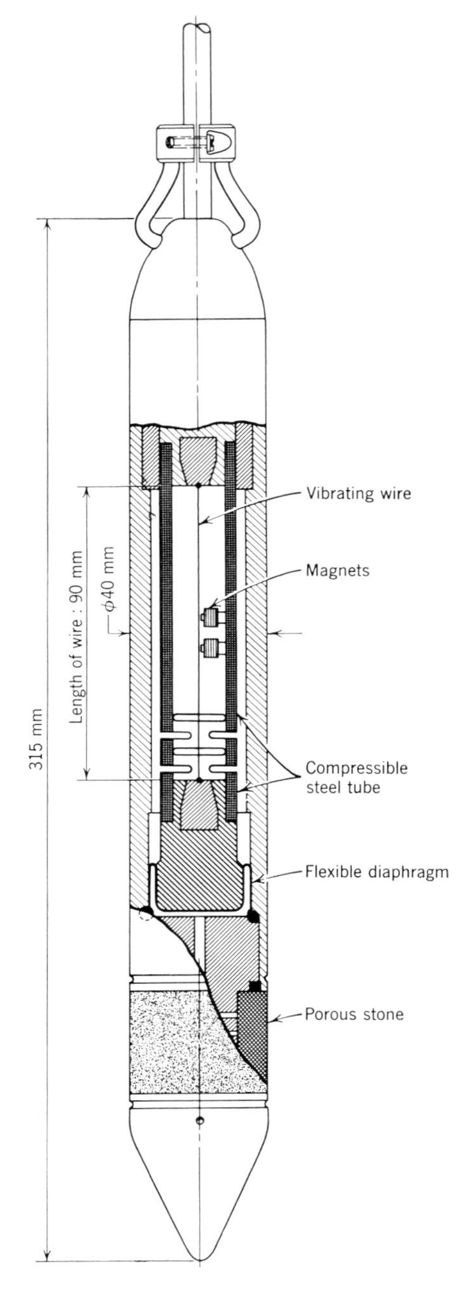

Fig. 8.1:8 Telemac vibrating wire piezometer (Type CL1)

eral advantages of electric cells as discussed above, it is probable that this type of piezometer will be much more widely used in the future. In one interesting application, the cells of the Telemac type were installed in the high Serre-Ponçon Dam which was completed in the French Alps in 1960. At several locations in the core of this dam, the vibrating wire cells were located directly adjacent to standard USBR hydraulic piezometers. Both sets of piezometers gave almost identical results during construction and the first years of reservoir operation.

8.1c Devices for Measuring Movements

HORIZONTAL EMBANKMENT MOVEMENT

Internal instruments for measuring horizontal movements within the embankment during construction have been installed at very few dams and there are no widely used apparatus for the purpose available. As a general practice on routine jobs, the cost of obtaining this information is difficult to justify. One device which has been used successfully for several dams in Europe (Swedish wells) consists of a vertical column of large-diameter reinforced concrete pipes, Fig. 8.1:9. The pipes are installed in pits during construction of the embankment and are first separated an inch or two with wooden wedges or shims. Then as the fill progresses upward, the wedges are knocked out so that each individual pipe section has no connection with the others and is held in place by the surrounding embankment material. With this arrangement the pipes are free to move with the embankment and the three components of the movement can be measured.[1]

At the high Trinity Dam in California (1959) the USBR engineers installed a special device for measuring horizontal strain within the embankment during construction. This device consists of two steel plates embedded in the embankment 20 ft. apart and centered on a crossarm compression measuring apparatus.[2] The change in the horizontal distance between the plates is measured with the same torpedo used for measuring embankment compression.

One of the most potentially valuable instruments though not yet widely used for dam embankments is the portable tiltmeter or slope indicator. These devices are constructed in the form of a torpedo which is run up and down a plastic or thin metal tubing placed in a vertical position in the dam foundation or embankment. The instru-

[1] Examples of the use of this device are given in Refs. 67, 69, 181, and 188.
[2] For details of the apparatus, see Ref. 274, p. 362.

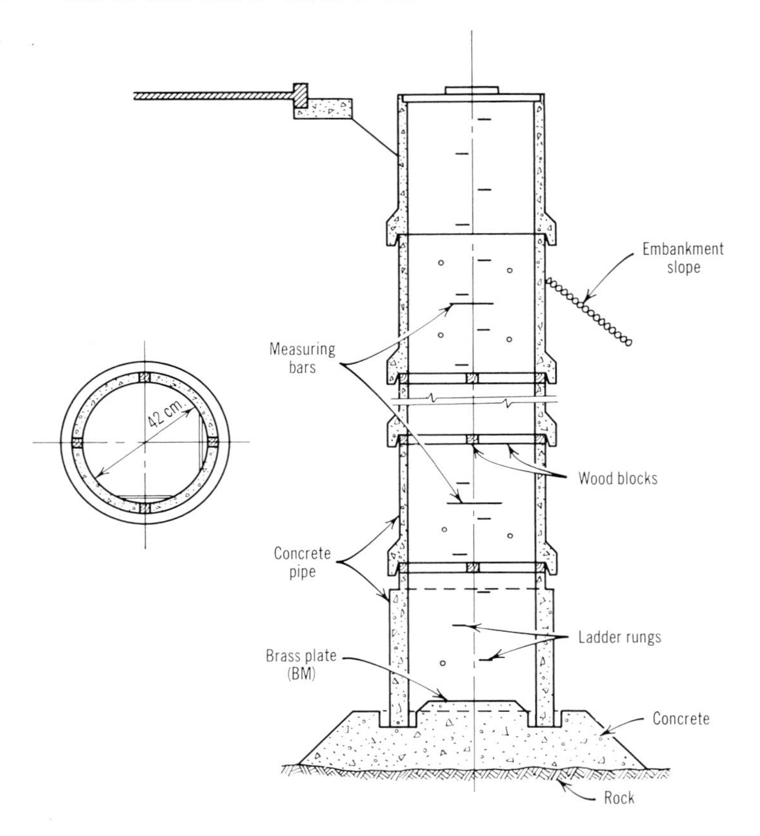

Fig. 8.1:9 Details of Swedish well for measuring embankment settlement and horizontal movement during construction (after Roberts, Ref. 452).

ment measures the slope of the tube at various elevations which allows the shape and the position of the tube to be obtained quite accurately. By making measurements at various times, the horizontal movement of the tubing can be obtained with an accuracy of a fraction of an inch in a long tube.

These portable tiltmeters have been most frequently used for study of natural land slides, and in this capacity have given the engineer a valuable analytical tool which was not previously available. The most commonly used apparatus in the United States is that developed by Wilson,[1] which uses a free-hanging pendulum and a variable elec-

[1] Details of the device with some applications are given in Refs. 117, 165, 509. Also, see Ref. 692 for a detailed recent review of slope indicators of various kinds.

tric resistance. An instrument using a pendulum on a thin steel-leaf spring in which the deflection of the spring is measured with SR-4 strain gauges was developed by the engineers of the Swedish Geotechnic Institute (Ref. 707), and was used satisfactorily in vertical holes in the embankment as well as in inclined holes in the rock foundation at the Messaure Dam in Sweden (1957–1962). Another device using the principle of a small camera which photographs a free-hanging pendulum is currently (1962) being used in both vertical and horizontal plastic tubing within the embankment of the 150-meter-high Gepatsch Dam in Austria. Recently, the authors have used a portable tiltmeter developed for them by the firm Telemac in Paris. This operates on the principle of the vibrating wire strain gauge. It has proved to be very reliable and accurate and may be the best type of all, since it is purely elastic without moving parts and the readings are not influenced by the temperature or the electric resistance of the cable.

FOUNDATION SETTLEMENT

At any dam where appreciable foundation compression is expected, it is advisable to install settlement plates at the bottom of the embankment. Knowledge of the magnitude and distribution of foundation settlement allows an analysis of the shear strains imposed on the embankment and also allows computation of the total volume of the embankment for payment purposes.

Steel plates, 2 ft. square or larger, have usually been used. In order to avoid damaging the riser pipe, its top should preferably be kept below the construction surface. Extensions to it should be added in pits excavated in the embankment, Fig. 8.1:10. This is a very rough type of instrumentation. If the dam is more than about 50 ft. in height and it is desired to measure the settlement within a precision of 0.1 ft. or less, an exterior pipe sleeve which is free to move must be used around the riser pipe to isolate it from the relative movement of the surrounding embankment.

VERTICAL EMBANKMENT COMPRESSION DURING CONSTRUCTION

The most generally used compression measuring apparatus is the telescoping pipe device with horizontal crossarms developed by the USBR, Fig. 8.1:11. The 6-ft.-long steel channels which protrude into

Fig. 8.1:10 Adding 2-in. pipe risers for foundation settlement plate measuring apparatus in hand-excavated pit.

the embankment are placed one by one in excavated trenches as the embankment is constructed. The cross arms are usually spaced 5 ft. on center, although they have been used at wider spacings. The vertical settlement of each segment of the telescoping pipe is measured with a specially designed torpedo, which is lowered into the pipe on a steel surveyor's tape, and which has collapsible fins allowing it to be drawn up tightly under the bottom of the inner telescoping pipe at the "measuring point."[1] This device is relatively simple and has been used by engineers all over the world. It has been adopted primarily in earth embankments but recently has been installed and has performed successfully in some rockfill dams built in layers up to 6 ft. in thickness.

A simple alternative device uses a series of plates or arms installed in various elevations as the embankment is constructed, each with its own vertical pipe or rod riser (Refs. 452, 505). At a few dams, steel plates have been buried during construction and located again later by drilling down through the embankment (Refs. 655).

Another device sometimes called the "Swedish water level settle-

[1] Detailed instructions for installation and reading the cross-arm measuring apparatus are given in Ref. 274.

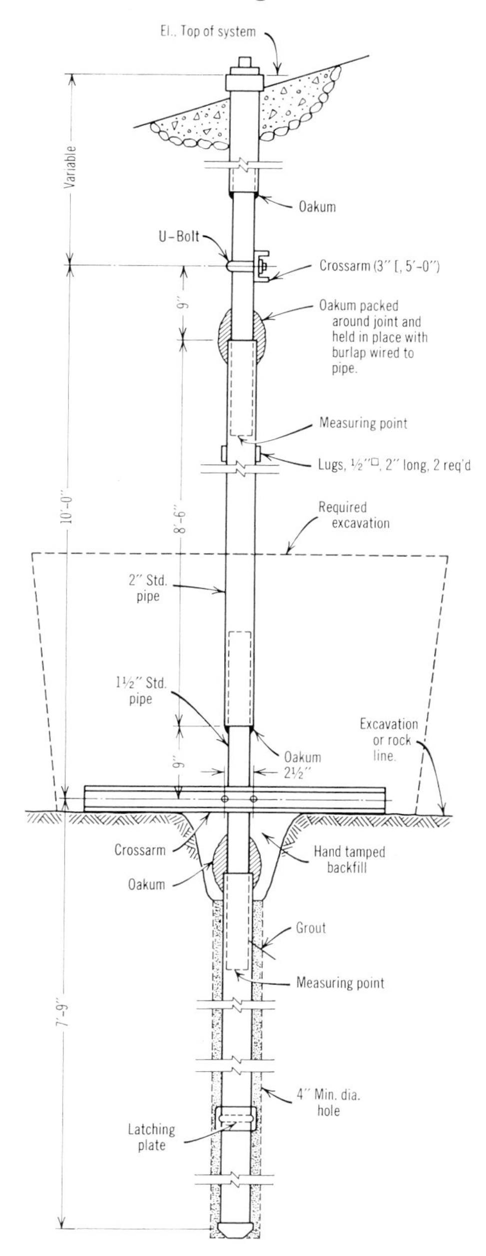

Fig. 8.1:11 Embankment compression measuring device (after USBR, Ref. 274).

ment measuring apparatus,"[1] used especially for rockfill dams, consists of placing an overflow pipe in a boxlike structure within the embankment at the point where the settlement measurement is desired (Ref. 106). The measurement is simply made by balancing the water level inside and outside the dam. Several tubes are needed to carry water in and out plus an air tube to be sure that the air pressure inside the dam is equal to atmospheric pressure. Apparatus of this type has been used in a number of large dams in Scandinavia and by French engineers in North Africa (Ref. 686). This apparatus is also currently being considered for the high Mont-Cenis Dam in the French Alps and the Djatiluhur Dam in Java. With careful measurements, settlements can be measured with an accuracy of about ½ in. The design of the apparatus and the technique of making the measurements have a large influence on the precision.

A recent development which appears very promising consists of burying doughnut-shaped steel plates in the embankment with a vertical riser pipe of plastic running up through the center of the steel plates. The elevation of the plates is measured by dropping a torpedo which contains a special electric coil into the vertical tubing. When the torpedo passes the steel plate, a noise is generated which allows the elevation of the plate to be determined with an accuracy of about ⅛ in. This device was developed and first used during the construction of the rockfill Biggetalsperre in Germany in 1962. It is also being used in both vertical and horizontal tubes in the high Gepatsch Dam in Austria, currently under construction (1962).[2]

SURFACE MONUMENTS FOR MEASURING EMBANKMENT MOVEMENTS

During construction it is easy to measure the vertical and horizontal components of surface movement of the lower slopes and foundation, and the results provide valuable information on performance. Such measurements should be made for all high dams, all dams underlain by a soft foundation, and all dams in which there is any question about the factor of safety against sliding. The monuments, consisting of concrete embedded steel rods, should be installed accurately along straight lines so that the horizontal movement can be obtained easily by measuring the offset from a transit line, Fig. 8.1:12. Usually it is sufficient to install the monuments only over that part of the dam where the embankment height is greatest. Too often, measurements of this

[1] Also called "teleniveau," Ref. 703.

[2] Patented device of Dr.-Ing. K. H. Idel.

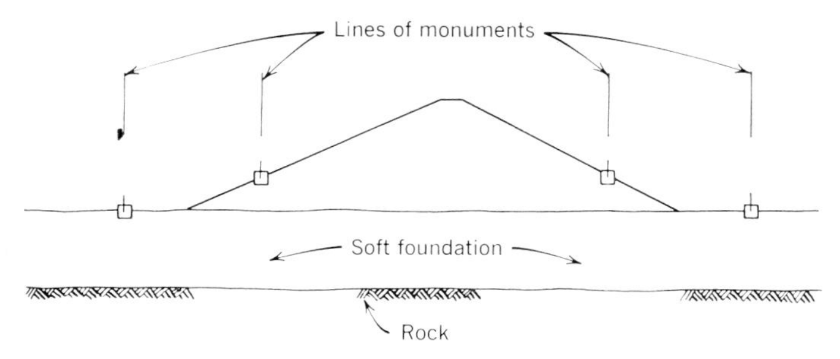

Fig. 8.1:12 Lines of monument for measuring horizontal movement of dam toes and foundation during construction.

kind have not been made at dams where they would have been helpful in analyzing subsequent performance. If a shear failure is imminent during construction, these measurements will indicate trouble long before any large or dangerous movements commence. Knowledge of the spread of the dam base which occurs during construction is also valuable in analyzing the stresses imposed on conduits passing through the foundation.

For dams of moderate or great height, it is always advisable to establish monuments on the crests in order to measure the postconstruction settlement and the upstream and downstream movement of the crest for the first few years. These measurements can be made easily and cheaply and are helpful in the analysis of any trouble which may develop.

For any dam where large settlements may develop, or where there is any reason to expect appreciable longitudinal movement, the engineer should arrange to have measurements taken of the crest movement in the direction parallel to the longitudinal axis of the dam. These measurements will reveal any tendency the dam has to develop tensile strains and transverse cracks (Sec. 2.3). Since little is known about longitudinal embankment movements (Sec. 3.2), results of the measurements on large dams should be published for the benefit of the profession.

8.1d Observation and Measurement of Leakage

Seepage through and under earth dams which emerges downstream may be perfectly safe (Sec. 6.4). However, such seeps must be carefully observed throughout the life of the dam (Ref. 98). Except for

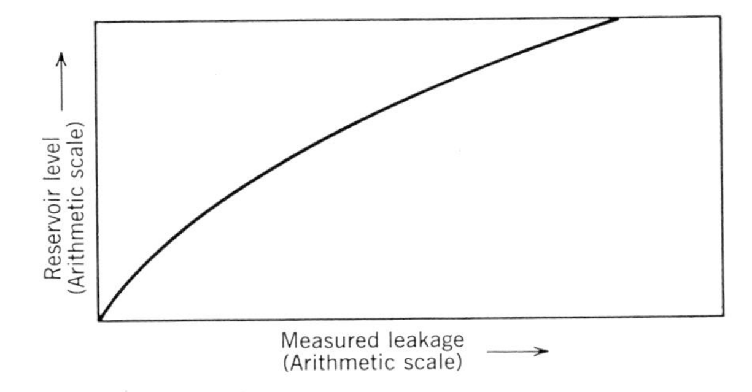

Fig. 8.1:13 Typical relationship between measured leakage and reservoir elevation.

small leaks located at large distances from the dam, all springs should be metered at intervals and permanent dated records kept of the volume of flow and the elevation of reservoir. Only by regular attention to these records can the engineer be sure that the seepage condition is not becoming worse. With time and reservoir silting, leaks should diminish in volume, if only slightly. Any increase in flow indicates that piping or solution is probably taking place. The frequency with which the flow should be metered depends on the volume and velocity of the flow and the engineer's estimation of its source.[1]

In the case of large leaks which develop soon after the reservoir is filled for the first time, or other leaks which appear to be increasing in volume, measurements should be made of the solids carried by the water (these can be measured with a stilling pond) and/or of the content of dissolved salts.[2] Even apparently clear leaks can carry away a considerable volume of fine-grained soil over a long period of time.

The total quantity of leakage commonly increases more rapidly than the reservoir head, Fig. 8.1:13. The volume of flow along any given seepage path increases as a direct function of the water head, but as the water level rises and the reservoir area increases, new seepage paths go into operation and the leakage volume increases more

[1] At the Cheakamus Dam in British Columbia (1955), which was founded on a deep deposit of erratic, pervious soil, a permanent gauging station was constructed with an automatic apparatus to measure and make a continuous written record of the leakage (Ref. 456).

[2] Initial leaks in the form of muddy water should always be viewed with apprehension. However, if the water is seeping through cracks in the bedrock and washing soil from these cracks, the initial muddy leakage may be of no significance. Clear leaks which suddenly become turbid always indicate potentially dangerous piping.

rapidly than a straight line relationship with the reservoir elevation. With the reservoir level at one elevation the measured leakage varies greatly with the temperature of the water. At some dams, the leakage in the winter months is as much as 30 percent less than in the summer (see, for example, Ref. 34).

8.2 DESIGN CONSIDERATIONS IN EARTHQUAKE REGIONS

8.2a Design Procedures

Although the limited experience available (Sec. 2.5) indicates that earth dams, as they are normally designed, have notable resistance to earthquake shocks, there can be little doubt that the shaking of a dam during a severe earthquake may still be the most dangerous condition to which it will be exposed during its lifetime. Consequently, it must be considered sound practice to provide special design details to increase the safety of a dam located in a seismic region. Enormous differences in the inherent safety of embankment type dams against earthquake can be obtained from differences in design details.

During the exploration process before the design is begun, as much information as possible should be collected about the earthquakes which have occurred in the area. The best geological and seismological advice should be obtained about the frequency and severity of possible future shocks (Sec. 8.2*c*). However, it is unlikely that these studies will enable the engineer to estimate the intensity of the shock for which he should prepare his design. Usually he can conclude only that the dam might be subjected to a severe earthquake.

Designing an earth dam to increase its safety during earthquakes is one of the most difficult special problems facing the engineer. In future years, when more information has been compiled on dam performance during earthquakes and as more basic research has been carried out, he will probably be better qualified to build earthquake-resistant dams. At present he can only fortify his experience by studying what has happened to dams in the past during earthquakes, and then he must bring his best judgment and ingenuity to the problem.

The accumulation of experience concerning the behavior or performance of dams when shaken by earthquakes is small because, at the present time (1963), we are at a special point in history where many major dams have been built in the worst seismic areas in the western United States, especially in California, but these have not yet been shaken. On the other hand, this absence of experience cannot be expected to last very long. Many major dams in California and other parts of the world will be shaken by earthquakes in the next few decades. As a substitute for this lack of experience at the present time, we must use our imagination to a considerable extent in picturing how dams may be damaged. We can speculate with some degree of assurance that most serious damages and failures will fall within the following general categories:

1. The shaking of the embankment may cause cracks to open through the core of the dam, leakage may develop through these cracks, and in the worst case failure may occur by piping.
2. Shaking of the dam may cause settlement of the crest due to compression of the foundation or the embankment, causing a decrease in the freeboard and leading in the worst case to overtopping.
3. Shaking of the reservoir bottom may cause all the water in the reservoir to oscillate in relatively large, slow waves (seiches) which may cause a considerable quantity of water to flow over the top of the dam (see Sec. 8.2*e*).
4. Slides of natural hillsides around the edges of the reservoir may displace enough water to overtop the dam or may damage appurtenant structures.
5. Deposits of loose sand in the foundation of the embankment might liquefy and move out from underneath the dam causing cracking, sliding, or actual horizontal movement of a large part of the dam.
6. Acceleration forces acting on the embankment as it is being shaken may cause a shear slide of an appreciable portion of the slope of the dam.
7. Crustal deformation associated with fault movement may lift the bottom of the reservoir with respect to the dam foundation, reducing the volume of the reservoir and possibly causing overtopping, Fig. 8.2:1.

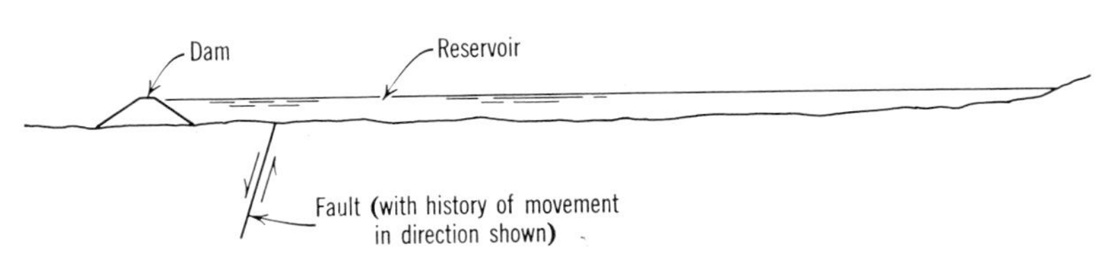

Fig. 8.2:1 Example of situation in which fault movement can decrease volume of the reservoir.

8. If a fault passes through the foundation of the dam itself, in addition to all the damages and potential modes of failure envisioned above, the dam may actually be sheared through, with the portions of the dam on either side of the fault moving with respect to each other, creating potentially large concentrated leakage.

In the ordinary process of design, the engineer generally studies a number of different embankment types and designs which could be built at the site. These vary in the use of different materials, different zoning, and different methods of foundation treatment. In an earthquake region, each of these possible embankment types and each of the individual possible design details should be scrutinized from the standpoint of their behavior during and after a severe shock. At the present time we do not know too well how to make this evaluation, but there are some very practical things which can be done.

As a first activity, the designer should study the special characteristics of his site with respect to the various types of potential trouble listed above; particularly, he should look at the foundation from the standpoint of settlement and liquefaction potential (Sec. 8.3) and at the walls of the reservoir basin for the possibility of landslides. Landslides might be dangerous under two main conditions: (1) if the volume of the material which slid was large with respect to the volume of the reservoir or (2) if the material which slid blocked the spillway[1] or outlet conduit.

As a minimum, at all sites, the designer must anticipate that the dam will crack and settle. Consequently, he must provide means to control leaks through the cracks and sufficient freeboard to prevent overtopping in spite of the settlement. By the excavation of soil from the foundation and the flattening of the embankment slopes, he can also attempt to reduce the cracking and settlement.

The degree of conservatism which he should use in selecting the details—and incurring the extra cost—of these safeguards depends on the size, location, and function of the reservoir and on the rapidity with which the water level can be lowered (Sec. 1.1*i*). The final decision, of course, can be made only on the basis of judgment; there are no rules which would meet with any general approval among experienced engineers. The design details discussed below are those from which the designer will choose in providing the amount of protection he judges necessary.

[1] The spillway, in particular, should be built in such a way as to minimize any possibility of blockage or damage from slides triggered by a shock, and side channel spillway flumes bounded by high, steep excavation slopes of earth or soft rock should be avoided.

COHESIONLESS GRADED FILTERS DOWNSTREAM

The use of graded filters of cohesionless materials in a zone located just downstream from the dam core is probably the best protection against excessive leakage and progressive piping through embankment cracks. Downstream filters of this kind are routinely provided in the designs of many dams not in seismic regions, but where filters are used also as safeguards against earthquake damage, they may be subjected to considerably more severe action than the ordinary graded filter and should be designed with more attention to gradation and dimensions. Since the upper part of the dam is likely to be subjected to the greatest cracking, the filter should be extended nearly to the top of the dam.

A crack which develops through the rolled impervious core of a dam during an earthquake will not extend into the downstream filter zone, because a crack will not remain open in a completely cohesionless material. Thus, if the reservoir water finds its way through the crack and discharges into the downstream graded filter, two results can follow: the core material can swell and seal the crack, or particles of the core material in the wall of the crack will be eroded by the flowing water and carried to the downstream filter. If the gradation of the filter is designed to prevent the eroded particles from entering the filter and the filter is intact, the particles will be stopped, the crack will be sealed, and no piping can develop.

EXTRA FREEBOARD

Since the crest of an earth dam may settle in an earthquake, the water level in the reservoir may be raised as a result of landslides into the reservoir basin or of the movement of the reservoir bottom, and great waves may develop on the reservoir, it is sound policy to provide extra freeboard in the design. No rules can be given for the amount of extra height that is desirable; it will depend on the nature of the foundation, the expected settlement, the size of the reservoir, and the degree of conservatism with which the problem is viewed. A dumped rockfill, for example, would probably settle considerably more because of shaking of the foundation than a well-compacted earthfill. Raising the crest is cheaply purchased insurance, especially since a higher crest can also frequently decrease the cost of the spillway for normal operating conditions.

VERY PERVIOUS DOWNSTREAM ZONES

In addition to controlling potential piping through embankment cracks, the designer protecting a dam against earthquake damage must make provisions for discharging the maximum anticipated leakage rapidly. If the water flowing through the cracks cannot get out of the drainage system, critically high water pressures may develop in the downstream portion of the embankment. Since the size of the cracks which may open during a severe earthquake can only be guessed at, the discharge capacity needed must be predicted conservatively.

If a good filter is provided downstream from the core, we can compute the theoretical maximum possible volume of leakage through the dam, since this would be limited by the permeability of the filter; that is, if the core of the dam was badly cracked and became very pervious, the maximum quantity of water which could get into the downstream drain is that which could get through the filter. However, the filter may be ruptured, and large quantities of water may get into the downstream drains from other sources such as leaks through cracks in the abutments and foundation or water flowing over the top of the dam. Consequently, even if good filters are used, the more leakage the downstream drain can handle safely, the greater the inherent security of the design.

Downstream zones of large quarried rock should be considered wherever the rock is available, since such zones offer the best combination of permeability and stability. There is little experience available to indicate how much leakage can be allowed to pass through a rockfill section without danger. It is probable, however, that an embankment of large rock on a sound rock foundation would be stable under any conceivable damage which might occur to the core, or even if the core were to be completely removed.[1] In addition, a massive downstream zone of large rock would probably provide the greatest security against the erosive action of large quantities of water passing

[1] There have been a number of experiences such as that described at Schofield Dam (Sec. 1.2*b*), in which earth-rock dams have developed large leaks through the core without causing movement or apparent threat of failure to the downstream rockfill section. In fact, there have been practically no failures of dams in which the main downstream section is of large quarried rock except for structures in which the rockfill was underlain by alluvium which eroded away and undermined the embankment (see, for example, Ref. 36). See Refs. 642 and 687 for some of the few studies which have been made of water flowing through coarse rockfill and the application to passing floods over partially completed rockfill dams.

over the crest. Consequently, the provision of a massive downstream section of large rock is one of the best ways to ensure the safety of a dam in a seismic region.

Other possibilities for providing rapid exit of leakage water are interior steel or concrete pipes or zones of coarse, screened gravels and cobbles. If it is not feasible to construct drains with large capacities, the downstream slope must be made flat enough to remain stable under the worst conceivable distribution of water pressures.

CORE TREATMENT

If there is a choice among several soils for material to construct the impervious core, the soil most resistant to piping should be selected (Sec. 2.2). Whatever material is used, the thicker the core is made, the greater will be the piping resistance. A thick core is also an advantage wherever an earthquake may tend to twist the dam alignment, because the additional width will minimize the influence of possible localized offsets.

What experience there has been with such dams subjected to earthquakes indicates that dams with central concrete core walls are apt to be more badly cracked than they would have been without the walls (Sec. 2.5). Consequently, interior concrete walls probably should be avoided in regions of earthquake activity.[1]

FLATTER SLOPES NEAR TOP OF DAM

Under normal conditions, the outside embankment slopes are often made steeper near the crest than in the lower elevations (Sec. 1.3*a*). From the standpoint of conventional stability analyses, this variation in the slopes provides the most economical distribution of the embankment material for any computed safety factor.

[1] An interesting dam with an articulated central concrete wall was used to provide earthquake resistance in Burma in 1937 (Refs. 254, 255). The 135-ft. high Gyobyu Dam was constructed at a site which had been about 30 miles from the epicenter of a major 1928 earthquake. The dam was designed with a central vertical wall varying in thickness from 8 ft. at the bottom to 4 ft. at the top. The concrete core wall was constructed in independent sections 25 ft. wide and 10 ft. high, connected by means of tongue and groove joints and asphaltic filler. Calculations indicated that the articulated wall could be deflected 16 ft. out of line without developing a leak.

On the other hand, it is known that during an earthquake shock the top of the dam is subjected to the worst damage and probably vibrates with considerably greater amplitude than the base. It seems prudent in a seismic region, therefore, to make the top of the dam thicker by increasing the crest width or by using flatter slopes at the top than would otherwise be required. There is no proved method to approach the problem analytically, and it is difficult to guess whether a thicker crest will be subjected to more or less of the "whipping" action which seems to occur than a thinner crest. In any event, however, the thicker the top is made, the longer is the path of seepage through cracks which may develop, and the safer is the dam in its most critical location.

FOUNDATION TREATMENT

Because a dam on a soft foundation will be more severely shaken in an earthquake than a dam on a hard foundation, consideration should be given to excavating loose or soft soil which might otherwise be left in place. As discussed in Section 8.2*d*, all our present knowledge demonstrates conclusively that steel and concrete structures at a given distance from an earthquake epicenter are much more badly damaged when founded on soil than when founded on hard rock. Both the acceleration and the amplitude of the ground movement are larger in soil. The magnitude of the difference—and the degree to which this difference may influence the damage to a dam embankment—is not known at the present time. This is one aspect of the problem for which it can be hoped that substantial advances in our knowledge will be made in the near future. At the present time, it can be speculated that the potential difference in damage to a dam is enormous, and that the cost of excavating a considerable volume of soil from the foundation may well be better justified than expenditures of a like amount to construct a more conservative embankment.

COHESIONLESS GRADED FILTERS UPSTREAM

Where the core of the dam is constructed from a very tough cohesive soil, it is conceivable that cracks which open during an earthquake will neither seal themselves nor erode. In order to prevent this, a cohesionless material which has an appreciable content of fine sand sizes can be placed in a zone directly upstream from the core of the dam.

The sand from this upstream zone will tend to wash into cracks of the core and partially seal them.[1]

8.2b Slope Stability Analysis Using Horizontal Acceleration Forces

In recent years it has become common practice to analyze the stability of slopes of earth dams in regions of seismic activity, using the common practice applied to building structures, by adding a horizontal acceleration force in the standard static slope stability analysis (Refs. 3, 138, 619). Depending on the reputation which the area has for seismic activity, the horizontal force is assigned a value which varies between $\frac{1}{20}$ and $\frac{1}{5}$ of the weight of the sliding mass.

In the absence of any other available analytical tool, this method came into general practice because it seemed better than nothing. No one believes that the procedure has any strong theoretical justification, and some engineers believe that it is completely unrealistic. Many others carry out the analysis in a perfunctory manner and, to avoid flattening the slopes greatly, accept low safety factors.

The main damage to earth dams subjected to severe earthquake shocks has consisted of cracking and settlement; records of slope slides are almost nonexistent, even though most of the dams which have been damaged were old and poorly compacted (Sec. 2.5). Compacted soil and rock as used for the construction of an earth dam does not lose its strength when it is momentarily stressed beyond the failure point.[2] The succession of main shocks during a severe earthquake normally lasts less than 60 seconds, and within this time the acceleration forces on any potential sliding surface in an earth dam embankment reverse themselves dozens of times. Therefore even if the average stress on the "critical" sliding surface should exceed the average strength for fractions of seconds, the movement along the sliding surface would be small. Unless the foundation consisted of loose sand or sensitive clay which might lose its strength when vibrated, it is not probable even in a severe earthquake that a shear slide with any appreciable displacement could occur in a well-constructed earth or rock dam designed with a reasonable safety factor against shear failure under static conditions.

[1] This was one of the main design provisions for the Coyote Dam constructed on the Hayward Fault (Sec. 8.2*c*).

[2] Recent research on the shear strength of soils under rapid or dynamic loading is given in Refs. 244, 245, 246, 306, 309, 310, 311, 514. See also studies of the influence of vibration on the pore pressure in compacted sand embankments in Refs. 75, 83.

The only possibility of a shear slide originating within the embankment itself would be in the circumstance where the material loses strength when subjected to a pulsating shear strain. Theoretically this can happen in a saturated granular soil if the soil structure decreases in volume under the strains imposed by the shock. We know that granular soils, when compacted by ordinary methods in dams of ordinary height, expand in volume when strained and, hence, the pore pressures decrease, creating a more stable condition. However, when dealing with very high dams where the embankment materials are under high stresses—outside the range of our normal experience with laboratory tests—we cannot rule out the possibility that the material might decrease in volume and have a tendency to "liquefy." The only way to approach this problem is to perform laboratory tests on undrained saturated samples with confining pressures of the order of magnitude of those expected in the dam, preferably with pulsating shear strains applied by a method which simulates those of an earthquake.

For these reasons, the authors do not believe that it is reasonable to make earthquake analyses by adding a hypothetical horizontal force in a standard static stability analysis in the ordinary method, and we have consistently recommended against the practice. An analysis in which special laboratory tests are performed, to simulate the conditions which might be expected to develop in the dam under an earthquake shock, might have some justification; however, in an analysis performed with the "effective stress" method in which the pore water pressures are assumed constant, or by the "total stress" method using ordinary laboratory test procedures, the inherent assumptions are too far from reality for the results to be capable of rational interpretation.

8.2c Fault Valleys as Dam Sites

Many prospective dam sites have been discarded as unsuitable because they were crossed by earthquake faults. On the other hand, many dams have been built on sites containing faults which were judged dead. Dams have even been constructed at a few sites on potentially active faults, sometimes because the true situation was not recognized. Since most of the best sites for earth dams will eventually be used up, the many problems connected both with deciding whether to build in fault valleys and with designing the safest dams for these sites will be encountered increasingly. Although, of course, an active fault in the foundation poses the most severe criteria on the dam designer, the authors believe that specially designed embankment type dams can be built safely at almost any site. In the following, some of the aspects of faults of most interest to the dam engineer are described.

In their every aspect earthquake faults vary over such a wide range that it is not possible to describe a typical one. The shear zone of a fault may consist of crushed or broken rock with a width of thousands of feet, or it may be essentially a plane surface with only a foot or two of broken material separating intact rock on either side. Both lengthwise and in depth, faults may be very short or may extend hundreds of miles. In general, the longer the fault—that is, the distance for which it can be traced on the ground surface—the greater the likelihood that it is active and the greater the potential movement when it slips.

With respect to movement, aggregate total displacements of dozens of miles are known to have occurred in some of the large faults. Fault movement may take place either as a continuous gradual creep which may or may not be accompanied by small shocks or, more commonly, as an alternation of sudden rather large shocks and periods of comparative quiescence.

During the periods of quiescence, very few attempts have been made to measure the relative movement of the ground on either side of faults. During actual earthquakes, there have been many more observations, and sudden fault movements of 10 to 20 ft. have frequently been measured.[1] These movements may be either horizontal or vertical: at the San Andreas Fault in California the relative movement between the two sides is essentially horizontal, but in other fault areas the displacement may be nearly or wholly vertical.

Little reliable information exists concerning movement in the direction perpendicular to the fault plane. Often there appears to be a compression on the fault plane after an earthquake is over. In some places on the main Hayward Fault during the earthquake of 1886, the fault remained open about 12 in. Only a few eyewitnesses have described the opening of faults. One account comes from the great San Francisco earthquake:

> . . . at the time of the 1906 earthquake, the line of rupture passed through a corral in which several men were milking cows. According to their statement, the fault fracture gaped during the movement and closed again. It opened wide enough to admit a cow which fell head first into the open fissure, where it remained permanently interred and after closure, invisible except for the tip of its tail (Ref. 261).

[1] See Ref. 257 for a summary of fault movements which have occurred in a number of severe earthquakes.

In the field, some faults can be recognized clearly from offsets of rock strata; others may be difficult or impossible to locate by any surface studies. Large and important faults at some dam sites have remained undiscovered even after extensive field explorations, and a number of sites have been abandoned when the faults were discovered during the foundation excavation. Some were discarded from fear that the faults might cause failure of the dams. Others were rejected even though the faults were judged dead, because the expense of treating the fault gouge zone in the foundation would have been excessive.

By definition, active faults are those which either are clearly undergoing movement or have undergone movement in historical or geological time. Dead faults are those which show no sign of having been active in recent geological time. Often it is not possible to distinguish between them with reliability, but the first step in investigating a fault valley is to make the attempt.[1] The engineer should secure the best professional help available and arrange for comprehensive studies. He will find it valuable to read the excellent papers on fault problems by Louderback (Refs. 257, 261).

In areas such as California where geologists have been studying the earthquake problems for decades, the extent and location of the major faults are well known, and rough estimates at least can be obtained concerning probable future movements. However, even in the well-studied areas, another several hundred years of observations and measurements of the type presently being made in California will be required before the amount and frequency of movement on a given fault can be predicted with reliability. In some other seismic areas of the world, very little or nothing of value to the dam engineer is known at the present time. Professor N. N. Ambraseys of the University of London has recently compiled a valuable Earthquake Engineering Reference Index (Ref. 701) which is designed to assist the civil engineer working in various parts of the world to find the literature and information which is available.

There is almost no information to indicate what might happen during an earthquake to an earth dam on an active fault. For two of the most pertinent cases available, see the descriptions of San Andreas Dam and Upper Crystal Springs Dam in Sec. 2.5. One of the few examples of the deliberate construction of a dam on what was considered to be an active major earthquake fault is that of the 125-ft.-high Coyote Dam constructed in 1936 about 25 miles southeast of San Jose,

[1] The Arvin-Tehachapi, California, earthquake of July 21, 1952 occurred as the result of movement on the White Wolf Fault previously supposed by many geologists to be inactive.

California (Refs. 193, 257, 261). The main Hayward Fault crosses the foundation.

The Hayward Fault moved last in 1886 and caused an earthquake with damage probably comparable to the 1906 San Francisco earthquake (which had an estimated Mercalli intensity of 10). It is believed that the fault has moved roughly 1,400 ft. since its beginning thousands of years ago. After careful study, the geologists predicted that the fault would move again within the lifetime of Coyote Dam. They decided that the probable movement might be about 20 ft. horizontally and several feet vertically, and that a 12-in. open crack might possibly develop.

After consideration of all the data, a consulting board of eminent dam engineers agreed that an earth dam could be safely constructed at the site. The board recommended these general design details:

1. The impervious section must be of such a type that a rupture will not take place. No rigid type of dam or concrete water face, concrete core wall, or narrow clay puddle may be used.
2. The core must be of rolled impervious material with a thickness equal to at least five times the water head.
3. Extra freeboard must be provided in the spillway.
4. A layer of gravel must be placed directly against the rolled impervious core so that if an opening should occur, the gravel would fill the void and prevent active erosion.

A section of the dam as built according to recommendations is shown in Fig. 8.2:2. The predicted movement of the fault has not yet occurred.

8.2d Ground Movements

Seismology, which is the science dealing with earthquakes and related phenomena, is a vigorously developing field and in future years it is to be expected that much will be learned of great value to the engineer building embankment type dams. In the last 30 years, approximately 60 seismographs capable of measuring the movement of the ground in the close vicinity of major faults ("strong motion" instruments) have been installed at various points in the seismic areas of the western United States (See Ref. 698 for a description of the program). The Japanese also have a similar program in operation. Up to the present time, however, only a relatively few measurements have been made with these instruments and none of these have been in

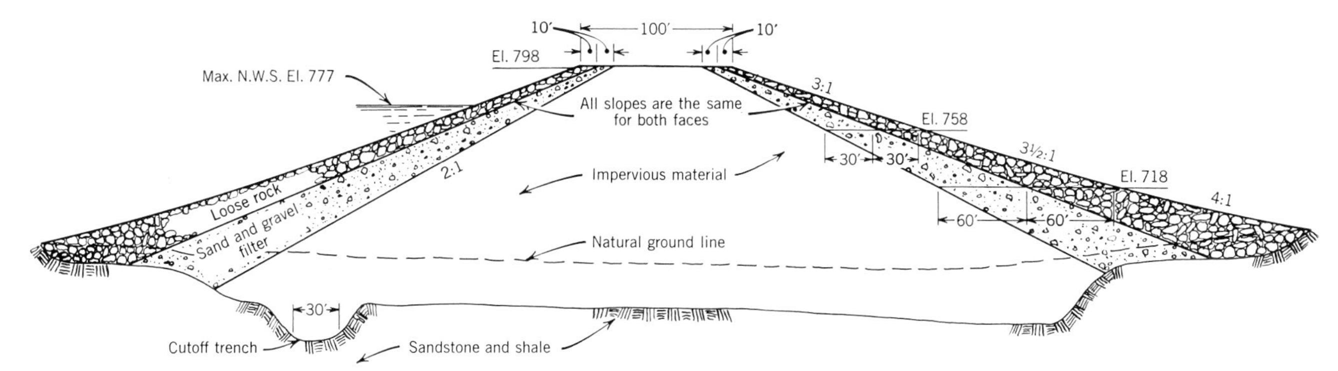

Fig. 8.2:2 *Cross section of Coyote Dam located on the Hayward Fault (after Tibbetts, Ref. 139).*

shakes of the most severe intensity. Consequently, in the next few decades we can expect to learn a great deal from measurements made on these instruments.

Strong-motion instruments have been installed on three major dams in the western United States—Cachuma Dam in southern California (an earth dam of the U.S. Bureau of Reclamation, Fig. 1.1:4) and on two concrete dams, Ross Dam in Washington and Hoover Dam in Nevada. Instruments are also being installed on two other very large earth dams presently under construction in northern California—Oroville and San Luis Dams. The purpose of the instruments located on dams is primarily research. Since a wide geographic distribution of these instruments is desirable in any event, it is hoped by placing instruments on dams that a double benefit will be obtained. Instruments are usually placed both on the dam crest and foundation to study the difference in the movement at these points.

The remaining paragraphs of this section describe some of the presently known aspects concerning ground movements which are of interest to the dam designer.

EARTHQUAKE INTENSITY SCALES

Long before instruments were available to measure earthquake movements, descriptive intensity scales were used to define the destructive violence of the ground motion in a particular area. These scales evaluate the earthquake violence in terms of the influence of the motion on people and things. The most widely known are the modified Mercalli intensity scale (1931) and the Rossi-Forel intensity scale, Fig. 8.2:3.

The device used to measure acceleration in the badly shaken area near the epicenter is the "strong-motion" seismograph, which records the ground surface movement at the point where the instrument is installed. If the accelerations measured on the strong-motion seismographs during an earthquake are plotted against the modified Mercalli intensity reported at or near the seismograph station, a primitive correlation is obtained. A plot of this type made by Neumann from data collected in ten earthquakes between 1933 and 1949 in the western part of the United States is shown in Fig. 8.2:4. As seen in the plot (which does not take into consideration the period of vibration), the maximum acceleration measured by the seismograph increases exponentially with the intensity. The curves shown are extrapolated beyond a modified Mercalli intensity of 9 because no earthquakes of

Rossi-Forel Intensity Scale (1883)
I The shock felt only by experienced observer under very favorable conditions
II Felt by a few people at rest recorded by several seismographs
III Felt by several people at rest, strong enough for the duration or direction to be appreciable
IV Felt by several people in motion; disturbance of moveable objects; cracking of floors
V Felt generally by everyone; disturbances of furniture; ringing of some bells
VI General awakening of those asleep; ringing of bells, swinging chandeliers, startled people run outdoors
VII Overthrow of moveable objects; fall of plaster, ringing of bells; panic, without great damage to buildings
VIII Fall of chimneys; cracks in walls of buildings
IX Partial or total destruction of some buildings
X Great disasters; ruins; disturbance of strata, fissures; rockfalls, landslides, etc.

Modified Mercalli Intensity Scale (1931, Wood and Neumann)
1. Detected only by sensitive instruments
2. Felt by few persons at rest, especially on upper floors; delicately suspended objects may swing
3. Felt noticeably indoors, but not always recognized as earthquake; standing autos rock slightly; vibration like passing truck
4. Felt indoors by many, outdoors by few; at night some awaken; dishes, windows, doors disturbed; motor cars rock noticeably
5. Felt by most people, some breakage of dishes, windows, and plaster; disturbance of tall objects
6. Felt by all; many frightened and run outdoors; falling plaster and chimneys, damage small
7. Everybody runs outdoors; damage to buildings varies depending on quality of construction; noticed by drivers of automobiles
8. Panel walls thrown out of frames; fall of walls, monuments, chimneys, sand and mud ejected; drivers of autos disturbed
9. Buildings shifted off foundations, cracked, thrown out of plumb; ground cracked; underground pipes broken
10. Most masonry and frame structures destroyed; ground cracked; rails bent; landslides
11. Few structures remain standing, bridges destroyed; fissures in ground, pipes broken, landslides, rails bent
12. Damage total, waves seen on ground surface, lines of sight and level distorted, objects thrown up into air

Accelerations

Cm/sec^2: 10, 20, 30, 40, 50, 60, 70, 80, 90, 100, 200, 300, 400, 500, 600, 700, 800, 900, 1000

Gravity Fraction: 0.01g, 0.05g, 0.1g, 0.5g, 1.0g

Magnitude (Instrumental): 2, 3, 4, 5, 6, 7, 8

Class E, D, C, B, Class A

Energy of Shock Ergs.: 10^{14}, 10^{15}, 10^{16}, 10^{17}, 10^{18}, 10^{19}, 10^{20}, 10^{21}, 10^{22}, 10^{23}, 10^{24}, 10^{25}

Fig. 8.2:3 Approximate relationships, earthquake intensity, accelerations, and magnitude (after Gumensky, Ref. 282).

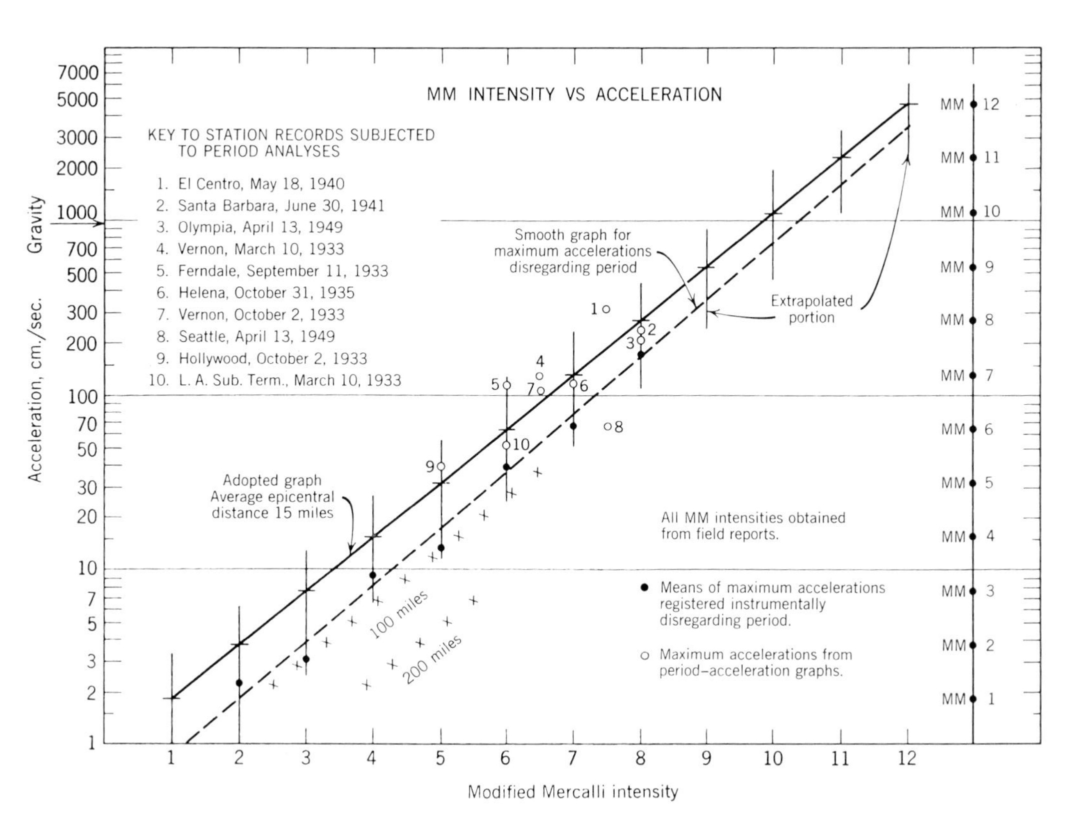

Fig. 8.2:4 Relationships between observed earthquake damage in terms of the Modified Mercalli scale and measured acceleration on strong-motion seismographs (after Neumann, Ref. 256).

this magnitude have been measured since the strong-motion instruments became available. It is interesting to note that the actual acceleration measured is much higher than the 0.1 g. commonly considered as a horizontal force in the design of engineering structures.

PERIODS AND AMPLITUDES OF GROUND MOVEMENTS

The influence that an earthquake may have on a structure cannot be defined by acceleration alone. The period and the amplitude of the ground movement are also important in interpreting the destructive action of an earthquake. These are arbitrary terms used only for convenience, however, since the actual movement of the ground surface is almost completely random and the periods between the crests are irregular.

Recent studies show that periods of ground surface vibrations are both longer and shorter than previously suspected. Some are so short that they can be heard with the ear and some may be as long as 5 minutes (Ref. 250). The very short waves are of no importance because the amplitude is too low to damage engineering structures. Similarly the very long waves, although their amplitude may rise high enough to be measured in inches, have no effect on engineering structures because the transition is too gradual. The ground movements which cause the most damage to building structures have dominant periods ranging between 0.1 and 1.0 second. In a severe shock the ground has usually been shaken between 30 and 90 seconds with near-maximum intensity.

The type of movement which may damage earth dams most severely is not known. There is some evidence that dams may be worse affected by ground movements with longer periods than those which cause the most harm to building structures.

It can be speculated that the amplitude of the movement of a point on the ground surface may be one of the most important factors influencing the damage to an embankment dam; however, at the present state of knowledge, very little is known about the amplitude. One of the most interesting measurements which has been made to date was obtained during the fairly severe El Centro shock of May 18, 1940. Figure 8.2:5 shows ground surface velocities and displacements computed from measurements made with an instrument located on deep alluvium about five miles from the fault. The maximum relative displacement across the fault was approximately 14 ft. and the portion of

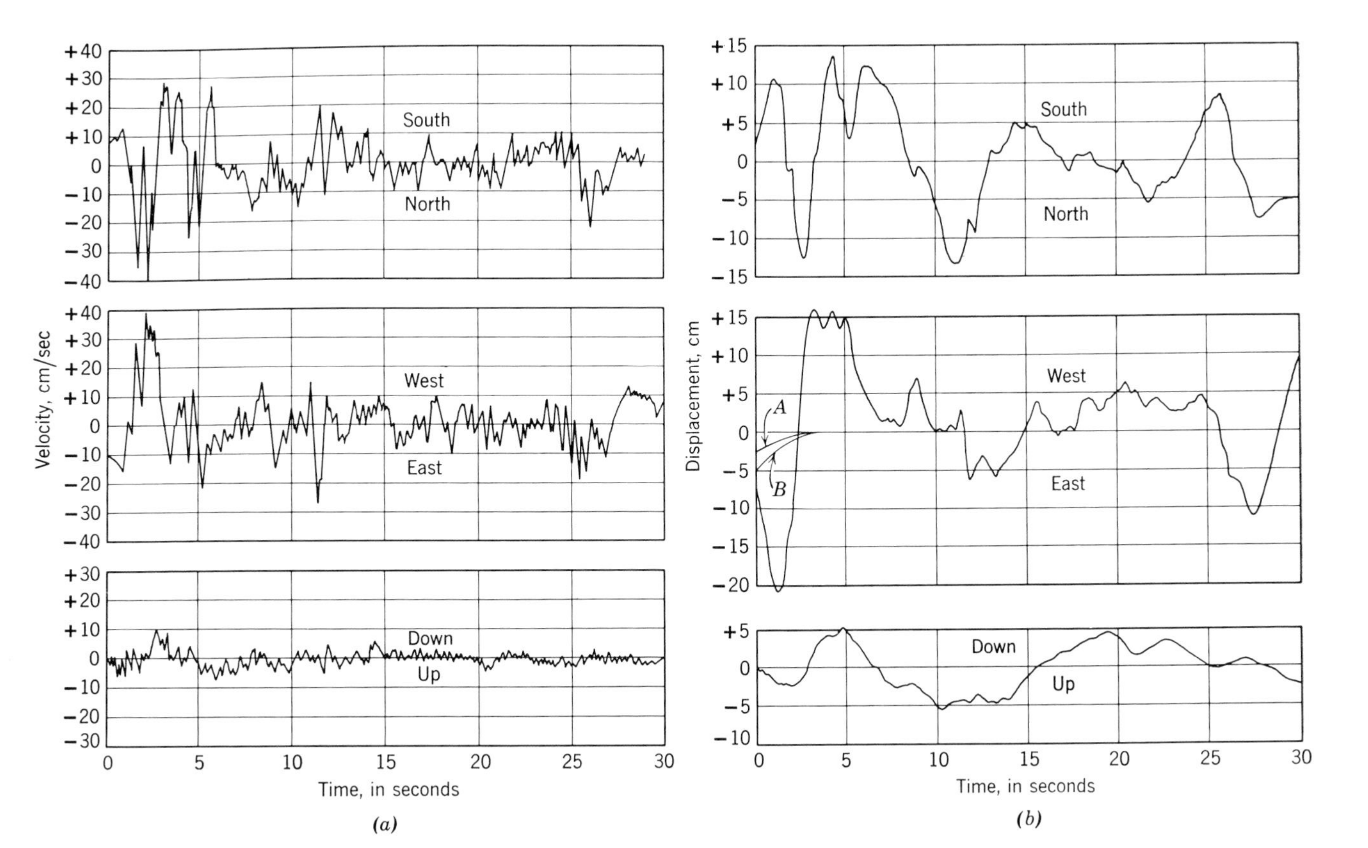

Fig. 8.2:5 Ground movement in El Centro earthquake (1940) (after Anderson et al., Ref. 269). (a) Velocity, (b) Displacement.

the fault which moved was about 40 miles in length. The maximum recorded acceleration was about 30% of gravity and the maximum displacement of the ground surface was about 20 cm. From this, other records, and eye-witness accounts, it appears likely that ground surface movements with vertical and horizontal components of several feet may be common in areas of severe damage and accelerations of 50% of gravity or more may be reached.

INFLUENCE OF FOUNDATION MATERIAL

When shock waves travel through rock, their velocity is relatively high while their amplitude is low. When they move from the bedrock to the overlying soil at the surface of the ground, their velocity decreases and their amplitude increases. For these reasons, structures founded on soild hard rock are less damaged than similar structures at a given distance from the epicenter founded on soil. With few exceptions all studies correlating earthquake damage with the characteristics of the underlying soil show this to be true (Ref. 228).

Fig. 8.2:6 presents such a study for the earthquake near Puget Sound, Washington (April 13, 1949). The curve forming the upper envelope of the observed intensities shows the damage suffered by struc-

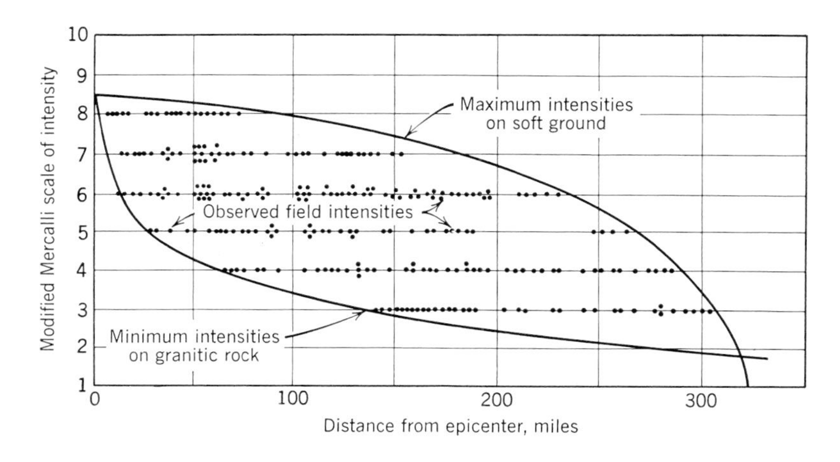

Fig. 8.2:6 Earthquake intensities from field surveys at various distances from epicenter—Puget Sound earthquake (April 1949) (after Neumann, Ref. 256).

tures founded on soft soil; the lower shows the damage suffered by structures on outcrops of the granitic bedrock in the area.

We do not yet have enough information to allow much to be said about the influence of the foundation conditions on the variation in the acceleration or amplitude of the ground surface movement. At present no strong-motion measurements have been made in badly shaken areas with instruments founded on rock. For relatively small earthquakes, in the range of period between 0.5 and 1.5 seconds, the ratio of acceleration (and amplitude), as determined with instruments underlain by soil and rock, has been as much as 8 to 1. Future measurements of this type can be expected to bring a great increase in our knowledge leading to a more rational evaluation of methods of foundation treatment for embankment type dams.

EARTHQUAKE WAVES

Seismologists have divided the types of shock waves into separate groups depending on the mechanics of the action by which the waves are propagated. However, the distinction between the wave types is of no interest to the engineer building earth dams. Dams which are close enough to the epicenter to be damaged receive all the waves in a jumbled mixture at essentially the same time. The wave pattern at any site is additionally deformed by reflection and refraction phenomena associated with discontinuities in the local geological structure.

Seismographic records of earthquakes have been studied in an endeavor to determine: (1) if the ground surface movements of a given earthquake as measured at different stations have similar characteristics; and (2) if the ground surface movements of several earthquakes as measured at one station have similar characteristics. With a few tenuous exceptions, the records of earth movements are so variable and so apparently random that no correlations have been obtained. It can be concluded that there is no "natural" ground frequency which can be expected for earthquake shocks at a given point on the surface, and therefore, that it is impossible to predict what type of earthquake waves may shake a given dam site.

At present we have no very good quantitative idea as to how the damaging action may diminish as a function of the distance between the dam site and the epicenter or the fault whose movement generates the shock. It is probable that the following relationship represents a conservative view of the attenuation which can be expected.

1. Within a distance of 0 to 10 miles measured perpendicularly to the fault, no dimunution of the damaging movement.
2. At a distance of 50 miles from the fault the damaging action may be 50 to 75% of the damage which could be expected near the fault.
3. At a distance of 150 miles from the fault little damage should be expected.

On occasion during the early stage of the design of a dam, the engineer has the opportunity to change the alignment of the longitudinal axis with respect to the direction of a neighboring fault whose movement can be expected to generate an earthquake. This can be done by selecting the dam site at different points in the valley if the direction of the river changes, or even by twisting the alignment of the dam at a given site. Consequently, it would be very valuable to know if the relative orientation of the dam axis and the fault might have a large influence on the damage which would occur when the fault moves. While future experience may throw some light on this subject, at the present state of our knowledge we cannot answer the question with any degree of confidence.

8.2e Seiches

Among the other great unknowns in this field is the effect of an earthquake shock on the motion of the entire mass of water in the reservoir. A reservoir having a large horizontal dimension has a long natural period of oscillation (Ref. 693) and it is suspected that some types of earthquakes, which contain a long period between fairly large ground displacements, may start the whole body of water in a relatively slow, rocking motion with appreciable magnitude. This type of motion is commonly referred to as a seiche. It is quite possible that under the worst combination of conditions, this action could cause a large quantity of the reservoir to be spilled over the top of the dam. There have been a number of examples, including the waves which passed over the top of Hebgen Dam (Sec. 2.5), which indicate that this is not simply a theoretical possibility. At this writing (1963), the State of California Department of Water Resources is considering making a thorough study of this problem and it is possible that within a short time we will be able to predict the maximum wave height as a function of reservoir dimensions and the type of ground surface movement.

8.2f Models and Vibration Theory

Experience with earth dams which have been shaken (Sec. 2.5) indicates that the upper portion of the embankment is subjected to greater forces than the lower. The most common damage due to foundation shaking is the development of nearly vertical longitudinal cracks on or near the crest; it seems likely that this cracking has been caused by a whipping action of the upper embankment in the upstream-downstream direction.

One of the main fears concerning the safety of an earth dam during an earthquake is the possibility that resonance associated with this whipping action might develop. Although most engineers have felt that the problem is too complex for solution, a few have approached it from the standpoint of vibration theory and scale models.

Heiland (Ref. 247) and Monobe et al (Ref. 267) both derived an equation for the fundamental period of vibration in terms of the height of the dam and the density and "shear modulus" of the embankment material.[1] Both found that calculations made with reasonable values for the embankment soil properties show the fundamental periods of vibration to be of the same general magnitude as the periods of large earthquake shocks. Therefore, while all the assumptions underlying these calculations may be questioned, the results do not allay the fear that high earth dams might develop resonance.

Both investigators then constructed models of earth dams on shaking tables and vibrated the models horizontally at various frequencies, using motion picture cameras meanwhile to measure the deflection of the crests with respect to the bases. Both studies showed that a resonance phenomenon occurred in the models under these conditions, and that at a certain narrow range of vibrating frequencies the crests of the models had vibration amplitudes as much as five times the amplitudes of the tables and the bases of the models, Fig. 8.2:7.

Therefore, both the theoretical approach and the model tests lend weight to the speculation that earth dams might resonate under earthquake shocks. Unfortunately we cannot evaluate the relationship between model and prototype or the influence of the approximations which were made to develop the theory. Neither can we determine the soil properties of an earth dam with sufficient reliability to calculate the fundamental period of the embankment. On the more encouraging side of speculations about resonance is the idea that, because of the relatively short duration of major earthquake shocks and the un-

[1] The latest and most detailed mathematical treatment of the problem is given in Ref. 485; see also Refs. 248, 249, 253, 265, 383.

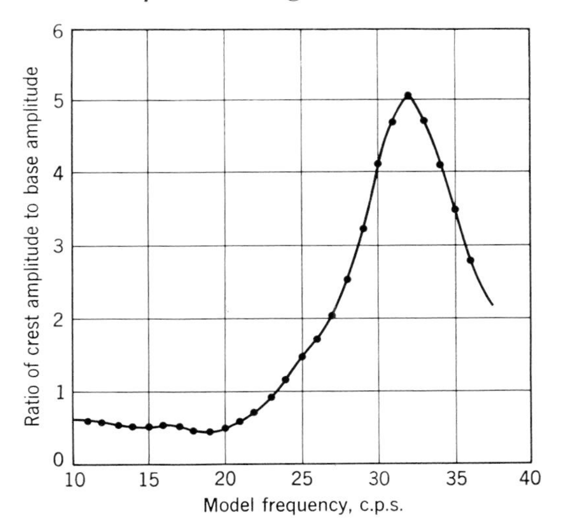

Fig. 8.2:7 Measured crest deflection of earth dam model subjected to sustained vibration at various frequencies (after Heiland, Ref. 247).

doubtedly large damping characteristics of a rolled-earth embankment, it is not likely that a resonant-like vibration could be sustained for more than a very brief time.

Another series of tests studying dam movement during an earthquake was made on sand models in connection with the design of the earth-rock Kenney Dam in British Columbia, Fig. 1.2:15 (Ref. 243). These tests were devised to study what damage might occur to the dam rather than to determine the natural frequency, and the table was struck by a heavy pendulum instead of being vibrated continuously. As an additional means of simulating an earthquake shock, the table was fitted with a heavy damping spring, so that the model, though subjected to strong shocks, was exposed to them for short periods of time.

Under the blow of the pendulum, the base of the dam model moved downstream and left the top momentarily stationary. After a time lag of 0.02 to 0.03 second, the top moved downstream somewhat further than the base. In the following several cycles of vibration caused by the heavy spring, the top came into phase with the bottom, and the model moved as a rigid unit. The investigators concluded that model studies of this kind can be reasonably used to study the problem,[1] that

[1] Grave questions have been raised by others concerning this extrapolation of conclusions from a small model; see, for example, the discussion of Spielman (Ref. 243).

rockfill dams with sloping earth cores were completely safe against slope slides during earthquake shocks, and that the worst circumstances which will arise for such a dam under the most severe conceivable shock is settlement of the crest and moderate flattening of the slopes. More recent and elaborate studies of a similar nature made in connection with the design of the high Portage Mountain Dam in Canada and the Oroville Dam in California have given the same conclusions (Refs. 511, 697).

8.3 LOOSE SAND FOUNDATIONS

FLOW SLIDES

A site with a thick deposit of loose sand in the foundation poses one of the most difficult problems for the earth dam designer. The difficulty arises not primarily because of low strength or high compressibility of the loose sand foundation material, but from the possibility that, under certain circumstances, the loose sand deposit may liquefy and flow out from under the dam. At the present state of our knowledge, the evaluation of the susceptibility of a deposit of loose sand under an earth dam to possible liquefaction for practical purposes is an intractable problem; neither theory nor experience offers much reliable assistance.

It is well known that "flow" or "liquefaction" slides have occurred in slopes of natural soil consisting of deposits of fine sand and silt, and in some types of clays. These slides are characterized by movement in which the material flows downhill under the action of gravity as though it were a heavy viscous fluid. The slides are caused when shear strains imposed on the saturated material result in a tendency for volume decrease, and consequent transfer of the stresses within the soil mass from the grain structure to the pore water. When a large part of the weight of the overlying soil is carried momentarily by the pore water, the shear strength of the mass is reduced literally to that of a liquid and the material flows downward; hence the term "flow" or "liquefaction" slide.

Flow slides in natural soil deposits have been triggered by earthquake shocks and by undercutting of the toe of the slope; in a few cases they have apparently been caused by increases in the ground water pressure. A number of major slides of this type have occurred in slopes of hydraulic fill earth dams during construction, and the Sheffield Dam probably failed as the result of liquefaction of the lower part of the embankment during an earthquake (Sec. 2.5). No failures have occurred as the result of liquefaction of natural sand foundations underlying rolled earth dams (Sec. 2.9).

Since no failures of this type have occurred, we are tempted to assume that sand foundations are safe from liquefaction regardless of the density; however, not many high earth dams have been constructed on sand foundations and this lack of experience cannot be considered conclusive. We must assume that, under some extreme conditions, liquefaction failures may occur.

DIFFICULTIES IN ANALYTICAL TREATMENT OF THE PROBLEM

Laboratory research on the problem, utilizing triaxial compression tests, has been carried out by a number of investigators for many years and this work is still continuing. At the present time, methods for the analysis of sand foundations through laboratory tests are still far from being routine procedures and it is beyond the scope of this book to recommend methods of laboratory investigation for this problem. Some of the principal difficulties which have arisen making the laboratory investigations difficult and of dubious reliability are discussed below.

The original concept of a "critical void ratio" below which cohesionless soil should be safe against liquefaction was introduced by Arthur Casagrande more than 25 years ago (Ref. 368). This was defined as the void ratio at which a specimen of cohesionless soil neither expanded nor contracted in volume when subjected to shearing strains. The method of testing and interpreting the test results has a large influence on the critical void ratio measured for a given material. Much subsequent laboratory research has been carried out for the purpose of determining methods of testing which would allow the most realistic evaluation of the liquefaction potential. Instead of coming closer to a solution of the problem, the primary result of these investigations has been to raise doubts as to whether any of the test procedures used to date ever will permit a reliable evaluation of the susceptibility of a given sand deposit to liquefaction.[1] Some of the principal difficulties

[1] The only notable exception to this is the work done by the U.S. Army Engineers since about 1948 in connection with studies of the stability of the bluffs of the Mississippi River. By painstaking field and laboratory research and study of many failures, it has been possible to develop empirical criteria for the special conditions existing along the Mississippi River which indicate the likelihood of flow slides within at least rough limits. At present the results of these studies are not published for general distribution. Some of the basic research has been summarized in Ref. 609.

encountered which make it difficult to apply the results of laboratory investigations are:

1. The critical void ratio as determined by triaxial tests using somewhat different but apparently reasonable procedures varies considerably and the most realistic procedure is not known (see for example, Ref. 213, Secs. 14.12 to 14.14).
2. Specimens of loose sand, with void ratio less than the critical void ratio, contract in volume but then the volume expands again at high strains.
3. Triaxial compression tests performed on saturated specimens of sand, prepared in the loosest state possible in the laboratory, cannot be made to liquefy even though the void ratio is much in excess of the critical void ratio. This is in spite of the fact that such tests have been run in the laboratory on specimens from natural soil deposits which have been known to have suffered flow slides in nature when existing at lower void ratios than the samples tested.
4. It may be probable that no laboratory test can be devised which represents the effect which sudden shocks of an earthquake may have on a large mass of saturated sand.
5. The volume change during shear in different parts of a triaxial test specimen varies considerably. By freezing samples, investigators at the U.S. Army Corps of Engineers Waterways Experiment Station showed that the overall volume changes commonly measured in the triaxial test were not at all indicative of what is happening in the shearing zone itself (Ref. 492).
6. Finally, in addition to the problems associated with test procedures and interpretation, very little information has been obtained concerning the densities of natural sand deposits which have undergone flow slides.

FACTORS INFLUENCING SUSCEPTIBILITY TO LIQUEFACTION

From these discussions, it can be seen that neither laboratory testing nor the available experience records of flow slides provide much tangible assistance for the evaluation of the susceptibility of a given sand foundation to liquefaction. The principal factors governing susceptibility to liquefaction are:

1. The change in void ratio of the sand caused by shear strains;
2. The effective permeability of the sand foundation relative to the

rate of imposition of shear strains, and hence, the rate at which pore pressures caused by shear strains can dissipate by drainage;
3. The magnitude and the rate of application of shear strains which may be imposed on the loose sand foundation by the construction of the dam, the operation of the reservoir, or by seismic shocks.

These factors, in turn, are dependent on the following properties of the foundation which can be determined by field and laboratory exploration:

1. The relative density of the sand;
2. The coarseness and the gradation of the sand;
3. The thickness and the permeability characteristics of the sand;
4. The strength characteristics of any soft layers of silt and clay located in or below the loose sand foundation which may determine the magnitude of the shear strains imposed on the sand layer.

The magnitude and the rate of application of shear stresses due to the dam and the reservoir depend on the weight and the shape of the dam, the rate of construction, and the rates of raising and lowering the reservoir level. The effective shear stresses imposed in the sand foundation mass by earthquake shock and the influence of these on volume changes and pore pressures cannot be estimated by any available means.

Even if the magnitude of each of the factors listed above could be reliably determined, their interrelationship and relative influence on the liquefaction potential remains a problem of formidable complexity. In future years, as the result of additional laboratory researches and experiences with flow slides, it will probably be possible to evaluate quantitatively the liquefaction potential at a given site at least within some rough limits. At the present time, the results of the laboratory and field investigations at a given site can only be used as a guide to judgment.

In the absence of better knowledge, the following tentative general conclusions concerning the factors involved appear to be a reasonable guide when evaluating loose sand foundations under rolled earth dams:

1. Sand with a relative density[1] of 50% or more probably cannot liquefy regardless of the gradation.

[1] Using the test procedures as described for test E-11 in the USBR Earth Manual. (Ref. 274).

2. The coarser and the better graded the sand, the less susceptible it is to liquefaction.
3. The materials most likely to liquefy are very fine, uniform sands and coarse silts consisting largely of rounded grains (see Ref. 126, p. 102 for gradation curves of some materials which have been unstable in natural slopes and excavations).[1]
4. Sand formations which are deposited by river action are likely to be less susceptible to the possibility of liquefaction than formations which have been deposited by wind, as slope wash, or as a still water deposit.
5. The greatest shear stresses are applied to the foundation material as the weight of the embankment is added during construction. Additional shear stresses due to the pressure of the reservoir are a small percentage of the stresses imposed during construction and usually are not added for many months after completion of construction. Consequently, if the foundation is saturated while the embankment is being built and, if a flow slide can occur, it should occur during construction.
6. We have practically no criteria which will allow an evaluation of the likelihood of a foundation flow slide at a given site as the result of an earthquake shock.
7. Loose sand foundations containing or underlain by horizontal layers of soft silt or clay will have larger horizontal movements and, hence, a greater strain in the loose sand foundation and greater potential for liquefaction.

METHODS OF FIELD EXPLORATION

Above the water table, suspect loose sand foundations should be explored with test pits. The "in-place density" of the sand at various elevations should be measured by field density tests. The visual appearance of cohesionless sand in the walls and the bottom of the test pit cannot be used as a reliable guide to the density even for experienced soil engineers.

It is more difficult to estimate the in-place density of the sand below the water table. In some major projects, where concern has arisen over the possibility of liquefaction, large expenditures have been con-

[1] In utilizing this data, it should be kept in mind that a flow slide of a natural slope consisting of loose sand will occur with much less provocation than a flow slide of the same material located in a horizontal layer under a dam.

sidered justified to dewater the foundation in order to sink test pits. Recently, investigators have been successful in obtaining apparently good undisturbed samples of fine to medium sands below the water table using piston samplers and drilling mud. In extreme cases, vertical columns of the sand have been solidified by freezing in order to obtain undisturbed samples for relative density determinations.[1]

Blasting tests, in which charges of explosives are jetted into the ground, have been used as an indication of the density. The settlement of the ground surface after blasting is an indication of the density of the deposit as well as the efficacy of blasting as a means for increasing the density.

More usually, the in-place density of the sand below the water table is estimated with dynamic or static penetration tests. Care must be taken when penetration tests are performed to avoid "quick" conditions developing in the soil at the bottom of the hole below the penetrometer. At a number of projects, where the penetration resistance obtained was low, the results were later determined to have been completely erroneous because the sand at the bottom of the exploratory holes was loosened by the upward flow of ground water into the hole. This phenomenon frequently occurs when the exploratory holes are drilled with casing and the water in the hole is allowed to drop below the adjacent ground water table. The trouble can be overcome by keeping the casing filled with water or through the use of drilling mud.

The resistance to penetration measured in sand layers depends on the depth of the overburden as well as the relative density. Results of the extensive research carried out by the USBR (Ref. 39) can be used to estimate the influence of the overburden pressure on the results of the standard penetration tests.

Field-bearing tests on 1-ft.-square steel plates have been used as a means of estimating the relative density. Above the water table, these tests can be performed relatively rapidly and economically. On important projects, they have been performed in dewatered pits below the water table. The relative density of the sand formation below the steel plate is estimated by comparing the "load-settlement" curve obtained in the test with families of curves representing different densities, Fig. 8.3:1.[2]

[1] For an important project in the Soviet Union, the sand foundation was explored with the use of a massive compressed-air caisson (Ref. 91).

[2] Reference 664 gives an excellent summary of the methods and difficulties of exploring sand foundations.

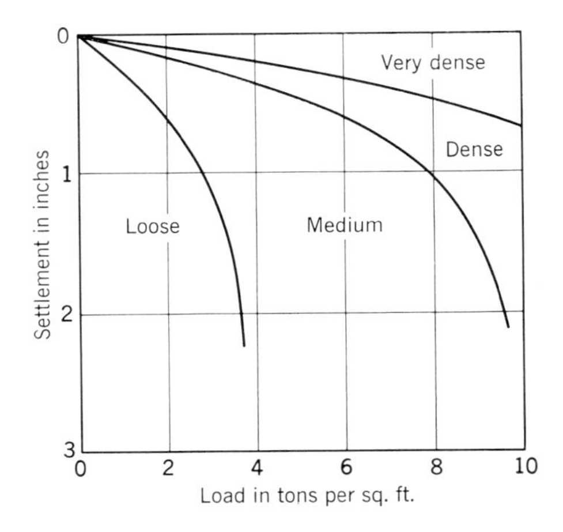

Fig. 8.3:1 Chart for estimating relative density of sand on basis of results of standard local test on 1-ft.[2] *bearing plate (after Terzaghi and Peck, Ref. 126).*

METHODS OF COMPACTING LOOSE SAND FOUNDATIONS

If the field investigations indicate that the foundation sand is so loose that there is reasonable suspicion that it might be subject to liquefaction, it must either be excavated or compacted in place by one of several means. On a number of dams, the surface layer of loose sand has been excavated and recompacted as a well-compacted, dense fill.

In only a very few cases have loose sand foundations been compacted in place. The principal means available are:

1. Compaction by setting off small explosive charges.
2. Vibroflotation.
3. Pile driving.

Explosives were first used for compaction of the loose sand foundation under an earth dam by the U.S. Army Engineers at Franklin Falls Dam, in New Hampshire, which was completed in 1942 (Refs. 17, 131). Though the technique proved quite successful and has been considered for use at a number of other sites, it has not been used to compact the natural sand foundation under a major dam since that time.[1] At the Franklin Falls Dam site, the upper 25 ft. of the founda-

[1] A procedure similar to that used at Franklin Falls Dam was used to compact about 60,000 yd.[3] of sandfill placed under water in a large pothole in the foundation of the Karnafuli Dam in Pakistan in 1958 (Ref. 672). The sand was fine

tion material consisted of loose, uniform, fine sand. This was compacted using 8-lb. charges of explosives jetted into the ground at a depth of 15 ft. on a 5-ft. grid pattern and exploded in a series of alternate shots. As a result of this work, the ground surface settled an average of 2.0 to 2.5 ft., and the void ratio of the material was decreased from a rough average of 0.95 to about 0.75. Approximately 400,000 yd.^{3} of sand were compacted in this manner at an average cost of roughly 9¢/yd^3.[1]

No cases are on record where loose sand foundations under earth dams have been compacted by vibroflotation (Refs. 89, 517) or by driving compaction piles (Ref. 6), although there is no doubt that both methods would be effective. The effectiveness of the vibroflotation method has been well demonstrated by the compaction of loose sand under the foundations for many major buildings. In all but the very finest silty sand, it is possible to compact loose deposits to a relative density of at least 75%. Successful tests were made for the use of the vibroflotation method for the compaction of large quantities of dune sand for the proposed construction of the High Aswan Dam (Refs. 79, 685). The Harza Engineering Company used the vibroflotation method for the compaction of sand backfill under a concrete dam in a narrow rock walled canyon for the Box Canyon Hydroelectric Project in Washington in 1957 (Ref. 369).

8.4 FOUNDATIONS OF SOFT CLAY AND SILT

8.4a Stability Analysis

Sites with thick deposits of soft and compressible soil generally have been avoided in the past, but as time passes and the better sites

and uniform and was washed to remove silty fines (after washing it contained less than 10% passing the No. 100 sieve). The maximum thickness of the compacted material was approximately 55 ft. Blasting charges (3 lb. and 8 lb.) were placed in plastic containers and jetted into the sand at three different elevations. The blasting was carried out from the bottom up and the charges at each elevation were set off at time intervals of approximately one day. Each blast caused some settlement. The average final settlement was approximately 5% of the thickness of the sand fill. The maximum surface settlement was 2.7 ft. The results were similar to those at the Franklin Falls site. The ground surface heaved up and settled back with each shot. Gas and water poured out of craters which developed over some of the holes through which the charges had been jetted into the fill.

[1] Recently similar blasting has even been used in the United States to compact loose sand in order to make it a better foundation to support small buildings, Ref. 691.

are gradually built upon, sites with soft foundations must be used with increasing frequency. When the height of the dam at a soft soil location approaches the maximum that the foundation material will allow, the embankment design is dominated by the need to insure against foundation shear failure.

Analysis of the stability of a dam at such a site is a problem for the soil mechanics specialist. The difficulties in testing soft, fine-grained foundation soils (and especially clays of very high plasticity and fissured or "slickensided" clays) are greater than for most embankment materials. Because of this and because almost all shear failures have occurred in dams underlain by clay foundations (Sec. 2.4), it is necessary at present to adopt a very conservative attitude toward the results of stability analyses in this circumstance.

Two conditions must be analyzed: (1) the stability during construction, when the stresses due to the weight of the embankment are added to the foundation for the first time, and (2) the long-time stability (full reservoir and drawdown conditions, Secs. 7.6, 7.7). The process of analyzing stability during construction is the same as for highway embankments, ore piles, and other heavy structures. Consequently more experience is available for this condition and more confidence can be placed in the results of current methods of analysis than can be given to analyses of long-time stability. Because of this, together with the fact that a movement or slide during construction is not as serious as a slide occurring during full reservoir, the long-time stability condition is usually the governing factor in the dam's design.

STABILITY ANALYSES FOR CONDITIONS DURING CONSTRUCTION

Stability during construction should be analyzed by the total stress method, in which the strength of the foundation soil is obtained from laboratory unconsolidated-undrained triaxial tests (Sec. 7.3*c*) performed on undisturbed samples or "vane" shear tests at the bottom of bore holes, Fig. 8.4:1. Studies made since 1945 of a large number of failures of embankments and other structures indicate that results of such analyses are reasonably reliable.

For the common situation where the foundation soil is very impervious and fully saturated, the analysis simplifies to the "$\phi = 0$" analysis in which only a "cohesion" component of strength is used (Refs. 124, 516). The strength may be constant throughout the foundation layer or it may vary.

In areas of flat topography a long, low dam is often needed on a

Fig. 8.4:1 Typical vane used for investigating in-place shear strength of soft foundations.

soft lake or marine deposit for the purpose of retaining a relatively large and shallow reservoir. In such circumstances the construction of a full-sized test embankment may be the cheapest, and certainly is the most reliable, way to test the during-construction stability.[1] For the USBR's 12-mile-long, 35-ft.-high Willard Dike, which was to be constructed on a deposit of soft organic silt of the Great Salt Lake in Utah, a 35-ft.-high test section of the dam with crest length of 240 ft. was put up for about $40,000 (Ref. 421). The test section made it possible to establish that considerable pore pressure dissipation oc-

[1] In fact, because of the difficulty in estimating the pore water pressure in seams and lenses of silt, full size test embankments are the only reliable means of analysis in some cases (see Ref. 521 for a detailed discussion of this subject).

curred in the upper parts of the soft foundation during construction, and that the dam therefore had a higher factor of safety than was indicated by the Total Stress stability analysis.

ANALYSIS OF THE LONG-TIME STABILITY CONDITION

If the dam embankment did not retain a reservoir, the stability condition during construction would be the most severe to which the foundation would be exposed. The construction pore water pressures in the foundation soil would gradually dissipate, and the material would become stronger with time. Actually, however, the construction pore pressures are replaced by seepage pore pressures, and in addition, the upper layers of the natural soil at the dam toes will swell and soften. Since the great majority of sliding failures have occurred after a number of years of operation (Sec. 2.4), it is evident that the worst stability condition frequently develops some years after construction.

It is difficult to evaluate the reliability of long-time stability analyses, because there have been few opportunities to study actual slides during the reservoir operation since the tools of modern soil mechanics have been available. However, such studies as have been made indicate that results of present methods of analysis do not always check the failure condition well.[1] Consequently, it is imperative under these circumstances to provide higher safety factors than are used for ordinary dams (Sec. 7.6), especially against the possibility of downstream slides where failure might result in a disaster.

In addition to the general limitations discussed in Chap. 7, the following factors, difficult or impossible to evaluate at present, all may detract from the dependability of stability analyses for dams on foundations of soft fine-grained soils:

1. *Strain incompatibility between embankment and foundation:* Typically, the stress-strain characteristics of the natural soft foundation and the stiffer rolled embankment are quite different.

[1] For example, see the detailed study reported by R. Peterson of the slides which occurred in the Seven Sisters Dike in Canada, founded on clay of high plasticity (Ref. 109). As the result of very careful laboratory and analytical studies using the latest testing equipment and methods currently available, it could only be shown that the average computed factor of safety in the embankment at the area of the slides was approximately 1.5. Higher factors of safety were computed using strengths obtained from ordinary 3-in. diameter Shelby tube samples or from vane tests. Reference 519 contains a comprehensive discussion of the difficulties inherent in present methods of stability analysis of embankments founded on soft clay of high plasticity.

The strain along the portion of the potential sliding surface which passes through the embankment may reach failure before the foundation is strained sufficiently to develop its full strength. At the present time, there is not sufficient information available to justify a definite recommendation on the best method of including the influence of this factor in the analysis. However, the conservative approach is to ignore the influence of the embankment strength on the stability.

2. *Relatively pervious horizontal layers:* Foundations consisting of soft, fine-grained soils are commonly very impervious, but they may contain continuous layers or isolated horizontal lenses of silt, which may or may not be observed in the exploratory borings. Although these layers often are not in themselves very pervious, they are many times more so than the surrounding clay and can transmit high pore water pressures horizontally for great distances. In this situation it is difficult to estimate what pore pressures might develop in the foundation soil. Continuous sampling with large diameter borings or test pits should be used to try to locate such layers.
3. *Thin, weak clay layers:* Natural deposits of clay of medium to high plasticity may sometimes contain thin, horizontal layers of clay of much higher plasticity than the average material in the deposit. These thin layers, which may be only an inch or less in thickness, may have the same appearance as the average material and go unnoticed even if they happen to be encountered in a sample. The effective stress strength parameters (c' and ϕ') of these thin layers are likely to be lower than those of the average material, and consequently will govern the stability.

The long-time stability should be analyzed by the effective stress method using pore pressures estimated for gravity flow (Sec. 7.6). In order to anticipate the existence of undiscovered, relatively pervious, silt seams in the foundation, the seepage study should be made on the basis of a high assumed ratio of horizontal to vertical permeability.

In order to locate thin layers of clay with different characteristics and possibly lower strength than the average material, it is highly desirable to take continuous samples, to inspect these inch by inch, and to determine Atterberg limits at intervals of depth of about 6 inches. The strength of the weakest material encountered should be used in the stability analysis.[1]

[1] The existence of layers of low strength (in terms of effective stress parameters) may not usually be determined with vane shear tests since the undrained strength of these layers may be as high or higher than that of the average foundation material.

8.4b Design Details

All dams on soft foundations should be provided with instruments carefully designed to allow the engineer to observe the movement of the lower slopes of the embankment and foundation as well as the pore water pressures which develop during construction (Sec. 8.1).

In order to obtain a safe and economical structure at a site with a soft foundation, one or more of the following design details are used:

1. *Excavation of soft material:* If the soft material in the foundation is not very thick, it can be excavated wholly or partially and replaced with compacted earth.
2. *The use of flat side slopes:* The embankment side slopes can be flattened to reduce the shear stresses in the foundation. This is the most common design procedure in dealing with fine-grained soils. Dams on soft foundations have been constructed with embankments containing many times the soil volume than would be otherwise required. Any type of material and construction method can be used for the outer slopes, because the overall stability of the dam is governed by the strength of the foundation material and the soil placed in the outer slopes of the embankment serves primarily to provide weight and mass. Lovewell Dam, Fig. 2.11:1, is a typical design in which flat upstream and downstream slopes were used to reduce the shear stresses in the underlying soft foundation. Another interesting design was used for the Jibacoa Dam in Cuba, which was underlain by approximately 150 ft. of soft, compressible, fine-grained soil, Fig. 8.4:2.
3. *The use of stage construction:* The embankment can be built slowly or in stages to allow the underlying soft soils to consolidate and gain strength during construction. At some dams the construction has even been broken into two or more separate contracts so that the engineer has the flexibility to allow him to wait for sufficient consolidation before work continues (Ref. 20).
4. *The use of drains:* Horizontal pervious blankets or vertical sand drains can be used to accelerate the foundation consolidation. Horizontal drainage blankets under the embankment slopes provide an exit for the excess pore water squeezed out of the foundation soil during construction. On a number of major dams they have been placed under both the upstream and the downstream slopes to accelerate consolidation, Figs. 1.4:6 and 1.4:7 These blankets also serve to control pore pressures in the embankment and foundation during reservoir operation.

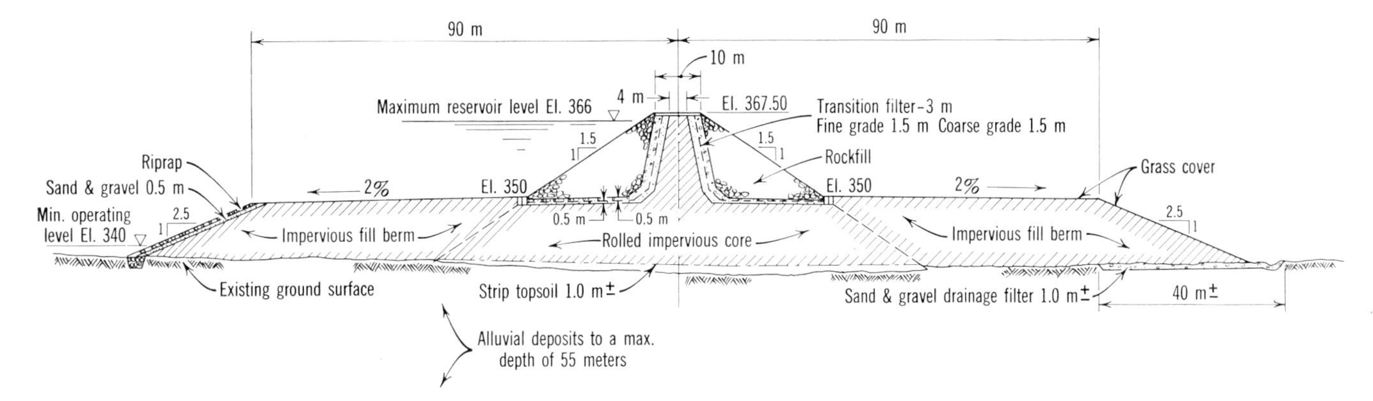

Fig. 8.4:2 Cross-section of Jibacoa Dam, Cuba, 1956, showing "Piggy-back" section on wide berm to allow construction of 38-m.-high dam on soft foundation. (Courtesy, Tippetts-Abbett-McCarthy-Stratton, Engineers and Architects)

Sand drains, although used very widely to accelerate consolidation under highway embankments, have been installed in only a few earth dams.[1] At the 90-ft.-high, 1,000-ft.-long Boundary Dam in Saskatchewan (Ref. 513) constructed in 1956 and 1957, both stage construction and sand drains were used under the portion of the dam constructed on a soft clay foundation, Fig. 8.4:3. Piezometers which were installed in the sand drains and in the clay at different distances from the drains, indicated that the sand drains were effective in keeping the pore water pressure low. In addition, sand drains have an important secondary function in acting as relief wells and thus controlling underseepage during the operation of the reservoir (Sec. 6.4).

Another method of strengthening the foundation which has not been used frequently but can be very effective under certain conditions is the construction of "shear keys" through the soft soil under the dam, Fig. 8.4:5. They consist of longitudinal trenches excavated through the soft soil and backfilled with well-compacted, rock or gravel. These trenches act as reinforcements for the foundation and carry a large part of the total shear stress imposed on the soft foundation soils. Because of difficulty in evaluating the relative stress-strain characteristics of the natural soil and the backfilled trench, it is not possible to compute the trench's quantitative influence on stability with reliability. Also, in order to obtain a reasonable estimate of the beneficial influence of such a shear key on the stability, it is necessary to use one of the methods of calculation which include side forces on the slices (Sec. 7.4) since the shear resistance in the key will depend to a large degree on the load which is transferred from the higher elevations of the trial sliding mass to the shear key.

Upward seepage of water at and below the downstream toe over long periods of time can soften foundations of even relatively stiff fine-grained soil. Even though the total quantity of underseepage may be very small, it is desirable to intercept the seepage water before it reaches the ground surface. For this reason, relief wells or drainage

[1] To the best of the authors' knowledge, at the present time (1961) sand drains have been used only in one dam in the United States (Rough River Dam, Louisville District, U.S. Army Corps of Engineers), and one dam in Canada (Boundary Dam, Prairie Farm Rehabilitation Administration). At least two low dams in England have been constructed with sand drains to consolidate soft foundations (Refs. 7 and 531). At the Rough River Dam which was completed in 1960, 12-in.-diameter sand drains spaced 13 ft. on centers were used under both the upstream and downstream slopes. Fig. 8.4:4.

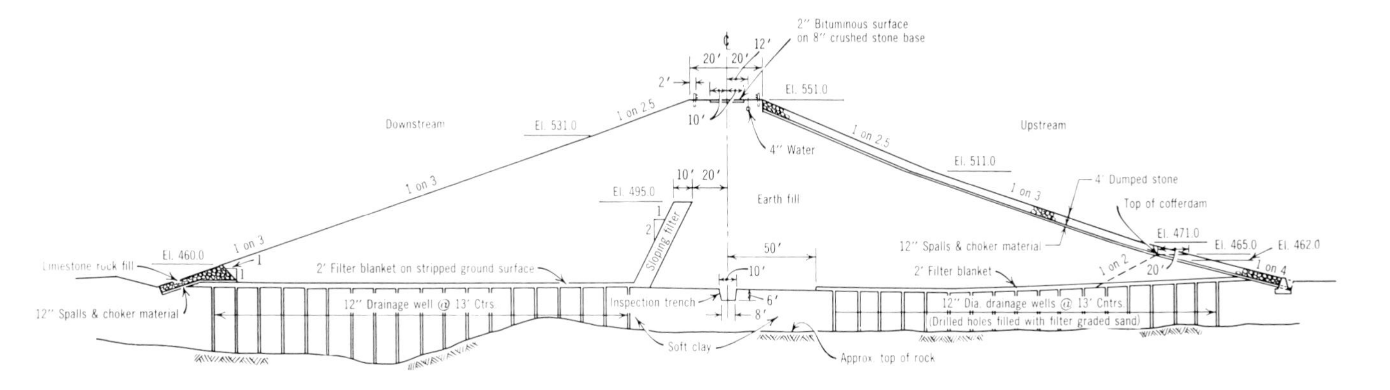

Fig. 8.4:4 Cross-section Rough River Dam showing foundation sand drains. (Courtesy Louisville District, U.S. Corps of Engineers)

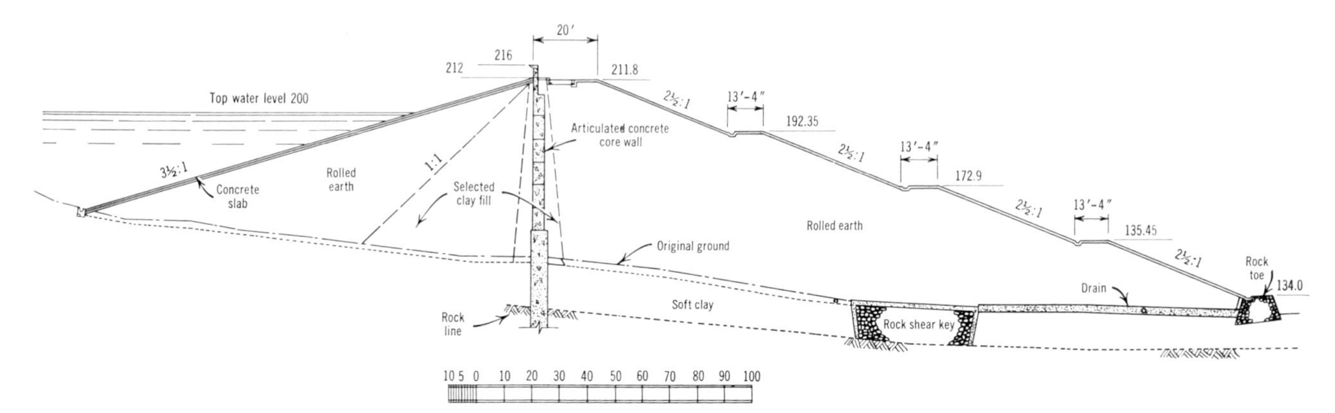

Fig. 8.4:5 Cross section of Waterfall Dam, Hong Kong, showing shear key of rock through soft clay foundation (after Little, Ref. 72).

trenches are always considered at the downstream toe of dams on foundations of fine-grained soil (Sec. 6.4).

FOUNDATIONS OF PARTIALLY SATURATED FINE-GRAINED SOIL

In arid and semiarid regions, fine-grained soils comprising the upper parts of dam foundations and abutments are often found to be partially saturated. These soils may be of any geologic origin, but often they consist of wind-blown deposits (loess) or slope wash deposits. If of low density and never before saturated, they will probably soften and compress under the weight of the embankment when they are wetted for the first time by seepage from the reservoir.[1]

While a typical earth dam can withstand considerable differential settlement if it occurs gradually over a long period of time, rapid differential settlements of much smaller total magnitude are likely to cause cracking. Consequently it is not good practice to permit the existence of these partially saturated soils in the foundation.[2]

In order to evaluate the "settlement on saturation" potential of the foundation material, undisturbed samples of the soil above the ground water table (preferably cut from test pits) should be tested in the laboratory for consolidation. Samples should be loaded in increments between dry loading stones to a maximum pressure approximately equal to the weight of the embankment. After equilibrium has been reached under the added pressure at the natural water content, the specimens should be saturated. If much additional compression occurs because of the increase in water content, the material should be considered unsuitable for the foundation. Unsuitable materials can be excavated and recompacted in the form of a rolled embankment, or they can be wetted by irrigation before construction so that most of the compression will occur during construction.

Prewetting of partially saturated soils under earth dams has not been common, but the USBR has irrigated the foundations of a num-

[1] See Refs. 40, 424 for a rough laboratory correlation between foundation soil properties (*in situ*) and probable "settlement on saturation" obtained for dams of various heights on partially saturated fine-grained soils.

[2] This is true even for small dams. The cracking which developed in the 40-ft. high dam shown in Fig. 2.3:1 resulted from rapid compression of an unsaturated, relatively low density clayey silt layer in the foundation when water was first added to the off-channel reservoir. If the settlement of this dam had taken place slowly, there is little doubt that the clay embankment would have been able to adjust to it without cracking.

Fig. 8.4:6 Aerial views of diked ponds used to irrigate loessial foundation material at Bonny Dam (1950). (Courtesy USBR)

ber of major earth dams consisting of the typical Missouri Valley Basin Loess.[1] The irrigation was carried out by ponding the flatter areas and by sprinkling the steeper slopes, Fig. 8.4:6. In most cases, after prewetting the soil did not measurably settle under its own weight, but the compression during construction was higher than would have occurred without irrigation, and the post-construction settlement was low.

At some other dams where the foundation consisted of low density loess with pronounced vertical soil structure, boards of engineering consultants have considered prewetting as being potentially dangerous. If the prewetting fills the soil voids with water but does not break down the soil structure, there is a possibility of a liquefaction-type failure when the shear strains from the embankment weight result in a sudden collapse of the soil structure.[2] At Fort Randall Dam in South Dakota, such a loess formation on the right abutment was excavated and replaced as compacted earth fill.

[1] See Refs. 42, 345, and 370 for descriptions of the USBR experience with prewetting foundations.

[2] See a discussion of the mechanics of liquefaction in fine-grained impervious soils in Ref. 126, p. 104.

8.5 UPSTREAM SLOPE WAVE PROTECTION

8.5a Types of Slope Protection

Any earth dam retaining a large reservoir, in which the upstream slope is composed of material other than cobbles or rock, must be provided with wave protection.[1] Present criteria governing the choice of type and design of wave protection are almost wholly empirical and based on past experience (Sec. 2.6). This experience demonstrates overwhelmingly that a properly proportioned layer of dumped rock riprap overlying a finer filter layer or layers provides excellent wave protection. Because of this and the fact that it is very durable and usually less costly than any suitable alternative, dumped rock riprap is used at the great majority of sites. In areas where suitable rock is absent, a monolithic, reinforced concrete slab is a suitable alternative and other manufactured materials have been used to a lesser extent.

Tightly packed layers of hand-placed rock in the form of a "dry masonry" blanket and articulated concrete slabs have been used for wave protection on many dams retaining reservoirs with large waves. It has been shown that layers of dumped rock riprap provide more effective wave protection than layers of carefully hand-placed rock of the same thickness; consequently, when rock is available for slope protection it should be placed in a dumped layer (Sec. 11.6) rather than in a carefully placed, hand-packed blanket. Similarly, experience has conclusively demonstrated that concrete pavements consisting of individual, separate concrete slabs or of articulated concrete slabs have been much less successful in resisting severe wave action than monolithic slabs. Therefore, until new methods are developed and tested, only dumped rock riprap, monolithic reinforced concrete, and asphaltic concrete should be considered except in special circumstances.

[1] At some small flood control projects such as those constructed by the U.S. Soil Conservation Service, where the reservoir is emptied a short time after flood occurs, it has been considered satisfactory to omit wave protection, especially where the upstream face is constructed of cohesive soils which have inherent resistance to erosion. For these structures it may be necessary to make some repairs at intervals but the cost of the maintenance work over the life of the dam may be less than the cost of providing wave protection. Slope protection has also been omitted frequently on dams retaining small municipal water supply reservoirs. For these, booms of floating logs tied together with cables and located a short distance from the embankment have been used satisfactorily to control the relatively minor wave action which develops.

Fig. 8.5:1 Hand-placed riprap gives pleasing finished appearance.

One such exception, for example, may be for dams in recreational areas or in other areas where the finished appearance of the structure is considered to have great importance. Here it may be desirable to place rock in a hand-placed layer because it provides a more pleasing finished appearance. Particularly, if the rock breaks with large plane surfaces, a very attractive riprap blanket can be obtained by hand placing, Fig. 8.5:1.

The cost of slope protection is an appreciable percentage of the cost of the embankment. For large dams, it frequently ranges between 10 and 30% and for small dams, retaining large reservoirs, the cost of slope protection can be 50% of the embankment cost or more. Because of its high relative cost, the fact that a slope protection failure is not a disaster, and the very rough and empirical character of design procedures available, the designing engineer has much freedom for the play of judgment and ingenuity in the selection of the slope protection at a given site. In the future, new methods and materials for slope protection will undoubtedly be devloped as an alternative to riprap at sites where rock is costly.[1]

[1] For example, see Sec. 2.6 for a description of USBR experience with soil cement and asphaltic concrete as wave protection, and Sec. 9.4 for a review of experience with the use of asphaltic concrete on the upstream slope of embankment dams.

When properly designed, layers of slope protection constructed from either rock or concrete are equally effective and have negligible maintenance costs. The primary factor governing the choice of the type of protection is initial cost. Under normal conditions, if there is a suitable rock within 30 road miles of the dam site, dumped riprap will always be cheaper.[1] Neither type of slope protection has a definite advantage from the standpoint of useful life expectancy.

Three advantages are sometimes cited for rock riprap over concrete.

1. It will be less damaged by postconstruction embankment settlement.
2. It is a more effective dissipator of wave energy (waves do not run as far up the slope, and therefore the necessary freeboard and height of the dam is less).
3. Extreme freezing weather and reservoir ice will cause more damage to concrete slabs than rock riprap.

The importance of these advantages is not great at most sites.

There is no doubt that rock riprap can withstand any conceivable postconstruction embankment settlement without damage; however, monolithic, reinforced concrete slabs can probably also withstand such settlement. Slabs of this type will deform greatly without developing large cracks. Except in the most extreme cases, where large, relatively rapid postconstruction differential settlement is expected, a reinforced, monolithic concrete slab which is called upon to perform only as wave protection is satisfactory.

It is true that waves run up the slope about twice as far on dams with concrete pavements as on riprap slopes. This disadvantage can be overcome to a large degree by the provision of a curved wave wall at the top of the slope which throws the wave back toward the reservoir. The detail shown in Fig. 8.5:2 which is used at Babcock Dam in Nebraska, where maximum wave heights of 6 to 7 ft. have been observed, has proved to be effective.[2]

[1] At many major dams, rock for riprap has been quarried and transported 50 to 100 miles. At Bonny Dam in Eastern Colorado, riprap was shipped 200 miles by train and hauled 24 miles by truck; a number of similar examples are available.

[2] Concrete wave walls or parapet walls at the upstream edge of the crest are useful to increase the freeboard against wave action during storms without increasing the dam height (Sec. 1.3*c*). This is true for dams with riprap blankets as it is for dams with concrete pavement. At some dams, continuous rows of piled rock have been placed at the upstream edge of the crest. This wall of piled rock serves the dual function of increasing the freeboard against wave action and as a stockpile of rock for maintenance of the riprap.

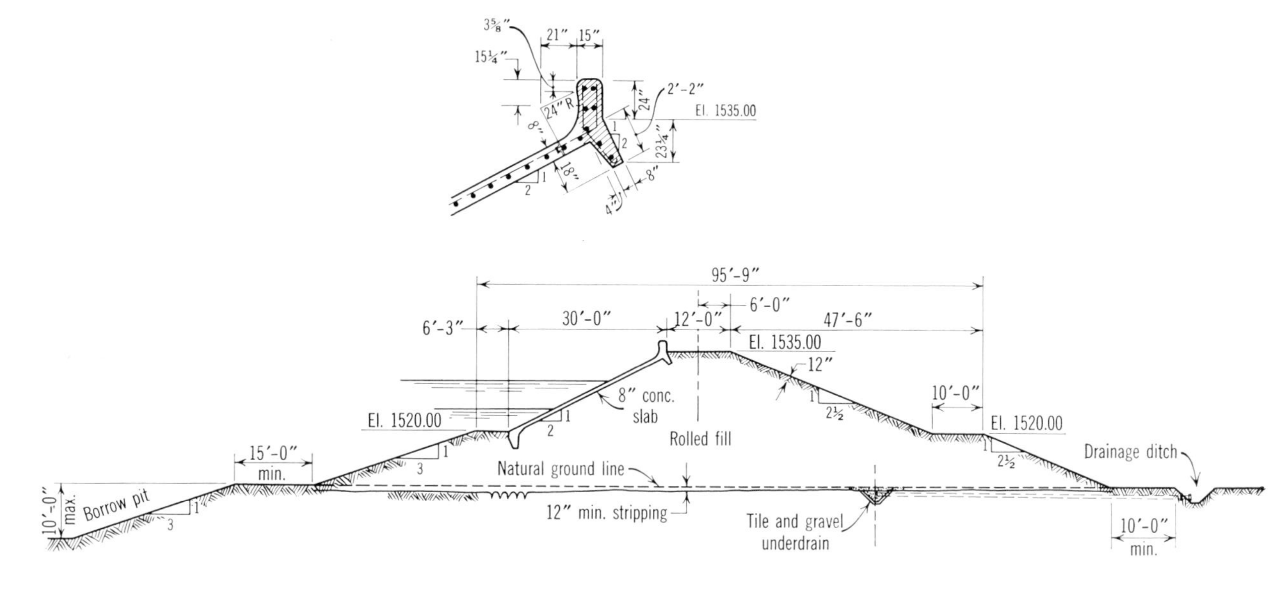

Fig. 8.5:2 Concrete wave wall at East Babcock Dike, Nebraska. (Courtesy Loup River Public Power District)

The advantage of rock over concrete due to its suspected better resistance to very severe cold winter weather is also difficult to evaluate. Several dams with concrete faces have been very successful after more than 25 years of operation in the mid-western part of the United States where the reservoir developed ice sheets of 12 in. or more in thickness. No trouble developed either from the influence of the ice on the reservoir or ice lenses in the embankment behind the slabs. These experiences lead to the tentative conclusion that monolithic concrete slabs are satisfactory in moderately cold climates; however, there is not sufficient number of examples available to give reasonable assurance that, under slightly different conditions, the concrete slabs will not be damaged by severe cold. On the other hand, experience tells us that cold weather can cause little serious damage to dumped rock riprap layers.

8.5b Design of Dumped Rock Riprap[1]

Rock riprap has been used as wave protection on the upstream face of earth dams since the earliest times. Until about 15 years ago, however, little progress had been made in establishing rational methods of selecting size of rocks or thickness of riprap layer as a function of wave intensity. In the mid 1940's, it was commonly considered that a dumped riprap layer of 3 ft. in thickness was satisfactory under any conditions of wave action and that a hand-placed riprap layer of 18 in. in thickness offered equivalent protection. The comprehensive study of actual riprap performance records made by the U.S. Army Engineers (Ref. 193) has led to a better understanding of these relationships. While these still leave much to be desired, the design criteria developed from the results of the U.S. Army Engineers' study are the most reliable guide for the design of dumped rock riprap available.

A successful riprap layer must be designed so that: (1) the individual rocks are not moved out of place by the hydraulic forces of the waves, and (2) the filter underlying the riprap will not be washed out through the voids in the riprap layer. Also, the filter must be successful in preventing any erosion of the underlying embankment material. The criteria for designing a riprap layer to satisfy these requirements are described in detail in the following sections.

[1] Section 11.6 covers construction of riprap layers.

SIZE AND GRADATION OF ROCK AND THICKNESS OF RIPRAP LAYER

The minimum suitable size and gradation of rock and thickness of riprap layer depend on the intensity of the wave action expected and on the steepness of the dam slope on which the riprap rests. The influence of the slope steepness is much less than that of the magnitude of the wave action and, at the present state of our ability to handle this problem, it is customary to assume that the influence of the steepness of the slope can be neglected over the range of upstream slopes normally used for earth dams; i.e., roughly between 2:1 and 4:1. The mechanics of wave action are complex and are the subject of much detailed study by hydraulics engineers; however, for the purpose of riprap design, it is sufficiently accurate to assume that the wave height is a direct measure of the intensity of its erosive action, and that the wave height, in turn, depends only on the velocity and the duration of the wind and the length of the reservoir over which it acts (fetch).[1] Since experience indicates that some of the highest waves observed have occurred in relatively shallow reservoirs, the influence of the depth of water on the estimated wave height is neglected.

Figure 8.5:3 is a chart developed by the U.S. Navy Hydrographic Office modified by the U.S. Army Engineers for short reservoirs relating wind velocity, reservoir fetch, and maximum wave height, which is sufficiently accurate and reliable for riprap design. The dotted lines in Fig. 8.5:3 also show the length of time during which it is necessary for a wind of a given velocity to blow in order to develop waves of maximum height on a reservoir of a given size. In the comprehensive Army study, it was found that observed wave heights agreed with those predicted from this chart for waves less than 4 ft. in height and that for higher waves, computed values were slightly lower than observed values. The maximum wave height observed in the 100 reservoirs studied was 8 ft.

In order to estimate the wave height which may develop using Fig. 8.5:3, it is necessary to estimate the maximum velocity and the duration of wind which may blow across the reservoir. To obtain a reliable and precise estimate of the maximum wind velocity and the average

[1] The fetch is defined as the open water distance over which the wind can blow and generate waves toward the dam. It is generally taken as the greatest straight line open water distance from the dam; however, the effective fetch may be much larger than the straight line distance for long reservoirs with a gradual curve toward the dam or it may be smaller where islands break up the wave action.

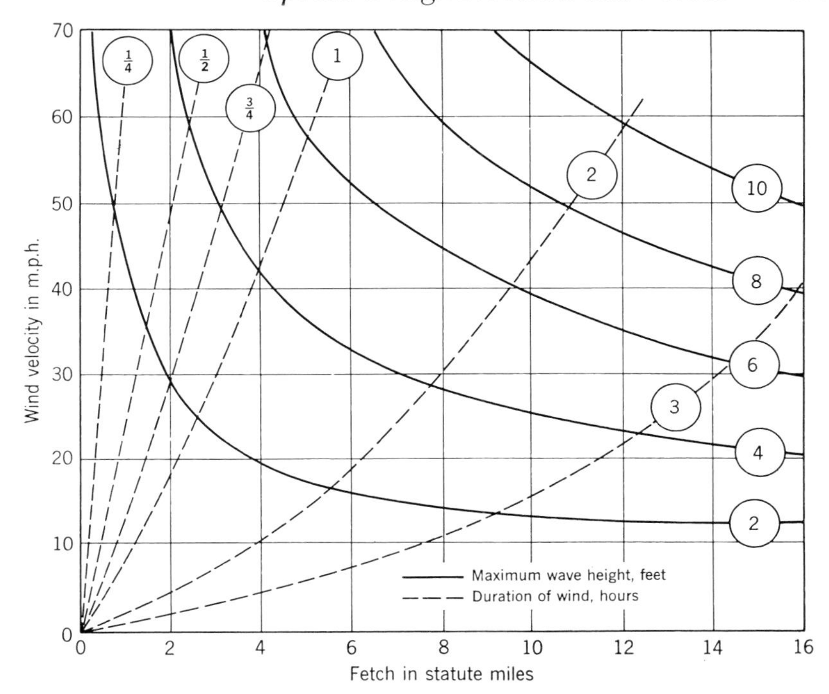

Fig. 8.5:3 Relationship of wave height to fetch, wind velocity, and wind duration (after Bertram, Ref. 193).

values which would exist over any given length of time is practically impossible. While some geographic areas are windier than others, high winds (up to 100 miles/hr.) can develop at any site on occasions, although the topography of the area may be such as to shelter the proposed reservoir surface. Wind velocities for design purposes are usually taken as the maximum measured values in the vicinity or simply estimated on the basis of the reputation which the area has for windiness. Estimated values of less than 50 miles/hr. should never be used, and values of 70 to 100 miles/hr. should be assumed in windy and flat areas. From the practical standpoint, estimated wave heights should never be in excess of 10 ft. since 8 ft. or a little more is about as high as waves have been observed to develop on inland reservoirs.[1]

[1] See Ref. 172 and the discussions for a summary of opinions about the prediction of wave heights.

From all the experience available, the authors believe that the riprap sizes given in Table 8.5:1 are reasonable and conservative for earthdams with ordinary slopes. In their most recent manual of practice, the U.S. Army Corps of Engineers have a more elaborate formula for riprap design which takes into account the specific gravity of the rock and the slope of the embankment (Ref. 362).

Table 8.5:1 Recommended Riprap Design Criteria

MAXIMUM WAVE HEIGHT (ft.)	MINIMUM AVERAGE ROCK SIZE (D_{50}) (in.)	LAYER THICKNESS (in.)
0–2	10	12
2–4	12	18
4–6	15	24
6–8	18	30
8–10	21	36

The most important criterion in Table 8.5:1 is the minimum average rock size (D_{50} of the riprap); e.g., for waves of 5 ft. in height, the riprap should be composed of rocks half of which by weight are equal or larger than a more or less equidimensional rock with average diameter of 15 in. The rock in the blanket should be well graded from a maximum rock size roughly equal to 1.5 times the average size varying down to 1 in. or smaller. The thickness of the riprap layer should be large enough to accommodate the largest size rock and should be equal to or a little larger than 1.5 times the average rock size.

In reservoirs in which the upper storage area is reserved for flood control purposes and the spillway crest and normal water level are well below the dam crest, the riprap thickness has been reduced in the upper portions of the dam since the water level is at the higher levels only for short periods of time. On many major irrigation-flood control projects in the western United States, a 3-ft.-thick riprap layer is used on the lower portion of the slope and a 2-ft.-thick layer on the upper portions. For dams without flood control storage, the full riprap should extend all the way to the crest where it may receive the greatest erosive action. For reservoirs which have a conservation pool or fixed minimum water level, it is sufficient to extend the riprap to a depth of 5 ft. below the minimum pool level.[1]

The criteria given in Table 8.5:1 are conservative for slopes flatter than 5:1 and can be relaxed appreciably. For the lower parts of high dams and special cases in which the upstream slope is flatter than 10:1, slope protection consisting of a 12-in. layer of reasonably well-graded gravel with a maximum size of 6 in. would probably be adequate at most sites.

UPSTREAM BERMS

It is customary and advisable to construct a horizontal berm at the lower edge of the riprap layer. The purpose of the berm is not to provide a foundation to prevent the riprap blanket from slipping down the slope, since each unit of area of the riprap has adequate resistance against downward slipping although this was probably the origin of its use. The primary function of the berm is to prevent undermining of the lower edge of the riprap, at the time of low reservoir, when the wave water which cascades back down the slope hits the bottom of the riprap and filter. A secondary function of the berm is to provide a horizontal working surface for construction equipment.

Although it is not customary to provide horizontal berms on the upstream slope above the bottom of the riprap, there are several good reasons for doing so. At a number of dams where the riprap was damaged and had to be reworked or replaced, the cost of this work would have been much less had the dam been provided with berms at convenient elevations from which equipment could have worked easily. Horizontal berms can be used also as truck roads during construction which will allow riprap placing without the necessity of working from the embankment construction surface.

QUALITY OF ROCK FOR RIPRAP

Ideally, rock for riprap should meet the quality specifications for concrete aggregate; however, in many cases, riprap blankets composed of much softer rock have given very satisfactory service. No minimum quantitative specifications can be given for the rock quality: the best available material should be used. Any rock is satisfactory if it does

[1] See Sects. 2.4*c* and 11.6 for descriptions of surface sloughing and treatment of portions of steep upstream slopes below the area protected by riprap.

not break up or soften when exposed to alternate wetting and drying and the temperature cycles which can be expected at the site. A good, quick indication of the suitability of the rock for riprap is simple visual observation of the condition of typical specimens which have been exposed to weather at the site. At a site where several rock types are available within an economical hauling distance, laboratory tests of the durability should be made on each and the soundest material should be chosen for the riprap.

Most igneous and metamorphic rocks and hard sandstones and limestones are sufficiently hard and durable for riprap on earth dams. Shales and rocks with shale seams often disintegrate when alternately wetted and dried.

Where possible, it is of course most desirable to use rock which is excavated during the construction of spillways or other structures. On a number of major dams riprap blankets have been constructed of relatively soft rocks from the required excavations with the knowledge that wave protection thus provided might have to be replaced during the life of the dam. Riprap layers constructed of rocks of this type are frequently made thicker than the minimum recommended design thickness of Table 8.5:1.

Some rocks, which are weak when they are first excavated from the ground, harden with time and exposure to the air. In some areas this is a common phenomenon and the difference in the rock hardness with time may be very great. At the site of the Sisga Dam, constructed in Colombia in 1952, the only rock available for riprap was a fine-grained, friable sandstone. When first excavated, pieces of this material would disintegrate into fine sand if dropped on a hard surface from a height of a few feet; however, when exposed to the air, the material case hardened to the point that, with care, larger rocks could be transported to the upstream slope where, subsequently, they continued to harden and proved to be adequately durable for the purpose.

High specific gravity is desirable for riprap since the heavier the material the more resistant it is to movement by the waves. The design criteria of Table 8.5:1 are derived for rocks of ordinary specific gravity; i.e., in excess of 2.60. Riprap comprised of rocks of much lower unit weight should be proportionately larger.

FILTERS UNDER RIPRAP LAYERS

A layer of filter material consisting of gravel or crushed rock is always required under riprap blankets to prevent the waves from eroding the underlying embankment material. Gradation requirements for the filter with respect to the embankment material are the same as the requirements for ordinary filters; i.e., the D_{15} size of the filter should not exceed 5 times the D_{85} size of the retained embankment material (Sec. 1.3*b*).[1] For the gradation relationship between filter and the riprap layer, it was well demonstrated by the comprehensive investigation of the U.S. Army Engineers (Ref. 193) that filters performed satisfactorily if the D_{15} size of the riprap does not exceed 10 times the D_{85} size of the filter. Also, experience indicates that the D_{85} size of the filter should not be less than 2 in.

Filters which are reasonably well graded between a maximum of 3 or 4 in. and coarse sand sizes are satisfactory for the great majority of dams. Filters are obtained as pitrun or screened sand and gravel or from crushed and screened rock. Many satisfactory filters have maximum sizes larger than 4 in. and appreciable quantities of medium to coarse sand. Filters with large percentages of fine sand are likely to be unsatisfactory.

On the great majority of earth dams, one layer of filter material has been adequate. No definite rule can be given for the minimum necessary thickness of filter layer and the problem is not subject to analytical treatment. Most filters are constructed with thicknesses varying in the range between 9 and 30 in. The following factors should be considered in the selection of filter thickness:

1. *Wave action:* The less the wave action, the less the need for a thick filter under the riprap.
2. *Gradation of the riprap:* If the riprap is well graded with plenty of quarry fines to fill the larger voids, there is less stress on the filter.
3. *Plasticity and gradation of the embankment material:* If the embankment material is a well graded granular soil with a tough clay binder, it needs erosion protection much less than if it is a fine silty sand.
4. *The cost of the filter:* If the material for the filter is obtained without washing or screening as a pit-run natural gravel and, consequently, is not very expensive with respect to the average embankment material, there is no reason to use minimum filter thicknesses.

[1] In order to verify that the filter criteria for steady-state seepage were applicable to the design of riprap filters which are called upon to withstand many reversals of flow under wave action, the U.S. Army Engineers performed experiments on models of sloping dam faces at the Beach Erosion Board Laboratory in Washington, D.C. (Ref. 172).

The U.S. Army Engineers use the following recommended filter thicknesses as a function of wave height (Ref. 352).

Table 8.5:2 Minimum Thicknesses of Single Layer Filters Under Riprap Blankets

COMPUTED WAVE HEIGHT (ft.)	MINIMUM FILTER THICKNESS (in.)
0–4	6
4–8	9
8–10	12

These values should be considered as absolute minimum thicknesses. Where the embankment material comprising the upstream slope directly under the riprap is composed of a very fine-grained soil, such as silt, it is often necessary to use two layers of progressively coarser material in order to meet the filter gradation requirements. These should not be less than 6 in. in thickness and preferably thicker.

THEORETICAL APPROACH TO RIPRAP DESIGN

Several attempts have been made to treat the problem of riprap design using the tools of theoretical and experimental hydraulics and the results have been used by some engineers as a guide to judgment.[1] While theoretical methods have already proved valuable for the analysis of dumped rock embankments which are used as seawalls (Ref. 343), the precision with which the theoretical equations represent reality has not yet been well enough confirmed by large scale laboratory tests or field observations on dams to justify their use for the computation of riprap size or thickness.

Using the theoretical approach, the first step is to calculate the maximum wave height which will develop on a given reservoir surface under the wind action expected. The estimated wave height is then converted, by one of several means, into an equivalent velocity of water which is assumed to be alternately rushing up and down the dam slope. The surge of water on the slope generates forces on the rocks due to the combined hydrostatic pressure and viscous drag forces which tend to move it down the slope. These forces are resisted

[1] Methods used for riprap analysis by the Tennessee Valley Authority are described in Ref. 52; also see Ref. 172 and discussions.

by the friction between the rock and the underlying embankment. As a result of the calculations, a "theoretical rock size" is obtained which will be stable (with a safety factor of 1.0) on a dam with a given slope under the action of waves of a given height.

Some of the main reasons that the theoretical analysis cannot practicably be used as a design procedure at the present time are as follows:

1. The theoretical equations are derived on the basis of simple and reasonable considerations of equilibrium; however, they inevitably contain constants which must be obtained experimentally either from the large-scale model tests or observations of prototype dams. At the present time, enough observations have not been made to give the formulas sufficient precision and reliability to justify small modifications of our empirical criteria for design.
2. The computations result in a theoretical rock size which will not be moved by the expected wave action. The relationship between the equilibrium condition for the individual theoretical rock and for a continuous well graded riprap blanket containing the theoretical rock size is not known; i.e., it is not known if the "theoretical" rock size should be the average size or the maximum size in the riprap blanket.
3. The theoretical rock size given is computed for a safety factor of 1.0. The safety factor which should be used in the design is not known.

8.5c Design of Monolithic, Reinforced Concrete Slabs[1]

The most important design requirement for a successful, concrete slab for wave protection is that it should not have or be able to develop any joints or other openings of sufficient width such that the underlying soil can be washed out. Reinforcing steel should be the same in each direction and the steel area should be such that when a section of the slab is stretched in tension, the strain will be distributed in the form of a large number of hairline cracks rather than in a few large cracks. Usually, steel reinforcement in each direction with an area of 0.5% each way of the concrete slab area, has been sufficient for this purpose.

[1] See also Sec. 9.2.

The steel should be run continuously through all construction joints in the slab. Experience indicates that it is not necessary to provide water stops of any type in the construction joints: if sufficient reinforcing steel is used, the construction joints will not open any wider than the other hairline cracks which can be expected to develop in the slab.

No definite rules can be given for the minimum permissible slab thickness. Undoubtedly, other things being equal, the thicker the slab, the longer it will last without deterioration; however, the quality of the concrete has a much greater influence on the durability than the thickness. The very best quality concrete should be used. An 8-in.-thick slab of good quality concrete should be adequate under the worst conditions of reservoir wave action. In order to provide a uniform subgrade for the slab, and to minimize subsequent cracking due to differential settlement caused by the reservoir water pressure, the embankment surface directly under the slab should be well compacted.

8.6 DOWNSTREAM SLOPE PROTECTION

Where the outer portion of the downstream slope consists of fine-grained soil; the surface of the slope must be protected from erosion. Sufficient erosion to necessitate considerable maintenance and repairs has been experienced on many dams, particularly in the first few years after construction before a good growth of protective grass is established, Fig. 8.6:1. In wet climates, the erosion is caused primarily by surface water runoff during rainstorms. In arid climates, wind causes most of the damage, and in very cold climates the frost action in the ground softens the surface and facilitates the erosion. In a few cases, wave water thrown over the top of the crest during storms has badly eroded downstream slopes of earth dams.

The worst damage is caused by the development of deep erosion gullies on the slope, occurring both at the abutment contacts and in the central portion of the dam. Gullies with depths of several feet are common; in severe cases, gullies 10 feet in depth have been eroded in a single rainstorm. Except for very low dams, erosion has not endangered the stability of the structure but has only required maintenance. Repairing and stabilizing the eroded slope can be difficult and expensive; consequently, the problem of protecting the downstream slope should be given the most careful attention and study by the designer.

The most severe erosion occurs on high dams with long slopes on

Fig. 8.6:1 Erosion of unprotected downstream slope by rainstorms during construction.

which rain water collects and runs off. On the other hand, severe erosion may be more serious on low dams where the depth of the erosion gullies may be appreciable with respect to the thickness of the embankment. In addition to the installation of an erosive resistant protective layer on the slope, every effort should be made to avoid the concentration of runoff water. Large flows of surface water from the abutments should not be allowed to strike the dam.

In almost all geographic areas, a properly cultivated cover of grass provides the most satisfactory and economical slope protection. Only in very dry areas, where there is not enough water to keep the grass cover alive, is it necessary to use other types of protection. Layers of rock riprap and coarse gravel also provide excellent slope protection.

Dumped rock riprap of the same type used for the upstream slope provides excellent protection for the downstream slope. Underlying blankets of filter material are not required. Oversized rocks obtained from the embankment construction material by raking or screening have provided an excellent supply of material for downstream rock blankets on many dams.

Layers of coarse sand and gravel with maximum particle size of 3 in. or more have resulted in good protection in most cases. If the slope below the gravel layer is impervious, rain water falling on the slope will not seep into the embankment but will run down the slope inside the gravel layer. If the gravel is not thick or pervious enough to carry off the rain water, shallow sloughing of the surface may occur in the lower portion of the gravel blanket.

The downstream slopes of the great majority of earth dams are protected by a growth of grass whose roots hold the surface soil in place. A good grass cover provides complete protection against the most severe erosive action which will develop. On many dams, satisfactory grass covers have grown naturally without any assistance from the dam builder. In the majority of cases, however, careful nurturing of seeded ground or hand-planted sprigs and a great deal of patience have been required to develop an adequate grass cover. The procedures to be used in establishing the grass cover should be decided upon in the design stage. Many different types of grass have been satisfactory for slope protection in different parts of the world. In the selection of a type of grass cover and the method of planting, the designer should avail himself of the experience of the highway engineers, landscape architects, or others in the locality who have faced the same problem.

Chemical fertilizers are excellent aids to the establishment of a good grass cover. Topsoil may or may not be spread on the downstream slope in layers of 6 to 12 in. or more in thickness. If the downstream slope consists of clean sand, it will be necessary to add some fine-grained soil in order to hold the grass roots. If the outer surface of the embankment material consists of soil with an appreciable content of clay or silt fines, topsoil is not needed provided fertilizer is used. The main disadvantage of using topsoil is that it contains the seeds of weeds which retard the development of a good growth of the superior seeded grass.

In areas of high precipitation, it may not be necessary to irrigate the newly planted grass; in hot and arid regions, it is essential to moisten the ground by sprinkling at periodic intervals to start the grass growing. Experience indicates that sprinkling at night is superior in very dry areas in order that a larger portion of the water will seep into the ground instead of evaporating. In very rainy areas, it is sometimes difficult to prevent the newly seeded soil from washing away. In some wet areas this problem has been satisfactorily overcome by covering the surface with a protective layer of hay or strips of burlap which are tied to stakes driven in the slope at frequent intervals.

Bushes of more than 2 ft. in height are not satisfactory slope protection. They kill the smaller grass whose roots are more effective in protecting the surface soil and they prevent inspection of the slope.

Horizontal berms have been used on the downstream slope of many

Fig. 8.6:2 Paved gutter at downstream abutment contact for erosion control—Mariposa Dam, Caracas. (Diego Ferrer, Civil Engineer)

dams to control the runoff water and erosion. The surface of the berm is sloped toward the dam and the water running down the embankment slope above the berm is caught and led in a gutter to the abutments or to vertical gutters or pipes in the slope. There is a considerable difference of opinion concerning the value of berms for the control of downstream slope erosion. At many dams, berms without proper slopes have served the opposite purpose of concentrating runoff on the lower portion of the embankment and accentuating erosion or causing local slides. Berms are satisfactory if designed in such a way that the water caught from the slope above is controlled in paved gutters; however, experience indicates that they are not necessary, even on high dams if the slope has a suitable protective coating.[1] Berms are also useful as access roads for vehicles for sprinkling grass and maintenance action.

The contact between the downstream slope and the abutments is almost always an area of concentrated surface runoff. The strip of embankment slope directly adjacent to the abutment should be especially well protected, preferably with a paved gutter or a strip of rock riprap, Fig. 8.6:2.

[1] The most recent practice of the U.S. Army Corps of Engineers has been not to use horizontal berms on the downstream slope of earth dams to control surface erosion.

nine

Dams with Impervious Membranes of Reinforced Concrete, Steel Plate and Asphaltic Concrete

A large percentage of all the major earth dams constructed before 1925 were provided with central concrete core walls or concrete slabs on the upstream face. These concrete membranes were used even in dams constructed of fine-grained soils, because the soil was not trusted to act as a permanent and reliable water barrier. Many of these old membranes cracked, and since adequate drainage often had not been provided, the cracking frequently led to poor dam performance.

When during the 1930's it became well established that cores of fine-grained earth were entirely satisfactory, the use of concrete declined rapidly. Today concrete membranes are considered, with a few exceptions, only for rockfill dams in the United States. In Europe they are occasionally used in combination with embankments of earth or gravel but the practice is not widespread.

Nevertheless completely safe earth dams can be constructed with impervious membranes of manufactured materials such as concrete, steel, and asphaltic concrete, and there are circumstances in which they are less costly or more easily and rapidly built than dams with earth cores.[1] In the authors' opinion, the advantages offered by manu-

[1] Although no dams of any size have been constructed with cores of the newly developed plastics, these appear to offer a fourth possibility which may, perhaps in combination with other materials, be used in future years. Reference 596 describes a very interesting experimental installation of an upstream slope membrane consisting of a sandwich construction with precast concrete elements and plastic sheet carried out by The Research Institute for the Building Industry of Bratislava (Czechoslavakia). In the United States, considerable research on the applicability of thin plastic sheets for use as the internal core for high rockfill dams has been carried out in the last few years by Mr. Erik Rettig, Consulting Engineer

factured membranes have not received the recognition they deserve. In the future it is probable that they will be used more often, on major as well as small dams.

Because there is not yet a large body of experience available on earth dams with manufactured membranes, the case histories given in the following sections include rockfill as well as earth structures. The requirements for successful membranes on rockfill dams are essentially the same as those for earth dams, except that the post-construction compression of rockfill is often greater. In addition, leakage which may develop through the membrane has little influence on the stability of an embankment of coarse and pervious rockfill but may cause trouble in earth fill if adequate drainage is not provided. If these differences are taken into account, the experience with manufactured membranes on rockfill dams can give useful assistance in the design of membranes for earth dams.

9.1 UPSTREAM IMPERVIOUS MEMBRANES OF MANUFACTURED MATERIAL

The impervious manufactured membrane can be placed either on the upstream face of the dam or as a core inside the embankment. Each location has advantages to recommend it. The authors prefer the upstream position except under special conditions, mainly because it is accessible and because it gives the dam greater stability. The relative merits of upstream and vertical membranes are compared in Sec. 9.5. The general advantages and disadvantages of upstream membranes with respect to standard earth cores are discussed below.

9.1a General Advantages

The major advantage of a dam with an upstream impervious membrane is that it has a greater margin of safety against shear failure than any other type of earth or rockfill dam. Consequently, for a

(San Francisco) in cooperation with the Dow Chemical Company. At the Mission Dam in British Columbia where great settlements were expected with possible cracking of the embankment, a thin plastic sheet was used as a temporary water barrier (Ref. 677). For experiments with plastic linings for ponds, see Refs. 356, 678. Also, see Ref. 694 for a description of the use of plastic sheet in the repair of a dam in California.

given safety factor the embankment slopes can be made steeper and the embankment volume smaller than for a dam with a core of rolled impervious earth or for a dam with a vertical impervious membrane of manufactured material. This advantage is the result of several characteristics of the upstream membrane:

1. With a suitable drain behind it, the membrane prevents seepage from entering the embankment and so eliminates the reduction in the stability which is associated with the development of seepage pore pressures. Consequently, the embankment strength is higher and the margin of safety against shear failure is increased for both the upstream and the downstream slopes.
2. The dam with an upstream membrane has the greatest possible mass of embankment acting to resist the pressure of the water in the reservoir.
3. Because the water pressure is resisted on the upstream slope, the total force exerted by the reservoir on the dam is directed downward into the foundation with as great an inclination as possible. This is the optimum condition from the standpoint of stability of the foundation.

In addition, the upstream impervious membrane has several other important advantages:

1. In contrast to internal cores, upstream membranes are exposed for inspection and repairs. Even if the reservoir cannot be drawn down, some work can be carried out by divers below water. When concentrated leaks develop through dams with rolled earth cores, however, it is difficult to do anything except control the leaks safely. Large leaks which develop through earth cores can be repaired by grouting or other extreme methods, but these are all expensive and cumbersone.
2. The membrane can serve a secondary function as wave protection.
3. Unlike a rolled-earth core, a membrane of manufactured material is not subject to progressive erosion under the action of water flowing through concentrated leaks.
4. With special designs, additional benefits may be obtained. For example, if the membrane and the underlying drainage layer are bulkheaded into sections and a drainage gallery is constructed, the locations of leaks which do develop can be determined. Locating leaks in dams with earth cores, on the other hand, is difficult and may be almost impossible. In another type

of design, laminated membranes which have a layer of stiff asphalt or other deformable material can be constructed so that leaks will be self-sealing.

9.1b Circumstances Favoring the Use of Upstream Membranes

At sites where the foundation is of hard rock or other strong material, dams with upstream impervious membranes may have a large cost advantage because the embankment can be designed with minimum volume. At such locations a dam with an upstream membrane should always be considered. This and other circumstances tending to favor the use of upstream membranes are discussed below.

FOUNDATION OF HARD ROCK

Where the abutments and foundation for a dam consist of strong rock, and where there is an ample supply of strong granular embankment construction material, the upstream impervious membrane may be the most economical design even if fine-grained soil is available for an impervious core. With a rolled earth core, the embankment would probably need slopes of the order of magnitude of 2.5:1 upstream and perhaps 2:1 downstream. By contrast, with the use of an embankment built of strong granular material and an upstream membrane there is no danger of shear failure or excessive crest deflection,[1] and the slopes can be made as steep as the construction will permit (assuming a competent foundation).

NO IMPERVIOUS SOIL

Where no impervious soil for an earth core exists within an economical haul distance, obviously a manufactured impervious membrane should be considered. This situation is common in high mountain country, where the little soil available often has a basically pervious nature.

[1] In studies made at the University of California on a 2-ft.-high model of a sand dam with an upstream impervious membrane, hydrostatic pressures were applied to the upstream slope with a magnitude equivalent to that which would be exerted by a fluid with a specific gravity of 15 without excessive deflection (Ref. 48).

WAVE PROTECTION EXPENSIVE

Where severe wave action is expected in the reservoir, and where rock for a riprap blanket involves an exceptionally long haul, the cost of any type of suitable wave protection may approximate the cost of an impervious upstream membrane which could also serve as wave protection.

POSSIBLE EMBANKMENT CRACKING

As increasingly higher earth dams are located in narrow rock-walled canyons, much concern has developed over the possibility of embankment cracking due to differential settlement (Sec. 2.3). In some cases intricate and costly drains and graded filters have been installed to minimize the danger of seepage through the anticipated cracks. All of this trouble can be essentially eliminated with a design using a flexible impervious membrane on the upstream slope.

CONSTRUCTION DURING WET WEATHER

With conventional methods of moisture-density control, it is not possible to accomplish much work with fine-grained soils in wet weather. In circumstances where long rainy seasons exist or where the construction period is limited, a dam with an upstream membrane and an embankment of rock or pervious soil which can be placed in wet weather may be the most economical and satisfactory solution[1] (see Secs. 1.1*c* and 1.1*g*).

EXTENSIVE GROUTING PROGRAM

If the foundation requires an extensive and time-consuming grouting program, the upstream membrane may allow a considerable

[1] During 1958 a severe shortage of municipal water for the city of Caracas, Venezuela made it imperative that a dam 80 ft. high and 400 ft. long (Rio Lagartijo Dike) be designed and constructed in a few months (Ref. 371). For this dam there was plenty of clay available; however, the designers were afraid that a clay core could not be constructed properly in the short construction period during the rainy season, and a rockfill dam with an upstream steel plate facing was used, Fig. 9.1:1. The total construction period for the steel plate facing was about 60 days.

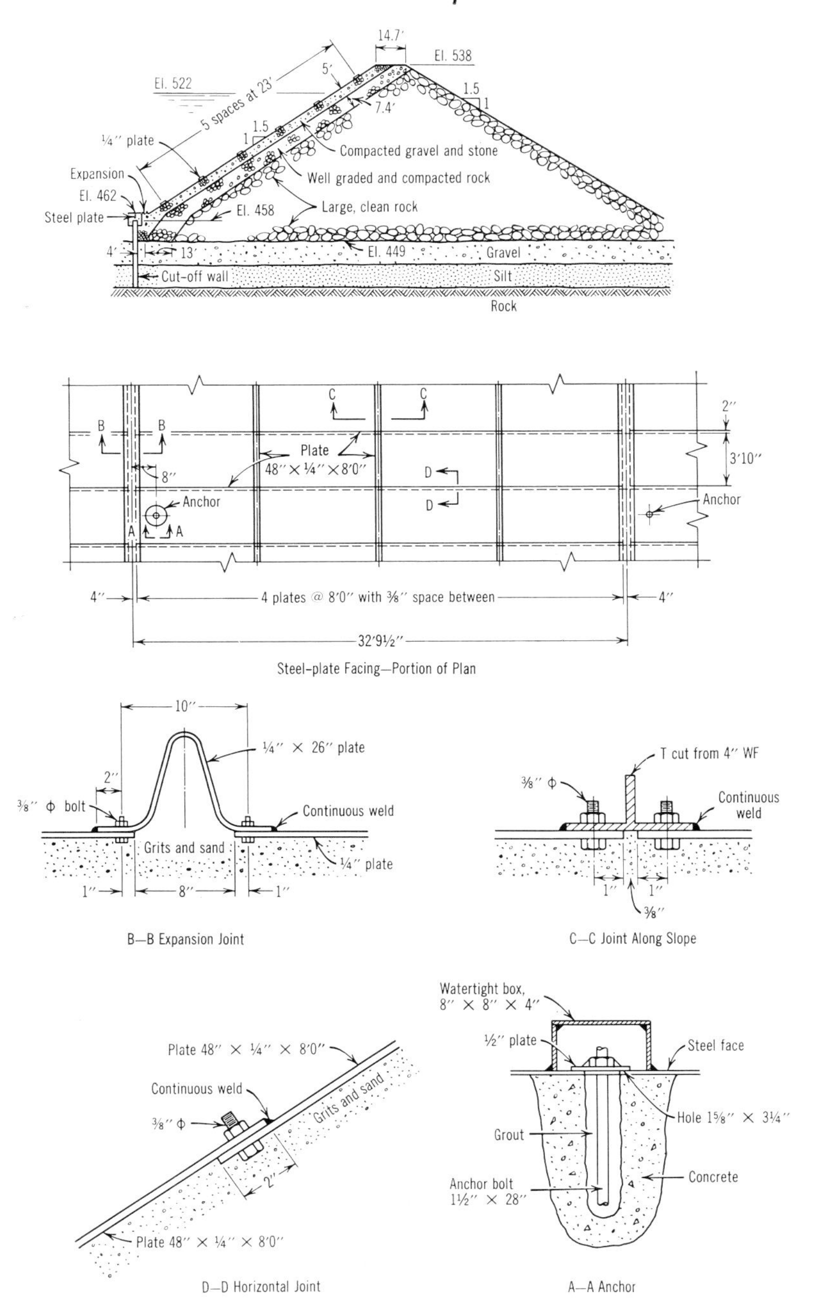

Fig. 9.1:1 Details of steel plate facing at Lagartijo Dam, Venezuela.

shortening of the construction schedule. With such a design the grouting can be carried out while the embankment is being constructed, whereas under ordinary conditions work on an embankment with a conventional rolled-earth core cannot be started until the grouting is completed, at least in the lower part of the valley.

EXPENSIVE APPURTENANT STRUCTURES

Where the outlet conduit or other appurtenant structures are imbedded within or abutted against the embankment, their length and cost often are directly related to the width of the dam at the base. If their cost is considerable with respect to the embankment, the most economical design for the total structure may be obtained with an upstream impervious membrane, which allows minimum base width.

9.1c General Disadvantages

In an approximate order of decreasing importance, the general disadvantages of upstream manufactured membranes compared with rolled-earth cores are:

1. *Shorter life.* Probably the principal deterrent to the wider use of manufactured membranes has been concern over the very real problem of deterioration. The basic constituents of a rolled earth core can be considered the final products of deterioration in nature, and consequently their properties will not be modified noticeably during the lifetime of a reservoir. The effective life of manufactured membranes depends on many factors. For Portland cement or asphaltic concrete, it is determined primarily by the mix design, the kind of aggregate, and the construction method. For steel plate, the presence or absence of elements tending to cause corrosion will dominate. Since these factors vary immensely, and since relatively few dams have been constructed of the different materials, there are no reliable criteria to use in predicting a given life span. However, from the limited experience available with such dams as well as from general knowledge of the performance of concrete, steel and asphalt when used in other structures, there is little doubt that well-designed and constructed membranes from any of these materials will last at least 30 years, and usually much longer.

2. *Higher cost.* The relative cost of membranes varies considerably, depending on the type of material, the thickness and other design details, the height of the dam, and the location. Other factors being equal, the steel plate is likely to cost most and asphaltic concrete least. The unit cost of membranes used under ordinary circumstances in the United States should fall between \$10 and \$25/yd.^{2} This cost is particularly a disadvantage for low dams in which the cost of the membrane may equal or exceed that of the embankment itself. For dams of increasing height, in which the width of the base and the embankment volume rapidly become larger, the cost of the membrane relative to the total cost becomes much less. Consequently, the higher the dam, the better the competitive position of the manufactured membrane. The location of the dam has an influence on the cost because the manufactured materials must be brought to the site. The farther the dam from the points of supply, the higher the cost and the greater the advantage of using a core of earth obtained nearby.
3. *Little experience.* Since relatively few dams have been built with upstream membranes, the designer does not have a large body of experience available to guide him in the selection of the best design details.
4. *Possibility of initial leakage.* There is the chance, especially when the reservoir is filled for the first time, that excessive leakage will develop through the membrane as a result of welding failures in steel, cracks in concrete, and tears or cracks in asphaltic material. The leaks may require that the reservoir be emptied for repairs, though some interim repair work can usually be done by divers. This is a benefit not offered by dams with compacted soil cores, which occasionally can also develop leaks or cracks large enough to make it necessary to drain the reservoir. In view of these factors, the possibility of having to draw down the water level constitutes only a slight disadvantage for a dam with an upstream membrane.
5. *Possibility of cracking at foundation.* Where the spreading of the dam base is appreciable during construction (Sec. 3.1*d*), the connection of the membrane to the foundation may be broken (see, for example, the experience cited by Terzaghi and Peck, Ref. 126, p. 498.) Therefore, if extensive spreading is anticipated, the engineer must construct the membrane and foundation cutoff after the embankment is completed and the base has already stretched, or must provide sufficient flexibility to avoid the cracking.

9.2 REINFORCED CONCRETE UPSTREAM SLABS

Very few recent dams with embankments of rolled earth or gravel have been constructed with upstream slabs of concrete, but a substantial number of rockfill structures have been designed in this way (Refs. 23, 24, 29, 30). Though the experience available is limited, it is adequate to demonstrate clearly that concrete slabs can be used successfully on earth or rockfill dams of any height, provided that the post construction settlement of the foundation is not so great that the slab cannot remain intact.

For earth dams up to a height of at least 150 ft. founded on reasonably incompressible materials, much experience is available to indicate clearly that the membrane can be constructed satisfactorily as a single monolithic concrete slab without expansion joints. The reinforcing steel should be the same in each direction and should be equal to at least 0.5% of the slab cross section.[1] Concrete slabs can undoubtedly be used successfully as impervious membranes for earth dams over 150 ft., but little experience is available as a guide to their design. For low dams as well as high dams there is much room remaining for ingenuity and improvement in present practice (i.e., over a single monolithic slab of reinforced concrete).

The performance of the concrete slab will be directly related to the quality of the concrete obtained. A slab can have a life, even in moderately severe climates, of at least 40 years if dense, low-slump air-entrained concrete is used and if close control is maintained during construction.

Where a single reinforced slab is adopted, some leakage will always occur through the inevitable hairline cracks, and it must be anticipated that some larger cracks or leaks due to construction inadequacies will develop soon after the first reservoir filling. Consequently, it is necessary to provide a drain behind the slab on an earth dam if the seepage water is to be kept out of the embankment.

Even though the earth embankment is not required to act as a water barrier, it should be well compacted in order to minimize post-

[1] Theoretically the reinforcing steel should be such that, when a section of the slab is stretched in tension, the strain will be distributed in the form of a large number of hairline cracks rather than in a small number of larger cracks. The cracking will be evenly distributed when the stress in the steel at the crack sections of the slab does not reach the yield point before the stresses in the concrete of the uncracked portion reach the tensile strength. However, it is not advisable to select the reinforcing steel wholly on the basis of such a calculation, since the slab is subject to bending as well as tension.

construction settlement of the upstream slope. The maximum settlement of the slab will probably occur near the midheight of the dam and can be expected to exceed the postconstruction crest settlement.

The following sections describe two important earth dams with upstream concrete membranes which have had long and satisfactory performance records and summarize the pertinent experience with rockfill dams.

9.2a Earth Dams

MCKAY DAM, OREGON

The 160-ft. McKay Dam, which was built in 1925 of gravels and cobbles in thin compacted layers, is one of the highest and most interesting earth dams constructed with reinforced concrete as the water barrier on the upstream slope, Figs. 9.2:1, 9.2:2. Even though the crest length is 2,700 ft., there are no expansion joints in the concrete slab. The dam retains a lake with a capacity of 74,000 acre-ft., and in the severe winters of the region ice sometimes forms 8 in. thick on the reservoir. For brief periods, temperatures drop as low as 30°F. below zero.

No maintenance work of any kind has been required on the slab since construction, and it is completely sound at the present except for a few isolated hairline cracks. The following description was given by the owner:

> The general appearance is similar to that of a rough terrazo finish; the wear on the aggregate and cement paste has been equal. Only occasionally has a pebble been plucked out of the surface or does the aggregate protrude above the surface. Although no expansion joints were provided in the slab originally, there has been little cracking or crazing of the concrete. Some hairline contraction cracks have occurred and some minor spalling has taken place along the construction joints.[1]

Another interesting feature of this dam is that the foundation cutoff wall was not extended down to bedrock under the left abutment because of the great depth to bedrock there. Leakage developed through this abutment when the reservoir was filled for the first time, but it has never increased. At full reservoir in 1959, the measured leakage was roughly 3.0 c.f.s.

[1] R. H. Harris, Acting Regional Director, USBR, Region 1, Boise, Idaho, 1960.

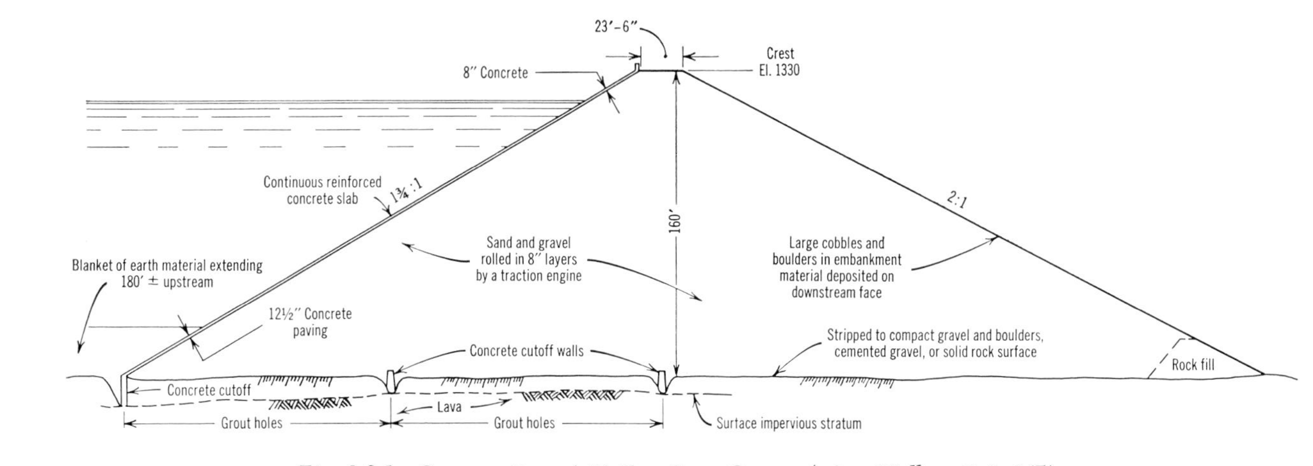

Fig. 9.2:1 Cross section of McKay Dam, Oregon (after Walker, Ref. 147).

Fig. 9.2:2 View of the upstream slope of McKay Dam at low water level in 1957. Note five steps in upper elevations designed as wave breakers. (Courtesy USBR)

DON MARTIN DAM, MEXICO

An embankment section and upstream concrete slab similar to McKay Dam was used in 1930 for Don Martin Dam in northern Mexico (Fig. 9.2:3), and the performance has been as satisfactory as at McKay Dam. Don Martin Dam has a maximum height of 100 ft. and a crest length of 4,000 ft.

Some cracking developed in the slab due to differential settlement at the right end of the dam where it connects with a massive concrete retaining wall at the spillway. The cracks were filled with asphalt in 1944, and no further trouble has developed. Over most of the length of the dam, the slab is in good condition (1960), Fig. 2.6:5.

9.2b California-Type Rockfill Dams

Design details for reinforced concrete slabs on rockfill dams have evolved empirically, primarily as the result of experience in California where a large number of major structures of this type have been built. The concrete slabs on the older dams were constructed monolithically without expansion joints and have been generally satisfactory. For the recent rockfill dams, which are much higher, the upstream concrete slabs have been divided in panels separated by expansion joints with water stops.

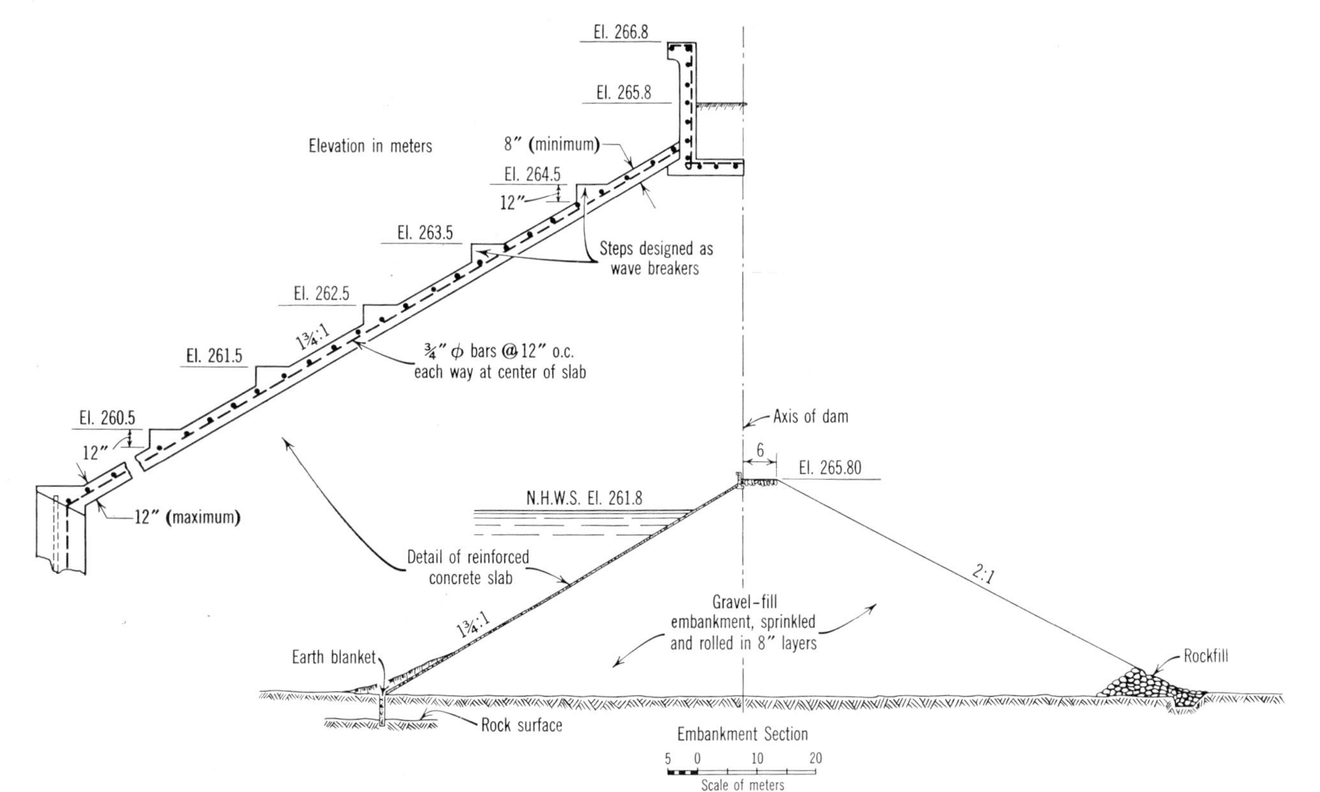

Fig. 9.2:3 ***Cross section of Don Martin Dam, Mexico, and details of concrete slab.*** (Courtesy Mexican Department of Hydraulic Works)

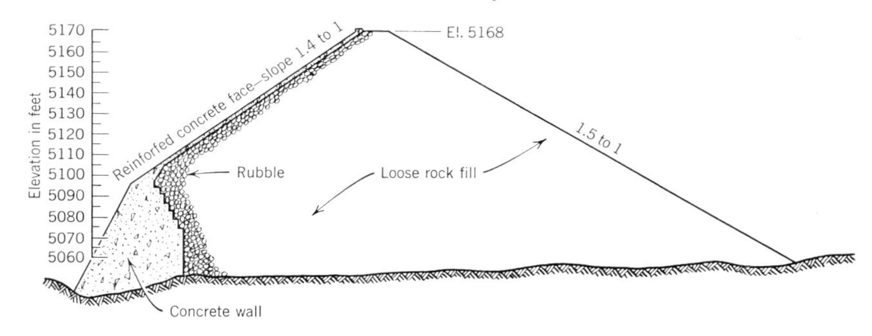

Fig. 9.2:4 Bucks Creek Dam, California (after Galloway, Ref. 262).

Among the older dams, one example which is especially useful in guiding the design of concrete faces for rolled earth structures is the Bucks Creek Dam built in California in 1928, Fig. 9.2:4 (Ref. 372). In this dam, which is 128 ft. high and 1,000 ft. long, the concrete face was built as a monolithic slab without expansion joints and with the reinforcing steel running continuously through the construction joints. The owner reports that "the face has developed no cracks, leakage is negligible and no maintenance has been required in its 30-year life to date" (Ref. 337).

On the recent dams the slab thickness is made approximately equal to 1% of the water head and usually has a minimum thickness of 12 in. Steel-reinforcing approximately equal to 0.5% of the total slab area is provided in each direction, and the steel does not run through the expansion joints. In most cases the steel has been placed at the center of the slab in one layer except where the slab is over 24 in. thick, in which event two equal layers are used, top and bottom.

Due to the relatively great postconstruction settlements, the concrete facings on most of the higher rockfill dams have developed some cracking and leakage in spite of the intricate system of expansion joints. Indeed, much of the leakage has been caused by torn water stops, and there is some question whether less trouble might have been experienced if the slabs had been built without the joints.

9.3 UPSTREAM MEMBRANES OF WELDED STEEL PLATE

Steel plate can be used at any site where reinforced concrete might be considered; and although steel is somewhat more expensive, it has two advantages over concrete: it is completely watertight, whereas fine cracks in the concrete will inevitably allow some leakage; and it is more flexible and better able to conform to differential settlements without rupture.

Experience with steel plate membranes on about a dozen earth and rockfill dams where it has been used indicates that steel facing has approximately the same life as reinforced concrete. In contrast to the fears which have existed on this point, in none of these dams has corrosion of the steel plate led to serious problems. On the other hand, the plate has never been placed directly on soil containing an appreciable percentage of clay or silt sizes.

Maintenance costs in general on the steel dams, which vary in age from 25 to more than 60 years, have been very low. The following examples give information on the performance of several dams with upstream steel plates, including two which are 60 years old.

ASH FORK DAM, ARIZONA

The oldest dam with a steel face in the United States is the 46-ft. Ash Fork Dam, which was constructed in 1898 with a sloping upstream face of 3/8-in. riveted steel plate supported on structural steel bents (Ref. 568). Since the weather in the vicinity is very dry and the reservoir is designed to store flood waters, there is a large reservoir drawdown each year, and the upstream face is alternately wet and dry. Analysis showed that the average chemical content of the water is (in parts per million) (Ref. 353):

Magnesium sulphate	8.6
Calcium and magnesium carbonate	113.0
Sodium chloride	26.0
Sodium carbonate	13.7
Alkalinity	100.0

The plate has been painted at average intervals of about 8 years. Except for unimportant minor pitting, it is in good condition at the present time.

SKAGUAY DAM, COLORADO

The Skaguay Dam is a steel faced rockfill structure built in 1900 in the Colorado Rockies. The maximum height is 75 ft., and the upstream slope is approximately 60° with the horizontal. The steel facing has a thickness of 1/2 in. at the bottom, reducing to 3/8 in. at mid-height.

It was constructed by riveting and caulking "in the same thorough manner as for boiler practice."

In the last 30 years the facing has been given two complete treatments of cleaning, chipping, buffing and painting. The last of these was in 1940. Concerning its present condition, the owner states:

> "Throughout the years, the steel face has become pitted, but not to the extent to become alarming . . . we are not able to examine the fill side of the steel, but have examined the portion extending above the fill at the crest of the dam and find that it has suffered less damage than the water side . . . the steel face has not received the maintenance it should have but . . . it has stood the years very well." (J. T. Van Derwalker, District Manager, Southern Colorado Power Company, 1959.)

EL VADO DAM, NEW MEXICO

Completed in 1934, El Vado Dam is a rolled fill embankment consisting primarily of gravel, Fig. 9.3:1 (Refs. 43, 45, 310, 374). It is the most important steel faced structure in the United States. The dam has a ¼-in. steel plate on the upstream face, a maximum height of 175 ft., crest length of 1,200 ft., and a 1.5:1 upstream slope. The dam retains a reservoir with a capacity of 200,000 acre-feet. The steel plate is underlain directly by the gravel fill embankment and was not painted on the under side. The dam has been in continuous operation and the maintenance forces report that "the steel plate is in excellent condition and, with minor inexpensive repair work from time to time, should have a life as long as the dam. (Recently an attempt was made in the laboratory to predict the remaining life, and a conservative estimate of 50 years was obtained.)"[1] Some wrinkling, buckling, and

[1] Leon W. Hill, Regional Director, USBR, Region 5, Amarillo, Texas, 1959.

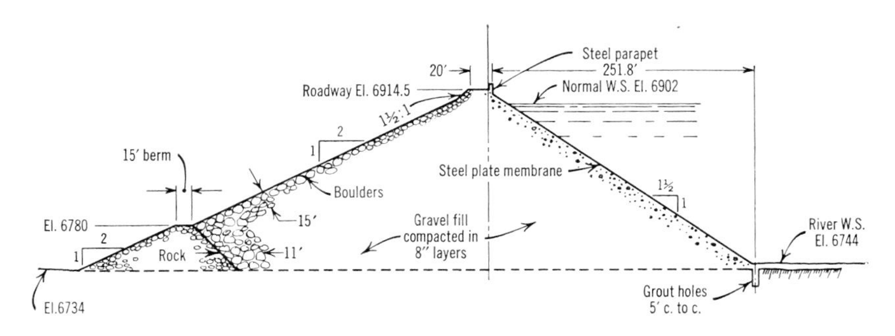

Fig. 9.3:1 El Vado Dam, New Mexico (after Segar, Ref. 314).

Fig. 9.3:2 General view of upstream steel face as El Vado Dam in 1954. (Courtesy USBR)

tearing of the steel plate has been caused on the left side of the dam by movement of the abutment, which is an old landslide. Figure 9.3:2 shows the steel facing when it was being cleaned and painted in 1954, and Fig. 9.3:3 shows a view of the typical embankment material and of the construction of the steel plate.

Fig. 9.3:3 Construction of steel plate facing of El Vado Dam and typical embankment material. (Courtesy O. O. Phillips)

Fig. 9.3:4 Construction of steel plate facing on Crystal Creek Dam (1935). (Courtesy O. O. Phillips)

CRYSTAL CREEK AND SOUTH CATAMOUNT DAMS, COLORADO

Stimulated by the design of El Vado Dam, two similar dams which were constructed for the water supply system of the city of Colorado Springs during 1935 and 1936 used upstream facing consisting of a ¼-in. plate of copper-bearing steel. Both dams were located at high elevations in the Rocky Mountains where the only available embankment material consisted of decomposed granite in the form of a coarse sand, which was considered too pervious for a conventional earth dam.

Fig. 9.3:5 Construction photo showing curved plate expansion joints in steel plate facing at South Catamount Dam. (Courtesy O. O. Phillips)

Fig. 9.3:6 Completed steel plate facing on South Catamount Dam (1936). (Courtesy O. O. Phillips)

Figures 9.3:4 and 9.3:5 show some details of the construction of the steel facing. As seen in the photographs, individual steel plates were first placed on frames consisting of light steel angles, and the space between the plate and the embankment was then backfilled with gravel. The legs of the frames were embedded in the upstream portion of the embankment so that they served also as anchors for the plate facing. Figure 9.3:6 shows a view of the face of South Catamount Dam at the end of construction.

Concerning the maintenance and present condition of the steel plate faces on Crystal Creek and South Catamount Dams, the owners state: "The steel plates show minor rust pitting at this time but are in what I would call 'excellent' condition . . . at intervals of 5 to 10 years, as reservoir levels permit, the exposed plates are sandblasted and repainted with red lead and metallic aluminum paint. The bottom portion, approximately one third of the steel decks, has had no treatment whatever."[1]

KHRAM DAM, RUSSIA

In 1935 a steel faced gravel-fill dam with a maximum height of about 110 ft. and an upstream slope of 1.5:1 was constructed in southern Russia (Refs. 354, 365, 651). The details of the upstream 6-mm. thick steel facing were very similar to those used for the above

[1] J. C. Nichols, Chief of Operations, Colorado Springs Department of Public Utilities, 1959.

three dams in Colorado and New Mexico. One principal difference was that the expansion joints running up and down the slope on the Khram Dam were constructed by curving the edges of the plates upward as shown in Fig. 9.3:7 instead of using a separate curved steel plate for the expansion joint. The Khram reservoir is emptied at least once a year, and the performance of the steel facing has been entirely satisfactory.

SALAZAR DAM, PORTUGAL

The 210-ft. rockfill Salazar Dam, constructed in Portugal in 1948 for an irrigation and hydroelectric project, is probably the highest and most important structure to be provided with an upstream impervious membrane of welded steel plate, (Refs. 216, 364). The main part of the dam is composed of quarried rock dumped in 9-ft.-thick layers.

An 8-in.-thick unreinforced concrete slab was provided under the steel plate as a levelling course. The plate itself is $\frac{5}{16}$ in. thick for the lower third of the dam and $\frac{1}{4}$-in.-thick above. Semicircular plates with a diameter of 1.3 ft. were used for vertical expansion joints placed approximately 50 ft. on centers. Drains installed in the concrete behind each expansion joint discharge in the concrete drainage gallery at the upstream toe. The concrete face was painted before the steel plate was laid, and the steel was painted with bituminous emulsion.

Crest settlement in the first 10 years was approximately 16 in., and the downstream crest movement was about 14 in. (Ref. 184). In the

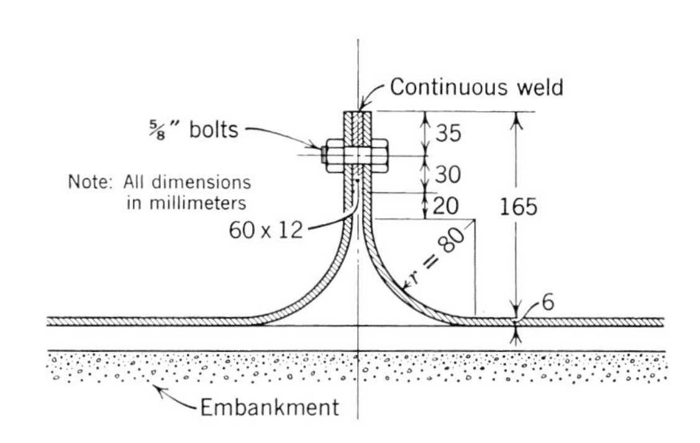

Fig. 9.3:7 Curved plate expansion joint used in upstream steel plate facing at Khram Dam, Russia (1935). (Courtesy C. Marcello)

first 2 years of operation, leakage developed in three cracks which opened in the welds of the expansion joints. The maximum measured leakage was 145 gal./min., of which 80 gal./min. were coming through the cracks in the welds. The drains behind each joint made it possible to locate the cracks in the welds and to cover them under water with tarpaulin until the permanent repairs could be made. After the welds were repaired at the end of the second year, the steel facing has been completely watertight.

Except for the minor repairs of the welds, the performance of the steel facing has been perfect and the owners consider the design very satisfactory from the standpoint of initial cost, performance, and maintenance. Only in 1959, after 10 years of operation, was it considered desirable to apply another coat of paint.

CARACAS WATER SUPPLY DAM, VENEZUELA

One of the latest[1] steel-faced dams is the 80-ft.-high Rio Lagartijo Dike built in 1958, Figs. 9.1:1, 9.3:8, 9.3:9. Steel was chosen in prefer-

[1] For other recent steel-faced dams, see the descriptions of the Sirinumu Dam in New Guinea (Ref. 651) and the Vielles Forges Dam in France (Ref. 667).

Fig. 9.3:8 General view of the upstream steel plate facing on the Rio Lagartijo Dike. (1960).

Fig. 9.3:9 Rio Lagartijo Dike—view looking down upstream slope from dam crest at location of bent-plate expansion joint.

ence to concrete primarily because large postconstruction settlements were expected as the result of compression of the soil foundation, and it was felt that steel plate could withstand the settlement better without rupturing.

The details of the plates are similar to those used at El Vado and the Colorado Springs Dam except that the bentplate expansion joint extends above the plate facing instead of below. Another innovation at the Rio Lagartijo Dike was the installation of a system of cathodic protection for the steel plate (Ref. 371). During the first four seasons of operation, the facing has been perfectly watertight, and there is no significant deterioration due to corrosion.

9.4 UPSTREAM MEMBRANES OF ASPHALTIC CONCRETE

For many years asphaltic concrete has been used here and there to protect the upstream slope of small dams against erosion;[1] however, only recently has it been used as a water barrier. Since about 1950, at least 15 large dams have been constructed with upstream im-

[1] See, for example, a description of the use of an asphaltic concrete slab in lieu of rock riprap by the USBR at Glen Anne Dam in California (Refs. 235, 594).

pervious membranes of asphaltic concrete. These have been in North Africa and in Europe with only one in the United States. During this brief period, opinion in the profession on the suitability of asphaltic concrete for this purpose has made an almost complete reversal from general skepticism to widespread recognition that asphaltic concrete upstream membranes cannot only be used safely but that they offer certain definite advantages over alternate materials and designs. There is no doubt that they will be used more widely in the future.[1] At the present a dam of over 300 ft. in height is in the design stage in Norway and others of this height have been seriously studied in the United States (Ref. 358).

Some of the main advantages of asphaltic membranes are:

1. They cost less than either concrete or steel.
2. They are more flexible than reinforced concrete slabs, and thus are better able to follow differential settlements without cracking.
3. They can be constructed rapidly.
4. Under certain circumstances, the leaks which develop are self-sealing.
5. The portions above the reservoir level are easier to repair than either steel or concrete.

Also, while the number of examples is limited, there is every reason to believe that a well-designed and constructed asphaltic concrete membrane will last as long as concrete or steel under ordinary conditions. The primary disadvantage is that the material is relatively soft and can be more easily damaged by falling rock, sabotage, or other activities of man and nature than either concrete or steel.

The asphaltic concrete used generally consists of very well-graded aggregate from a maximum size of about 1 in. to fine sand sizes and contains approximately 10% by weight of ground rock dust (filler) which is finer than the No. 200 sieve. Pure asphalt binder of 8 to 10% by weight of aggregate is used. The material is mixed and compacted hot. All the field and laboratory experience shows that the degree of compaction obtained and the resultant air content of the asphaltic concrete have an important influence on the principle properties of the membrance, i.e., the permeability, the rate of aging, and the resistance to alternate freezing and thawing. An air content of 2 to 3% appears to be optimum.[2]

[1] References 206 and 591 give an excellent summary of criteria and practice for the design and construction of asphaltic concrete membranes; also, see Refs. 197, 546, 547, 552 for general information on the use of asphalt for dam membranes.

[2] Reference 708 gives a valuable summary of the properties of the asphaltic concrete used on most of the major dams.

Although there is a great deal of information about asphaltic concrete for highway and airfield pavements, few criteria are available to guide the design of an upstream membrane for a dam. The criteria for the selection of the thickness, mix, drainage provisions, and surface protection must be considered still in the developmental process. No doubt, different and better design details will be devised and used in the next few years.

A more thorough review of current experience with asphaltic concrete membranes is provided in Ref. 563. In addition, an excellent paper by Scheidenhelm et al. (Ref. 358) gives a summary of such major dams constructed before 1957. Of these, most have been in North Africa and Europe. Only one major dam, Montgomery Dam, has been built in the United States with asphaltic concrete on the slope.

Details of the construction and performance of some important dams with asphaltic membranes are given below. It should be noted here that all of these dams except Montgomery Dam have control galleries at the upstream toe similar to that described below for Ghrib Dam.

GHRIB DAM, ALGERIA

When completed in 1935, this was the first major dam in modern times constructed with an upstream impervious membrane of asphaltic concrete (Refs. 220, 544, 588, 589, 593). It is still one of the highest (210 ft.), and it retains the largest reservoir (230,000 acre-ft.) and has the steepest upstream slope (steeper than 45°) of any structure with such a membrane, Fig. 9.4:1. The construction of the Ghrib Dam is a milestone in the art of dam building and there have been few comparable single advances.

The successful performance of Ghrib Dam for more than 20 years under extreme conditions of temperature is one of the strongest supports for the contention that asphaltic concrete is suitable for impervious membranes on embankment dams. While there is little doubt that the development of the use of asphaltic concrete for membranes for high dams has been unduly delayed by the natural reluctance of the profession to use untried methods for such potentially dangerous structures, there can also be no doubt that the acceptance of this method of construction would not have reached its present, albeit tentative, status if it had not been for the precedent of Ghrib Dam.

For design purposes, it was estimated that the temperature of the outer surface of the membrane might go as high as 70°C. (160°F.)

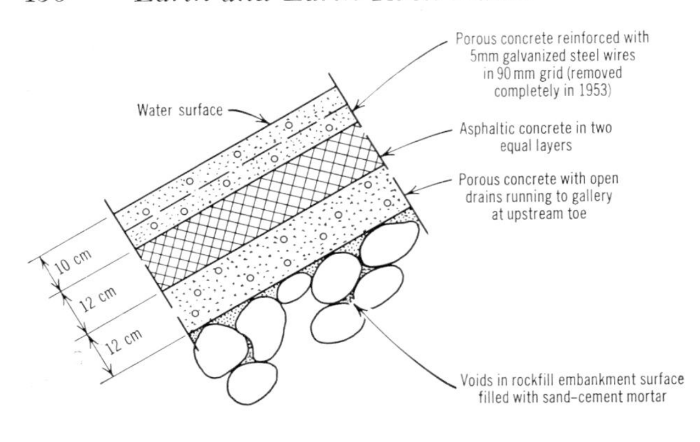

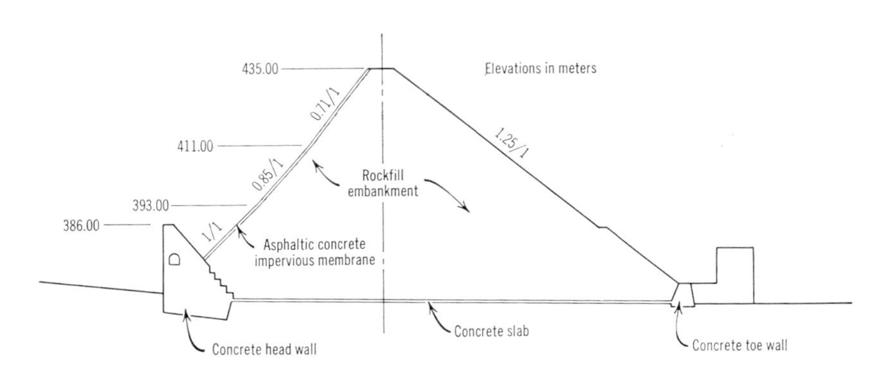

Fig. 9.4:1 Ghrib Dam, Algeria, with details of asphaltic concrete upstream membrane (after Thevenin, Ref. 220).

under the normal summer sun and it was necessary to provide an asphaltic concrete to withstand such temperatures without dangerous softening and downward creep on the steep slopes. As shown in Fig. 9.4:1, the asphaltic concrete slab finally chosen was only 12 cm. (about 4.8 in.) thick. It was sandwiched between an 8-cm. lower layer of porous concrete, which acted as a drain, and a 10-cm. outer layer, which acted as both thermal and physical protection, and which was also made of porous concrete reinforced with galvanized steel wire mesh.

The embankment was constructed of large, bulky limestone rocks fitted together carefully with cranes in such a way that the large rocks were in contact with each other and the smaller rocks filled the interstices. They were so tightly and meticulously packed that the post-construction settlement of the dam crest in the first 20 years of operation was only about 12 in. The horizontal component of the movement of the crest during this time was also roughly 12 in. in the downstream direction.

The underlying porous concrete drainage layer was placed for the purpose of catching and controlling any water which found its way through leaks in the asphaltic concrete. It contains interior open drains which run vertically down the slope at intervals, discharging into the reinforced concrete gallery which runs along the upstream toe of the dam. The locations at which the water enters the gallery give some indication of the position of leaks in the membrane.

The reservoir capacity is so large in comparison to the stream flow and the water use requirements that several years were required for the reservoir to fill the first time. In 1939 when the water level reached a new high, some rupturing of the asphaltic concrete evidently took place due to differential settlement. This was manifested by concentrated leakage appearing in the control gallery. No repairs were made and, even though the reservoir remained at the same level, the leakage gradually decreased until at the end of several months it had practically stopped, indicating a remarkable facility for self-repair in the asphaltic concrete. No further leaks developed afterwards and at present the membrane is completely watertight.

Soon after construction, the reinforcing mesh in the outer layer of porous concrete began to corrode. Subsequently the layer deteriorated and gradually peeled off, until in 1953 the engineers completely removed it from the surface of the dam. Parenthetically, this is only one in a number of experiences which indicate that porous concrete is not a suitable material for use on the upstream face of a dam (see, for example, Ref. 148).

AGING AND THERMAL PROTECTION TESTS

In 1954–1955, after the outer layer of the membrane had been taken off, and approximately 20 years after the asphaltic concrete had been installed, an extensive series of tests were performed on it. From laboratory test results on samples cut from the membrane at various locations, the following conclusions were obtained:

1. Aging of the asphaltic concrete was characterized by a moderate loss of plasticity and a concomitant increase in the compressive and shearing strength.

2. Aging was most pronounced in sections which had been above the high water level, and insignificant in parts which had been continuously immersed.
3. Aging was greater where the asphalt had been less well compacted and more porous.
4. The permeability was not influenced by aging.

On the basis of laboratory slope creep tests it was decided that the asphaltic concrete membrane could be left bare if its temperature did not rise above approximately 50°C.

To determine the insulating properties of several prospective materials for a new protective layer, field tests were performed in place on the membrane. The tested materials included were: those, such as "aerated" concrete, which have high coefficients of thermal insulation; light-colored paints, which reflect heat; and synthetic resins of light colors, which do both. Temperature measuring probes were installed at the surface of the membrane and at different depths in the asphalt, and the temperatures which developed under the August sun were recorded. The conclusions were:

1. The asphaltic concrete in itself had a considerable thermal inertia. On a hot day when the surface of the unprotected asphaltic concrete reached 60°C., the temperature at the middle of the 12-cm. slab reached a maximum of only about 40°C.
2. The addition of white paint kept the temperature on the surface at a minimum of 16°C below the value which occurred when no protection was used, and well below the temperature at which flow might be expected.

As a result of these studies, the surface of the asphalt was given only a coat of white paint, and up to the present the performance of this thermal protection has been excellent.

IRIL-EMDA DAM, ALGERIA

The Iril-Emda Dam, completed in 1954, was the third of four Algerian dams built using asphaltic concrete membranes (Refs. 187, 434, 549, 550).[1] With a maximum height of 245 ft. and a membrane

[1] The second was the Bou-Hanifia Dam, for which the design and construction were similar to Ghrib. Performance has been satisfactory and uneventful (Refs. 556, 557). The fourth was the Sarno Dam, a rolled earthfill constructed in 1955, which has also given satisfactory performance (Refs. 107, 548).

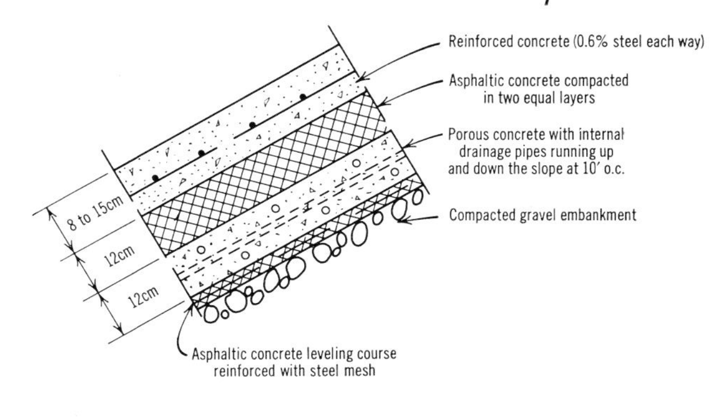

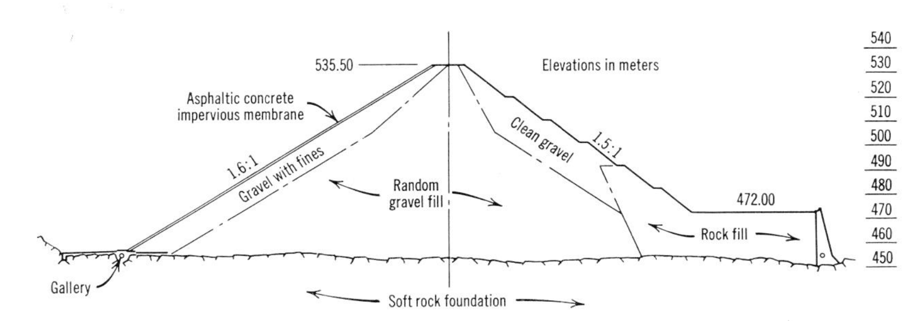

Fig. 9.4:2 Iril Emda Dam, Algeria, with details of asphaltic concrete upstream impervious mambrane (after Guilhamon and Castelnan, Ref. 187).

area of more than 66,000 sq. mtr., it is the largest dam of this type which has been constructed, Fig. 9.4:2.

A compacted gravel-filled dam, it was built of broken limestone in the form of a well-graded mixture of cobbles, gravels, and fines, and was compacted in layers with vibratory compactors. In the upstream third of the dam, under the membrane, the layers were made 12 in. thick, and the thickness was increased to a maximum of about 3 ft. in the downstream third of the dam. The upstream slope was made approximately 1.6 : 1—the inclination at which it was estimated that the material could be held without raveling during construction while the asphaltic concrete was being placed.[1]

[1] In 1959, the Radoina rockfill dam which had a laminated upstream membrane very similar to that of Iril-Emda Dam was constructed in Czechoslavakia to a height of 130 ft. The average upstream slope was approximately 0.8 (horizontal) to 1.0 (vertical), and the asphalt layers were placed and compacted satisfactorily with specially designed spreaders and vibrating rollers suspended from the crest operating up and down the slope (Ref. 598).

The membrane is essentially the same as that used at the Ghrib and Bou-Hanifia Dams. Because of the unsatisfactory performance of the porous concrete at Ghrib Dam, the outside protective layer at Iril-Emda was constructed of monolithic reinforced concrete varying in thickness from 8 cm. at the crest to 15 cm. at the toe and reinforced with 0.6% steel in each direction. The construction of the asphaltic concrete was handled with standard road-building equipment, and the layers were compacted with 15 passes of a 5-ton smooth steel roller running up and down the slope.

In addition to reviewing the performance of the two previous dams, the designers of Iril-Emda performed a detailed series of laboratory tests on the asphaltic concrete (Ref. 549). Among the tests, an attempt was made to study the self-repairing properties of the asphaltic concrete with specimens which were essentially complete cross-sections of the whole membrane, including the upper and lower concrete layers. These specimens were placed in a steel cylindrical permeability testing apparatus of the type usually employed for earth specimens. Holes were gouged in the asphaltic concrete layers, and the rates of seepage through the specimens as a function of time were measured under various hydraulic gradients.

In all cases during the tests the leakage decreased with time. This was evidently due to the plastic flow of the asphaltic concrete under the pressure caused by the water gradient, which had a tendency to close the leakage channel by squeezing. At a water pressure equivalent to a head of 160 ft. or more, nearly complete self-sealing occurred on all samples.[1]

Also as a part of the testing program, tension tests were performed on full-sized reinforced concrete slabs to determine the amount of steel which would be necessary in the outside protective layer to force the slab to crack in many hairline fissures rather than in a few larger cracks which might rupture the asphaltic concrete membrane. The results, as shown in the sketch, led to the use of 0.6% steel, which is a little more than would normally be considered necessary.

The performance of the dam has been completely satisfactory. The maximum postconstruction settlement of the crest has been only 2.5

[1] Other investigators have found this same phenomenon. For example, in connection with the studies for their dams, the engineers of the Norwegian Geotechnical Institute cut a sample of asphaltic concrete membrane from a small test dam which they had constructed and tested this sample in a laboratory permeameter under a head of 6 km./cm.^{2} At the beginning of the test the permeability of the specimen was approximately 1×10^{-6} cm./sec., and after the test had run for six months the coefficient of permeability had decreased to 1×10^{-9} cm./sec.

in., indicating the efficacy of the vibratory compaction. The leakage measured in the control gallery when the reservoir reached maximum height for the first time was about 0.3 c.s.f. (through the entire 66,000-m.^{2} surface). With time, this amount decreased to approximately half.

GENKEL DAM, GERMANY

Genkel Dam (Refs. 201, 553) completed in 1952, is the first of a series of five German rockfill dams which have been built in the last decade using essentially the same details for the impervious membrane. As shown in Fig. 9.4:3, the membrane design was a departure from that used by the French engineers in Algeria in that the asphaltic concrete layer is exposed directly on the surface with no protective covering. The membrane is also more complex, with a gravel drain under the upper slab and a second layer of asphaltic concrete under the gravel.

Because of the novelty of the membrane, the embankment was given a 2¼ : 1 exterior slope even though it was believed that a steeper slope might have been feasible. It was constructed of 400,000 m.^{3} of quarried rock, which broke into relatively small pieces mostly in the range between 1 and 15 in. The rock was hauled to the embankment with rubber-tired hauling equipment and spread in layers approximately 18 in. thick, and the hauling equipment was routed over the construction surface in order to obtain as much compaction as possible.

The layers were subsequently compacted with 2½-ton vibrating tampers suspended from cranes and, to a lesser extent, with smooth steel rollers. This type of compaction with heavy tamping and vibrating plates has been used for a considerable number of rockfill dams in Germany since World War II, Fig. 9.4:4. In the case of Genkel Dam, the maximum postconstruction settlement at the crest of the embankment was less than ½ in. Such low settlements have been consistently obtained on the German rockfill dams compacted with the vibrating tampers (though these dams have not been very high).

The outer impervious asphaltic membrane (9 cm. total thickness) was placed in three equal layers. These were compacted with light vibratory plates, which were believed to give as good compaction as heavy static rollers and which did not require heavy equipment on the crest to act as anchors. The upper surface of the final membrane was covered with two layers of mastic placed with a special spreading machine to a total thickness of about 0.5 cm. to "seal the pores."

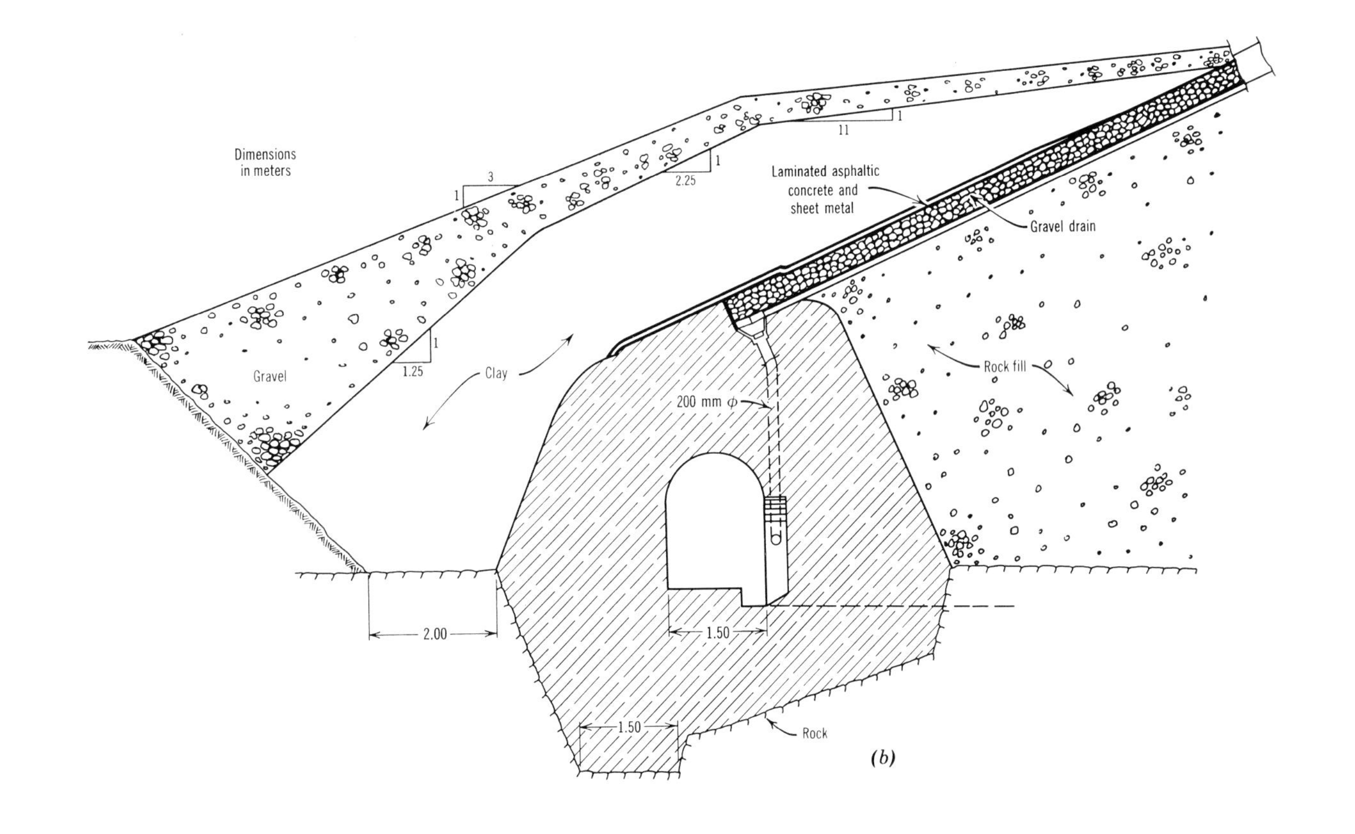

Dimensions in meters
3
1
2.25
1
11
1
Laminated asphaltic concrete and sheet metal
Gravel drain
Gravel
1
1.25
Clay
Rock fill
200 mm φ
2.00
1.50
1.50
Rock
(b)

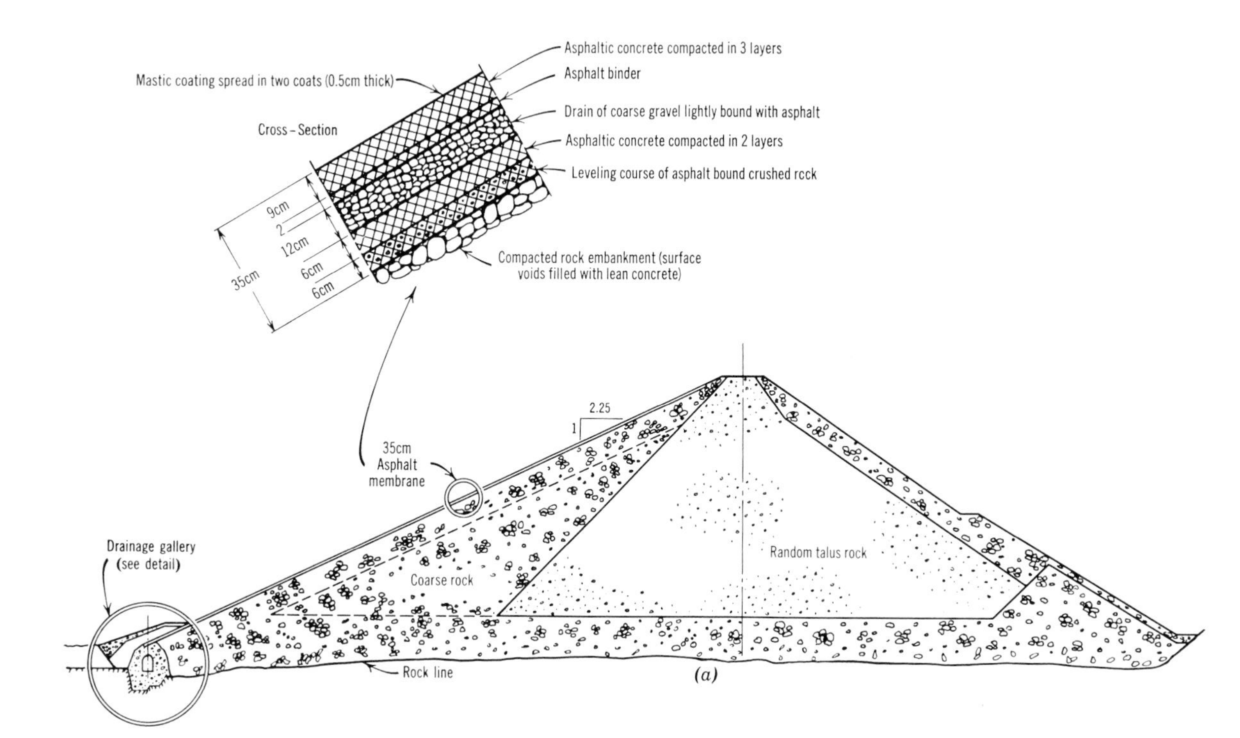

Fig. 9.4:3 Genkel Dam (after Schmitz, Ref 553). (a) Typical cross-section with membrane details. (b) Detail of drainage galley at upstream toe.

The drainage layer immediately under the outer asphaltic concrete layer consisted of gravel graded roughly between 1½ and 3 in., with enough asphalt to bind the stones together but not enough to reduce the permeability. This layer was bulkheaded into sections with strips of asphaltic concrete, with each section drained to the gallery at the toe with a pipe so that any leaks through the membrane could be located.

Concerning the performance of the dam, the owner reports that the membrane has been satisfactory in all respects. No maintenance or repairs of any kind have been required. The reservoir has been in operation since the spring of 1953, and the total leakage which discharges into the gallery from the membrane (covering 11,000 $m.^2$) consists only of drops of water probably due primarily to condensation of moist air in the drainage layer.

Ice on the reservoir reaches a maximum thickness of about 10 in. but does not adhere to the membrane, probably because warm air rising from the gallery keeps the membrane sufficiently warm. During most of the winter, the ice is observed to stop short of the face of the dam, and there is a space of several inches of water between the ice sheet and the membrane, indicating that a considerable quantity of heat is supplied from the interior. Similar performance has occurred in winter at all the other German dams. In the case of the Henne, which has a similar upstream membrane (Refs. 201, 554, 590), the maximum ice thickness reaches about 17 in., and no strong bond forms between the ice and the dam.[1]

WAHNBACH DAM, GERMANY

The 170-ft.-high Wahnbach Dam is the last of the series of 5 dams which have been completed in Germany with essentially the same asphaltic concrete upstream membranes.[2] Both the embankment and

[1] In 1962 the authors inspected the 190-ft.-high Henne Dam which had been in operation at that time for 8 years. At the time of this inspection, the surface of the upstream membrane could only be described as having been in excellent condition. The only sign of deterioration was a slight pealing of the surface layer of mastic to a depth of 1 to 2 mm. No leakage whatsoever was being obtained through the membrane—and all the leakage is being measured since it is channeled into the gallery at the upstream toe. The appearance of this dam left no doubt that, with occasional painting of the upstream slope with additional coats of mastic above the water level, its life span would be essentially indefinite.

[2] Another very similar rockfill dam with upstream membrane of asphaltic concrete is currently under construction (1962–1963), Ref. 394.

(a)

(b)

Fig. 9.4:4 German plate compactors. (a) Horizontally vibrating plate. (Courtesy Losenhausenwerk, Düsseldorf.) *(b) Three-ton tamping plate raised and dropped about 10 ft.*

the membrane were constructed in essentially the same manner as at the Genkel Dam described above (Ref. 201). However, the upstream slope of the embankment was steepened to 1.6:1, which the designer considered the maximum on which the work could be carried out in an economical and safe fashion.

The reservoir has been in operation since 1957 and the dam has required no repairs or maintenance.

MONTGOMERY DAM, COLORADO

Montgomery Dam, completed in 1957, is a dumped rockfill structure with a height of 113 ft. and a provision for being raised to 150 ft., Fig. 9.4:5 (Refs. 84, 355, 358). It is located at an elevation of approximately 11,000 ft. in the Colorado Rockies. Its 12-in.-thick asphaltic concrete face was an extreme departure from any previous practice in the United States, although it was not radical when compared with the thinner facings used on the higher dams in Algeria and Germany. The embankment was constructed from large hard quarried granite dumped and sluiced in 30-ft. layers following the American practice for construction of dumped rockfill dams. The problems of design and construction are given in detail in Refs. 84 and 358.

Economy and speed of construction were the main incentives for the use of asphaltic concrete. The designers estimated that the cost of the asphaltic facing, at the isolated mountain site, was approximately half the cost of a concrete slab and that its use may have saved a year of construction time.

An extensive test program, probably the most thorough of its type ever made, was carried out in the asphalt laboratories of the USBR

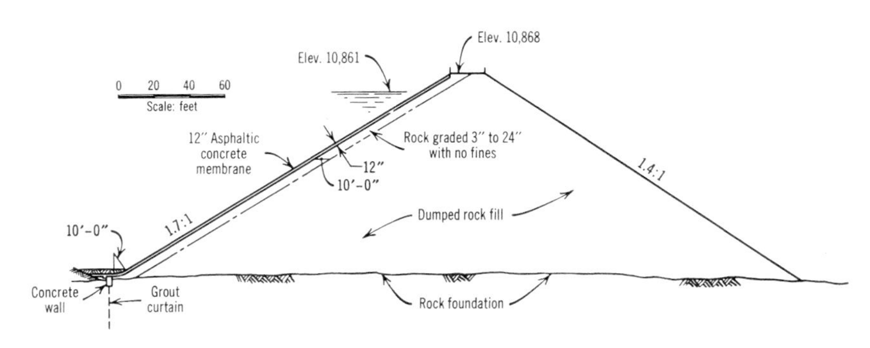

Fig. 9.4:5 Cross section of Montgomery Dam (after Scott, Ref. 84).

in order to determine a mix which would be suitable for the extreme range of temperatures expected as well as for the possible influence of a thick ice sheet on the membrane (Ref. 355). The tests also included an attempt at an evaluation of the influence of waves beating on the upstream slope.

Montgomery Dam is unprecedented in a number of respects. It is the only dam in which a single membrane of asphaltic concrete has been used without an exterior protective layer or underlying drainage layer, and it is the only dam with an asphaltic concrete face which has been constructed as a dumped rockfill. It does not have the drainage collection gallery which can be entered to observe and evaluate the seepage which gets through the membrane, and in addition to these innovations, it is constructed at a site where the asphaltic concrete membrane will be exposed to much more severe cold than at any similar dams. In fact, the winter conditions at the site are probably as severe as those to which any dam will ever be exposed.

The asphaltic concrete was mixed and spread with equipment commonly used for highway construction. The membrane was constructed in three layers 3 to 4 in. thick on an underlying leveling course. The material was compacted with a 3,000-lb. steel drum vibrating roller which was run up and down the slope, and the whole membrane (approximately 28,000 yd.^{2}) was completed in a single two-month construction season.

After 3 years of reservoir operation, the total crest settlement was about 0.1 ft. (Ref. 375). No noticeable quantity of leakage through the membrane developed and the effect of the ice on the surface of the pavement had been negligible. Winter temperatures have probably dropped as low as —25°F. A few minor surface cracks, which had a maximum depth of about 4 in. and widths varying between ¼ and ½ in., developed above the high water level along the construction joints, at the end of the first season. They were undoubtedly due to the fact that the paving machine could not operate as effectively on the extreme upper portion of the membrane at the crest. They were cleaned out and filled with sand and asphalt using the local practice for repair of highway pavements.

9.5 INTERNAL IMPERVIOUS MEMBRANES

As discussed in the introduction to this chapter, vertical concrete cores were once installed in almost all large earth dams, even those constructed of clay. In current practice, internal cores of concrete or other material are not used to any large extent in the United States, though a few main dams of this type have been constructed in Europe in the last 20 years. As stated earlier, the authors believe that there still are circumstances in which a dam with a membrane of manufactured material should be used and, while they generally prefer the upstream slope membrane, special conditions may make an internal membrane more advantageous. The principal relative advantages of the two locations are listed in the following.

PRINCIPAL ADVANTAGES OF INTERNAL MEMBRANES (CORES)

1. For a given dam the area of the membrane is smaller than that of an upstream facing, so that less manufactured material has to be used.
2. The length of the grout curtain in the rock below the wall is shorter.
3. The manufactured material comprising an internal membrane is protected by the surrounding embankment from weather and accidents such as falling rock.
4. The core membrane can be made almost completely watertight, even if cracking develops, by placing a thin layer of clay upstream.
5. A vertical extension of the core membrane below the base of the dam can be used as a seepage cutoff through soil deposits in the foundation. In such a situation a vertical cutoff extended down from the upstream facing is less satisfactory, since the spreading of the base of the dam (Sec. 3.1*d*) creates forces which may crack the cutoff wall.

PRINCIPAL ADVANTAGES OF UPSTREAM MEMBRANES

1. Where the membrane is on the upstream slope, the optimum stability conditions are produced (Sec. 9.1*a*) and the volume of the embankment can therefore be made smaller.
2. Since the upstream slab is exposed, damage can be inspected and repaired easily.
3. The upstream membrane can be built after the embankment is completed, whereas a central core must be brought up in relatively short lifts during the embankment construction.
4. Foundation grouting can be carried out while the dam is being built.

Another advantage frequently cited for the vertical core is that the membrane is less influenced by embankment settlement, and less likely to crack as a result. Actually, it is problematical whether the central core or the upstream facing has the advantage from this standpoint. The upstream membrane certainly undergoes larger post construction vertical settlement in most cases. On the other hand, the central core is subjected to very high stresses during construction because of the strains in the lower part of the embankment. It is probable that many vertical concrete walls are badly cracked but the damage is never seen. The principal disadvantages of the upstream reinforced concrete slab with respect to earth cores as discussed in Sec. 9.1*c* (i.e., shorter life and higher cost) apply also to internal concrete cores.

Most vertical membranes for earth and rockfill dams have been constructed of concrete. Steel plate has been used in only a few cases, and while asphaltic concrete has been considered for a number of dams[1] it has never been finally selected for any important structure, Fig. 9.5:1.

Methods have been proposed for designing central reinforced concrete cores on the basis of the estimated strains which may be imposed by the movement of the embankment (Refs. 13, 44, 189, 376). The results of calculations of this type are not reliable since we cannot estimate the magnitude of movements which may occur with sufficient precision.

As in the case of concrete membranes on the upstream slope, the design of vertical concrete cores must be based largely on experience. On the other hand, since most vertical cores are never exposed for inspection, very little reliable performance experience is available from which to form an opinion concerning optimum thicknesses and reinforcing for central reinforced concrete walls in dams of various

[1] For example, the designers studied the possibility of using a narrow core of compacted asphaltic concrete for the 440-ft.-high Goeschenenalp Dam in Switzerland (Refs. 5 and 597). At the Henne Dam constructed between 1952 and 1955, an internal core with a slight upstream slope was placed in the embankment of quarried rockfill. This core was constructed of small rock (varying approximately between 2 and 12 in.) which was pounded into a previously placed layer of pure asphalt with heavy tampers supported from cranes (see Fig. 9.4:4) in such a way that the asphalt and some added sand filled the voids between the rocks and created an impervious wall. At the Henne Dam, which also had an impervious asphaltic membrane on the upstream slope, the internal core was provided as a second line of defense or as a "braking zone" which would be called on to act only in the circumstances that extensive damage were to occur to the upstream membrane by some action (Refs. 201, 554, 595).

Fig. 9.5:1 Thin central core of asphaltic concrete compacted in 3-inch layers in 50-ft.-high Bremke-Vordam (Germany), 1962. Core is flanked on both sides by gravel zones as shown. Main dam of rockfill placed in three foot compacted layers.

types and height. In the absence of better information, the thickness and reinforcing for a central concrete core wall should be made about the same as would be used for a reinforced concrete slab on the upstream slope (Sec. 9.2). Though used in many of the older dams, thick unreinforced concrete cores have generally been unsatisfactory.

A few examples of concrete and steel cores are given in the following.

TIETON DAM: CONCRETE

When completed by the USBR in 1923, Tieton Dam was one of the highest earth dams in the world and had a concrete core wall with a maximum height above bedrock of 320 ft., Fig. 9.5:2 (Refs. 377, 378). Though the performance has been completely satisfactory,[1] the USBR

[1] The total maximum measured seepage is about 0.2 c.f.s. The crest gradually moved downstream until it had a maximum deflection of about 1.2 ft. in 1947. Each time the reservoir is lowered (about 100 ft.), the crest rebounds upstream approximately 0.1 ft.

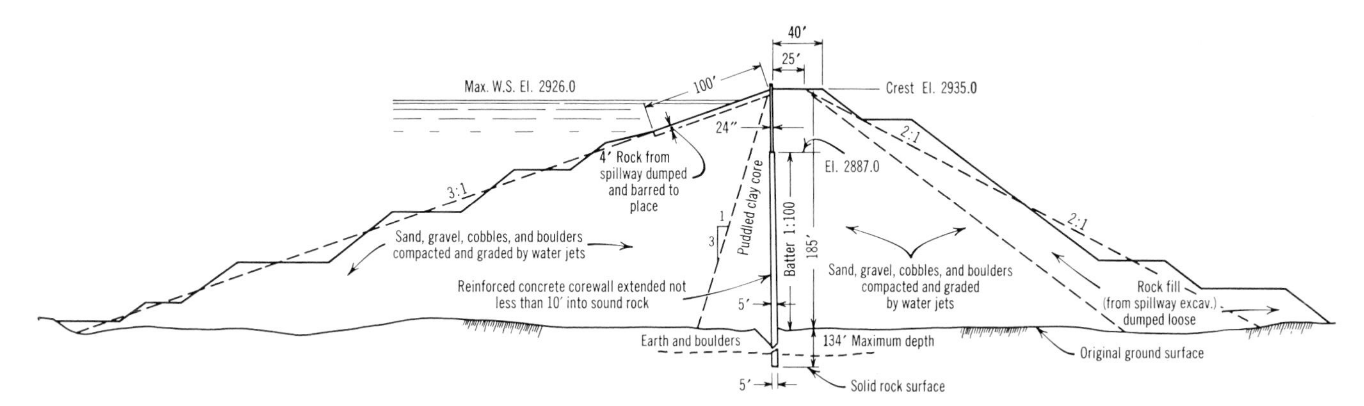

Fig. 9.5:2 Cross section of Tieton Dam, Oregon (after Walker, Ref. 147).

never used a design of this type again, partly because of the difficulty in constructing the high concrete core wall.

SWEDISH DAMS: CONCRETE

A number of major earth and rockfill dams with heights up to 170 ft. and central reinforced concrete core walls have been constructed since 1940 in Sweden, and the performance has been uniformly satisfactory (Refs. 188, 217). The walls are generally between 12 and 16 in. in thickness and are heavily reinforced, and a thin layer of fine-grained soil is placed just upstream for the purpose of sealing any cracks which may develop in the concrete.

MERRIMAN DAM: CONCRETE

One of the last major dams constructed with a central concrete core in the United States is the Merriman Dam for the New York City water supply system, which was completed in 1947 (Ref. 167). A view of the concrete wall and the embankment during construction is given in Fig. 9.5:3.

Fig. 9.5:3 View of construction of the central core wall at Merriman Dam (1946).

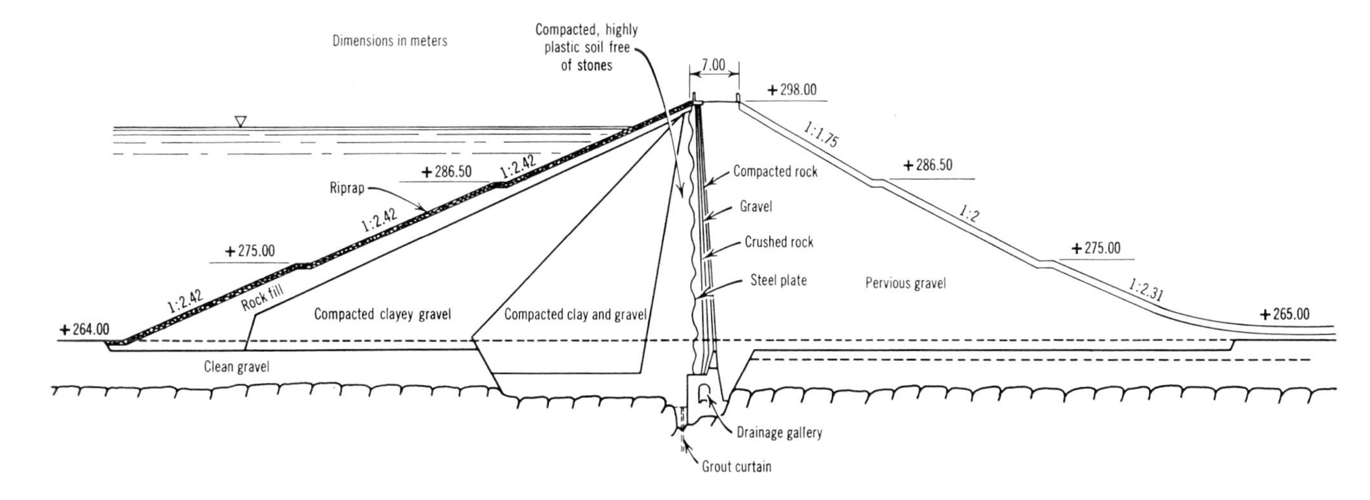

Fig. 9.5:4 Bever Dam, Germany, with central steel plate core, 1938 (after Mohle, Ref. 185).

BEVER DAM AND HEART LAKE DAM: STEEL PLATE

Few dams with vertical membranes of steel plate have been constructed. One exception is the 140 ft. Bever Dam in Germany (Ref. 185), completed in 1938 with a steel core $\frac{5}{16}$ in. thick, Fig. 9.5:4. The core was corrugated horizontally to permit it to adapt easily to the embankment settlement, and as in the case of the Swedish dams with vertical concrete walls, an upstream layer of clay varying in width from 8 m. at the bottom to 2 m. at the top was placed upstream from the steel plate to seal any leaks which might develop.

Fig. 2.7:3 shows installation of an "asbestos-bonded" corrugated steel sheet core being placed in the center of the 70-ft.-high Heart Lake Dam in the Colorado Rockies in 1954. This core was chosen for the dual purpose of controlling leakage in the semipervious embankment constructed of silty sand and for deterring burrowing animals.

ten

Treatment of Rock Foundations

10.1 DESIGN PROBLEMS

In recent years, the increasing economic advantage of rolled earth over concrete has resulted in the use of earth dams at sites with rock foundations and abutments where concrete dams would have been automatically selected a few years ago. Although rock is superior to earth as a foundation from the standpoint of strength and compressibility, the engineer still has several major problems at sites with rock foundations. First, the impervious section of the dam must be bonded to the rock foundation and abutments to prevent excessive seepage from developing along the contact surface between the embankment and the rock. The second main problem is to analyze the seepage which will occur under and around the dam in the cracks and joints in the rock and to provide means to control the seepage in such a way that it will not create a piping hazard and the water loss from the reservoir will be tolerable.

10.1a Types of Rock Foundation

There are many different types of rock foundations for earth dams. From the standpoint of planning the treatment of the foundation, the rock has several main characteristics of interest to the engineer:

1. Shape or surface topography.
2. Hardness.
3. Permeability and nature of crack systems.

Each of these individual properties varies between wide limits. The shape (or topography) of the rock may be smooth with flat, gradually changing slopes against which it is relatively easy to compact earth embankments. At the other extreme, the rock may be steep, irregular, or jagged, in which case it is more difficult to obtain a good bond between the rock and the dam embankment.

The hardness of rock foundations varies from the most solid, unweathered igneous rock, which can be excavated or shaped only by blasting, to soft, sedimentary or badly weathered rocks which can be excavated easily with earth-moving equipment. Unless the natural rock surface is very regular an earth dam embankment can be bonded to soft rock with less difficulty than to hard rock. Soft rock can be shaped to a smooth surface against which it is possible to use the heavy rollers and literally fuse the embankment material with the abutments and foundations. Where the rock is hard, there is always a definite surface of demarcation between the embankment and foundation and it is difficult to compact the embankment in the zone immediately adjacent to the hard rock, especially with a sheepsfoot roller.

Rock foundations range from practically impervious to very pervious. Except for a few types of sandstones, which may be fairly pervious because of an open porous structure, the high permeability of pervious rock formations is caused by fractures, bedding planes, joints, or cavernous voids which permit the passage of water. Individual, intact pieces of rock are impervious.

In the case of stratified rock, the structural attitude (strike and dip) is important because it can influence the direction of water movement through the rock. Stratified rock may be alternately hard and soft, pervious and impervious, and the topography of the rock surface may be radically different in each layer. The strata may be horizontal or may be dipping in any direction. The infinite combinations of rock conditions which are possible assure that each site has its special problems requiring individual study.

10.1b Sealing the Impervious Section of the Dam to the Rock

The best way to obtain a tight bond between the impervious core of the dam and the foundation is to make the rock surface regular enough to permit each embankment layer to be compacted directly against the rock with the heavy roller. Soft rocks can be trimmed to a suitable surface with earth-moving equipment. Overhangs and promontories in hard rock foundations and abutments are removed by hand labor with compressed air tools (or by wedging or boring) or by blasting.[1] Local depressions in the rock surface are filled with slush grout, gunite, "dental" concrete, or hand-compacted impervious soil.

Steep abutments should be cut back to the flattest practicable slope. The flatter the average slope, the easier it is to obtain good bond between the dam core and the rock. On the other hand many dams have been constructed successfully against very steep rock abutments, and no definite criteria can be given for the maximum permissible abutment slope.

The greater the width of the contact area between the core of the rolled embankment and the rock, the less is the likelihood that a leak will occur along the surface of contact. Consequently, dams with thin impervious cores are more susceptible to trouble than those with thick cores. This disadvantage of the thin core dam can be overcome to some degree by increasing the width of the core at the juncture with the rock in order to give a wider contact area.

Where the contact area between the impervious embankment core and the rock is relatively narrow, the downstream filter assumes more importance. Theoretically, if a properly designed graded filter is provided downstream from the core, any unexpectedly large leak developing at the abutment contact will discharge harmlessly into the filter. Since the gradation of the filter material is such that the soil from the embankment core cannot wash into the filter, progressive erosion cannot start and the leak should seal itself. Where very thin filter bands are used for the main embankment it is desirable to increase the width of the filter at the rock contact. Increasing the width of the core and filters in thin core dams, at the contact with the rock abutments and foundations, does not create difficult construction problems and may have a large beneficial influence on the control of leakage.

CONCRETE CUTOFF WALLS ON STEEP ABUTMENTS OF HARD ROCK

A few years ago, it was common practice to construct short, vertical, reinforced concrete walls as a connection between the core of the dam and the rock, Fig. 10.1:1. These were poured into trenches excavated in the rock of the foundation and abutments and extended into the earthen embankment. The walls varied in height between 5 and 20 ft., depending on the height of the dam, and usually extended up both the abutments to the high water elevation. These walls were

[1] Blasting must be carried out very carefully in order to avoid shattering the rock and increasing its permeability.

Fig. 10.1:1 Typical foundation cutoff wall. (Courtesy USBR)

intended to provide a seepage barrier for that water which flows along the contact surface between the compacted earth fill and the rock. The footing for the wall was often used as a grout cap.

Cutoff walls were constructed in the foundations of some of the largest dams which have been built including the 450-ft.-high, Anderson Ranch Dam (constructed 1941–1947 in Idaho), the 330-ft.-high Granby Dam (constructed 1946–1948 in Colorado), the 220-ft.-high Pactola Dam (constructed 1952–1956 in South Dakota), and many others. A typical wall of this type is shown in Fig. 10.1:2.

In the last few years, there has been a definite change of opinion concerning the value of these cutoff walls, and since 1950, they have been used much less frequently. At the present time, the use of concrete cutoff walls has been practically dropped from the practice of the U.S. Corps of Engineers (Ref. 3) and the USBR (Ref. 147). The primary objection to the concrete walls is that they prevent the use of heavy rollers for the compaction of the soil against the rock abutment in the vicinity of the wall. It is generally believed that the desirable function of the concrete wall, which serves to force any water seeping along the abutment contact to detour around the wall into the rolled earth fill, is less important than the disadvantage of having to compact the soil around the wall by mechanical tampers and hand compaction (Sec. 10.3). Secondary objections to the use of cutoff walls are the cost and the fact that blasting to excavate a foundation trench may open cracks which will make the abutments and foundations more pervious.

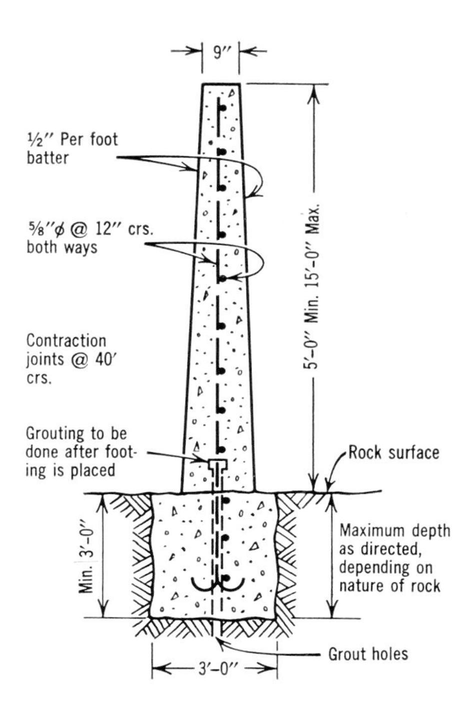

Fig. 10.1:2 Typical concrete cutoff wall used on rock foundation and abutments in major USBR dams before 1945–1950 (Ref. 35).

None of these objections are important when compared with the possible consequences of leaving out the cutoff wall at a site where it is needed. The cost of the wall is not an item for consideration when it reduces the possibility of failure. The problem of compacting the soil in the confined space around the walls can be handled easily. The foundation trench for the wall can be cut into the abutments with such small charges that the blasting will not disturb the character of the rock abutment; if sufficient concern develops about this problem, the trench can be excavated by drilling closely spaced holes along both edges ("line drilling") and removing the rock by barring or wedging.

At the great majority of sites where the rock surface is not excessively steep or smooth, the value of concrete walls is small. On the other hand, the authors believe that under some circumstances concrete walls on hard rock abutments of high earth dams are advisable and that it may be dangerous to leave them out. This is particularly true where the rock is smooth and where the abutments are steep or have sharp breaks in slope.

CUTOFF TRENCHES EXCAVATED INTO ROCK

The second principal method which has been used to break the path of possible seepage along the contact surface between the earth embankment and rock is the construction of a cutoff trench which is backfilled with compacted impervious soil. The cutoff trench in the rock is similar to that used to cut off seepage through foundations consisting of pervious soil (Sec. 6.2) but the dimensions of the trench are smaller. Cutoff trenches into the rock are especially useful for creating a seal between the dam core and foundation when the rock is relatively soft and decreases in permeability with depth below the surface.

At sites where the rock consists of alternate layers of hard and soft rock, such as alternately bedded sandstones and shales, oftentimes the softer rock has eroded to a greater extent than the harder rock and the valley walls are ribbed. In such a case it is desirable to excavate heavily in order to provide an even surface on the abutments so the earth can be compacted directly against the rock with heavy rollers. Sometimes the work can be lessened by the excavation of a wide, smoothly sloping cutoff trench over the whole area of contact between the core and the foundation, Fig. 10.1:3.

In a few dams, cutoff trenches have been excavated in foundations

Fig. 10.1:3 View of wide trench excavated in schist bedrock to provide smooth regular abutment against which the dam core can be compacted with heavy roller, Caribou Dam, California (1958).

and abutments consisting of hard rock. Hard rock usually is best treated, however, by grouting and surface layers of slush grout or concrete, Sec. 10.3. A trench of approximately 4 ft. in depth below the general level of the foundations and abutments was blasted into the hard rock at the site of the 300-ft.-high Cherry Valley Dam (constructed in the Sierra Nevada Mountains in 1951–1954). The trench averaged 20 ft. in width and varied in depth from 20 ft. at the abutments to 4 ft. in the stream bed, Fig. 3.2:5.

10.1c Planning Rock Grouting

Design methods for controlling seepage along the contact between the rock and the core of the dam have been discussed in the foregoing. The second major problem at sites with rock foundations is to evaluate the permeability of the rock, to decide whether it needs improvement by grouting and, if so, to work out grouting details. This is a problem which requires considerable judgment and experience. The desirability of close cooperation with an engineering geologist experienced in grouting cannot be overemphasized.

Except in rare circumstances rock foundations are too hard to be eroded by the water seeping through the cracks in the rock under and around a dam and, consequently, are not susceptible to piping failure. Grouting is used to make rock foundations and abutments more impermeable for two primary reasons: (1) to reduce the seepage loss from the reservoir; and (2) to control the water pressure in the fractures of the rock foundation at the downstream portion of the dam which might have a deleterious influence on the performance of the earth dam itself.

A third major incentive for grouting—which is the real reason for a great deal of grouting carried out in cases where it might not otherwise be considered essential—results from the security it affords from the standpoint of exploration of the rock. By the time that a closely spaced line of holes has been drilled and grouted across the valley, the engineer has obtained considerable assurance that he has located all major permeable strata, and large localized cracks and faults within the depth of the grout curtain.

EVALUATING NECESSITY FOR GROUTING

One of the most difficult problems facing the engineer at a given site is to decide whether or not it is necessary to grout and this is a subject of considerable debate among experienced engineers. There are no general or reliable rules which can be given stating that grouting must be done for the foundation of dams over a certain height or for dams founded on rock of certain types.

From a *historical viewpoint,* it can be stated that dams with a height of 50 ft. or less founded on rock foundations which are not excessively fractured generally have not been grouted. On the other hand, foundations for dams which are retaining a water head of more than 100 ft. usually have been grouted in some way.

The desirability for grouting depends primarily on the height of the dam, the permeability of the bedrock, and the value of the water which might be lost from the reservoir as a result of seepage through the rock foundation. It depends to a lesser degree on the zoning of the dam. In cases where the results of the field investigation show that the rock is very pervious and where the water in the reservoir is valuable, grouting obviously should be done. In the more general case where a reasonable quantity of water loss from the reservoir due to foundation seepage is not important, and where the results of preliminary explorations indicate that the rock is not excessively pervious, the decision is not as easy.

The decision as to whether or not to grout should depend also on the nature of the cracks in the rock. For some kinds of rock foundations, curtain grouting cannot have any appreciable influence on the quantity and pressure of the underseepage. Grouting cannot be of much value unless the permeability of the rock decreases with depth. If the permeability of the rock mass in the center of the valley and on the abutments is as great for hundreds of feet below and beyond the lateral extent of the grout curtain as it is near the rock surface, it can be readily shown theoretically that the grout curtain will have negligible influence on the underseepage. In this case, the situation is directly analogous to a partially penetrating cutoff wall through a pervious soil foundation (Sec. 6.3*a*). Fortunately, the permeability of most rock foundations does decrease with the distance from the surface since the joints are generally opened wider near the surface due to stress relief and weathering. However, in some types of rocks, such as columnar basalt and some limestones, the cracks do not become appreciably smaller with distance from the surface, and in these rocks grouting may be almost completely ineffective, even if the grout curtain extends hundreds of feet below the bottom of the valley and beyond the abutments.

Grouting is also of doubtful value in rocks where the permeability results from a great number of very fine cracks. Since with ordinary cement grout it is not possible to seal cracks which have a width much finer than 0.2 mm. (Ref. 240), in such a finely cracked rock, the ratio of ungrouted cracks to the total will be high and the grouting is likely to be almost useless. (See interesting discussions of this phenomenon in Refs. 384, 618.)

A conservative rule in common use is "When in doubt—grout, especially on large dams." On the other hand, the converse is almost equally true. If there is reasonable doubt in the mind of an experienced engineer that grouting is necessary, its omission will not endanger the safety of the dam but will result only in high foundation leakage under the worst circumstances which can develop. In borderline cases, if the maximum estimated quantity of seepage through the foundation is not likely to create a serious economic problem, or if it is desired to construct the dam as rapidly as possible, the grouting can be omitted and drains installed which will safely control the foundation seepage.

The Vigario Dam, which was constructed in Brazil in 1951, is an instructive example of a foundation in which the grout was omitted, Fig. 1.2:6. The Vigario Dam is a homogeneous embankment of clay with a maximum height of about 130 ft. The following description of the foundation investigation and underseepage control is given in Ref. 103.

The underlying gneiss contains many fissures and it was felt, at first, that a grout curtain would be required to prevent seepage under the dam. As a test, grout was pumped into a hole drilled into the foundation, and another hole was drilled only 1 meter away. No grout was encountered in the second hole during drilling. The second hole in turn required about as much grout per linear ft. as the first hole—an indication that a grout curtain would be very expensive.

It was decided, instead, to intercept the underground seepage by vertical filter wells spaced 6½ ft. apart and going down to sound bedrock at a depth of about 46 ft. These vertical filter wells were placed directly below the vertical sand filter in the dam and are thus directly connected to the horizontal filter and sump pit . . . . Rate of seepage will be measured in suitable observation wells and, if it becomes excessive, grouting may be resorted to.

Measurements of the seepage through the foundation of this dam when the reservoir was filled indicated that the seepage was very low (Ref. 457). This was in direct contrast to the fact that the upper portions of the foundation rock appeared badly cracked and large quantities of grout could be pumped into it. Many other similar experiences are available which leave little doubt that much of the rock grouting which has been carried out under earth dams has been a waste of money and effort.[1]

PLANNING GROUTING DETAILS

At the great majority of earth dams, a single line of closely spaced grout holes is used to create a narrow zone of rock in the foundation in which all the larger cracks are filled with cement. This relatively impervious wall of grouted rock is commonly called a "grout curtain." The grout curtain is placed at the center of the contact area between the impervious core of the dam and the foundation or slightly upstream from the center. Occasionally two, parallel grout curtains, located 50 ft. or more apart have been used under major dams. No rules can be given for the circumstances under which it is desirable to use more than one grout curtain. The authors believe that it is almost always preferable to devote the grouting energies to obtaining one very good grout curtain (even if two or more closely spaced rows of grout holes are used) and to seal the cracks in the horizontal

[1] See, for example, the comprehensive discussion of this subject by A. Casagrande, Ref. 618.

surface of the rock with slush grout rather than to construct two independent grout curtains at wide spacing.

In grouting alluvium (Sec. 6.2*d*), the efficiency of the grout curtain increases in direct proportion to the number of lines of grout holes used. In grouting rock in which the cracks are very fine, closely spaced, and erratic, undoubtedly the only way one can hope to create an effective grout curtain is to use several lines of holes. However, for foundations in which the rock consists of large solid blocks, and relatively large, widely spaced cracks, the grout travels a long distance and one line of holes may be almost as effective as multiple rows of holes.

Grout curtains are usually constructed in a vertical plane. Where the nature of the rock indicates that it is desirable, however, grout curtains are constructed at an inclination in order that the grout holes will cross more of the potential leakage cracks than could be intersected with the use of a vertical curtain. The decision to provide an inclined grout curtain is usually made in the design stage but may be modified in the field during construction as more information concerning the nature of the foundation is obtained. Holes can be drilled and grouted with standard equipment at inclinations up to about 60° (with the vertical) with little or no extra cost. Grout curtains are seldom inclined more than 30° from the vertical.

Almost all rock grouting under earth dams has been done with Portland cement. In a few cases, where large cavities were to be filled, sand-cement grouts have been used. Clay and clay-cement grouts have been used to a very limited extent up to the present time. (See, for example, Refs. 239, 639.) Because clay-cement grout is cheaper, less viscous, and adequately strong and durable when set, it will probably be used more widely in the future. Laboratory researches indicate that clay-cement grouts containing as little as 25% cement will satisfy all the conditions of stability and permanence required even for grouting large rock cracks (see, for example, Ref. 599).

The depths of grout holes should be chosen primarily on the basis of the results of the engineer's and geologist's study of the condition of the bedrock and on the purpose of the grout curtain. Depths of grout holes should never be chosen on the basis of some predetermined formula or on the basis of the average depths of grout curtains used under other dams. In the design stage the engineer determines the approximate depth of grout holes, the initial spacing of the holes, the location of the grout line in plan, and sometimes the details of a grout cap. He should normally wait until construction starts to determine the rest of the details of the grouting program, since the information which becomes available during construction concerning the permeability of the foundation and the details of the geology is more comprehensive.

The preparation of construction specifications and methods of payment for grouting is difficult because the extent and details of the work to be carried out are not known well in advance. Reference 238 contains a discussion of the main problems involved in writing grouting specifications with suggested guide paragraphs and a breakdown of pay items.

BLANKET GROUTING

"Blanket" grouting as opposed to "curtain" grouting (in which the holes are drilled on a line) consists of drilling and grouting a number of relatively closely spaced, shallow holes on a grid pattern over an area. Blanket grouting makes the upper portions of the bedrock stronger and less pervious. It is commonly used for the upstream portion of the foundation area under concrete dams and is used to a much lesser extent under earth dams.

Since even badly fractured rock formations normally have sufficient strength to support earth dams, blanket grouting is used most commonly only where the upper portions of the rock are so pervious that a single line of grout holes is not considered sufficient to create an impervious barrier at the surface. In some cases, the upper portions of the bedrock for a number of feet may be very pervious as a result of extreme weathering or fracturing. In these circumstances the engineer has a choice of either excavating down to the more impervious rock or blanket grouting an upper strip of the more pervious rock. The thinner the impervious core, the greater the desirability of blanket grouting the upper zone at the rock foundation. Some type of blanket grouting has been used under most of the high rockfill dams with thin impervious cores, at the juncture of the impervious core and the bedrock, Fig. 10.1:4.

DRAINAGE HOLES IN ROCK FOUNDATIONS

Until recently the drilling of open holes in the rock downstream from the grout curtain for the purpose of catching and controlling seepage has not been widely used for the foundations of earth dams. They have been used more frequently in the last few years, however, particularly on large dams. They should definitely be considered at

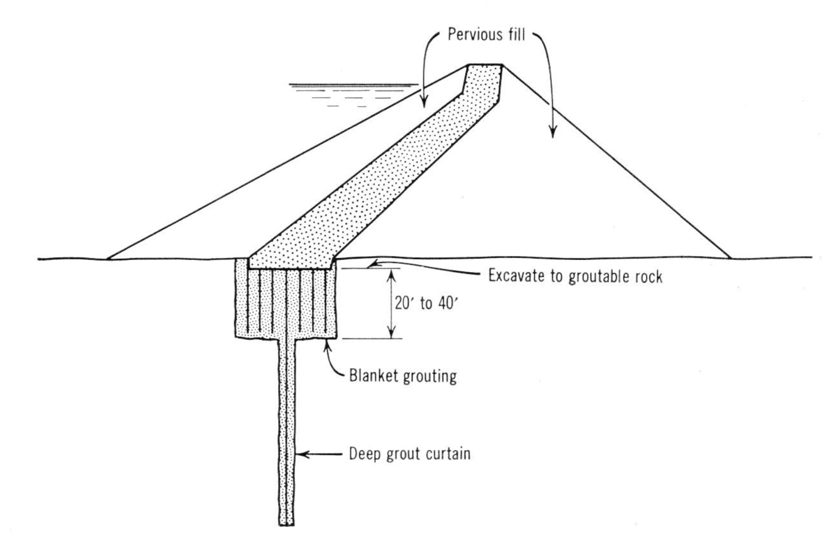

Fig. 10.1:4 Blanket grouting commonly used under thin core dams where upper surface of rock is fractured.

any site where the engineer has concern that high water pressures may develop in the fractures of the foundation or abutments near the downstream portion of the dam—and especially in horizontally bedded sedimentary rocks. A line of closely spaced drainage holes can serve the same important purpose of providing detailed foundation exploration for which the engineer frequently relies upon the grout holes.

In many rock formations, holes only 2 or 3 in. in diameter will be adequate to catch all the water which may seep past the grout curtain and to reduce the water pressures in rock seams to tolerable magnitudes. The drainage holes can be located underneath the dam at locations where the intercepted water can be discharged into the internal drainage system, or they can be drilled in the abutments and foundation downstream from the dam.[1] The advantage of the former method is that the water is intercepted further upstream, which is preferable from the standpoint of stability. The advantage of the latter method is that the leakage coming out of each hole can be observed and analyzed.

Under certain circumstances, both of these advantages can be

[1] See typical installation at Hirfanli Dam, Fig. 1.2:16, and Mammoth Pool Dam, Fig. 1.2:5.

obtained by using inclined holes drilled from the downstream toe. At Oahe Dam on the Missouri River, inclined holes with a diameter of 8 in. and lengths of up to 1,000 ft. were drilled to reduce pressures in fault zones in the Pierre Shale Formation, thus increasing the stability of high cut slopes around the stilling basin for the outlet works. For these drains an internal, plastic pipe (2.5-in. diameter) with circumferential drainage slots was used for the full length. The annular space around the plastic pipe was backfilled with graded gravel which was blown into the hole with air pressure by using a technique which allowed a comparison of the volume of gravel installed and the volume of the annular space.

In cold climates, consideration must be given to the probability that drainage holes which are discharged downstream from the dam may become sealed with ice and inoperative in the winter. Also, drainage holes may be plugged by a number of other actions such as an accumulation of deposition from the discharging leakage water and the squeezing together of the walls of the hole—especially in shale formations and rocks with clay seams. Consequently, if the drainage is a vital part of the design, means to measure the pore water pressures must be provided in order to study the long-time effectiveness of the drains and the owner must be prepared to drill supplementary holes if necessary.

GROUTING AND DRAINAGE GALLERIES

Internal galleries large enough for men to enter and work, provided for the purpose of grouting and collecting and discharging drainage water, are a common feature in concrete dams, and similar galleries or rock tunnels have been provided under many major earth dams by European engineers (Refs. 190, 219, 303, 307, 427); however, they have not been used to the authors' knowledge in the United States.[1] Some of the major possible benefits to be obtained from the use of galleries in earth dams are as follows:

[1] Perhaps the only major earth dam in the United States with such a gallery is the USBR's Alcova Dam constructed in Wyoming in 1938, Fig. 10.1:5. The foundation at Alcova Dam consisted of tilted sedimentary strata. Drainage holes were drilled to tap expected high pressures, and they discharged into the drainage gallery. In the event that it was necessary, additional grouting was planned from the gallery. The cost of the gallery was so great that the design was not used again by the Bureau Engineers (Ref. 147). Since the gallery at Alcova Dam is located below the tail water level, it stands full of water except when pumped empty.

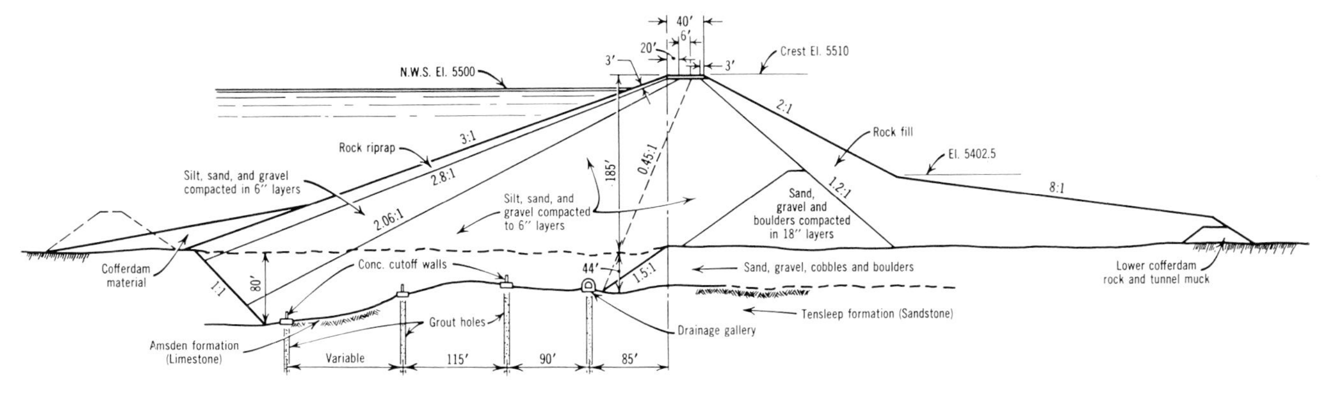

Fig. 10.1:5 Cross section of Alcova Dam showing internal reinforced concrete gallery (after Walker, Ref. 147).

1. The grouting of the rock can be carried out at the same time that the earth embankment is being placed without raising the primary objections which many engineers have about grouting rock foundations through the embankment itself. Often the use of the grouting gallery can appreciably shorten the time of construction and make the grouting timetable independent of the general construction plan. Drain holes or "bleeder wells" drilled in the rock foundation downstream from the grout curtain can be discharged into the gallery. Observation of the quantities of seepage in these drain holes indicates where foundation leaks are occurring.
2. By having access to the foundation under the dam when the reservoir is filled, the engineer can install any additional grouting or drainage required.
3. Higher grout pressures can be used because of the weight of the overlying embankment.
4. The drainage gallery can be used to house the outlets for piezometer lines and other types of measuring apparatus in a more convenient fashion than running them to the downstream toe of the dam.
5. If the gallery is constructed in the form of a tunnel below the rock surface along the longitudinal axis of the dam, it provides another great advantage as an exploration tunnel for the rock foundation. Some engineers recommend that such an exploration tunnel be made for all large dams on rock foundations in any event, simply because it is the only way to obtain a complete and detailed understanding of the nature of the rock along the grout curtain (see, for example, Ref. 648).

Drainage and grouting galleries have been used in perhaps half of the major earth and rockfill dams built in Europe in the last 15 years.[1] They are constructed either as a reinforced concrete conduit founded on the bedrock surface or as tunnels excavated inside the rock of the foundation and abutments, or of a combination. In order to allow the use of standard drilling equipment, the galleries should have a minimum vertical dimension of about 9 ft. However, in some cases galleries with a minimum dimension of 6 ft., 6 in., have been used with a slight increase in the time required for and the cost of grouting.

Most American engineers have not considered the use of galleries in or under earth dams. This is primarily due to the fact that there is no precedent for them in the United States. Secondarily, they would

[1] See Figs. 1.2:44–1.2:47 for some examples.

be somewhat more expensive because of the higher relative cost of hand labor and concrete. It is probable that, if the advantages offered by the use of galleries as listed above were carefully evaluated, it would be found that galleries would be desirable in some earth dams in the United States.

10.2 PROCEDURES FOR GROUTING ROCK FOUNDATIONS

While the construction of a grout curtain consists only of drilling the holes in the rock and pumping them full of grout, differences of opinion among engineers and engineering geologists on the best methods of grouting are probably as great as on any other aspect of earth dam design or construction. Until recently, there has not been much communication among the proponents of various grouting methods, and most knowledge about grouting was confined to a very few construction engineers.

In the last few years there has been sufficient experience published on the more controversial points to clear the air a good deal.[1] While there is still no unanimity of opinion on "correct" grouting procedures, the experience of the last decade or two has provided a common fund of knowledge which has dispelled much of the mystery formerly surrounding the subject.

Regardless of the methods used, the successful installation of a grout curtain is a process of trial and experimentation. The rock is systematically probed, washed, water tested, and grouted; the depths of the holes, the spacing of the holes, and the procedures for putting in the grout must be decided upon in the field as the work progresses. It is essential that the construction of the grout curtain be carried out under the continuous and immediate supervision of a competent engineer who is familiar with grouting theory and procedures.

At the start of the grouting program, no one can accurately predict the number of drill holes which will be required and the amount of grout it will prove desirable to pump into the rock. The program must be kept flexible and the engineer must have full control and authority to modify the procedures on the basis of the results obtained as the work progresses. During the grouting program, from the results of the large number of pressure tests and grout take records, the engineer learns much more about the foundation than he knew as the result of the exploratory program carried out during the design stage.

[1] See, for example, Refs. 231, 236, 237, 242, 491, 633.

The process of grouting normally consists of four elementary steps: (1) drilling the holes in the rock foundation; (2) washing out the drilled cuttings and all loose material possible from the sides of the holes; (3) pressure testing the holes with water to obtain an indication of the permeability of the rock; and (4) pumping grout into the holes under pressure to force it out in all directions into the cracks and crevices of the rock. In all the commonly used construction methods for grout curtains, it is customary to drill the first holes (called "primary" holes) at fairly wide spacing along the line of the proposed grout curtain. The initial spacing between primary holes is selected in advance by the designer on the basis of his preliminary knowledge of the rock conditions. In most dams the spacing of the primary holes ranges between 20 and 60 ft., although it might be somewhat more or less. The primary holes may or may not be drilled to their full proposed depth at the first drilling.

After the primary holes are drilled, washed, pressure tested, and grouted, secondary holes are drilled on the line of the grout curtain mid-way between the primary holes. Again, the secondary holes also may or may not be drilled to their full depth immediately. After the secondary holes are grouted, depending upon the results of the pressure tests and the grout "take," "tertiary" holes may be drilled at the midpoints of the distances between the existing primary and secondary holes. On most projects, the final spacing of grout holes at the completion of the job will be something between 5 and 10 ft. It has seldom been necessary to put down grout holes more closely spaced than 5 ft. except in areas of badly cracked rock such as fault zones. On the other hand, it is usually necessary to finish grouting holes spaced not further apart than 10 ft. This process of drilling primary holes and then intermediate holes at the midpoints is known as "splitting." This splitting process is done on almost all jobs.[1]

10.2a Stage Grouting and Packer Grouting

There are two main methods in common use for the construction of grout curtains. These are called the "stage method" and the "packer method." While each of these methods has advantages for use in cer-

[1] At reservoirs designed wholly for flood control, the water loss due to seepage through the cracks in the rock is not important, since the water retained in the reservoir is discharged rapidly in any event. Nominal shallow grouting is carried out more for the purpose of exploration for large cavities than for sealing small rock cracks. For such dams, grout holes are satisfactorily drilled in consecutive order without "splitting."

tain types of rock conditions, either can be used at most sites. The choice of the method used at a given site is made primarily on the basis of the preference of the individual engineer in charge, although in some cases the rock conditions may make it necessary to use one of the methods.

The principal distinguishing characteristic of the stage method is that the hole is drilled and grouted in stages of depth from the top with the grout being inserted through a pipe nipple which is fixed into the rock at the top of the hole.[1] The hole is usually drilled in two or three increments of depth. The depth increments are chosen as roughly equal (though not always) and are decided upon in advance on the basis of the preliminary knowledge of the rock conditions, and subsequently are modified according to the experience obtained. In the first operations the upper stage of the hole is drilled and grouted at a relatively low pressure. After the first stage of the hole is grouted completely, the hole is redrilled and extended down to the second stage. The new, greater length of hole is now grouted under a higher grout pressure; and if a third stage is used, the process is repeated.

Packer grouting is done almost in the reverse fashion. The holes are drilled at once to the full proposed depth. Subsequently, a "packer," which is a mechanical device for sealing off the hole at any elevation, is inserted and sealed against the walls of the hole at some distance above the bottom. After the packer is in place, grout is injected under relatively high pressure into the lower portion of the hole below the packer. Grouting of the hole is completed by gradually raising the packer in successive stages (10 to 20 ft. at a time) and grouting at successively lower pressures.

There are three types of grout packers in general use, as shown in Fig. 10.2:1. The leather cup packer seals itself against the walls of the grout hole when the pressure of the grout pushing up below the packer expands the leather cups. This type of packer can be used only where fairly hard rock is being grouted and where smooth drill holes of nearly constant diameter are obtained. It can be used for pressures up to approximately 750 p.s.i. The solid rubber packer is expanded and sealed against the walls of the grout hole by turning the adjusting nut which compresses and expands the rubber ring. This rubber packer can be used in a hole which is more irregular than that which is suitable for use of the leather cup packer. The solid rubber packer has been widely used and can stand high grout pressures. The pneumatic packer cannot stand such high grout pressures as the other two types.

[1] Because of the use of this nipple at the surface, stage grouting is sometimes known by the name "nipple" grouting.

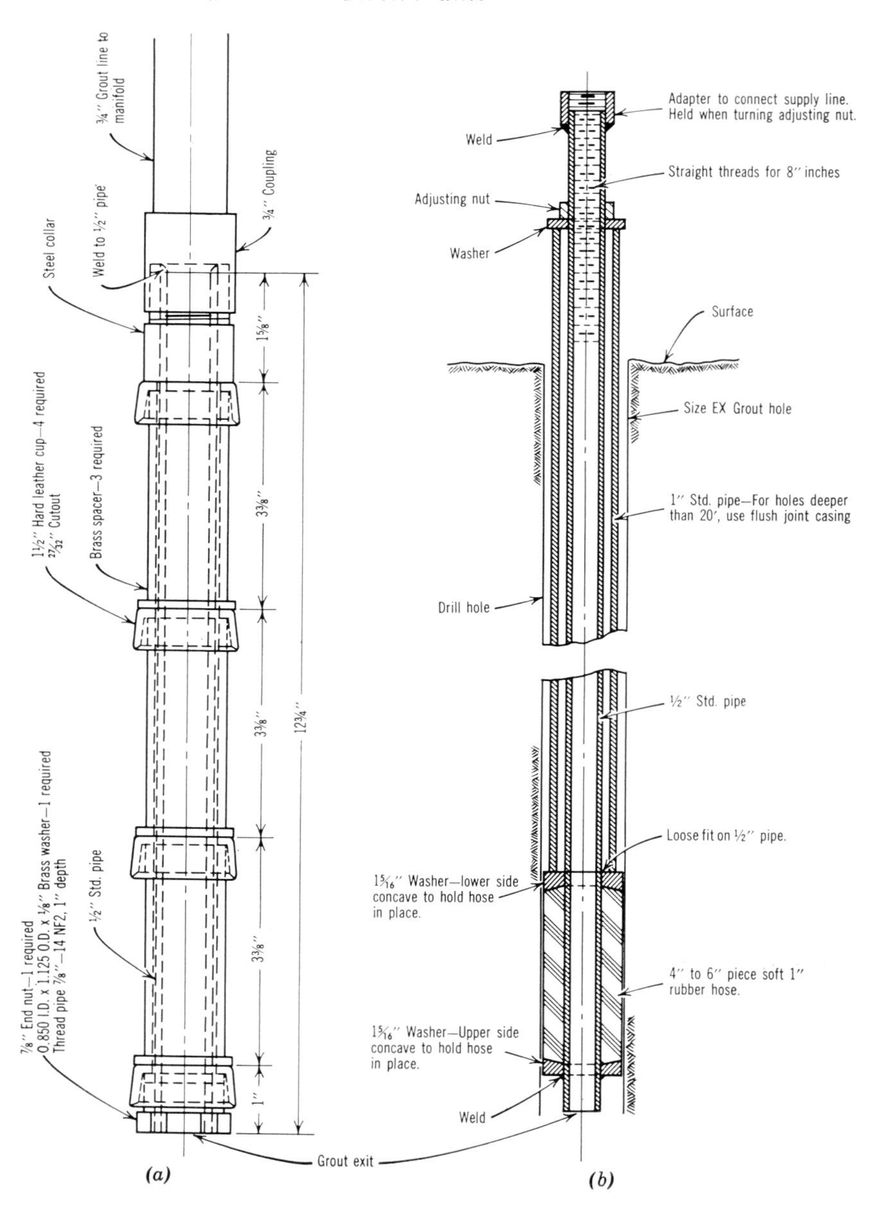

Fig. 10.2:1 Types of grout hole packers (after USBR, Ref. 231). (a) Leather cup packer. (b) Solid rubber packer. (c) Pneumatic packer.

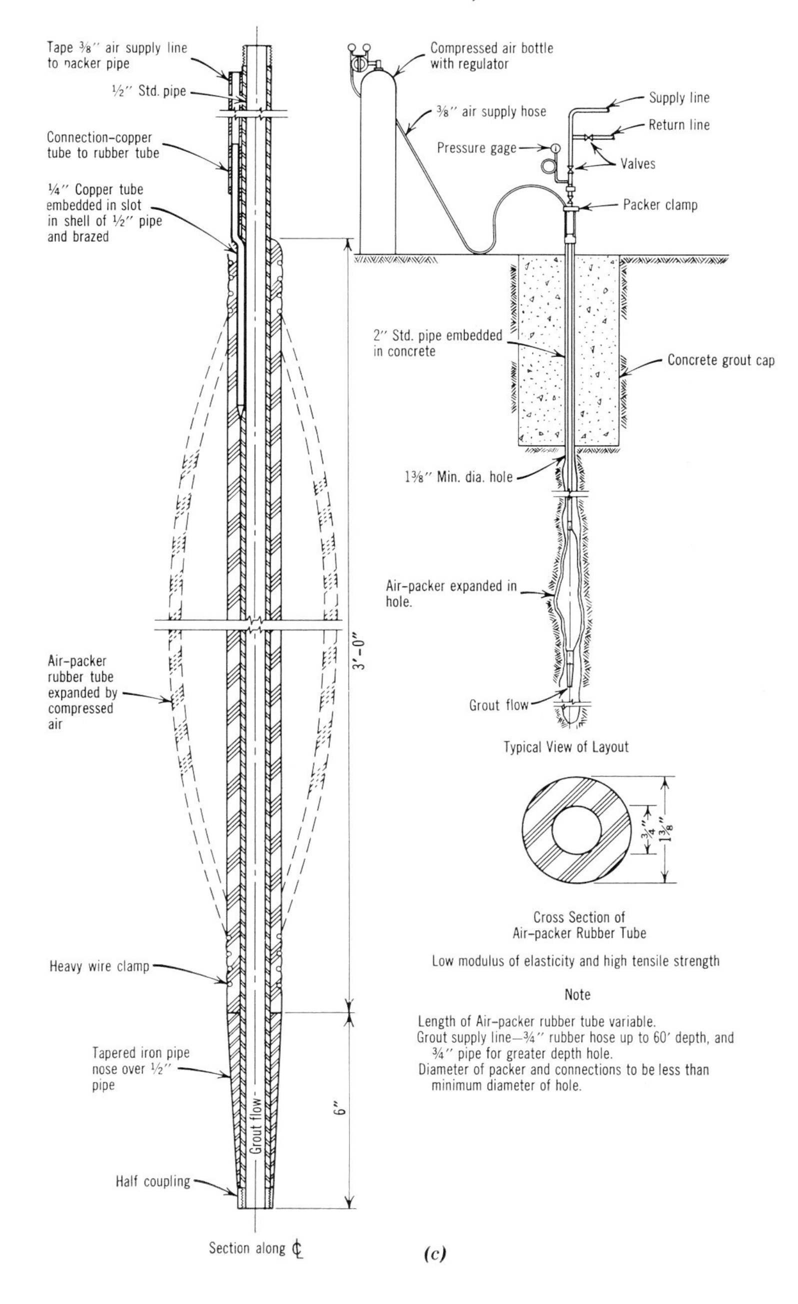
Tape 3/8″ air supply line to packer pipe
1/2″ Std. pipe
Connection–copper tube to rubber tube
1/4″ Copper tube embedded in slot in shell of 1/2″ pipe and brazed
Air-packer rubber tube expanded by compressed air
Heavy wire clamp
Tapered iron pipe nose over 1/2″ pipe
Half coupling
Grout flow
3′–0″
6″
Section along ℄
Compressed air bottle with regulator
3/8″ air supply hose
Pressure gage
Supply line
Return line
Valves
Packer clamp
2″ Std. pipe embedded in concrete
Concrete grout cap
1 3/8″ Min. dia. hole
Air-packer expanded in hole.
Grout flow
Typical View of Layout
3/4″
1 3/8″
Cross Section of Air-packer Rubber Tube
Low modulus of elasticity and high tensile strength
Note
Length of Air-packer rubber tube variable.
Grout supply line—3/4″ rubber hose up to 60′ depth, and 3/4″ pipe for greater depth hole.
Diameter of packer and connections to be less than minimum diameter of hole.

(c)

It is best adapted for use where the walls of the holes are irregular in shape. It is used widely for grouting in sedimentary formations with alternating layers of shale, sandstone or limestone but is generally limited to grouting where the pressures do not exceed 200 p.s.i.

Packers perform more satisfactorily in small grout holes because the total area and vertical pressure on the packer is less. They are usually used in EX (1½ in.) holes but are sometimes used on NX (3 in.) holes.

Some of the main advantages of the two grouting methods are listed below:

ADVANTAGES OF STAGE GROUTING

1. The equipment is simpler. No packers are needed and troubles with setting packers and leakage past packers are not encountered.
2. The rock is progressively grouted at higher pressures as the hole progresses downward and leaky zones are grouted as they are encountered.
3. The upper portions of the rock are grouted at higher pressures.
4. Less trouble is experienced with caving holes since the grout is always applied at the surface.

ADVANTAGES OF PACKER GROUTING

1. The work is usually carried out more rapidly since the hole is drilled only once and it is not necessary to move the drill rig on and off the hole.
2. Better information is obtained concerning the elevations of the hole which take the largest quantities of grout.
3. Better control of grout leakage through cracks in the surface of the rock is obtained.

Although strong preferences exist between groups of engineers who advocate one or the other of these methods as the most suitable for general foundation grouting, both methods are frequently used with some modification on many jobs. The foundations of most earth dams constructed by the U.S. Corps of Engineers are grouted primarily by the stage method. They do, however, occasionally use packers near the surface, when grouting the first stage, to control surface leakage. Other engineers who favor the use of stage grouting often use parkers to isolate and grout especially leaky zones. The engineers of the USBR favor the packer method; however, they use stage grouting methods where caving holes are met or in other conditions where it is difficult to set the packers.[1]

10.2b Drilling Grout Holes

SIZE OF HOLES

Most grout holes are drilled with an EX bit which gives a hole of approximately 1½ in. in diameter. In previous years larger holes were drilled as it was felt that the hole size might be a factor in obtaining better grout injection. Later experience indicated that the diameter of the hole had little effect within reasonable limits on the amount of grout that the hole would take. At some dams where 1½-in. diameter grout holes have been specified by the engineer, the contractor has drilled larger holes at his own expense in order to avoid difficulties and delays due to the hole caving and squeezing.

It is frequent practice to require that a small percentage (in some cases 10%) of the holes be cored in order to explore the results of the grouting program.[2] In some cases these exploratory holes are used for later grouting and in some cases they are not grouted. For exploratory holes, an NX bit which gives a core of about 2 in. in diameter and drills a hole of about 3 in. is the minimum size ordinarily used. Larger core barrels are used if necessary in order to obtain good core recovery in rock which is difficult to core.

DIRECTION OF HOLES

Holes are drilled for a grout curtain to form a plane in the rock foundation. Grout curtains are usually vertical but may be inclined to

[1] Grouting rocks in which the holes cave or squeeze is especially difficult with the packer method since the holes become too small to allow the packer to be installed or the squeezing prevents the packer from being withdrawn. Rocks of this type may be difficult to grout with the stage method also since the hole may not remain open long enough to be grouted the full depth. In severe cases, with the stage method, a ½-in. grout pipe is extended to within a few feet of the bottom of the hole and a stuffing box is used at the surface of the ground. This procedure, which is called "circuit grouting," assures that the grout is reaching the bottom of the hole at full pressure even if the rock squeezes in around the ½-in. pipe.

[2] At a few dams, all grout holes were cored in order to learn as much as possible about the progress of the work.

intersect more rock fractures. For maximum efficiency the holes must be drilled at the same inclination in order to form a plane surface. If the deviation of any individual grout hole from the plane surface of the grout curtain exceeds approximately one-third the minimum distance between grout holes, the continuity of the grout curtain may be broken.

Holes on sloping rock abutments may be drilled either vertically or inclined at an angle which is approximately normal to the slope of the abutment. Where the latter system is used, it is sometimes necessary to add some additional grout holes at the juncture of a horizontal rock foundation and the abutment in order to compensate for the "fanning" effect. Holes drilled normal to the abutment slope are usually preferred since shorter lengths of grout holes are needed for a given depth of grout curtain. The steeper the abutment, the greater the advantage of the perpendicular grout hole. At some dams when difficulty was experienced in forming a tight grout curtain with the inclined holes (perpendicular to the abutment surface), additional vertical holes which crossed the inclined holes have been drilled and grouted on the same line. In badly fractured areas, two or more closely spaced rows of holes may be required to seal the grout curtain using both vertical and inclined holes.

DRILLING EQUIPMENT

Holes for a grout curtain are usually drilled with rotary drilling rigs using diamond bits. Holes drilled with percussion drilling equipment are much less satisfactory. Some engineers prohibit the use of percussion drilling and others limit its use to shallow holes in rock formations of the type which produce granular cuttings rather than slimes. Experience has indicated that when grout holes are drilled using compressed air and percussion drilling equipment, the drill cuttings frequently plug and seal cracks in the rock adjacent to the hole. Nevertheless, much grouting has been done through holes drilled with percussion type equipment, especially for shallow blanket grouting. Percussion equipment is seldom used for holes in excess of 40 ft. in depth.

Most grout holes are drilled with rotary rigs using plug bits which grind up all the rock without taking a core. The primary advantage of plug bits is speed since the drill tools do not have to be removed from the hole until the final desired depth is reached. In some very hard rocks, contractors find it more desirable to use standard core barrels for drilling the holes since there is less area of rock to be cut and greater speeds can be obtained even though the cores have to be pulled from the holes at periodic intervals.

Rotary holes are usually drilled with clean water. Careful observations are kept of losses or gains of drilling water which indicate leaks or artesian flows. Where large losses or gains of drill water are encountered, the pervious length of hole should be tested and grouted before drilling is continued. Rotary drilling using compressed air to remove the cuttings has been very satisfactory in some cases. Drilling mud should never be used.

Drill rigs of all kinds have been used satisfactorily for grout holes. Portability of equipment and fast drilling speed are always essential to an economical grout program. Truck-mounted drill rigs are satisfactory where the terrain is such that they can travel easily. Special small rotary rigs with air motors have proved very adaptable to drilling grout holes, Fig. 10.2:2. On steep rock abutments, wooden staging may be required for the drilling and grouting operations, Fig. 10.2:3.

In past years, some contractors have coated the drilling rods with grease to reduce the wear on the rods. At the present time the use of rod lubricants of this type is considered very poor practice and is generally prohibited. Where compressed air has been used to remove the

Fig. 10.2:2 Typical light-weight, air-powered rotary drill rig most commonly used for drilling grout holes.

Fig. 10.2:3 Typical staging constructed for drilling and grouting on steep abutments and grouting plant.

drill cuttings, rather than drilling water, water soluble rod lubricants have been used with some success.

10.2c Washing and Pressure Testing Grout Holes

Before grouting, the holes are washed and pressure-tested with water. This work may be done in one operation but is usually done in two separate operations using different equipment.

The purpose of washing the hole is twofold: (1) to clean the drill cuttings out of the hole; and (2) to remove, as far as possible, soil and other loose material from the cracks in the rock adjacent to the hole. In the construction of grout curtains under earth dams, where the grouting is intended only to make an impervious zone of rock, no attempt is made to wash the soil in the cracks of the rock from one hole to another.[1] Most engineers believe it is preferable to leave the natural silts or clays in the fissures of the rock and compact them by the pressure of the subsequent grout injection. In curtain grouting, washing is intended only to remove soil from the fissures in the rock in the immediate area of the hole being washed.

The washing procedure is the same for both stage and packer grouting. The wash water is pumped to the bottom of the hole through the drill rod or other pipe and no pressure connection is used at the rock surface. The cuttings and the soil from the crevices in the rock adjacent to the hole are washed to the surface by the upward flow of the water discharging from the hole. The water is pumped into the hole either with the pump on the drill rig or with the pump on the grouting apparatus. In some cases, the pressure on the construction water supply line is sufficient for washing, so that the contractor does not have to tie up his other equipment.

Washing should be continued until the discharge water is running clear. The extent to which soil can be removed from the crevices of the rock adjacent to the hole depends on the character of the joint filling material. Tough, hard clays cannot be removed easily, and it is usually not desirable to try to do so. On the other hand, cohesionless silts, silt-

[1] In shallow holes used for blanket grouting under concrete dams, where the grouting is carried out for the purpose of strengthening the rock, frequently an attempt is made to force the wash water from one hole into adjoining open holes or out through the surface cracks in the area. Using this procedure, which is called "pressure washing," much higher pressures are used for wash water than are used for washing holes for the construction of a grout curtain under an earth dam.

like materials, and sandy materials may be washed from the crevices in the rock to a distance of many feet from the grout hole.

Where considerable soil is being removed from the cracks in the rock adjacent to the hole, the washing process can be assisted by pumping air into the wash water with an air compressor. This causes a boiling action in the hole which in some cases is more effective in removing the soil from the rock fractures than the water alone.

After washing, the next step is pressure testing in which water is pumped into the hole under pressure and the volume of inflow measured. The results are used in a given hole to plan the best method of grouting the hole. There are no standard procedures for pressure testing. Before starting, the engineer already knows something about the permeability of the rock from his examination of the rock cores and from the observations of loss of drilling water in the grout hole. From the preliminary knowledge of the permeability of the rock foundation, the procedures for the pressure testing should be planned in such a way as to give the most information possible.

Somewhat different methods of pressure testing are used for stage and packer grouting. For the stage method, the pressure is applied to the top of the hole through the nipple and each stage is tested before grouting, Fig. 10.2:4. Where packer grouting is used, the pressure testing is commonly carried out from the bottom of the hole upward by setting the packer at successively higher elevations. Using either method, zones which are suspected to be especially pervious are isolated with packers and pressure-tested individually.

Fig. 10.2:4 Typical hookup for pressure testing in the stage method where the pressure is applied through the nipple at the top of the hole showing piping connections, flow meter, and pressure gauge.

The pressure testing is usually carried out with the grouting plant immediately before the hole is grouted. This has several advantages. It permits testing the grout pump lines and connections. Also, the water pumped into the rock during the pressure test wets all the fractures into which it is desired to pump the grout, which minimizes problems caused by the tendency of some dry rocks to absorb water from the grout.

Pressure testing is usually carried out with a pressure approximately equal to the maximum grout pressure which is subsequently used. The water pressure should be at least equal to the maximum reservoir head expected at the location and preferably 150% the maximum reservoir head. The tests are continued at full pressure for 10 to 15 minutes and the volume of water intake measured. Sections of hole which take less than about 1 gal. of water/min./10-ft. of length will take very little grout and grouting can be omitted.[1]

When pressure testing the upper sections of the bedrock, considerable water leakage may develop through fractures in the rock surface adjacent to the hole being tested. If the leakage exceeds the capacity of the pump, and it is not possible to build up the pressure desired, it is necessary to seal the major surface leaks. Water leaks are sealed in the same manner as grout leaks as discussed in the following section.

10.2d Grouting Procedures

OBJECT OF GROUTING

The object of the grouting is to seal with grout all the open cracks in the bedrock for a narrow width adjacent to the plane of the grout curtain. The most effective grouting in a given hole is that in which all the open cracks in the rock within a large radius from the hole are uniformly filled with grout. In general, to avoid drilling an excessive number of grout holes, the best grouting procedures are those in which the maximum quantity of grout is pumped into each hole. The only

[1] The only widely quoted criterion relating the results of water pressure tests and the desirability of grouting is that proposed by Lugeon who recommended that water tests be carried out using a pressure of 10 atmos. and that grouting be done if the water loss exceeded 1.0 liters/meter of hole/min. This latter unit is sometimes called the "lugeon" and it corresponds roughly to a coefficient of permeability of 1×10^{-5} cm./sec. This criterion was proposed for application to the foundations of concrete dams over 30 m. in height and should be considered of only historical interest by the earth dam engineer. Refs. 8, 581, 632.

exceptions to this arise under those circumstances where the grout pressures are so high that the ground surface is lifted and/or where the grout may be traveling through some rock crevices a long distance from the grout curtain.

GROUTING EQUIPMENT

As shown in Fig. 10.2:3 and 10.2:5, a grouting plant consists simply of a grout mixer, a grout agitator, a grout pump, and piping with connections to the grout hole. Two types of piping systems are used for the grout. The circulating system (solid lines in Fig. 10.2:5) has a return line to the grout agitator tank through which all grout which does not go into the hole is returned. With the single-line system (dotted lines), all the grout pumped goes into the hole or is wasted through the blow-off valve shown. The circulating system is used almost universally at the present time in the United States, though a few engineers still strongly prefer the single-line system. In Europe, in the last few years, a number of major dams have been grouted using single-line grout systems. The primary advantage of the circulating system is that the grout never has an opportunity to set up in the supply pipe. Engineers who prefer the single-line system believe that the high surges of pulsating pressure in the grout supply line coming directly from the piston stroke of the pump are important in the process of getting the maximum amount of grout into the rock fissures. These peaks of pulsating pressure generated in the fluid grout in the hole are largely dissipated through the return line when the circulating system is used. Since the value of this attribute of the single-line system is problematical and since it is much more difficult to use and control, the authors prefer the circulating system.

The agitator in the storage tank is used only for keeping the grout in uniform suspension. Various types of pumps have been used satisfactorily for grouting and in recent years many improvements have been made in grout pumps. Air-driven, positive displacement type of pumps with special fittings for cement grout service have been the most popular and satisfactory. In a few cases centrifugal pumps have been used for low-pressure grouting and, at present, there are centrifugal pumps available specially designed for high-pressure grout service.

Grout mixers have usually been simple tanks with slowly rotating, mechanically operated paddles into which the cement and water were dumped. Recently, high-speed, impeller-type "colloidal" mixers,

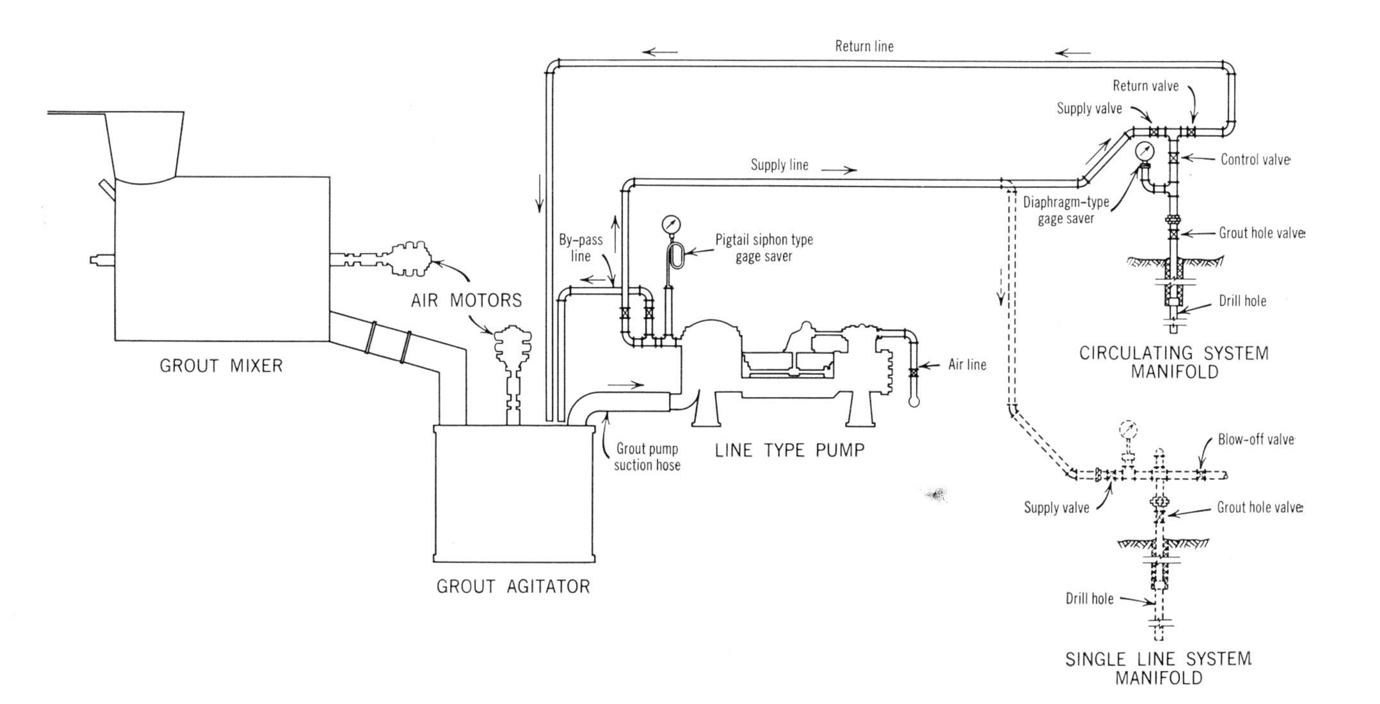

Fig. 10.2:5 Elements of typical grouting plant (after USBR, Ref. 231).

which are much more effective in eliminating lumps of cement from the grout, are coming into wide use and should be specified. A water meter is needed in the system to measure the water coming into the grout mixer. Pressure gauges must be protected from direct contact with the grout. This is done by connecting the gauge and the grout line with an oil-or grease-filled pipe. Grout lines exposed to the sun should be covered with wet cloths to prevent them from getting too hot, which sometimes causes early setting of the cement and poor grouting results. In cold weather, grout must be kept from freezing before it is pumped into the rock.

Where the stage method of grouting is to be used, the grout supply line is connected to the grout hole through the nipple which is sealed into the surface of the rock as shown in Fig. 10.2:6. The grout nipple is usually a 1½-in. diameter pipe with one end threaded and the other end flared for caulking. The length of the nipple depends upon the soundness and permeability of the upper part of the rock and normally varies between 2 and 5 ft. The pipe nipples are usually sealed into the surface of the rock with thick cement grout. Where the hole is to be drilled immediately, before time can be allowed for the grout

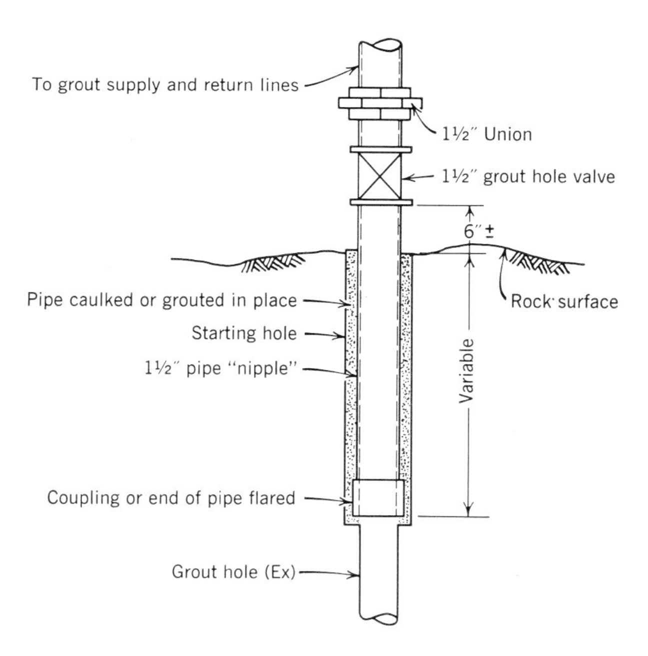

Fig. 10.2:6 Typical "nipple" used for stage grouting.

to set, the nipple can either be caulked into the hole with lead wool, or sealed with molten sulfur.

PRESSURE AND CONSISTENCY OF GROUT

The maximum permissible grout pressure should be used in each hole and at each depth grouted. The maximum grout pressure is limited to that value which will not cause appreciable displacement of the rock formation and "spring" its fracture system. No reliable, quantitative rules can be given for the determination of the maximum grout pressures which can be safely used because there are so many indeterminate influencing factors. For horizontally bedded, sedimentary rocks, it is usually not safe to use a grout pressure at any depth which is much greater than the weight of the overlying rock. The maximum permissible grout pressure depends on the type and the continuity of the cracks in the rock, the strength of the rock, the consistency of the grout being used, the permeability of the rock, the shape of the valley, the length of hole being grouted, the previous grout history of the rock, and on a number of lesser factors. For massive hard rocks, grout pressures equal to ten or more times the weight of the overlying rock have been safely used.

The only direct and reliable method to establish maximum permissible grout pressures at a given site is to make a series of tests in which the grout pressure is raised in increments until failure occurs. The tests are made with a grout that is sufficiently thin that it could be pumped indefinitely without sealing the hole. Several tests are made with holes of different lengths and packers at different elevations. At low pressures, the grout is forced to flow out through existing fissures in the rock and the rate of grout take is roughly proportional to the pressure. At a pressure above which the rock is being broken or lifted by the grout, there will be a sudden increase in the rate of grout take because of the new fissures which are opened. By plotting the results as shown in Fig. 10.2:7 a fairly good idea can be obtained of the grout pressure which should not be exceeded.

Considerably higher grout pressures frequently are used in Europe than would be considered necessary or safe practice in the United States, and differences of opinion on this subject sometimes arise when American and European engineers have occasion to collaborate on a project. After observing several of these, the authors have concluded that the difference in practice is probably due primarily to the fact that the grouting procedures and standards in Europe are left

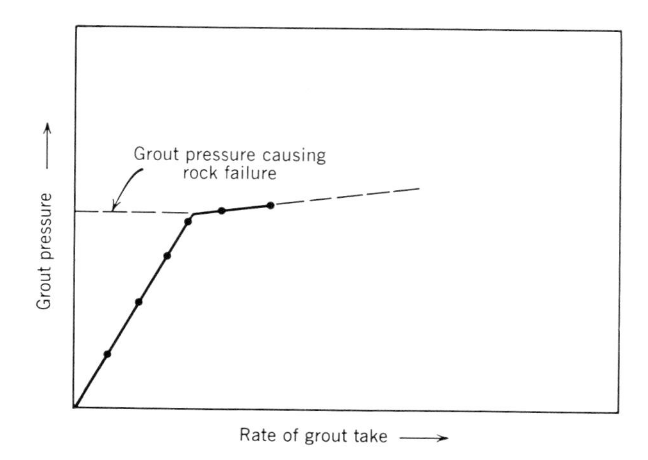

Fig. 10.2:7 Plot of results of grout tests made to determine maximum safe grout pressure.

more often to the specialized contractor. The authors definitely favor the arguments for low pressures since the consequences of using pressures which are too low are much less serious than the possible results of using excessively high pressures.

Experience has demonstrated that the maximum amount of cement grout can be pumped into a grout hole when the maximum pressure is applied at the time that the grout is first introduced into the hole and that this maximum pressure should be maintained throughout the grouting as well as possible. Conversely, the amount of grout which a hole will take is minimized by starting the grouting at the lowest pressure at which the hole will accept grout and then allowing the pressure to build up as the resistance increases. High pressures in the early part of the grouting are obtained by speeding up the grout pump.

Grout mixes or consistencies are given in terms of the water-cement ratio on the basis of a volume measure. The consistency of the grout mix is varied throughout the program in an effort to get the greatest possible quantity of cement into the rock adjacent to each hole. At the great majority of sites, grout consistencies ranging between 10:1 (10 volumes of water to 1 volume of dry cement) and 1:1 are used. A consistency of 20:1 is a very wet or thin grout, and 0.5:1 is a relatively thick grout.

For most rocks it is desirable to start grouting with a consistency between 3:1 and 5:1. If the pressure develops rapidly and the hole takes little grout, the grout mix is thinned. If it is difficult to build up grout pressure with the maximum speed of the pump, the mix is thickened. The object is to keep the grout feeding into the hole at a uniform rate and at not less than about 75% of the final pressure. It is always better to err by starting a hole with grout which is too thin, since a grout which is too thick may block the main cracks and cause the loss of the hole.

If, because of the size and continuity of the fractures in the rock, the pump cannot build up the desired pressure even with a thick grout, grouting should be stopped and the grout allowed to set. Subsequently, the hole should be redrilled and regrouted, or additional grout holes should be drilled and grouted in the immediate vicinity.

If a hole takes a large quantity of grout over a long period of time, it is desirable to stop pumping grout occasionally and pump in a quantity of clean water (5 to 10 ft.^{3}) at intervals in order to keep the grout hole and fissures open.

The object of the grouting is to fill the cracks in the strip of rock along the grout curtain to a width of 5 to 10 ft. Consequently, by estimating the volume of the groutable cracks in the rock for this width, one can obtain a rough estimate of the grout which would be needed if it could be confined to this zone. Actually, of course, the grout travels further and does not fill the cracks uniformly so that an estimate based on such a calculation can only be considered a very rough guide. For most jobs, average grout quantities vary between 10 and 600 lb. of cement per lineal foot of grout hole, depending on the permeability of the rock.

One of the most common errors in grouting is to pump thousands of sacks of cement continuously into a single hole. Such an action is always wasteful since it can only indicate that the grout is traveling long distances from the grout curtain. In some jobs, 50,000 sacks of cement and more have been pumped into single holes. It is always advisable to stop grouting after several hundreds of sacks of thick grout have been pumped, and to continue after the grout has set and to use nearby holes and other techniques to confine the grout to the zone of rock adjacent to the grout curtain.

The grout "take" sometimes does not correlate with the results of the water pressure testing. At sites where the rock takes large quantities of water but will not take grout, the addition of a small quantity of silicate gel under pressure (Ref. 25) sometimes has a lubricating effect on the fissures which subsequently permits larger grout takes. Also, the addition of small quantities of bentonite (usually less than 5% by weight of cement) has an appreciable influence in decreasing the viscosity of the grout mixture and sometimes has increased greatly the grout take in rock with fine cracks.

Various criteria are used to determine when the rock has been adequately grouted. Some specifications require that grouting shall continue to refusal, which means that the grout pressure is applied until the grout sets up in the hole. A reasonable criterion is that grouting shall be considered completed when the hole or section of hole being grouted will not take more than 1 $ft.^3$ of the grout mixture in 5 minutes under the maximum grout pressure.

STOPPING SURFACE LEAKAGE

In grouting the upper portions of the rock, there is a tendency for the grout to flow upward and seep out through cracks in the rock surface adjacent to the hole. Where the fractures are large, this surface leakage may be so great as to prevent the development of the desired grouting pressures. The following methods have been used successfully to control this surface leakage:

1. Constructing a concrete "grout cap" or concrete wall in a trench excavated in the bedrock.
2. Caulking the leaks in the surface of the bedrock with wooden wedges, dry-packed cement grout, oakum, burlap and other materials, Fig. 10.2:8.
3. Covering the cleaned rock surface with a thin layer of gunite or slush grout.
4. Pumping very thick grout into the hole until it spills out on the rock surface and then letting it set up before redrilling and grouting the hole.
5. Constructing a temporary earth embankment (often called "grouting platform") with height of 5 to 20 ft. along the line of the grout curtain and grouting through the embankment.[1]

Where the condition is not too severe, a concrete filled trench (grout cap) of dimensions approximately as shown in Fig. 10.2:9 usually suffices to control the leakage in the upper portion of the bedrock. The grout cap serves several desirable functions: (1) It constitutes an impervious wall through the upper few feet of broken rock which are the most difficult to grout; (2) it forces the surface leakage to travel a longer distance; and (3) it eliminates troubles with caulking

[1] In a few cases, this same effect was obtained by grouting the rock before the cutoff trench was excavated and the procedure proved quite satisfactory.

Fig. 10.2:8 Caulking surface grout leaks with wooden wedges driven into the rock surface in vicinity of grout curtain. (Courtesy Rene Marchand)

pipe nipples into the broken surface rock. Because of these advantages some engineers (including the USBR) use grout caps for almost all grouting and include the minimum dimensions in the design drawings, Fig. 10.2:10. Often pipe nipples are cast directly in the concrete of the grout cap as it is poured; however, problems are sometimes ex-

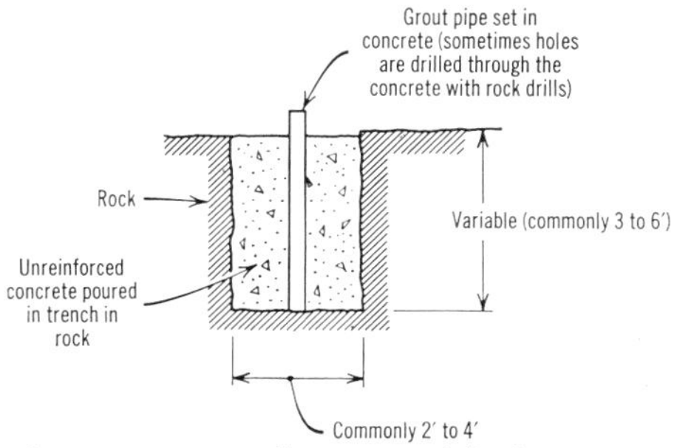

Fig. 10.2:9 Typical grout cap used to control leakage of grout through upper few feet of broken rock.

Fig. 10.2:10 Typical grout cap used in USBR earth dams—Cachuma Dam, California. (Courtesy USBR)

perienced in keeping these pipes vertical. As a consequence, many engineers prefer to drill holes through the concrete grout cap and either caulk in a pipe nipple or use a packer at the top of the hole and no nipple.

Where the surface rock is badly broken to a considerable depth, a gunite blanket performs very satisfactorily to control surface leakage. At the surface of badly broken rock, the grout tends to flow upward within a relatively confined area around a grout hole. When the surface of the rock in the vicinity of the grout hole is covered with an impervious layer of gunite, the grout is forced to move horizontally under the gunite blanket and spread out to create a much wider zone of grouted and impervious rock, Fig. 10.2:11. The surface of the rock is thoroughly cleaned with hand labor and water jets (Sec. 10.3) for a distance of 10 ft. or more on both sides of the grout curtain. This cleaned surface is covered with a continuous layer of gunite (or sanded slush grout) having a minimum thickness of about ½ in. and greater thickness as required to fill local surface depressions. While this gunite layer has little weight and theoretically cannot withstand large uplift forces, actually (because it clings tenaciously to the rock surface and the area over which the grout pressure can act is small) the gunite layer can resist high grout pressures in the rock cracks without heaving.[1]

Where the surface cracks are not too large, they sometimes can be filled by pumping thick grout from the hole. The grout is pumped under low pressure and allowed to spill out of the main cracks, to fill the surface depressions, and to set up before continuing.[2] For bad leaks, intermittent grouting may be used. It may sometimes be necessary to grout the upper reach of the hole several times in this manner before sufficient pressure can be built up.

The construction of an earth embankment along the line of the grout curtain has two major advantages: (1) it provides weight on the rock surface which allows the use of higher grout pressures in the upper part of the hole and (2) it confines the grout and prevents it from spilling freely out of the rock surface cracks. This procedure,

[1] See Ref. 49 for a description of the successful use of this method for grouting the upper 30 ft. of badly cracked and jointed basalt with high pressures at the Kenney Dam in British Columbia.

[2] Grout overflowing at the surface must be removed before compacted earth of the dam is placed against the rock unless the rock surface is carefully cleaned in advance to assure that the grout spillage is not underlain by loose, pervious material.

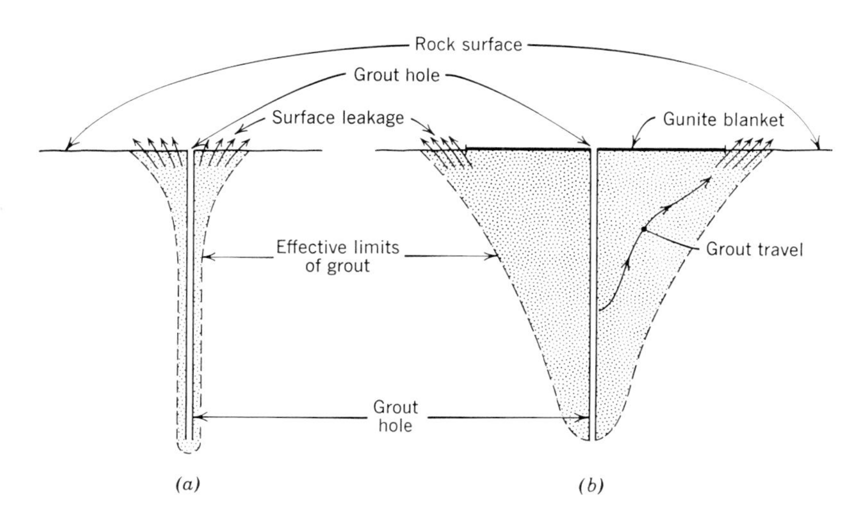

Fig. 10.2:11 Influence of a surface gunite blanket in obtaining a wider grouted zone at cracked surface. (a) Without sealing rock surface cracks (b) With surface sealing.

which is also used for grouting alluvium (Section 6.2*d*) has been used with increasing frequency in recent years for rock foundations and has proved to be both practical and economical. Most frequently temporary embankments of fine-grained soil (at least at the bottom) have been used and then removed after the grouting is completed. In a few cases, the grout platform has been constructed as part of the dam and then left in place as discussed in the following.

10.2e Grouting Through Completed Earth Dam Embankments

For earthfill dams, the grout curtain is usually completed before the dam is constructed, and most engineers strongly prefer this procedure. However, the foundations of many dams have been grouted through the completed or semicompleted embankment. This is usually done as a construction expediency, or because the need for the grouting became apparent only after the reservoir was filled. A few engineers prefer to grout foundations after completing the embankment. Some of the main advantages and disadvantages of these two methods are as follows:

ARGUMENTS AGAINST GROUTING THROUGH COMPLETED EARTH DAM EMBANKMENTS

1. Bore hole washing and pressure testing cannot be carried out as vigorously because of the danger of eroding the soil at the contact between the foundation and the embankment.
2. The use of grout pressures which are too high is more dangerous since the embankment may be heaved and cracked.
3. Less control can be exercised over the travel of the grout and, if there is any underseepage, the grout will tend to be carried downstream where it is not desirable. In some cases, the grout has found its way up through the rock crevices into drains in the dam, reducing the effectiveness of the drainage system.
4. Higher cost.

ARGUMENTS FOR GROUTING THROUGH COMPLETED EMBANKMENTS

1. The total construction time period may be shortened.
2. The grout will enter and seal any spaces or holes between the fill and the rock foundation which have been left.

3. The grout may find its way into any cracks or shear planes in the lower part of the embankment which have developed during construction and seal them.
4. In a site where there is doubt whether a grout curtain is needed, it can be left out until the reservoir is filled and the leakage can be observed. If the leakage is too high, the rock can be grouted through the dam. If the leakage is tolerable, the whole cost and construction time for grouting will be saved.

When grouting is done through earth dam embankments, steel casing should always be used and sealed well into the surface of the bedrock. Packers should always be used.

10.3 SURFACE TREATMENT OF ROCK FOUNDATIONS AND ABUTMENTS[1]

10.3a Purpose and Methods of Surface Treatment

After the construction of the grout curtain (if any) is completed, the rock surface, against which the impervious core of the dam is to be compacted and sealed, must be treated. The proper execution and control of the foundation treatment are among the most important aspects of the construction of a safe earth dam.

Treatment of soft rocks under the dam core usually involves fewer problems and less difficulty than treatment of hard rocks. If the rock is soft enough to be excavated with earth-moving equipment, treatment of the abutments and foundation consists simply of excavating a smooth surface against which the impervious soil of the embankment core can be compacted with heavy rollers, Fig. 10.3:1.

Some soft rocks, which are suitably hard and impervious in their natural state, deteriorate rapidly to less than satisfactory materials when exposed by excavation to the atmosphere. This problem has been solved satisfactorily under the dam core by excavating the deteriorated surface layer in small strips directly before the rolled earth embankment is placed or by protecting the exposed surface with sprayed coats of asphalt.

For earth-rock dams, the deterioration of soft rock foundations under the rockfill shells may create a more important and difficult problem than the treatment of the foundation under the core. The

[1] An excellent discussion of this subject with illustrations is given by T. F. Thompson (Ref. 223).

Fig. 10.3:1 Compacting the rolled-earth core against the smoothly excavated surface of a horizontally bedded shale abutment with 50-ton rubber-tired roller.

soft rock under the rolled-earth core will not continue to deteriorate significantly after the foundation is covered since it will be protected from the atmosphere. However, the portions of the foundation under the rockfill embankment sections are exposed to the air circulating in the voids of the rock and may continue to soften during the lifetime of the dam. This has been an especially important and difficult problem for high earth-rock dams in which the development of a weak plane on the surface of the foundation may require that the embankment side slopes be made as flat as those for an earth dam. No satisfactory solution for all sites of this type is available. Each problem must be studied individually and extensive field tests of methods to prevent the rock from deteriorating may be justified. There is, however, enough experience from such tests to indicate that even a thin coating of asphalt is generally much more satisfactory than gunite or concrete slabs in protecting the soft rock from deteriorating during construction.

The surface of hard rock foundations must also be smoothed up sufficiently so that the heavy rollers used for compacting the core of the dam can operate as close as possible to the rock and a minimum of "mechanically tamped" earth embankment is necessary at the abutments. Hard rock surfaces are treated by removing promontories and overhangs by light blasting or jackhammers, filling depressions with concrete or hand tamped earth and, if the rock is fractured, cleaning out and filling the surface cracks with cement grout.

The purpose of cleaning and grouting surface fractures is to protect the earth core from possible erosion caused by leakage water flowing through the rock cracks. Secondly, by creating an impervious skin on the surface of the rock, the total quantity and pressures of the underseepage through the rock is reduced. Many engineers do not hesitate to spend much construction time and considerable money providing a tight vertical grout curtain under the dam core but are much less careful about sealing the surface of the rock which is equally important and much less costly. As seen in Fig. 10.3:2, if the rock surface (*AC*) is not made impervious by sealing the cracks, a concentration of seepage with a very high hydraulic gradient develops through the earth core directly above the top of the grout curtain (seepage path no. 1).

This creates all the hydraulic conditions necessary for potential piping of the earth core material into the cracks in the rock downstream from the grout curtain. If the cracks in the rock surface are sealed with cement grout (over area *AC*, Fig. 10.3:2), seepage water is forced to flow through the dam core as the designers intended (seepage path no. 2).

The vertical grout curtain and the sealed rock surface create an impervious "Tee" in the rock foundation. The primary purpose of the vertical leg of the "Tee" formed by the grout curtain (*BD*) is to reduce the leakage under the dam, and the secondary purpose is to reduce seepage pressures and gradients of the water flowing through the rock cracks downstream from the grout curtain. The primary purpose of sealing the surface of the rock (*AC*) is to control hydraulic

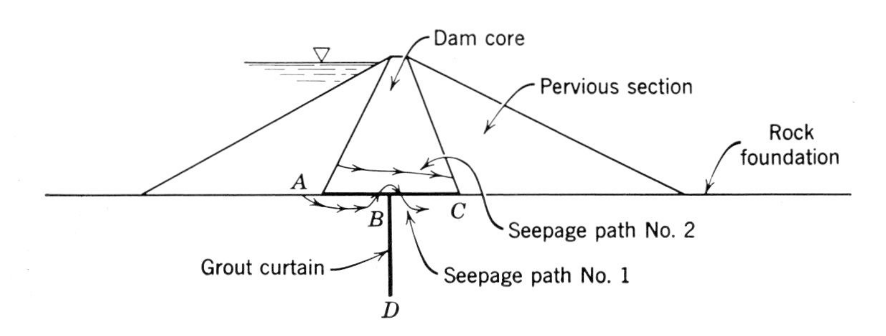

Fig. 10.3:2 The influence of sealing the surface of a rock foundation with slurry grout or other means on the seepage pattern.

gradients and piping through the lower part of the dam core, and the secondary purpose is to reduce the quantity of seepage through the earth core.

Treatment of the rock surface under the impervious core of the dam is carried out in four steps:

1. Cleaning all loose and erodible material from the rock surface and shaping the rock to as smooth and regular a surface as practicable.
2. Washing the exposed rock surface with water and/or air jets.
3. Sealing the cracks in the rock surface.
4. Compacting the first few lifts of earth embankment against the prepared rock surface.

Some of the principal methods which have been used successfully for foundation treatment are described in the following subsections.

SHAPING AND HAND CLEANING THE ROCK SURFACE

Before the grout curtain is installed, the area of the rock surface which is to be in contact with the impervious core of the dam is exposed with rough excavation. When the slopes are not too steep, the rough excavation is carried out with earth moving equipment. Hard rock promontories and overhangs are shot off using careful blasting procedures and small charges to avoid serious shocks to the main mass of the rock foundation and abutment.[1] Rough excavation on steep abutments is carried out by scaling with hand labor or water jets. It is preferable to expose the rock with rough excavation over the whole contact area of the foundation and abutments at one time in order to be able to examine the rock surface characteristics and plan the method of foundation treatment which will be used.

After the grout curtain, if any, is completed, the rock surface is painstakingly cleaned with hand labor. Pockets of sand and gravel and other soil are removed by hand shoveling, and soft, erodible seams of rock resulting from localized decomposition are cleaned out by hand labor ("dental" excavation). Loose rocks are removed by hand picking or wedging. Loose layers of grout spillage from the curtain grout construction are chipped out and removed. Finally, the hand cleaned surface is thoroughly washed with powerful water jets

[1] At a few major dams with steep abutments, extensive rock bolting has been used to increase the stability of the slope during construction.

Fig. 10.3:3 Final cleanup of rock foundation under Trinity Dam (1956), showing sluicing with water jets and hand-shoveling of soft material. (Courtesy USBR)

from hoses to obtain as clean a surface as possible and to remove fines which have been worked into the seams of the rock, Figs. 10.3:3, 10.3:4, 10.3:5. Washing is carried out from the higher elevations to the lower elevations. Compressed air jets are frequently used as a final step in the cleanup, Fig. 10.3:6.

SEALING FRACTURES AND FILLING DEPRESSIONS IN THE ROCK SURFACE

The final decisions regarding the methods of rock treatment are made only after the surface is exposed by the hand cleaning and washing. The purpose of the final treatment is twofold: (1) to smooth

Fig. 10.3:4 Pressure washing and "dental" excavation of rock abutments at Folsom Dam (1953). (Courtesy T. F. Thompson)

Fig. 10.3:5 Hand-picking rock surface, Mormon Island Dam (1953). (Courtesy T. F. Thompson)

out the local irregularities and, consequently, to obtain a good surface against which to compact the embankment, and (2) to seal the surface cracks. Both purposes are accomplished by the same treatment. Some of the principal rock surface characteristics which need treatment are: (1) potholes or depressions of limited areal extent; (2) very jagged surfaces caused by the joint and fracture systems which cannot be made smooth by excavation; and (3) open cracks.

Deep potholes or pockets of limited areal extent can be filled with hand compacted soil or concrete, Figs. 10.3:7, 10.3:8. If the rock sur-

Fig. 10.3:6 Final cleanup of rock abutment under impervious earth core with compressed air jets, Success Dam (1960).

face in the bottom and sides of the potholes is cracked, the cracks should be sealed with cement grout if the holes are to be filled with compacted earth. If the rock surface contains a great number of closely spaced potholes, the best solution may be to pour a slab of concrete over the entire rock surface area, Fig. 10.3:9.

If rock with a very jagged surface is not very pervious, the easiest method of treatment is to compact the earth core against it at a high water content in thin layers with mechanical tampers, Fig. 10.3:10. When the soil is compacted wet, it can be forced into all the irregular depressions in the rock surface to create a complete bond. If the jagged rock has open cracks, they should be sealed with cement grout.

Fig. 10.3:7 "Dental" concrete used to fill pot holes in bedrock surface under core trench at McNary Dam. (Courtesy T. F. Thompson)

Fig. 10.3:8 Filling local depressions in the rock surface with hand-compacted clay, Pomme de Terre Dike (1959).

Fig. 10.3:9 Irregular rock surface in bottom of cutoff trench completely covered with plain concrete. (Courtesy USBR).

Gunite is especially well adapted for treatment of hard, jagged rock with open cracks, Fig. 10.3:11.[1] For a rock of this type the gunite layer seals the cracks and smooths out the surface irregularities.

[1] Gunite should only be used to treat the surface of relatively hard rock, which will not be displaced under the pressure and impact of the heavy rollers used to compact the soil against the rock.

Fig. 10.3:10 Compacting impervious earth-core material against jagged rock abutment. Concrete cutoff wall in background.

Surface fractures should be sealed with a sand-cement slurry grout which is poured or broomed into the previously cleaned fractures. Large cracks can be filled individually, Fig. 10.3:12, and fine, closely spaced cracks are usually filled by brooming the slurry grout over the whole rock surface, Fig. 10.3:13. Depending on the nature of the cracks, the slurry grout may penetrate distances varying between inches and a few feet. Fault zones or large cracks are dug out to a depth of several feet and backfilled with concrete, Fig. 10.3:14.

EARTH COMPACTION AGAINST ROCK FOUNDATIONS AND ABUTMENTS

If the rock surface after treatment is sufficiently uniform, the soil should be compacted directly against the rock with heavy compaction equipment. When compacting the soil against steep rock abutments (or walls of concrete structures) the construction surface of the embankment should be sloped away from the rock or walls for a distance of 8 to 12 ft. at an inclination of 6:1 or steeper. The purpose of sloping the construction surface is to allow the roller to act more directly to compact the earth against the rock surface. If the surface is too irregular to allow the use of a large roller directly against the rock, the roller should be used to compact the soil as close to the rock as possible, and the portion of the embankment directly against the rock should be compacted with smaller equipment in thin layers,

Fig. 10.3:11 Before and after guniting jagged rock abutment at Isabella Dam (1952). (Courtesy Sacramento District, U.S. Army Engineers)

Fig. 10.3:12 Large cracks in rock foundation filled with sand-cement "slurry" grout, McNary Dam (1950). (Courtesy T. F. Thompson)

Fig. 10.3:13 Sand-cement "slurry" grout being broomed into closely spaced, fine rock cracks, Mormon Island Dam, (1949). (Courtesy T. F. Thompson)

Fig. 10.3:14 Excavated fault zone filled with concrete under dam core, Miramar Dam (1960). (D. C. Moorhouse, Civil Engineer)

Fig. 10.3:15 One ft.-wide clay layer used to obtain seal at rock abutments, Mission Dam (1959). The special clay layer was compacted with compressed air tampers and protected against drying with wet burlap cloth.

Fig. 10.3:10. The heavier the tamper used, the less likely that poorly compacted embankment layers will be constructed in the dam.

Sheepsfoot rollers are not very satisfactory for the compaction of soil directly against irregular hard rock surfaces. The feet of the roller ride on the higher rock projections and the soil in the depressions in the rock surface is not compacted. There is also the danger that the feet of the roller will loosen protruding rocks or rupture impervious membranes of gunite or slush grout. On the other hand, sheepsfoot rollers may be superior to rubber-tired rollers for compacting the first layer of earth against soft rock foundations. If the rock surface is weak enough that it can be chewed up and compacted by the sheepsfoot roller, the first layer of embankment soil may be mixed and fused with the rock surface creating an excellent seal.

The soil used directly adjacent to hard rock foundations and abutments (and concrete structures) in the core area should be relatively fine-grained. If the typical core material of the dam is coarse and contains only the minimum necessary quantity of fines to make it impervious, special finer soil should be used for the core zone directly adjacent to the rock for a width of at least a few feet. The water content of the soil placed against the rock should be high enough to allow it to be compacted into all the irregularities of the rock. It is more important for this soil to be wet and plastic and capable of being compacted to form a good seal against the rock than it is to compact it to a dense and strong condition. The more irregular the rock surface, the more desirable it is to compact the soil at a high water content.[1]

[1] At several major dams in recent years, trenches with a width of 12 to 24 in. have been filled with water at the abutments to condition the soil to the consistency of a stiff mud. At the narrow core Mission Dam in British Columbia a 1-ft. wide layer of very wet clay was hand tamped against the rock abutments for a distance of several feet above the general construction surface, Fig. 10.3:15. An almost identical technique is being used at the high Gepatsch Dam currently under construction in Austria (1963).

eleven

Embankment Construction

11.1 CONSTRUCTION SUPERVISION

The skill, experience and judgment required of the engineer in charge of construction is in no way less than that needed by the designer. The information provided by the explorations of the foundation and borrow areas during the design stage is limited compared with that obtained during construction. Conditions different from those anticipated are often encountered. The differences may be small and require only a minor modification of the originally proposed design or construction details, or they may be so important as to require major changes. For this reason, as discussed in more detail in Chapter 1, it must be generally considered that the design process is not completed until the dam is completed and the reservoir is in successful operation. The engineer who directs the construction must have a full understanding of the considerations which led to the design, so that he can make rapid and technically sound decisions for modifications to cope with conditions as they are exposed.

The necessity for competent supervision commences at the earliest part of the construction during the foundation treatment. Adequate foundation treatment is one of the most critical and difficult tasks. Since this work occurs when both the engineering and construction forces are new on the job and relatively unorganized, the resources and judgment of the engineer are likely to be taxed from the beginning of construction.

The size of the crew needed for adequate construction supervision varies depending on the size and difficulty of the job. As a minimum even on a small job, one inspector should work continuously on the construction surface for each shift. Larger jobs require a number of inspectors, each with responsibility for some specific aspect of the construction such as the control of compaction at abutment contacts or the supervision of foundation grouting. On projects where large volumes of material are being moved, a full-time inspector should be assigned to each borrow pit to inspect the excavation and the moisture control operations. Unless the distances are short or the rate of earth moving is low, all inspectors should be provided with vehicles to allow them to carry out their functions without delaying the work.

If construction is to continue at night, the same close inspection is required as during the day shifts. Good lighting is essential, both in the borrow pits and on the construction surface. Even so, it may be desirable to confine to the daylight hours certain critical work, such as compaction of the impervious embankment against concrete structures and abutments.

The supervisor must provide a soil laboratory on all but the smallest projects. The size and the number of laboratory personnel will vary with the size of the dam and the control procedures being used.

Except for the smallest jobs, it is absolutely imperative that detailed written records be kept of all activities and observations made during construction. In order to assure that these records are properly kept, it is advisable to instruct the field inspection force that they will be required to prepare a formal report at the end of the job. If the field forces know that they are going to have to make such a report, it will be a strong incentive for them to keep better current records with drawings and photographs than they might if this task were not hanging over their heads or if they felt that the daily records were only to be filled away and never again used.

The records are primarily of value in the analysis and correction of any unforeseen difficulties which may require design changes during construction or treatment of some condition after construction. Secondarily, they provide a precaution against unjustified claims by the contractor. These reports should include all pertinent observations made, such as depths of excavations, details of dewatering and foundation treatment, drains installed, and descriptions of the materials and construction methods used for the different embankment zones.

During construction much more is learned about the foundation than was known in the design stage, and it should always be considered that the foundation exploration is continuing through the construction. This additional exploration is desirable to correct and refine the results of the previous studies. It is used also to assist in planning the details of the foundation treatment. Additional exploratory borings should be made as required to supplement the observations made in the excavations. On jobs of moderate and large size, a full-time resident engineering geologist should be used, even though a consulting geologist may be retained to make frequent visits to the site. The resident geologist should make inspections and records with drawings and photographs of all conditions exposed by excavations in the foundation for the dam and for appurtenant structures and, especially, in tunnels, spillway excavations, etc. He should record all details of the rock including clay seams, faults of all kinds, and the joint systems. In a large majority of the cases, these detailed records may not be used, but in the few cases where they are needed, they may be invaluable since, once covered up, the conditions seen during construction can never again be exposed for inspection.

The supervising engineer's relationship with the contractor is very important to the success of the project. It is generally unwise to award a construction contract for a price which is much below the engineer's estimate of the cost required to do a good job. In any event, such low bids should be carefully examined and understood before letting the contract. Once the contract is let, the engineer should use all his experience and judgment to help the contractor plan the most satisfactory and economical methods of construction. Lack of cooperation or antagonism between engineer and contractor has resulted in unnecessary delays and higher costs on many jobs.

While the engineer should supply all the assistance possible, the specific procedures for doing the work are left to the judgment of the contractor. One of the principal reasons for the relatively low cost of earth embankments is the fact that earth-moving contractors have consistently devised superior construction methods and stimulated the development of better equipment. The engineer should not insist arbitrarily that the work be done in one way when a contractor believes that he could do it better by another, unless the contractor's method is demonstrated to produce less than satisfactory results. On the other hand, many unnecessary problems and delays have arisen primarily as the result of poor planning or inexperience on the part of the contractor. The engineer cannot abdicate all responsibility for planning of the methods and order of construction. He should require from the contractor well in advance of each phase of construction a plan and schedule for carrying out the work. By studying these plans, the engineer can be reasonably sure that the contractor is not going to create a situation from which it will be necessary to extricate himself by less than satisfactory construction procedures.

11.2 EMBANKMENT CONSTRUCTION EQUIPMENT[1]

The construction procedure consists of excavating the material, hauling it to the dam, mixing it (either in the borrow pit or on the

[1] See Refs. 399, 400, and 605 for more detailed descriptions of embankment construction equipment.

embankment surface) to predetermined water contents and uniformity of properties, spreading it in layers, and compacting it to the desired density. The rate of embankment construction depends on the size of the project and the amount and types of equipment used. It may be limited to a maximum practicable value by the area of the construction surface. Rates of embankment placement ranging from 10,000 to 30,000 yd.^{3}/day are common for large dams. On a few major projects more than 70,000 yd.^{3}/day have been achieved.

On most projects the contractor is given considerable latitude in the types of equipment he will employ. In some cases, however, the equipment has such a large effect on the final embankment properties that the engineer should set some restrictions on the type to be used. This is particularly true for excavating equipment, since the method by which the material is dug determines how it is broken down and mixed, and consequently, governs to a large extent the characteristics of the embankment.

Over the last three decades increasingly larger and faster construction equipment has been devised, and there is no indication that the process of development is slowing down. The basic pieces used on earth dams are:

1. Excavating equipment.
2. Hauling equipment.
3. Spreading equipment; i.e., bulldozers, graders, etc.
4. Rollers and special compacting equipment.
5. Rock separation equipment.
6. Discs, harrows and plows.
7. Watering equipment.

In addition, special equipment is frequently devised for individual construction problems.

11.2a Equipment for Excavating and Hauling

Excavation is usually carried out with power shovels, draglines, scrapers or side-delivery loaders. Each offers certain advantages, and frequently several types are used on the same job.

POWER SHOVELS

Power shovels have been the most common excavation equipment. They provide the best means for mixing and blending soils of different properties and water contents in the excavation. When soils are being mixed, the depth of cut can be varied to obtain the percentage of materials desired (see Fig. 1.1:1). Where more extensive mixing is required, it can be achieved by running the open bucket through the mixture several times before loading the material. No other type of equipment is as effective or versatile for mixing layered materials in the excavation, and on many jobs it is desirable to specify that shovels be used.

Shovels with 2 to 3 yd.^{3} dipper capacities are most common, although larger capacities have been used at many large dams. Rates of excavation vary considerably, typical rates for average construction conditions and efficiency ranging between 70 and 90 yd.^{3}/hr./yd.^{3} of dipper capacity. Excavation faces of 10 to 25 ft. in height are considered most efficient, but shovels have been used in some circumstances to excavate much higher faces.

DRAGLINES

Draglines have been used less frequently, although they can be worked at almost any location where a shovel is suitable. They are a little slower and not quite as satisfactory for mixing soils in a vertical cut. Moreover, they will not excavate such hard materials as can be cut with power shovels. The primary advantage of the dragline is that it works from above the pit and can be employed where the material is too wet to support a shovel.

Draglines are especially effective in excavating pervious sands and gravels below the water table without the necessity for dewatering the excavation, Figs. 11.2:1, 11.2:2. In many cases the material from below water has been loaded directly into trucks for transportation to the dam surface; however, for some materials it may be necessary to stockpile while the excess water drains out.[1]

Dragline buckets of the same size as power shovel dippers are customary, although on a few large jobs bucket sizes up to 17 yd.^{3} have

[1] See Sec. 11.4*b* for description of problems which arise when compacting pervious materials excavated below the water table.

been used economically by dumping the material into loading hoppers with volumes equal to several times the capacity of the hauling units.

SIDE-DELIVERY LOADERS

In the early 1940's, side-delivery loaders of the type shown in Fig. 11.2:3 came into wide use for earth dam construction.[1] When the conditions are right, i.e., when the borrow areas are large, with flat

Fig. 11.2:1 Excavating below water and stockpiling sandy gravel, Folsom Dam (1950).

topography and uniform soil conditions, these loaders provide the most economical and rapid means of excavation. On large projects the cost has ranged between 2 and 4¢/yd.^{3} at rates between 800 and 1,200 yd.^{3}/hr.

Depending on the hardness of the material, one or two tractors may be required to push the loader. It can excavate almost any soil, though it is not too efficient for borrow pits which contain large cobbles and

[1] Loaders of similar design drawn by horses were adopted for the construction of some earth dams in the western part of the United States before 1920.

Fig. 11.2:2 Excavating sandy gravel below the water table for direct loading in trucks, Cachuma Dam (1952). (Courtesy USBR)

Fig. 11.2:3 Side loader excavating weathered shale at Oahe Dam (1957). (Courtesy Omaha District, U.S. Army Engineers)

boulders. At some sites there has also been difficulty with side loaders in borrow pits containing layers of cohesionless sands. Since most side loaders excavate by making a cut 3 to 4 ft. deep, they have not been as satisfactory as power shovels for mixing horizontal strata of different soil properties and water contents.[1]

SCRAPERS

Scrapers are combination excavating and hauling units of two kinds: (1) slow-moving types pulled by crawler tractors and (2) rapid-moving types pulled by rubber-tired tractors. The slow-moving scrapers serve primarily for stripping and cleanup work and only for excavation when the haul distances are very short. Rubber-tired scrapers compete with trucks as hauling units and can be used for any haul distance. Generally rubber-tired scrapers require the assistance of "pusher" tractors in the borrow area during loading.

Scrapers have some advantage over trucks, since they spread the material on the embankment surface in a layer which is approximately the correct thickness for compaction. The thickness of the spread layer can be controlled within very close limits, and scrapers have been employed at some projects in lieu of bulldozers for spreading material dumped in piles or windrows by trucks.

Since scrapers excavate by cutting a relatively thin strip of soil, they are not efficient mixers and are best suited for borrow pits where the soils are uniform or where it is desired to excavate horizontally stratified deposits selectively. They have been used to mix stratified deposits by developing the excavation in such a way that the scraper is loaded on an incline cutting across several horizontal strata of different materials; however, they are never as effective for mixing as a shovel or dragline. Scrapers can excavate conveniently on slopes up to 3:1 (horizontal to vertical).

Because they excavate in thin strips, scrapers are especially valuable for chewing up stiff clays and soft rock. At some sites where power shovels have been used, the material has arrived at the construction

[1] At a few dams side loaders have been fitted with specially adapted cutters which have allowed satisfactory excavation and mixing on vertical faces up to 10 and 12 ft. in height. Also see Ref. 601 for a description of an interesting mechanized excavator in which the cutting element consisted of a number of shovel dippers mounted on a large rotating wheel used at Abique Dam in New Mexico (2,000 to 3,500 $yd.^3$/hr.). At this writing, the contractors for two major earth dams in California plan to use similar excavating wheels with capacities to 5,000 $yd.^3$/hr. in combination with conveyor belts.

surface in such large chunks that it has been very difficult to break it down into a suitable form for compaction. In such cases it has been found desirable to change to scrapers, which have then delivered the material to the embankment surface in a much finer form.

Sizes and speeds of scrapers have been gradually increased over the last 20 years. Those with capacities of 30 $yd.^3$ are in wide use, and, at some dams, scrapers with a capacity of 50 $yd.^3$ and speeds of 45 miles/hr. have been practicable.

TRUCKS

Trucks of all kinds (bottom-dump, end-dump, and side-dump) are employed to haul materials from the excavation to the dam. The type of truck has little influence on the properties or the processing of the embankment material. Trucks with a capacity of up to 50 tons are common on large earth dam projects; probably the biggest capacity to date is an 80-$yd.^3$ end-dump truck with 18 wheels and speed of 35 miles/hr. which was used at the huge Oahe Dam on the Missouri River.

BELT CONVEYORS

Belt conveyors for hauling material from borrow areas have been set up at relatively few dams, though some of these have been major ones. They are economical in areas of very rough topography where there is a large difference in elevation between the borrow pit and the dam and where the cost of building and maintaining haul roads is high.

An electrically operated conveyor belt 30 in. wide and 925 ft. long was used as early as 1912 on the Lahontan Dam in Nevada. At the 450-ft.-high Anderson Ranch Dam in Idaho, constructed in 1941–1947, an extensive system of 36-in. and 42-in. belts was installed with seven flights between 500 and 1,600 ft. in length and transfer points between the belts. The material from the end of the last belt was dumped directly on the embankment construction surface, where it was distributed with scrapers. A conveyor belt system at the Youghiogheny Dam, completed in 1942 in Pennsylvania, had a 65-ft.-long final belt flight which was pivoted and could swing in an arc to distribute the material at various points on the construction surface (Ref. 406).

Fig. 11.2:4 View of conveyor belt at Trinity Dam (1959).

At the 550-ft. Trinity Dam in northern California, where the main borrow pit was located about 1500 ft. above the dam crest, a 42-in. conveyor with a total maximum length of nearly 12,000 ft. and a design capacity of 2000 cu. yd./hr. was constructed for about $2,000,000 (1960) (Figs. 11.2:4, 11.2:5). The material, which was a weathered rock, was excavated with scrapers in order to chew it up and was dumped into a combination screening and crushing plant. The rock passing a 5-in. screen was dropped onto the belt, and what was left was fed back through a rotary crusher. Instead of carrying the material directly to the dam with the conveyor belt, the contractor provided large steel hoppers at the terminal end with automatic facilities for adding water and loading the trucks (Fig. 11.2:6).

Rollers and Other Equipment for Spreading, Blending, and Compacting

The hauling equipment dumps the material onto the construction surface in layers, piles, or windrows. Piles and windrows are spaced in such a way that they can be spread with a bulldozer, grader, or other spreading equipment into a layer of the desired thickness without moving the material long distances.

When the material in the layers has a suitable average water content and uniformity, it is then compacted. If the water content is not at the value desired or the material has not been sufficiently blended by the excavation and spreading operation, it is watered, dried, or mixed with harrows or plows (Fig. 11.2:7) before compaction. Procedures for changing the water content and for moisture-density control are covered in Sec. 11.3 and 11.4.

Either sheepsfoot or rubber-tired rollers are used for the compaction of impervious and semipervious materials, while pervious materials

Fig. 11.2:5 Closeup of conveyor belt used at Trinity Dam, at transfer point (1959).

Fig. 11.2:6 Automatic loading hoppers used at the terminal end of long conveyor belt at Trinity Dam (1959).

are compacted by sluicing and by crawler tractors, rubber-tired rollers, or vibratory rollers.[1] Roller equipment is described briefly below, and sheepsfoot and rubber-tired rollers are treated in more detail in Sec. 11.3*c*.

[1] Smooth steel rollers and wobble-wheel rollers are not used to any extent for earth dam cores. See Refs. 483, 605 for detailed accounts of rollers in common use in the United States.

Fig. 11.2:7 Disk and tooth harrows used to blend fine-grained core material before compaction.

SHEEPSFOOT ROLLERS

As the name implies, the sheepsfoot roller evolved as the result of observations that the travel of animals over earth surfaces resulted in excellent compaction.[1] More than 100 years ago in England, herds of cattle and sheep were actually used for earth compaction. Probably the first sheepsfoot roller was developed in California about 1905, when a wooden log, 3 ft. in diameter and 8 ft. long with railroad spikes for feet, was adopted for road surface compaction. A steel sheepsfoot roller was employed during construction of the Drum Dam in California in 1907 (Ref. 95), and subsequently the use of similar rollers increased until in the period between 1930 and 1950 the majority of earth dams were constructed with them.

The weights and dimensions of sheepsfoot rollers increased gradually until about 1940 when, following the development of rollers with ballasted weight of about 4,000 lb./ft. at the San Gabriel Dam No. 1 in California (Ref. 419), a standard heavy roller was adopted by the USBR. This has been used on all USBR dams since then, and has been adopted widely by other engineers, Fig. 11.2:9. A lighter roller is commonly employed for small dams, although the standard heavy roller is being used more often even for them. The current specifications for the USBR roller are (Ref. 100):

1. The drums must be not less than 5 ft. in diameter and between 4 and 6 ft. long, with the space between adjacent drums from 12 to 15 in.
2. Each drum must be free to pivot about an axis parallel to the direction of travel.
3. One tamping foot must be provided for each 100 in.^{2} of drum surface.
4. The space between the feet must be equal to or greater than 9 in.
5. The length of the feet must be maintained at a minimum of 9 in.
6. The cross section of the feet must be equal to or less than 10 in.^{2} at a distance of 6 in. from the surface of the drum. It must be equal or greater than 7 in.^{2} but not greater than 10 in.^{2} at a distance of 8 in. from the surface.
7. The weight of the roller when fully loaded with sand and water must be not less than 4,000 lb./ft. of drum length.

[1] The term was originally used only for a roller with a knob on the end of the tamping foot, but it now is applied generally to all steel drum rollers with tamping feet.

8. Cleaners must be provided to keep the spaces between the tamping feet clear of material that would interfere with the compaction.

RUBBER-TIRED ROLLERS

Since the end of World War II, a number of major earth dams have been constructed both here and abroad with rubber-tired rollers. They have been more popular in the western United States but now are becoming widely used in all parts of the world. Although many sizes have been tried, most dams have been constructed with a "standard" 50-ton roller,[1] Fig. 11.2:8, with specifications approximately as follows:

1. A minimum of four wheels located abreast and equipped with pneumatic tires.
2. Tires capable of operating at air pressures varying between 80 and 100 p.s.i. under full wheel load.

[1] The first major modern dam constructed with this roller was the Lookout Point Dam, Oregon (1949).

Fig. 11.2:8 Typical "standard" 50-ton rubber-tired roller.

3. The body segmented in such a way that it can be ballasted to a load of at least 25,000 lb/wheel, and so that the wheel loads will be equal when the roller is pulled over rough ground.
4. Wheels spaced so that the distance between the tires will not be greater than one-half the tire width when fully loaded.

IMPERVIOUS COMPACTION IN CONFINED AREAS

Neither sheepsfoot nor rubber-tired rollers can be maneuvered easily into confined areas adjacent to concrete structures or rock cliffs. The heavy rollers are used as close as possible—usually within several feet of these obstructions, Fig. 11.2:9—and the rest of the impervious material is compacted with special equipment.

Many types of compaction equipment are satisfactory for confined areas. These include loaded trucks with long rear axles and wheels extending beyond the body; wheeled tractors with tamping feet; small sheepsfoot rollers; heavy weights dropped by cranes; compressed air tampers of various types; modified pavement breakers and pile drivers. Light compressed air tampers, Fig. 11.2:10, will compact earth embankment materials to adequate densities in thin layers (3 in.), although if this equipment is used, very careful construction control is necessary to avoid burying poorly compacted layers. For this reason heavier tampers which compact the soil to a greater depth are preferable. Various rolling and tamping equipment which has been found satisfactory is shown in Fig. 11.2:11. Usually the same moisture content and density is required for material tamped in confined areas as for the material rolled in the main dam.[1]

COMPACTING PERVIOUS SANDS AND GRAVELS

It is usually much easier to compact pervious sand and gravel embankment sections than the impervious core. Clean, coarse material can frequently be placed in an adequately dense state by spreading it in thin layers, wetting it well, and routing the hauling equipment over it. A common specification, which is satisfactory for nearly all pervious soils, is to require that the material be spread in

[1] See Sec. 10.3 for the special compaction required in impervious sections at the contact with rock abutments. Heavy drop weights mounted on tractors or other mobile equipment (see Fig. 10.3:10) are among the best pieces of equipment for this work.

Fig. 11.2:9 Views showing compaction of embankment as close as possible to a concrete outlet conduit with a large sheepsfoot roller.

about 12-in. layers, sluiced with hoses or water wagons, and then compacted with two to four passes of a heavy crawler-type tractor (40,000 lb.). The tractor is run in high gear, and the pressure together with the vibration imparted to the soil layer are sufficient to densify it. A 50-ton rubber-tired roller is also effective in compacting sluiced cohesionless material unless the soil is too clean to support the heavy wheel loads.

Fig. 11.2:10 Hand-tamping backfill around concrete outlet conduit with lightweight compressed air tampers sometimes results in unsatisfactory compaction.

It is known that vibratory compactors can compact sands and gravels to higher and more uniform densities than either tractors or rubber-tired rollers. Although these compactors have not yet been widely used in the United States for earth dam construction, they have been common in Europe for some time and will undoubtedly be more widely adopted here in the future. Three basic types are available:[1] (1) smooth steel drum rollers with vibratory attachments, Fig.

[1] While there is little quantitative information available on the subject at the present time, a number of recent experiences indicate that vibratory rollers can sometimes be used successfully for compacting the impervious sections of dams which consist of basically cohesionless silty sands or gravels. At a small dam in New Jersey (Glendola Reservoir), where the core was composed of fine, silty sand (10 to 15% finer than the No. 200 sieve), the contractor used a vibrating sheepsfoot roller recently manufactured by the Vibro-Plus Corporation of New Jersey, (Ref. 690) Fig. 11.2:12. Tests indicated that four passes of the vibrating roller gave about the same density as ten passes of a standard USBR sheepsfoot roller. At another recent low dam, a smooth steel drum vibrating roller was used to compact a well-graded silty sand. The material was compacted in layers 9 to 12 in. thick at a water content near Standard Proctor Optimum. The resulting embankment was uniformly dense and showed no sign of stratification or lack of bond between the layers. See also the description of the core of Slottmoberget Dam in Norway built in 1959 from well-graded glacial moraine (silt, sand, gravel and cobbles) which was compacted satisfactorily in 3-ft. layers with smooth-drum-vibrating rollers (Ref. 542).

Fig. 11.2:11a Tamping soil around outlet conduit with heavy drop weight suspended from a crane (behind clam shell bucket).

Fig. 11.2:11b Small, single-drum sheepsfoot roller used to compact soil in confined area.

Fig. 11.2:12 Vibrating sheepsfoot roller used to compact core of silty fine sand at Glendola Dam, New Jersey (1961).

11.2:13; (2) rubber-tired rollers with vibratory attachments; and (3) flat, vibrating plates. There is as yet not much reliable information concerning the optimum combination of roller weight and vibration frequency and energy for compacting soils of various characteristics.

11.2c Equipment for Separating Oversized Rocks and Cobbles

Oversized rocks or cobbles must normally be removed from coarse embankment materials in screening plants or on the construction surface to allow proper compaction. In some cases the choice of the method is left to the contractor; in others the designer specifies that a screening plant be used (Sec. 1.1*a*). Four principal types of screening plants have been constructed:

1. Horizontal or sloping screens which are stationary and act by gravity.
2. Vibrating screens, Fig. 11.2:14.
3. Rotating trommels.
4. "Wobblers" (or "rotating cams"), Fig. 11.2:15.

Fig. 11.2:13 Smooth steel drum vibratory rollers used to compact sand and gravel zone in Navajo Dam, New Mexico (1959) (Courtesy, Construction Methods and Equipment, Ref. 418).

Plants have been designed for capacities ranging from 100 to 2,000 yd.^{3}/hr., although the majority process 300 to 500 yd.^{3}/hr. Materials with an appreciable content of wet clay are the most difficult to process in a separating plant, since the clay blocks the screen openings.

Fig. 11.2:14 Sloping vibrating screens, Cachuma Dam, California (1950). (Courtesy USBR)

At some dams, coarse (12-in.) stationary screens are first used to scalp off the boulders before the material is dumped into the main screening plant. In a few cases rock crushers have been built into the plant to crush the oversized cobbles. Sprays are often installed at one

Fig. 11.2:15 Wobbler-type screening unit, Casitas Dam (1957). Rotating action of cams causes fines to fall through and rocks to discharge over front end. (Courtesy USBR)

or more locations in separating plants, since this is one of the most efficient ways to add water to dry soils.

In order to avoid rehandling the material, some attempts have been made to construct mobile screening plants and move them around in the borrow pit so that they can be loaded directly with the excavating equipment. These have usually been unsuccessful, both because they

(a)

(b)

Fig. 11.2:16 (a) Raking oversize cobbles from embankment material on the construction surface. (b) Rockfill section from oversized rocks raked out of construction material on the construction surface. Note absence of small rocks.

could be loaded with only one piece of excavating equipment and because it is necessary to keep a definite relative elevation between the screening plant and the excavating unit. As a result, most plants have been constructed at a fixed site located on the shortest haul road between the borrow areas and the dam.

A number of techniques have been adopted to remove oversized rocks on the construction surface. At some dams, large crews of men have been used to throw out the cobbles. More frequently, rocks have been segregated into piles with bulldozers or rakes and loaded into trucks, Fig. 11.2:16. In some cases they have been pushed with bulldozers directly into pervious zones in the embankment. The effectiveness of rakes on tractors varies considerably with the soil type and the experience of the operator. With such equipment there is a tendency to leave many large rocks in the fill, and continuous attention by the inspectors is required to keep this to a minimum.

At a few dams it has been possible to eliminate most of the oversized cobbles in the borrow excavation by providing a grizzly over the mouth of the power shovel dipper. This procedure is practical only where there are relatively few oversized rocks and where the soil separates easily from the rock.

11.3 CONSTRUCTION PROCEDURES: IMPERVIOUS AND SEMIPERVIOUS EMBANKMENT SECTIONS

11.3a Mechanics of Compaction

Although the cost (usually 2 to 6¢/yd.^{3}, 1960 prices, of compaction with heavy rollers after the soil has been dumped and spread into layers is a small fraction of the cost of the embankment, the method of compaction is one of the most important aspects of construction.[1] It governs to a large degree the uniformity of the completed embankment and also influences the important embankment properties, such as compressibility, shear strength and permeability.

For a given method and effort of compaction, a soil is compacted to a unique density at each water content. As seen in Fig. 11.3:1, the dry density of the compacted soil increases with increasing water content until a "maximum dry density" is obtained at an "optimum water content." At this point the dry density begins to decrease as the water

[1] For more detailed discussions of the mechanics of soil compaction, see Refs. 140, 402, 404, 408, 412, 416, 438, 605.

content continues rising. For each method of compaction, there is always a single maximum density at an optimum water content.

For any soil, the optimum water content, the maximum density, and the shape of the curve vary considerably with different methods of compaction. On the wet side of optimum water content, the compaction curve becomes roughly parallel to the zero air voids curve (100% saturation), and the distance between the two curves *d*, (Fig. 11.3:1) represents the amount of air contained in the compacted soil.

If a soil is compacted by the same method but with different compactive efforts, families of compaction curves with similar shape are obtained, Fig. 11.3:2. The higher the compactive effort, the higher is the maximum density and the lower the optimum water content. In such a family of water content-density curves, a line connecting the maximum points for the individual curves approximates a straight line which is roughly parallel to the curves of constant percent saturation. On the wet side of optimum water content, the curves obtained for all compactive efforts merge into a single line indicating that, at a given

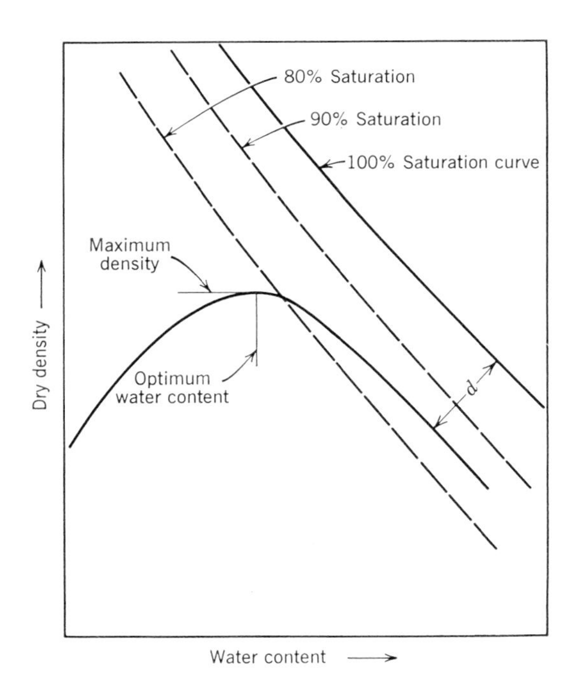

Fig. 11.3:1 Typical relationship between dry density and water content for one compactive effort.

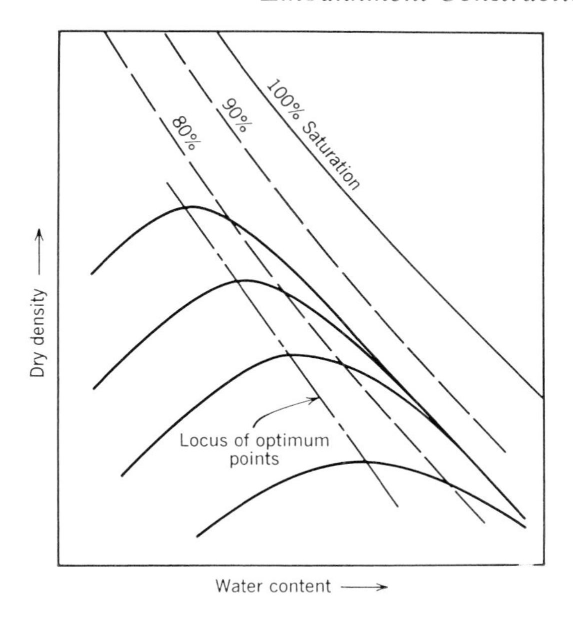

Fig. 11.3:2 Typical family of density-water content curves for one soil obtained using the same compaction method and equipment but different compactive efforts.

water content, there is a maximum degree of saturation (or maximum density at a given water content) which can be obtained practicably with a given type of compaction equipment regardless of the compaction effort expended. On the dry side of optimum water content, the curves are approximately parallel. From these results it is seen that an increase in compactive effort is more effective in increasing the density of the soil when its water content is on the dry side of optimum than when on the wet side.

Figure 11.3:3 shows qualitatively the relationship between increased compactive effort and increased density for soils at the same initial density but at different water contents. Curves 1 and 2, Fig. 11.3:3*a*, represent water content-density curves for the same soil compacted with the same equipment but using two compactive efforts. Points *a* and *b* represent two conditions of the soil compacted with the lower compactive effort to the same dry density, but at two different water contents (below and above optimum). As the compaction is increased to the higher compactive effort (curve 2), the dry density of the material at *a* increases more rapidly than that at *b*, Fig. 11.3:3*b*. This difference in the influence of the increased compactive effort on

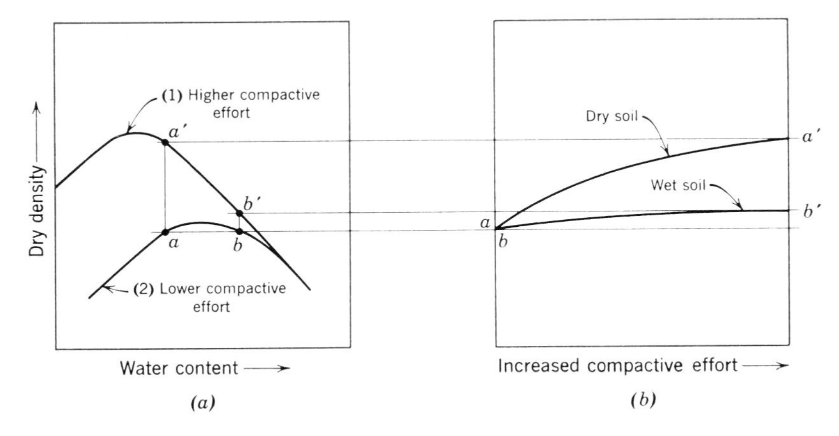

Fig. 11.3:3 Influence of increased compaction effort on a compacted soil at the same density and different water contents (using the same compaction equipment).

the density is due to the fact that the wetter material is soft and the shear stresses imposed on the soil during the compaction process are greater than the shear strength, so that the compaction energy is dissipated largely in shearing the compacted material without much additional densification. On the dry side of optimum water content, the material is stiffer and more of the energy of the compaction equipment goes into compressing the soil to a denser state.

For a given soil each different method of compaction results in a somewhat different family of water content-density curves of the type shown in Fig. 11.3:2. In the laboratory, where compaction tests can be performed under controlled conditions, the curves for a given soil can be plotted easily and reliably. Curves for compaction of embankments in the field for various types of rollers are more difficult to obtain because of the difficulties in controlling the variables which influence the test results.

11.3b Laboratory Compaction

In the design stage, compacted specimens of the soils from which the dam is to be constructed are tested in the laboratory, and assumptions are made concerning the strength and the permeability of the various zones of the embankment. During construction, laboratory compaction tests are used as a control standard of the compaction being obtained. Laboratory tests are performed by compacting the soil in layers in cylindrical steel molds. Three general types of tests have been developed:

1. Dynamic tests in which the soil is compacted by the free fall of a hammer. The compactive effort is varied by varying the weight of the hammer, the height of the hammer drop, the number of hammer blows per layer, and the thickness of the compacted layers.
2. Kneading tests in which the soil is compacted in layers with a kneading piston or tamper.
3. Static tests in which the soil is compacted in layers by static pressure of a piston with area equal to the area of the compaction mold.

For earth dam projects dynamic tests are used almost exclusively. Kneading compaction tests have been performed occasionally, either for research or for the preparation of samples for strength or permeability tests, but very rarely for embankment construction control. Static compaction tests have not been employed to any significant degree.

DYNAMIC TESTS

The dynamic test most commonly used for earth dam studies is the standard Proctor, which has been adopted by the ASTM and the AASHO. It is performed in a 4-in.-diameter mold with a volume of $\frac{1}{30}$ ft.^{3} The soil is compacted in three equal layers with a 5.5 hammer dropped 12 in. 25 times on each layer, resulting in a total compactive effort of 12,375 ft.-lb. of energy per cubic foot of compacted soil. The USBR has adopted a slightly different standard test in which a $\frac{1}{20}$-ft.^{3} mold is used and the hammer is dropped 18 in. instead of 12 in., so that the compaction effort is the same.

A "modified Proctor" test with much higher compactive effort, developed primarily for the control of heavily compacted bases for airfield pavements, has also been used to some extent for earth dams. This test is made in a $\frac{1}{30}$-ft.^{3} mold, where the soil is compacted in five equal layers with a 10-lb. hammer dropped 18 in. on each layer. The resulting compactive effort amounts to 56,250 ft.-lb./ft.^{3} of compacted soil.

Experience indicates that the field compactive effort exerted by rollers commonly used for earth dam embankments is slightly higher than the laboratory standard Proctor test and much lower than the modified Proctor. For this reason an intermediate laboratory test with

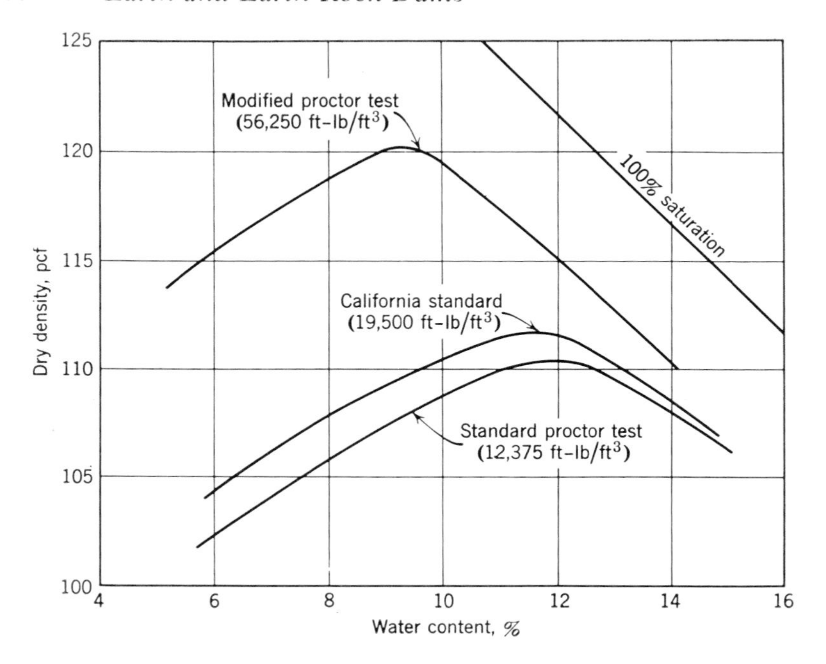

Fig. 11.3:4 Typical results of laboratory compaction tests on sandy clay of medium plasticity.

compaction effort of 19,500 ft.-lb./ft.^{3} is sometimes chosen for construction control, particularly for dams in the state of California. Figure 11.3:4 shows water content-density curves obtained for these three laboratory compaction tests on a sandy clay of low plasticity. The difference between the maximum densities obtained in the Standard and Modified Proctor tests varies considerably as a function of the soil type. For fine uniform sands the difference is only 2 or 3 lb./ft.^{3}; for well graded silty or clayey sands or gravels the difference is commonly 8 to 10 lb./ft.^{3}, and it may be as much as 15 lb./ft.^{3} for some clays.

In addition to the compactive energy imparted to the soil (in ft.-lb./ft.^{3}), the results of the laboratory dynamic compaction test are influenced to some degree by the details of the test; i.e., the size of the molds, the height of drop, and the weight of the hammer, the relative dimensions of hammer and mold, and the thickness of the compacted layer. However, the researches which have been carried out, although few in number, all indicate that these details can be varied over wide ranges without appreciably influencing the test results (see, for example, Refs. 414, 433). Within the limited range of dimensions, weights and procedures commonly used for laboratory dynamic compaction tests, approximately the same water content-density curve is obtained for tests on the same soil performed with a given compaction energy. For example, the standard Proctor and the USBR tests give practically the same results even though both the height of hammer drop and the volume of the mold is 50% greater for the USBR test. Details of the four most common dynamic laboratory compaction tests are summarized in Table 11.3:1.

KNEADING TESTS

It has long been observed that the water content-density curves for actual embankments compacted with rollers have maximum densities at higher degrees of saturation than the curves for the same soil compacted in the laboratory with dynamic tests. Developed in an effort to simulate the field procedure more closely, kneading compaction tests are made by compacting the soil through the kneading action of a tamping piston or foot. Such tests do provide water content-density curves closer to field results than the curves from dynamic laboratory tests. However, because the differences are not great, and because the kneading tampers are generally more expensive and complicated than the equipment required for the simple dynamic test, kneading compaction tests have not been widely used for earth dams.

On the other hand, recent researches are showing that the basic properties of specimens compacted to the same water content and density by dynamic and kneading procedures may be quite different for some soils (see, for example, Refs. 140 and 441). Consequently, although kneading compaction is not likely to be adopted widely for construction control, it may be used more frequently for the preparation of specimens in the design stage to study the properties of the material.

11.3c Field Compaction with Sheepsfoot and Rubber-Tired Rollers

The density to which an impervious embankment section of a given soil type is compacted will depend primarily on the following factors.

1. Roller-type, gross weight, and soil contact pressures.
2. Thickness of compacted layers.
3. Number of roller passes.
4. Soil water content.
5. Roller speed.

Table 11.3:1 Details of Laboratory Compaction Tests in Common Use

NAME OF TEST	MOLD SIZE		HAMMER		TEST PROCEDURE				COMPACTIVE EFFORT (FT.-LB/FT.^{3})
	Diameter × *Height* (*in.*)	*Volume* (*ft.*^{3})	*Diameter* (*in.*)	*Weight* (*lb.*)	*Hammer Drop* (*in.*)	*No. of Layers*	*No. of Blows Layer*	*Maximum Particle Size*	
Standard Proctor (ASTM D-698-58T and AASHO T99-57)[2]	4.000 × 4.584	1/30	2	5.5	12	3	25	No. 4[1]	12,375
Modified Proctor (ASTM D-1557-58T and AASHO T180-57)	4.000 × 4.584	1/30	2	10	18	5	25	No. 4[1]	56,250
USBR	4.281 × 6.000	1/20	2	5.5	18	3	25	No. 4	12,375
State of California Department of Water Resources	4.250 × 6.094	1/20	2	10	18	5	13	No. 4	19,500

[1] For coarse soil, 3/4-in. maximum particle size may be used. (See ASTM and AASHO Test Procedures for "Method C".)

[2] In common usage the tests described by the terms "standard Proctor," "standard ASTM" and "standard AASHO" are the same; however, this is not the same laboratory compaction test proposed by Proctor in his original papers. In Proctor's original test, the soil was compacted with a "firmly struck" blow from a height of 12 in. rather than by the free fall of the hammer. Because of the difficulty in duplicating the results, the original test of Proctor has not been widely used and the name "standard Proctor test" has been generally used for the standard test procedures adopted by the ASTM and AASHO.

Because of the expense and difficulty of making full-scale controlled field compaction tests, the influence of these factors is known only in a general way. It is not practicable to obtain water content-density curves of the type shown in Figs. 11.3:1 and 11.3:2 except with extremely uniform soils, close construction control, and literally hundreds of field density tests.[1] Consequently it is impossible at the present time to predict accurately from the laboratory compaction curves the field compaction characteristics of a given roller on a given embankment material.

The experience of the USBR indicates that the average field water content-density curve is similar in shape to and slightly higher than the standard Proctor laboratory compaction curve (Ref. 100). The dams on which this experience is based were compacted, according to USBR practice, in 6-in. layers with 12 passes of the sheepsfoot roller which is described in Sec. 11.2*b*. Average field and laboratory curves for three USBR dams with embankments of varying gravel content are compared in Fig. 11.3:5.

ROLLER ACTIONS

Both rubber-tired and sheepsfoot rollers compact soil layers by a kneading action: that is, the pressure causing compaction varies from zero to a maximum value and then to zero again as the roller passes over the soil. Aside from this similarity, however, the two rollers work on quite different principles.

The rubber-tired roller applies the force at the surface of the soil layer, and the compaction is obtained by the repetition of pressure with successive roller passes. The pressure, although it is slightly lower for the first few passes because the tires sink deeper into the soil, is essentialy the same for all roller passes. It is roughly equal to the air pressure in the tires for any roller of a given weight, so that the air pressure directly influences the embankment density obtained (see Table 11.3:2, page 595).

In a rather different compaction process, the tamping feet of the sheepsfoot roller first penetrate the loosely spread layer. For the first few passes the roller weight is carried by the soil under the steel drum, which exerts relatively low unit pressures and by the tamping feet bearing on the next lower layer with relatively high unit pressures. Consequently, on the first few passes of the roller, a large part of the

[1] For one of the few large studies which have been made under such conditions, see the account at the end of this section of the Army Vicksburg roller research

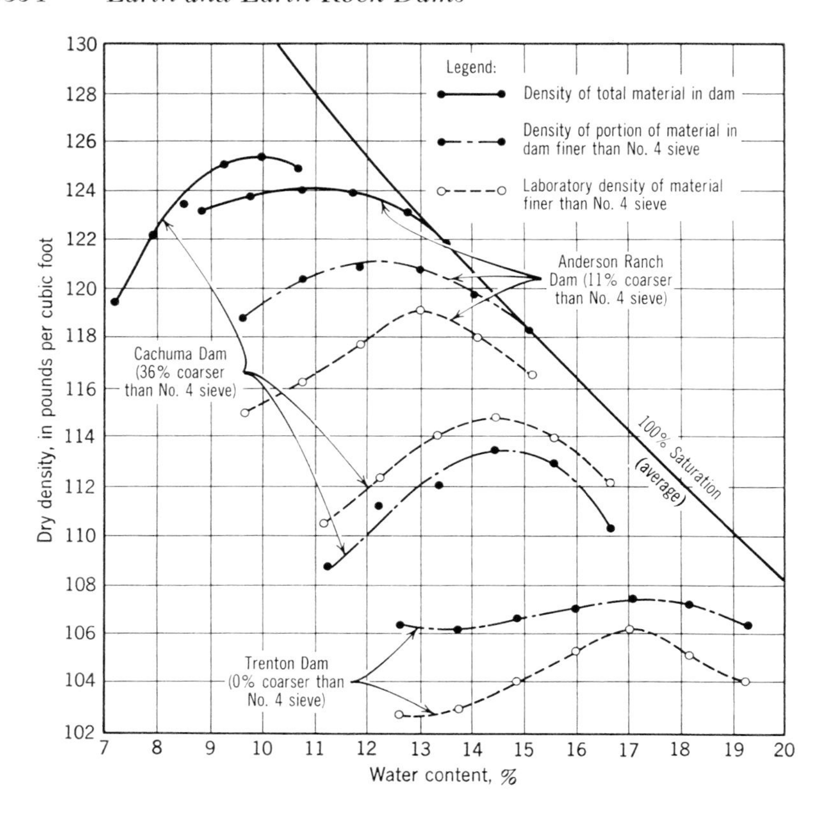

Fig. 11.3:5 Typical field and laboratory compaction curves obtained as statistical averages for USBR dams compacted with 12 passes of standard sheepsfoot roller (after Hilf, Ref. 100).

compactive energy is transmitted to the next lower layer. Gradually, as the upper layer becomes denser and stronger, provided that the water content is not too high, it becomes able to support the high unit pressures exerted by the tamping feet. The most efficient utilization of the compactive energy occurs when the roller drum just lifts off the surface ("walks out") for the last few passes.

Sheepsfoot rollers are often rated in terms of pressure exerted by the feet, assuming one row of feet in contact with the embankment. However, any description of the compactive ability of the roller must include other properties such as the weight per ft. of the drum and the spacing and area of the feet. When fully ballasted, the standard USBR roller exerts a foot pressure of about 490 p.s.i. computed on the assumption that 5% of the 240 feet are in contact with the earth at any one time.

Layer Thickness and Roller Passes. No hard and fast rules can be given for the optimum thicknesses of layers compacted with either type of roller. The common procedure is that embankments are constructed in 6-in. layers (after compaction) with sheepsfoot rollers and 9-in. layers (after compaction) with the 50-ton rubber-tired roller.

Table 11.3:2 Summary of Field "Optimum" Water Contents and Maximum Densities for Rubber-tired Rollers on Silty Clay (Vicksburg Research, Ref. 415)

	150-PSI TIRE PRESSURE		90-PSI TIRE PRESSURE[1]		50 PSI TIRE PRESSURE	
	Optimum Water Content (%)	*Maximum Density (lb/ft.³)*	*Optimum Water Content (%)*	*Maximum Density (lb/ft.³)*	*Optimum Water Content (%)*	*Maximum Density (lb/ft.³)*
4-roller passes	16.0	113.5	17.7	110.5	19.7	107.0
8-roller passes	15.3	115.0	17.2	111.2	19.0	107.5
16-roller passes	14.7	116.5	17.0	111.5	18.8	108.0

[1] "Standard" 50-ton rubber-tired roller.

Soil properties: Liquid limit = 38%; plasticity index = 13%; standard Proctor optimum water content = 17.8%; standard Proctor maximum density = 107.5 lb/ft.^{3}

These dimensions are satisfactory for most soil types, although thicker layers could probably be used safely for some soils.

For a 6-in. layer compacted with a sheepsfoot roller, satisfactory densities can be obtained with 6 to 12 passes when the water content of the soil is right. For most soils little extra density is obtained after six passes of the sheepsfoot roller except where the water content is well on the dry side of the optimum value for the equipment. Nevertheless, because the extra rolling is relatively inexpensive and may improve the uniformity of the embankment, many engineers specify a minimum of 10 or 12 sheepsfoot roller passes. With the 50-ton rubber-tired roller three to six passes are usually sufficient to provide comparable densities.

For the purpose of construction control, each embankment layer is considered to be compacted only by the roller passes on its surface, and specifications require that the desired densities be obtained before additional soil can be placed. However, experience indicates that the density of the embankment may be increased to a depth of several feet below the construction surface by repeated passes of a heavy rubber-tired roller. Even with sheepsfoot rollers, which have an approximate weight of 4,000 lb./ft., some additional compaction to a depth of 3 ft. has been measured; but rubber-tired rollers weigh about 15,000 lb./ft. and so are far more effective in compacting lower layers of the embankment. The lower the water content of the soil, the greater is the compactive influence of rolling at depth.

Roller Speeds and Rates of Compaction. The influence of the roller speed on the compaction obtained is not well known, but within the range of speeds at which rollers normally operate, and probably even at higher speeds, most of the evidence makes it seem doubtful that roller velocity has any great effect on the density. On the other hand, one of the few studies which have ever attempted to assess the influence of roller speed clearly demonstrated that 22 passes of a grid-type roller at 8 to 10 miles/hr. were required to obtain the same density which was achieved with eight passes at 2 miles/hr. (Ref. 401). In the absence of less contradictory information on the subject, roller speeds are usually restricted to those which have been customary in the past.

Rubber-tired rollers are towed either by crawler tractors or by pneumatic tired tractors, and in the former case the speed is commonly 3 to 4 miles/hr. In the latter case, it is usually limited to a maximum of about 8 miles/hr. unless tests are performed to demonstrate that adequate compaction can be obtained at higher speeds. The rate at which embankment can be compacted with one rubber-tired roller usually falls between 500 and 1,000 yd.^{3}/hr., depending upon the roller speed and upon normal delays due to turning and stops.

Sheepsfoot rollers are customarily pulled with large crawler-type tractors at a maximum of about 4 miles/hr. The typical rate at which embankment can be compacted with one sheepsfoot roller varies, depending upon the size of the working area and normal delays, between 100 and 500 yd.^{3}/hr. per double-drum roller.

Special self-propelled sheepsfoot rollers, which considerably accelerate the rate of compaction, have been developed and used at several large dams, Fig. 11.3:6. At the USBR's Twitchel Dam, completed in

Fig. 11.3:6 Recently developed self-propelled sheepsfoot roller meeting USBR specifications. (Courtesy USBR)

1958 in southern California, the contractor designed and built on the job a four-drum self-propelled roller meeting the USBR specifications. It traveled at 6 miles/hr., maneuvered easily, and operated in either direction, thus reducing lost time for turns. The supervising engineers concluded that somewhat more uniform and slightly higher densities were obtained with the self propelled roller and that it did approximately as much work as three standard double-drum rollers pulled by tractors in the normal fashion.

"Walking Out" by Sheepsfoot Rollers. For most efficient compaction, the sheepsfoot roller should walk out enough to lift the steel drum off the construction surface at least an inch or two during the last few roller passes. However, it is not necessary for the drum to rise more than several inches.

When sheepsfoot rollers do not walk out, it means that local shear failure is taking place because the soil water content is so high that the embankment below the bottom of the feet will not support the high unit pressures. Even in this circumstance, tests may show that the pressures which have developed under the tamping feet are quite adequate to secure the compaction desired. In fact, a number of major dams have been compacted to satisfactory densities although the drum has not visibly risen from the construction surface.

During the construction of some dams the rollers have been lightened by removing ballast in order to make them walk out of the fill on the last few passes. Slightly higher densities have been reported with this procedure, but since there is a maximum practical density which can be obtained at any given water content (Fig. 11.3:2), substantially higher densities can only be achieved by decreasing the water content of the soil.

Almost all sheepsfoot rollers are provided with feet which are about 3 in. in diameter, although it is probable that larger feet would be more effective in certain types of fine-grained soils. Proposals have been made to construct rollers with feet of varying sizes so that at a given job a roller could be used which would walk out of the construction surface when the embankment was compacted at the water content desired. However, rollers with large feet have not been used except at a few dams on an experimental basis. The U.S. Army Engineers have conducted some research on the problem of selecting optimum sizes of roller feet for various soil types from the results of field penetration tests (Ref. 413). This work indicated clearly that sheepsfoot rollers with larger feet than those commonly used would be more efficient for certain types of soils, but the study was not extensive enough to allow any general correlations between soil type and most desirable foot sizes.

RELATIVE ADVANTAGES OF SHEEPSFOOT AND RUBBER-TIRED ROLLERS

At the present a spirited debate[1] is in progress over the relative merits of rubber-tired and sheepsfoot rollers. Because the sheepsfoot roller generally produces a better bond between compacted layers, the USBR engineers will not permit the use of rubber-tired rollers on impervious embankment sections at all. On the other hand, a larger group which includes the U.S. Corps of Engineers believes that embankments built with rubber-tired rollers under the proper construction controls can be both safe and cheap.

Most engineers agree that there are some special soils which are best compacted under the high unit pressures and the mixing action of the sheepsfoot roller. For the majority of impervious soils, however, rubber-tired rollers are receiving increasing favor. Rubber-tired rollers are more effective in compacting soils containing large cobbles and hard rock fragments since the feet of the sheepsfoot roller ride on the

[1] See Ref. 100 for an example.

cobbles thus preventing the adjacent feet from compacting the soil between the cobbles. On some jobs, both rollers in some combination have been used by contractors to produce the specified compaction with minimum effort. The principal advantages of the two types are as follows:

Sheepsfoot Rollers.

1. Since the sheepsfoot roller leaves the upper few inches of the compacted layer in a loose state, successive layers are fused together in a continual process. Consequently the embankment constructed with sheepsfoot rollers has a relatively uniform density and rarely contains horizontal surfaces which are not intimately fused together. In contrast, the upper surface of an earth layer compacted by rubber-tired rollers is hard and smooth and must be scarified in order to bond it to the succeeding layer (Fig. 11.3:7).
2. Sheepsfoot rollers are much more effective than rubber-tired rollers in breaking down large pieces of soft rock, so that a

Fig. 11.3:7 Ripper used to scarify the upper surface of layers of clay core compacted with rubber tired rollers before the next lift of soil is spread, Pomme de Terre Dam, Missouri, 1960.

residual soil containing gravels and cobbles of soft decomposing rock will result in a finer, more impervious embankment if constructed with a sheepsfoot roller.

3. Because the churning action of the sheepsfeet mixes the soil and breaks up clods, less blending with harrows and plows is needed. In addition, wet soils can be dried by rolling, and dry soils can be sprinkled while rolling. For rubber-tired rollers, on the other hand, the moisture content of the soil must be just right before compaction.
4. Since the unit pressure on the soil is low at first and only gradually increases as the soil becomes strong enough to support the tamping feet, sheepsfoot rollers can compact over a wider range of water contents than rubber-tired rollers. Consequently they are able to work on fine-grained soils at water contents that would cause the heavier rubber-tired machines to bog down on the construction surface.

Rubber-Tired Rollers.

1. The basis of the increasing popularity of rubber-tired rollers is that they can compact the majority of embankment materials at a lower cost and in less time than sheepsfoot rollers. This is true for a number of reasons. Rubber-tired rollers compact the soil in thicker lifts than sheepsfoot rollers, and larger sized gravels or cobbles can thus be used in the lifts. Rubber-tired rollers require fewer passes than sheepsfoot equipment, and the tractors pulling them can often travel faster. Due to these savings, fewer rubber-tired rollers than sheepsfoot rollers are needed to compact an embankment in a given period of time, and this in itself will speed construction because the embankment surface will be less cluttered with the rolling equipment. All of these advantages combine to offset substantially the additional trouble of adjusting the water content of the soil, scarifying the surfaces of the layers, and taking the other steps necessary to achieve a satisfactory bond between layers.
2. In some coarse, granular soils the feet of sheepsfoot rollers are rapidly worn out requiring continual maintenance. For these soils the rubber-tired roller, which rides on top of the layer, may be much less rapidly damaged by the abrasive character of the soil and therefore more economical.
3. Construction during a rainy season proceeds more rapidly with rubber-tired rollers since there is less trouble with water soaking

into the upper soil layer. The sheepsfoot roller always leaves a loose surface, which becomes excessively wet from rain unless it is sealed beforehand with smooth wheel rollers (Sec. 11.3*f*). Even at sites where the smooth roller is used, sudden storms may soak the soil before the sealing has been completed.

4. Higher densities, especially in granular soils, can be obtained with the rubber-tired roller because of its heavy weight of approximately 15,000 lb./ft. While the sheepsfoot roller exerts higher unit compactive pressures on relatively small areas, its total weight is only about 4,000 lb./ft.
5. The rubber-tired roller furnishes a guide to construction control in that the high individual wheel loads indicate the location of soft spots in underlying layers. The sheepsfoot roller sometimes bridges over these localized spots so that they go unnoticed.
6. Rubber-tired rollers obtain better compaction than sheepsfoot rollers against hard rock abutments and foundations and at junctures between the earth core and concrete structures (see Sec. 10.3).

VICKSBURG ROLLER RESEARCH

One of the best studies of the factors which influence compaction by rubber-tired and sheepsfoot rollers was the comprehensive research program by the Army Vicksburg Waterways Experiment Station (Refs. 409, 410, 411). Most of the work was carried out with a single very uniform silty clay, of which many individual embankment sections were compacted at different water contents, by different numbers of roller passes, and with different total weights and unit pressures for both types of roller.

Typical results with rubber-tired rollers (Figs. 11.3:8, 11.3:9) show that the air pressure in the tires has a major influence on the density, and that the influence is greater when the soil water content is on the dry side of optimum. As seen in Table 11.3:2, the field optimum water content rose from 14.7% to 19.7% when the compactive effort was decreased from 16 passes of a 150-p.s.i. roller to 4 passes of a 50-p.s.i. roller. For the 90-p.s.i. roller, which is approximately equivalent to the standard 50-ton model, the optimum water content was slightly below standard Proctor optimum while the maximum density was several pounds higher than standard Proctor maximum.

The sheepsfoot rollers used in the study had foot areas of 7, 14, and 21 in.^{2}, each roller being ballasted to give a foot bearing pressure of

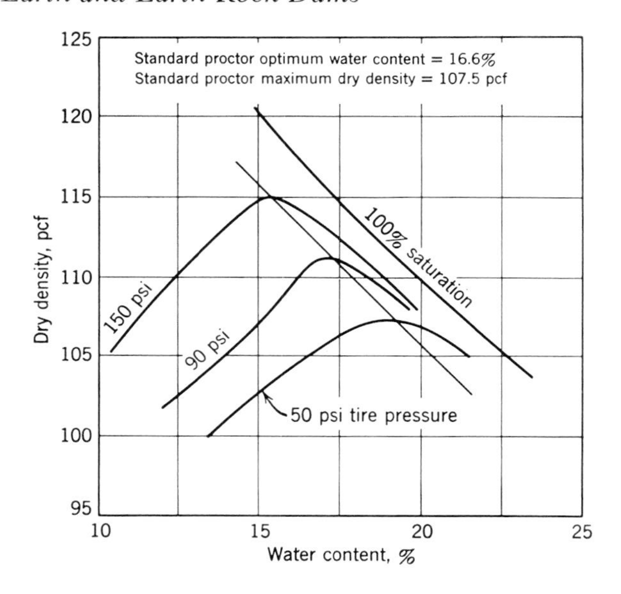

Fig. 11.3:8 Water content-density curves for eight passes of rubber-tired roller with different tire pressures from test embankments on silty clay (after Turnbull and Foster, Ref. 415).

250 p.s.i. (gross roller weights of 14,000, 28,000, and 42,000 lb. respectively). As shown in Figs. 11.3:10 and 11.3:11, the results indicate that the field optimum water content decreases with the number of roller passes, and that the larger the foot size, the higher the maximum density and the lower the optimum water content.

One of the most interesting findings of the Vicksburg research was that the sheepsfoot rollers compacted more air into the embankment than the rubber-tired rollers at their respective optimum water contents. Figure 11.3:12, which summarizes the various densities obtained, shows that the optimum conditions for the sheepsfoot roller occurred at less than 80% saturation, whereas those for the rubber-tired roller were at about 90% (Ref. 564).

The results of the Vicksburg Research Project, as summarized in Figs. 11.3:8 to 11.3:12, were obtained from studies on only one silty clay; however, it is probable that the general relationships observed would be similar for all fine-grained soils of low to medium plasticity. Although there is little and fragmentary data published, it appears that for the great majority of soils from which the impervious section of an earth dam would be constructed, the field optimum water con-

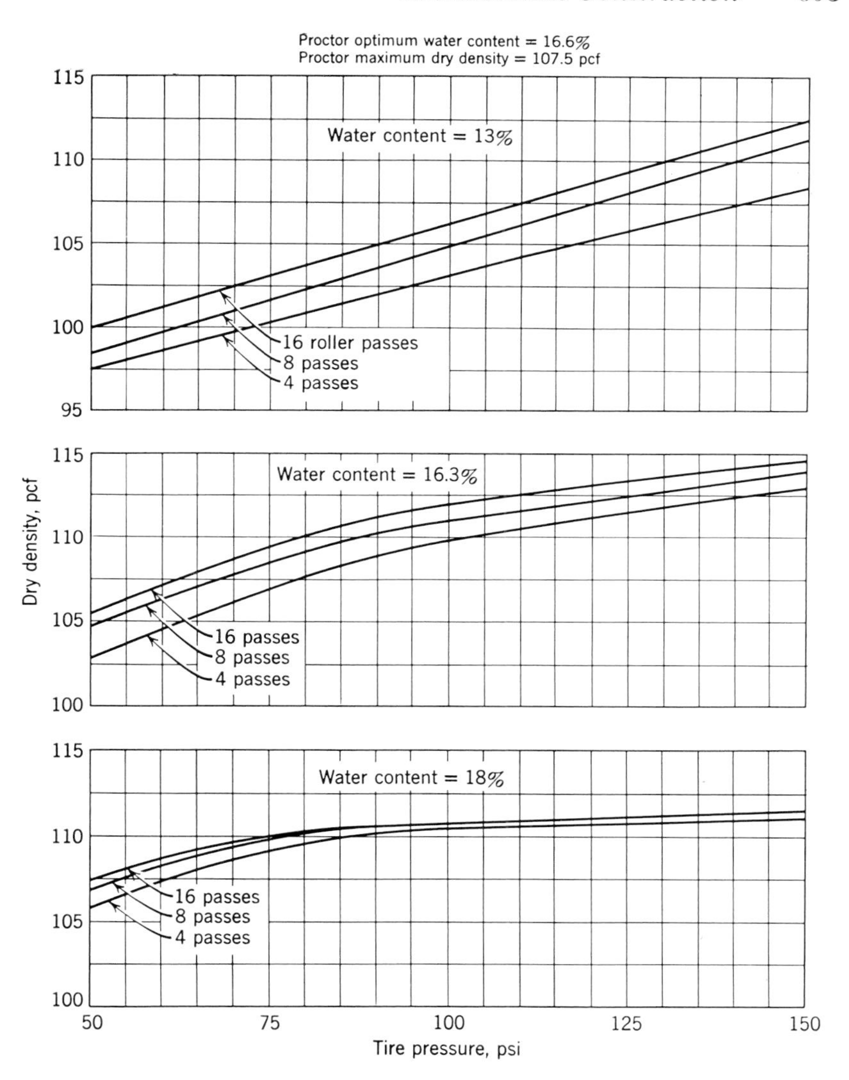

Fig. 11.3:9 Influence of tire pressure and number of roller passes on dry density of silty clay (after Turnbull and Foster, Ref. 415).

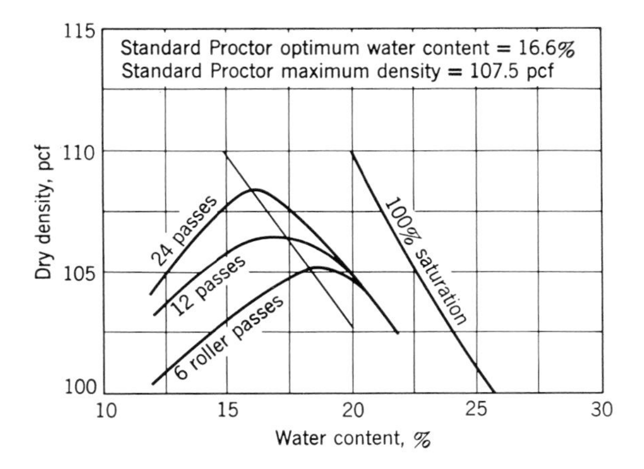

Fig. 11.3:10 Water content-density curves for 250-p.s.i. sheepsfoot roller from test embankments on silty clay (after Turnbull and Foster, Ref. 402).

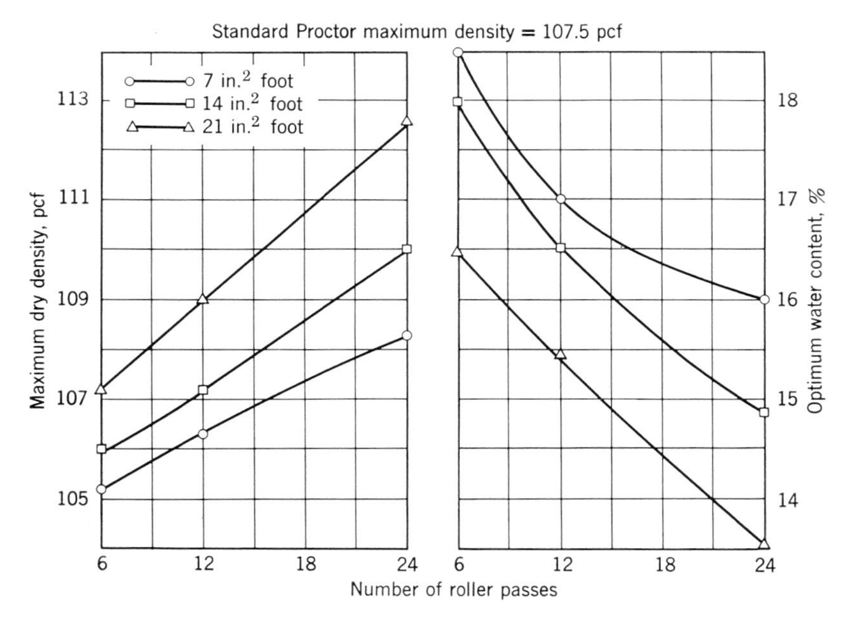

Fig. 11.3:11 Influence of number of roller passes and foot size of sheepsfoot roller from test embankments on silty clay. In all cases the roller foot pressure is 250 p.s.i. (after Turnbull and Foster, Ref. 402).

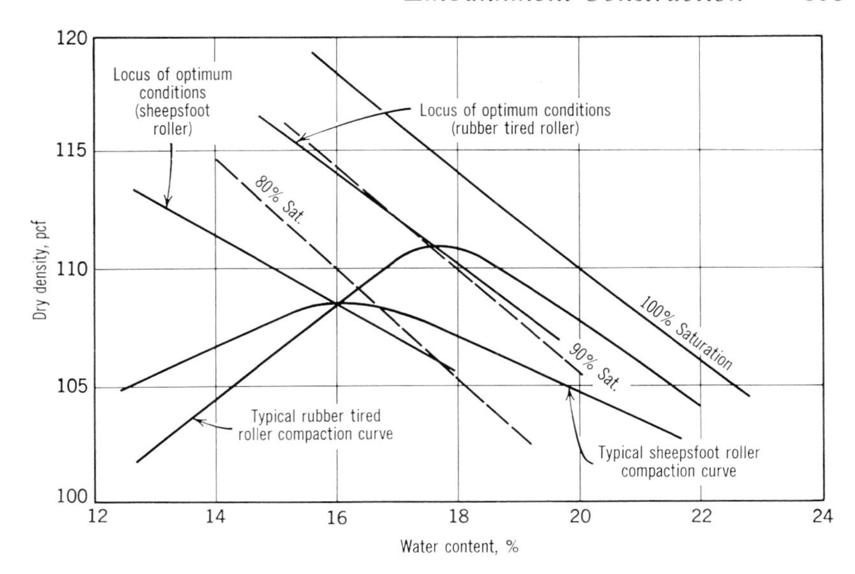

Fig. 11.3:12 Relationship between sheepsfoot and rubber-tired roller compaction from test embankments on silty clay (after Zegarra, Ref. 564).

tent for both the 50-ton rubber-tired roller and the standard sheepsfoot roller is the same general order of magnitude as the standard Proctor optimum water content.

11.3d Embankment Quality Control

PURPOSE AND METHODS OF CONTROL

The construction of an embankment must be controlled in a manner to assure that the zones are relatively homogeneous and that the average properties are at least equal in quality to the values assumed in the design. Basic construction control is exercised by visual evaluation of: (1) the uniformity of the water content and the properties of the soil layer before compaction; (2) the thickness of the compacted layers; and (3) the action of the roller and heavy hauling equipment on the construction surface.

In addition to its role as a compactor, the heavy roller also can be considered a testing apparatus. If the water content is uniform and the layer is not too thick the action of the roller indicates whether the water content is satisfactory and whether compaction is being obtained. If after several passes a heavy rubber-tired roller ruts excessively or a sheepsfoot roller continues churning up the construction surface and shows no sign of walking out, the water content is too high for the roller. If the water content is too low for good compaction, the sheepsfoot roller will walk out several inches after only three or four passes, and a rubber-tired roller will leave a smooth, hard and stiff surface. At water contents which result in good compaction, there will always be some penetration of the compaction surface by heavily loaded rubber tires, Fig. 11.3:13, and sheepsfoot rollers will walk out and lift the drum off the surface at least a small distance after eight passes.[1] At a good water content, there may or may not be a noticeable undulation of the embankment construction surface adjacent to the passage of heavy construction equipment. This will depend largely on the soil type.

The layer thickness frequently has an important influence on the

[1] As discussed in the previous section, many dams have been constructed to satisfactory densities by heavy sheepsfoot rollers which did not walk out. However, in such circumstances higher densities could probably always have been obtained at a lower water content.

Fig. 11.3:13 Appearance of clayey soil at good water content being compacted with 50-ton rubber-tired roller.

ease of obtaining the desired compaction; however, the contractor naturally is interested in placing the material layers which are as thick as he is allowed to use. Because of this and because it is difficult to control the thickness by visual inspection alone, arguments frequently arise on this point. Consequently, as a minimum practice it is necessary to take levels on the same point on the construction surface after every few layers in order to see what the average thickness "after compaction" has been and to calibrate the inspector's judgment. On jobs where considerable difficulty arises with the contractor, it is necessary to resort to the measurement of each layer of spread material. This is most frequently done by driving stakes in the ground with marks on them showing the top of the spread layer. Another procedure, which isn't as cumbersome as it might seem, is to set up an instrument and have the contractor spread each layer to a given surface elevation. On some jobs, the disputes which have arisen over the layer thickness and particularly over the difference between the thickness "before" and "after" compaction have been so troublesome that the engineer has preferred on subsequent jobs to specify the thickness "before compaction" and to specify a layer several inches less in thickness than he might consider necessary.

To supplement the visual evaluation of the embankment construction and provide the engineer with guides to judgment, tests of the density and water content are also performed. Except in special cases, laboratory strength tests are not used for routine construction control; the relationship between embankment density and strength is established in advance, and the strength is controlled indirectly through tests of the embankment density. In a few dams the embankment strength has also been controlled by limiting the plasticity of the fines to predetermined maximum values.[1]

Another useful technique is to keep track of and control the rate of roller production in terms of average values of cubic yards of material compacted per roller per unit of time. This figure will often correlate very well with the average compaction obtained. Consequently, when a trend toward a higher frequency of substandard test results is ob-

[1] The strength of a compacted clay is roughly related to the liquid limit. At some dams where the principal construction material consists of clay, and the liquid limit of the prospective borrow materials varies considerably, the strength of the embankment has been controlled to a pre-established value by restricting the liquid limit of the construction material to a maximum value. While there is not sufficient information available to allow any reliable generalization, in some clay deposits in which the material is all of the same origin but varies in plasticity, the shear strength in terms of effective stresses may be decreased 10 to 15% by an increase in the liquid limit from 50 to 60%.

served, one means for reversing the trend is to decrease the rate of roller production.

FIELD DENSITY TESTS

The compaction water content of impervious embankment sections is controlled in order to achieve the desired densities. The water content is also controlled within pre-established limiting values in order to minimize the danger of embankment cracking or to control construction pore water pressures (Sec. 1.4).

The density and the water content of the compacted embankment are measured by "field density" tests and are then compared with the results of laboratory compaction tests (usually standard Proctor, Sec. 11.3*b*) on the same sample. If the measured density and/or water content of a section of the completed embankment fall outside the pre-established limits, the section may be rerolled or broken up, reworked to an acceptable water content, and then rerolled.

Two types of field density tests are made: routine tests, and tests on portions of the embankment which the inspector believes may be inadequately compacted. The routine tests are performed at regular intervals (frequently for each 2,000 to 3,000 yd.^{3} of embankment material), and after the initial compaction problems are solved, the results are employed primarily as a quantitative documentation of the construction record. Results of tests on suspect areas are used by the controlling engineer to decide whether or not to require reworking and additional compaction.

The most common field density test is the simple "calibrated sand" test, in which a measured weight of compacted soil is excavated carefully by hand-digging a small hole with smooth walls from the compacted soil layer being tested. The volume of the hole is then measured by filling it with calibrated dry sand.[1] Other types of field density tests have been used successfully but less frequently; for example, water-filled rubber balloons or oil can be used instead of calibrated sand to measure the volume of the hole, or, in fine-grained soils, undisturbed samples can be taken by thin tube drive samplers.[2]

The results of field density tests vary considerably with the test

[1] For details of field density test procedures, see AASHO or ASTM standard test procedures.

[2] For a study of the accuracy of various methods of field density tests, see Ref. 447 and discussions.

procedures, even within the relatively limited range of procedures which are generally considered good practice. On a number of dams in which investigations of the embankment density have been made (with rotary or drive samples from drill holes, or with handcut block samples from test pits) after completion of construction, the average density measured has been quite different from the average of the tests performed during construction. One of the most frequent causes of error in field density tests, and one which may go unnoticed throughout construction, is that the compacted material adjacent to the hole excavated for the field density test tends to squeeze into the hole and reduce its volume. Well-documented evidence exists that this action has caused measured densities to run as much as 5% higher than the actual density without being suspected.

The results of field density tests depend to a large degree on the depth below the construction surface at which they are made. Usually specifications require that they be performed on the uppermost compacted layer. However, for many soil types considerable additional compaction will occur to depths of at least 12 to 18 in., so that the results of tests made at a depth of 18 in. below the compacted surface are likely to be higher and more uniform than those made at a depth of only 6 in. For construction control on any given project, it is desirable to investigate the influence of compaction at various depths on the embankment soil and to adopt a suitable test procedure to be used throughout the job.

Field Water Content Determination. The standard method of determining the water content of the embankment material taken in the field density test is to dry a representative specimen in an oven at a temperature of about 110°C. for at least 24 hr. Since the laboratory tests are used primarily as a guide and a supplement to the inspector's visual evaluation of the material, a delay of 24 hrs. in obtaining the results is usually not important. However, several rapid water content tests have been developed. The most satisfactory, which has been used at a great number of dams, is carried out by heating the soil in an open pan over a hotplate or gas flame. For many soils the open pan procedure gives the same results as the standard oven test, although often this can be achieved only by cooking the soil on the hotplate for a definite length of time which is determined in advance by experimentation.[1] By running parallel water content tests in the

[1] Especially for fine-grained soils, "water content" is not an absolute value. The water content measured in a test depends on the temperature and the length of time the soil is heated. See, for example, "How dry is a dry soil," *Proceedings of the Highway Research Board*, vol. 29, p. 491.

conventional oven, the engineer can determine and reliably control the factors which influence the results of the hotplate tests.

A number of other rapid methods of water content determination have been devised. In his original papers (Ref. 209), which had such a dominant influence on the development of modern control testing procedures, Proctor introduced a penetration "needle," a cylindrical rod of small diameter, which tested water content by measuring the resistance to penetration of the soil. Although some earth dam engineers have found it to be a convenient supplement to construction control procedures, it is not commonly used because (1) it does not work satisfactorily in the coarse soils from which the impervious sections of many dams presently are being constructed, and (2) even for fine-grained soils, the results are not very precise.

Other more recently formulated methods of determining water content have not yet been widely practiced for earth dam construction, so that their accuracy is not established (Refs. 394, 641).[1] One very promising method, which was developed by J. W. Hilf in 1957 is based on the statistical relationship between optimum water content and maximum density (Refs. 405, 422). The procedure allows the determination of the relationship between the embankment water content and dry density and the laboratory optimum conditions without the necessity for measuring the water content. Consequently, results can be obtained in an hour or less. The Hilf rapid method has been adopted for compaction control on USBR dams, and its procedures are described thoroughly in the USBR's Earth Manual (Ref. 274). It is probable that the method will be widely used in the future (Refs. 322, 429).

CONTROL PROCEDURES FOR MATERIALS CONTAINING GRAVELS

Influence of Gravel on Compaction Curves. Coarse materials consisting of mixtures of gravel and sand with sufficient fines to fill the voids (GM and GC soils) make excellent impervious cores for earth

[1] In one method, for example, calcium carbide is mixed with the soil in a closed container and the pressure developed by the gas generated is measured. The gas pressure is directly related to the quantity of water mixed with the carbide and hence is a direct measure of the water content (Refs. 445, 515). For soils which do not contain high contents of clayey fines, this method has proved quite reliable and convenient. Another new development is a testing apparatus for measuring both the water content and the density of the compacted embankment using "nuclear methods" (Refs. 357, 443, 444). This is being tried out by a number of agencies.

dams. In a natural deposit of soil of this type, in which the index properties of the fines may be quite uniform, the content of coarse gravel-sized particles (usually defined as particles coarser than the no. 4 sieve) is likely to vary considerably. By using the same compaction effort on such a material, different water content-density curves are obtained for portions of the soil with different percentages of gravel. A typical family of compaction curves for one soil deposit is given in Fig. 11.3:14*b*.

The density of the total material increases and the optimum water content decreases with increasing percentages of gravel sizes up to a maximum of about 60 to 75%. Above this value the density decreases again (curve 1, Fig. 11.3:14*a*). Soils containing more gravels than the value necessary to result in maximum dry density (60 to 75%) do not have sufficient fines to fill the voids and rapidly become pervious materials with a small increase in the gravel content.

Experience indicates that for the material with less than approximately 30% by weight coarser than the no. 4 sieve, the compaction of the fine fraction is not affected by the presence of the gravels. For material with gravel content in the range roughly between 30 and 50%, the presence of the gravel interferes somewhat with the compaction of the fines; consequently, the maximum density of the minus no. 4 material is likely to be lower and the optimum water content higher than they would be if the gravel were not present. This is true for both field compaction with rollers, Fig. 11.3:5, and for laboratory compaction tests on the total material. Laboratory researches indicate that the higher the plasticity of the fines, the higher the quantity of coarser material which will be tolerated before the compaction of the fine fraction will be influenced (Ref. 414).

For soils containing more than about 50% by weight coarser than the no. 4 sieve, the gravel content has a large influence on the water content and the density of the minus no. 4 material. For these very coarse soils, the water content of the minus no. 4 fraction necessary to achieve maximum density for the total material is higher than the normal optimum water content. In extreme cases, this water content may even approach the liquid limit of the material as was experienced during the construction of the Hills Creek Dam (Sec. 11.3*f*). For this reason different methods of water content-density construction control are used for materials containing large quantities of gravels.

Control Procedures. For soils with a gravel content up to 25 or 30%, the most widely used and satisfactory control method is to compare the density of the minus no. 4 (or ¾ in.) fraction of the compacted embankment with the results of the laboratory test performed on the

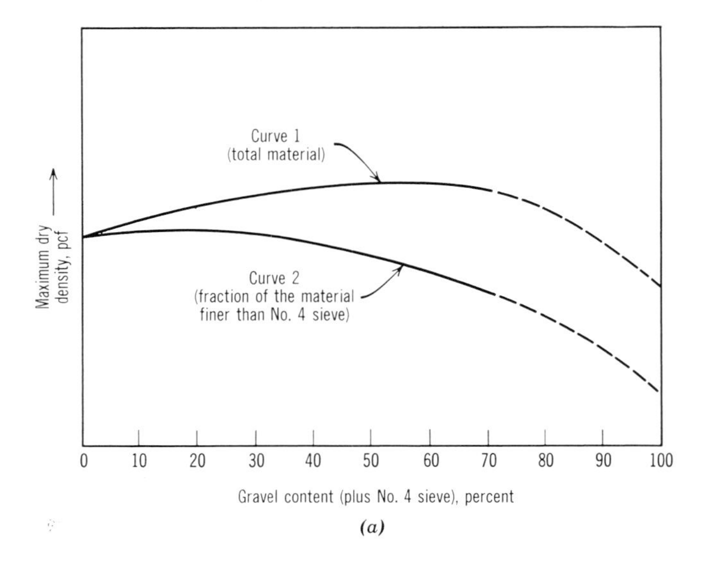

(a)

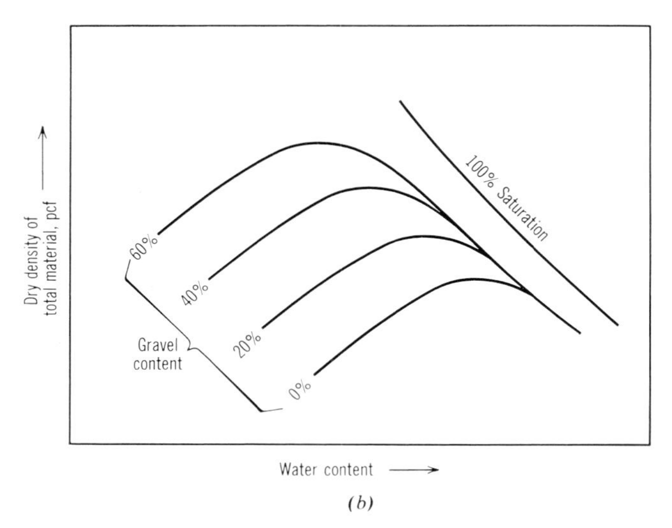

(b)

Fig. 11.3:14 Relationship between density and gravel content for coarse, impervious soil with single compaction effort.

same fraction. This is done by computing the density of the fine fraction of the embankment material by measuring and subtracting the weight and volume of the gravel particles encountered in the field density test.[1] For some soils testing may indicate that this procedure can be used up to a gravel content of 50%.

For coarser soils, considerably more testing effort is required to obtain the same degree of density control during construction as obtained for fine-grained soils. There is no widely accepted "best" method of compaction control for the coarse soils.[2] Each soil deposit presents a somewhat different problem and control methods are improvised to suit the conditions. The most straightforward method is to use a laboratory test with a compaction cylinder which is large enough to allow the compaction of the total material including gravels. Laboratory tests using compaction cylinders with a volume of 1 $ft.^3$ or larger have been used satisfactorily to control material with maximum gravel sizes of 3 to 6 in. The work required for these large laboratory tests is much greater than for standard small tests and, consequently, fewer tests can be performed by a laboratory staff of a given size. However, since the tests are intended only to supplement the primary construction control which is carried out by visual inspection of the construction activities, many engineers prefer to use fewer large scale tests performed on the whole material than a greater number of tests performed on a fraction of the soil.

Another method of density control which has been used with reasonable success for coarse materials is to establish a set of standard curves for the soil in each borrow pit which relate the maximum laboratory density of the total material to the percent of gravel (Ref. 449). For a given soil with varying quantities of gravel, the percentage of gravel sizes has the major influence on the total density and this relationship can be established within fairly close and reliable limits. Curves of the type shown in Fig. 11.3:14*a* (upper solid curve) are obtained. Often one or two curves are sufficient to cover the range of soil in a single borrow pit. After these basic laboratory curves are established, the control procedure consists of measuring the density of the total embankment material (field density test), determining

[1] In soils where only a few pieces of gravel are encountered in the field density test, the influence of the existence of the gravels on the computations can be obviated by replacing the gravel particles in the field density test hole before pouring in the calibrated sand to measure the volume of the hole.

[2] Reference 704 contains an interesting recent summary and analysis of methods which have been used by different engineering organizations for control of coarse grained soils.

the percentage of gravel, and comparing the field and laboratory densities. The basic laboratory curves are obtained by compacting representative specimens of the soil with different percentages of gravel in large sized laboratory compaction molds.[1]

11.3e Handling Dry Materials

ADDING WATER ON THE CONSTRUCTION SURFACES

In arid regions the natural water content of soils in prospective borrow areas is usually too low, and water must be added. If the average water content is only a few percent below the desired value for compaction, the extra water can be applied by sprinkling the soil after it is spread on the construction surface and before it is rolled. Except where the soil is very sandy and is being rolled with a sheepsfoot roller which acts as a mixer, the additional water must be blended into the soil with harrows or plows, Fig. 11.2:7. The amount of blending necessary depends on the fineness and plasticity of the soil. If the layer to which water is added is not thoroughly mixed, and if sufficient time is not allowed for uniform absorption, the water is retained in pockets of wet soil and poor compaction results. Poor mixing on the construction surface is one of the most common deficiencies observed during the construction of earth dam embankments.

The quantity of water which can be added practicably to the soil on the fill with a reasonable amount of harrowing and plowing also depends on the fineness and plasticity (or permeability) of the soil. For the great majority of soils used for impervious sections (silts and clays of low to medium plasticity, and coarse soils with silty or clayey fines), it is possible to add at least 4% water (by weight) without excessive working. The coarser and less plastic the soil, the greater the quantity of water which can be worked uniformly into it. Little or no water can be added satisfactorily to clays of medium and high plasticity unless they contain a large amount of sand or gravel. Regardless of the blending effort expended on very fine clayey soils, a minimum of several days of "curing" is required for the water to be absorbed by the individual clay lumps.

Where it is difficult to add water uniformly to the soil on the construction surface, water sprinkling trucks equipped with air com-

[1] An interesting method for computing the basic laboratory curves for this purpose has been developed by the Washington State Highway Department (Ref. 448).

Fig. 11.3:15 Water trucks with pressure-sprinkling systems are superior to gravity feed sprinklers for increasing the soil water content on the construction surface.

pressors are much more satisfactory than sprinkling trucks with gravity-feed only. The pressure tank sprinkler spreads a larger quantity of water and distributes it more evenly than other sprinkling methods in common use, Fig. 11.3:15.

IRRIGATING BORROW PITS

In arid regions the average natural water content of soils in borrow areas is often 10 to 15% below the desirable value for compaction. Under such conditions, borrow pit irrigation always results in a more satisfactory and uniform distribution of the water in the soil and is usually much more economical than adding the water on the construction surface.[1] In addition to its lower cost and greater efficiency, irrigating the borrow pit results in less work being required on the construction surface so that earth placing can proceed more rapidly.

[1] Comparisons of the cost of adding about 5% water to silty clay during the construction of Oahe Dam on the Missouri River indicated that sprinkling on the construction surface was more than twice as costly than borrow pit irrigation (Ref. 74).

Fig. 11.3:16 Typical sprinkler system used for irrigating borrow pits before excavation. The piping consists of light-weight aluminum which can be moved to a new area in a few hours by hand labor.

Borrow pits are irrigated by ponding water on the surface with low dikes or with a pressure sprinkler system, Fig. 11.3:16. The length of time required may vary from a few days to several months, depending on the permeability of the natural soil deposit and the depth of soil being moistened. After irrigation, a curing period is also desirable to allow the added water to be absorbed uniformly by the soil. The time needed for irrigation and curing and the most desirable method of adding the water at any given site can best be determined by experimentation.

The water content of dry soil deposits is seldom raised by irrigation to values which are too high for compaction. Experience indicates that the great majority of natural soils, even if irrigated over very long periods, will take only a certain amount of water which approximates a good construction water content. In very few cases has it been necessary to limit the irrigation period or to change from ponding to sprinkling in order to avoid adding too much water. On dry days, in fact, several percent of water may be lost by evaporation during the process of excavation, spreading, and rolling; and supplementary sprinkling will be desirable on the construction surface.

Irrigation by ponding is best adapted to relatively flat areas and tight soils for which long periods of irrigation are needed. Sprinkling can be used successfully at almost all sites, but it is especially advantageous on sloping ground and in large borrow areas where a relatively shallow depth of soil is to be irrigated. In some borrow areas it has been found desirable not to strip the top soil before irrigating, since the stripping operation may tend to seal the natural holes and cracks in the ground surface which allow the penetration of water. Ripping the surface to a depth of 2 to 3 ft. with rippers on the back of tractors is effective for opening any impervious layer which may exist on the surface, Fig. 11.3:17. When sprinklers are used, contour plowing is often found desirable to prevent surface runoff from hilly sites.

Borrow areas are frequently wetted to depths of 5 to 15 ft. by surface irrigation, and in some cases it has been possible to raise the water content successfully to a depth of 30 ft. or more. The main deterrent to irrigation at great depths is the existence of relatively tight horizontal soil layers which prevent the downward percolation of water. In extreme cases it has not been possible to make the water penetrate such layers, and the effective practical depth of irrigation is limited.

In cases where the borrow pit irrigation is less than satisfactory, supplemental water can be added by sprinkling the face of a shovel excavation. With this procedure the water is mixed into the soil by the shovel dipper and blends with the loosely excavated soil as it is

Fig. 11.3:17 Ripper used to dig furrows in the surface of dry borrow areas to accelerate the penetration of irrigation water.

Fig. 11.3:18 View of borrow excavation face in dry silty sand after 72 hr. of sprinkler irrigation showing penetration of water to a depth of about 3 ft. A power shovel excavation on the face shown resulted in a mixture of dry and moist soil with water content near optimum.

being hauled and spread on the embankment, so that it is mixed more uniformly than it could be if it were added on the construction surface.

ADDING WATER IN ROCK SEPARATION PLANTS

At projects where the soil is processed through a rock separation plant to remove oversize cobbles (Sec. 11.2*c*), a considerable quantity of water can be blended into the soil by sprinkling within the plant. Since most of the large lumps in the soil are broken down by the extra handling and the action of the plant, the material is in a condition to absorb water readily. After the rocks have been screened out, the water is added to the soil by sprays, which are located at the terminal end of the chutes or conveyor belts, just before the soil is dumped into the hauling equipment or loading hopper. With a little experimentation, the quantity of water added can be adjusted to bring the water content of the soil within very close range of the desired value.

ADDING WATER TO COARSE RESIDUAL SOILS

A special problem arises when controlling the construction water content of residual soils or other soils which break down and become appreciably finer grained during the process of excavation, spreading and rolling on the dam surface. Such materials have different gradation curves and a higher content of fines at progressive stages of the construction process. Moreover, they have a higher optimum water content after each roller pass, particularly when sheepsfoot rollers are used. Consequently a soil which may appear too wet before rolling commences may be too dry when finally compacted. Each soil of this type creates a different moisture control problem which must be studied and solved. Because the optimum water content of such materials increases as they are being rolled, it is necessary as a general rule to add water by sprinkling between roller passes in order to avoid compacting the material excessively dry. For treatment of the extreme case where the material consists primarily of disintegrating rocks, see Section 11.5*d*, "Embankment Sections of Soft Rock."

11.3f Handling Wet Materials

It is usually easier to add water to a dry soil than to reduce the water content of a wet soil. The amount of difficulty in lowering the water content will depend both on the fineness and plasticity (permeability) of the soil, and on the rainfall during the construction season, and the influence of the rain will depend as much on the type of rain as on the total precipitation. When rain comes down in a few hard, short storms, for example, the air may be dry during the rest of the period and the soil will have a chance to dry. On the other hand, if the rain falls continuously in a light drizzle, little if any drying will be possible.

In most parts of the world there is a dry season of sufficient length that the water content of the impervious material can be controlled without excessive work and cost. In areas with very wet climates, however, it may be extremely difficult to reduce the water content of the available impervious soils enough to allow them to be compacted in the normal manner with heavy rollers. Under such conditions the engineer has two choices:

1. Use conventional construction methods and fight the water content down. This procedure has been used at many dams in wet areas in spite of considerable difficulties and delays in construction.
2. Place the impervious material at its high natural water content even though it cannot be compacted with heavy rollers to a high density. This method has been used for a number of recent Swedish and Norwegian dams, though it is infrequent in other parts of the world.

PROCEDURES FOR DECREASING THE WATER CONTENT

Experience has shown that for wet soils as well as dry ones it is more effective and economical to process the soil to an acceptable water content in the borrow pit before excavation. If the borrow material is below the ground water level, the level must be lowered by drainage. The water should also be drained out in soil deposits which are located above the ground water table, but which contain layers or lenses of pervious material with free water in the form of perched water tables.

If the construction season is not excessively rainy, wet materials can sometimes be dried by ripping or plowing and aerating the soil to a depth of several feet. This is relatively easy for silty and sandy soils but difficult for clays. Even in dry climates it can be a formidable undertaking to lower the water content of tough clays to an acceptable uniform value by aeration, because they dry in hard chunks which are difficult to process.[1]

In very wet climates it may only be possible to prevent the material from getting any wetter during construction. This is done by providing surface drainage in the borrow area[2] and by using excavating

[1] At a few dams with wet borrow materials, heroic battles have been fought to reduce the water content. The soil at the Mud Mountain Dam, Washington (1940) was dried in rotary kilns, and a huge canvas tent was erected over the construction surface to shelter it from rainfall (*Engineering News Record*, March 27 and August 28, 1941). See Ref. 222 for the description of the use of an aggregate dryer at the Beechwood Dam, Canada, (1955). At the Sasumua Dam, Kenya (1950), wet clays in the borrow areas were aerated in 4 in. layers with 6 passes of a rotary cultivator (Ref. 68). Similar problems were experienced at the Dorena Dam, Oregon (1949) (Ref. 407).

[2] At the Yale Dam which was constructed in the very wet coastal area of Washington in 1953, contractors surfaced portions of the borrow area with asphalt to prevent penetration of rain water (Ref. 51).

equipment which minimizes the tendency of the borrow material to soak up additional water. Excavation with power shovels on a vertical face is more effective in preventing soil from being wetted by showers than excavation with scrapers.

When the material is spread on the embankment surface in dry weather, the water content can be reduced further by plowing or discing before rolling. Impervious materials should not be placed on the dam while it is actually raining except in special circumstances, though embankment construction often can be carried out successfully in the dry periods between intermittent rains. In such cases rubber-tired rollers are superior to sheepsfoot rollers because the rubber tires leave a hard construction surface, which soaks up rain less than the loose surface left by the sheepsfoot. When a sheepsfoot roller is used at a dam where the construction is shut down frequently by rain, smooth wheel rollers are brought in to seal the surface, Fig. 11.3:19. Whichever roller is used, the surface should be sloped to allow the rain water to run off instead of standing in puddles and seeping into the upper layers.

A few soils which are suitable for the construction of the impervious section of earth dams have been placed successfully in the rain. At the site of the 512-ft.-high Swift Creek Dam, which was completed in Washington in 1958, the average annual rainfall is about 120 in., and

Fig. 11.3:19 Smooth steel roller used to seal embankment surface to prevent excessive wetting during rainstorms.

there is no dry season—only periods of lesser precipitation. The design of the dam and the selection of the materials was more or less dominated by considerations of weather. It was decided that the core would be of gravelly, silty sand, Fig. 3.3:4, which could be placed even during heavy rain provided that certain construction precautions were taken. In the design stage, a test embankment was actually made from the sand, and rainy conditions were simulated by a sprinkler system.[1] The principal procedures developed for the wet weather construction, which were completely successful, consisted of (Refs. 77, 159, 160):

1. Excavating the borrow pit on a vertical face.
2. Grading and compacting the surface of the borrow area to prevent surface water from puddling and seeping into the fill material.
3. Sloping the construction surface (8 to 10%).
4. Rolling the compacted layers in the direction transverse to the longitudinal axis of the dam to facilitate drainage on the construction surface.
5. Compacting the material as soon as possible after placement to minimize the time that loose fill would be exposed to rainfall.

With these adaptations of conventional procedures, the embankment construction surface was sufficiently stable to support compaction by a 50-ton rubber-tired roller even when the soil was very wet, Fig. 11.3:20.

METHODS OF COMPACTING AT HIGH NATURAL WATER CONTENTS

Swedish "Wet Compaction." Because of the long, cold winters in Sweden, earth dams can be constructed only in the 6 months when the temperature is above freezing, which are usually so rainy that there are only a few days when the wet soils available can be dried. For the impervious sections of moderately high earth dams, the Swedish State Power Board has used successfully since 1951 a wet compaction method which makes it possible for work to continue in the rain (Refs. 1, 76, 217).[2] A few years later the technique was

[1] In order to demonstrate the difficulties which would be encountered, moving pictures taken of the construction of the test section were shown to prospective contractors bidding on the job.

[2] The construction of wet and very soft cores in earth dams has a long-time precedent in the "puddled" cores designed by English engineers. For these, the

Fig. 11.3:20 Typical view of 50-ton rubber-tired roller working on wet embankment construction surface at Swift Creek Dam. (Courtesy Pacific Power and Light Company)

adopted by engineers in Norway and has been used to date on about a half dozen dams.

In these Swedish dams the cores, which have consisted usually of glacial moraines in the form of gravelly sands with silty fines, are compacted at a water content so high that the construction surface is too soft to support the travel of heavy rubber-tired hauling equipment or any kind of heavy rollers. The impervious material is dumped on the construction surface adjacent to the core and pushed to the core zone with bulldozers. It is then compacted by crawler-type tractors, which in some cases have to travel rapidly in order to avoid bogging down. For construction control the material is considered too wet if a rapidly moving, heavy tractor penetrates the construction surface more than about one foot and too dry if it does not penetrate more than 1 in., Fig. 11.3:21*a*.

---

clay is processed to a soft puttylike consistency in pug mills, spread by hand labor in thin layers, and knit into an impervious mass with the feet of laborers. It is interesting to note that this practice was used for a few of the early dams constructed in California in the mid 1800's.

(*a*)

(*b*)

Fig. 11.3:21 Swedish wet-compaction method (a) Typical view of construction surface. (b) compaction, of soil at abutment contact with rock surface using the boots of laborers (Arstaddalen Dam, Norway, 1962).

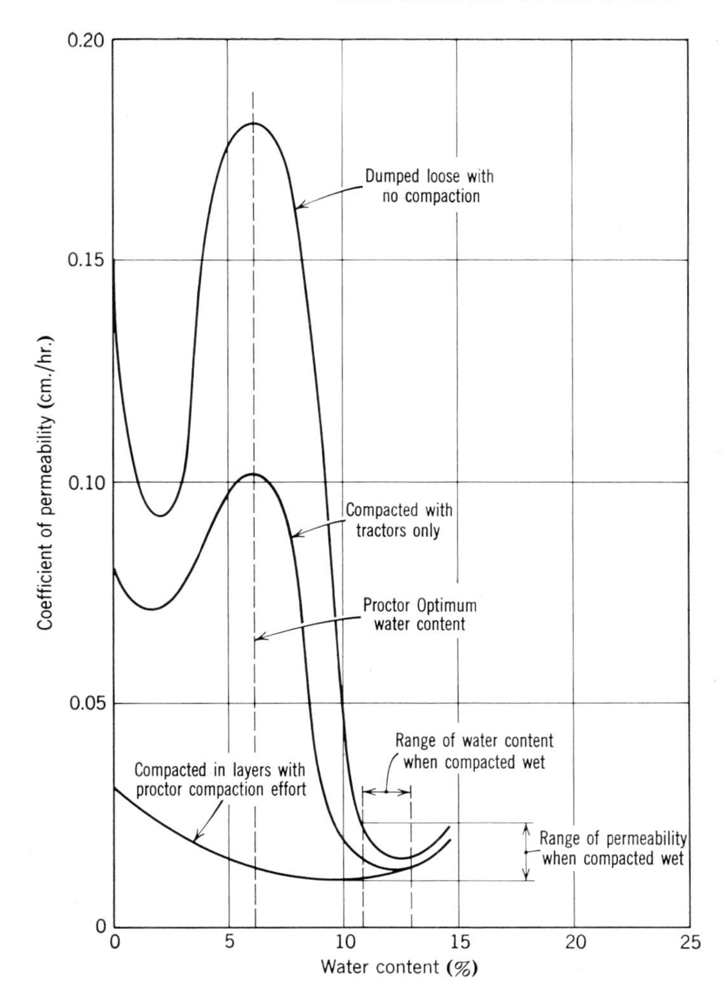

Fig. 11.3:22 Influence of compaction effort and water content on permeability of typical sandy moraine used for dam cores in Sweden. (after Swedish State Power Board, Ref. 709).

The density of the core material using this method of construction is, of course, considerably lower than the Proctor maximum density; however, the permeability is practically the same as if it were constructed with heavy rollers at the optimum water content (Ref. 442). Fig. 11.3:22 shows the influence of water content and compaction method on the permeability of embankments constructed from the

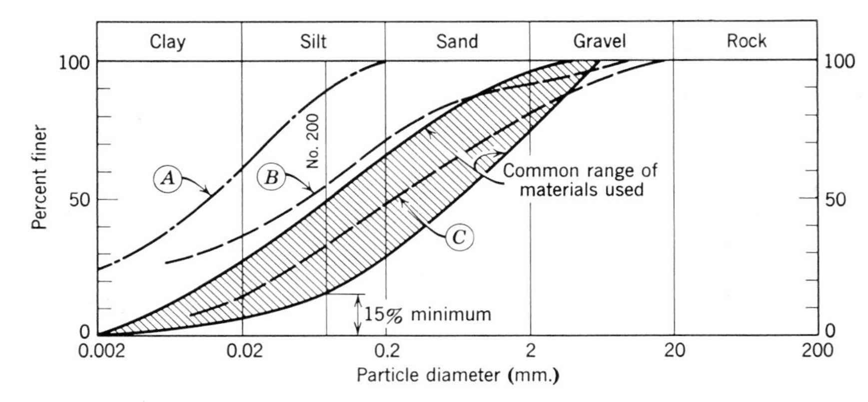

Fig. 11.3:23 Gradation curves for the common and finest materials which have been used with the Swedish wet-compaction method.

Average Curves for Specific Dams:

Curve (A) Tustervatn Dam, Norway (1956–1957) (L. L. ≈ 31; P. I. ≈ 12)
Curve (B) Arstaddalen Dam, Norway (1962–1963) (L. L. ≈ 25; P. I. ≈ 7)
Curve (C) Messaure Dam, Sweden (1958–1962) (Cohesionless)

typical sandy moraine in Sweden. While the relationship shown in Fig. 11.3:22 may be a little startling at first to some, it would be suspected by any child making a dam of silty sand as he soon finds that he must place the material in a wet sloppy condition if he is going to make his dam hold water. Fig. 11.3:24 shows a typical excavated face of the material used for the core of the high Messaure Dam in Sweden. (See cross-section of dam, Fig. 1.2:49.)

The wet compaction method has been used most often for dam cores of silty, gravelly sand though, in a few cases, fairly fine silt has been used. The range of gradation for most of the materials used, as well as curves for two of the finer materials, is shown in Fig. 11.3:24. The Tustervatn Dam, which was built of a very fine silt (see dam cross-section Fig. 1.2:49), is only about 50 ft. high. During the construction of this dam, the surface of the core was too soft even to support a crawler type tractor, and the material was spread and compacted with a light, rubber tired farm tractor with chain treads on the front and rear wheels, Fig. 11.3:25.

Some general details and comments concerning the Swedish wet-compaction methods follow:

1. An effort is usually made to spread the material in layers 9 to 12 inches in thickness, though it is difficult to control the layer

ig. 11.3:24 Typical view of excavated face in borrow pit of glacial moraine core naterial placed by wet compaction in Messaure Dam, Sweden, 1958–1962 (see ypical gradation curve, Fig. 11.3:23).

ig. 11.3:25 Light farm tractor used to compact wet silt impervious core at Tus-ervatn Dam, Norway. (Courtesy Mr. B. Kjaernsli)

thickness since tractors sink 12 in. or more into the constructio surface. Specifications usually call for three to five coverages c the tread of a D-7 tractor over each point on the constructio surface.

2. Little or no control of the density of the core is attempted. It i impossible to use calibrated sand tests since the soft materia squeezes together and reduces the volume of the hole by a unknown amount. At some of the larger dams, no density test were made whatsoever. At the 60-meter-high Arstaddalen Dar in north Norway in 1962, tests were made by digging an un disturbed sample from the construction surface, trimming this i the laboratory, and weighing it in and out of a bath of oil to ge its volume.
3. Primary emphasis in the construction control is on the permea bility of the core material. Generally, a specification is used tha the material shall not contain less than 15% by weight finer tha the No. 200 sieve.
4. One major design problem is the settlement of the core durin and after construction. The magnitude of the settlement depend to a great extent on the soil type. At the high Messaure Dam i Sweden, in which the core was constructed from the typical non plastic glacial morraine, the postconstruction settlement is ex pected to be less than 0.5% of the height. At the 50 ft. hig Tustervatn Dam, where the core was of fine silt, the post con struction settlement was nearly 2% in the first 5 years after con struction, Ref. 706.
5. Pore water pressures equaling nearly 100% of the overlying em bankment weight generally are measured soon after the installa tion of the piezometers. However, because the cores are gener ally made thin, fairly rapid dissipation occurs.
6. In Sweden, zones of the embankments near concrete structure and steep rock abutments are compacted by the more conven tional techniques with the water content near Proctor optimun in order to minimize differential settlements. In Norway, at th 60-meter-high Arstaddalen Dam, the wet silt core was compacte against the cleaned rock abutment surface with the boots o laborers, Fig. 11.3:21*b*.
7. When the weather is wet, the water content of the material i raised naturally to the high value desired by the rain. When th weather is dry, the material is wetted with hoses or water truck on the construction surface. Seldom is it necessary to stop worl during wet weather for more than a few hours.

8. In a few cases, contractors have tried to make roads on the construction surface to cross from one side to the other by placing temporary pavements of dry material. This generally has not been satisfactory, and it has been proved preferable to cross from upstream to downstream by using roads cut into the abutments. This is definitely a disadvantage of the method.

The authors believe that this Swedish wet-construction technique ın be used to advantage in many other parts of the world where iny weather is a major problem in construction of earth cores. This probably especially true in some tropical areas. In fact,[1] the conruction procedures described for the Hills Creek Dam in Oregon in e following section are very similar and were adopted for the same asons.

acing Wet Core at Hills Creek Dam (Ref. 439). Problems similar those encountered in wet weather arise at dams where the imperviıs sections consist of coarse, gravelly soil with a barely sufficient ıantity of fines to fill the voids and make the material impervious. ɔr these soils the water content of the minus no. 4 fraction must be nsiderably higher than standard Proctor optimum in order to result the maximum density of the total material (Sec. 11.3*d*). The higher ater content may also be necessary to insure that the voids are cometely filled with fines and that no "honeycombing" occurs.

At the 340-ft.-high Hills Creek Dam in Oregon, constructed in 59–1960, the core is a very coarse clayey gravel with barely enough ıes to fill the voids and render the compacted material impervious. he material was screened to remove the plus 6-in. cobbles, so that e gradation was as follows:

PARTICLE SIZE	PERCENT FINER THAN
3 in.	70 to 85
No. 4 sieve	30 to 40
No. 200 sieve	10 to 20

he Atterberg limits of the fines fell on or slightly below the A-line the plasticity chart, Table 4.2:1, and the liquid limit usually ranged tween 30 and 50%. The clay mineral was predominantly a halloysite

[1] lso in Japan, clay cores with water content of 15 to 20% above standard Proctor timum have been compacted by specially devised rollers with short (4-in. ıg), wedge-shaped feet (Ref. 71).

with some montmorillonite.[1] The standard Proctor optimum wat content of the minus no. 4 fraction varied from approximately 24 28%.

In the early part of construction, attempts were made to compa the core material with a 50-ton rubber-tired roller at a water conte which was at or slightly above standard Proctor optimum (of th minus no. 4 material). At this water content, the fines were so stiff th the material could not be worked into a homogeneous mixture, an segregation and honeycombing occurred, Fig. 11.3:26.

In order to make the fines soft and plastic enough to allow them flow or be squeezed into the voids, it was necessary to place the so at a water content (of the minus no. 4 fraction) 10% or more abov standard Proctor optimum. At the higher value, which was used f most of the core, it became possible to compact the material in homogeneous, impervious mass in which the fines filled the void Because the construction surface was then too soft to support th heavy roller or the large earth moving equipment, the material wa dumped on the surface adjacent to the core section and pushed in

[1] The liquid limit of samples which have been dried before testing is much low than for tests in which the sample was not dried below the plastic limit.

Fig. 11.3:26 Segregation and honeycombing of coarse, well graded clayey gra used for the impervious section of Hills Creek Dam when placed at water conte near standard Proctor optimum (of the minus no. 4 material.)

Fig. 11.3:27 Views of the construction surface of very wet, clayey gravel impervious core after compaction with crawler tractor, Hills Creek Dam. This is the same material as shown in Fig. 11.3:26, except that it is being placed at a higher water content which minimizes the tendency for segregation of the coarser gravels.

the core and spread with bulldozers. It was compacted to 8-in. layers with two passes of a heavy crawler-type tractor, which sank into the construction surface 6 to 8 in., Fig. 11.3:27.

Readings of piezometers located in the wet compacted core indicated that the initial pore water pressures were nearly equal to 100% of the weight of the overburden and that moderate dissipation due to drainage occurred during construction (see Fig. 8.1:6). Laboratory tests on compacted specimens of the total material showed that the coefficient of permeability ranged roughly between 1 and 300 ft./yr. with an average of about 70 ft./yr.

11.3g Problems Caused by Cold Weather

In cold climates it is necessary to halt the construction of impervious embankment sections during the winter because good compaction cannot be obtained if the soil is frozen.[1] Often the contractor desires to keep working as long as possible, and unless a large snow storm

[1] Even at temperatures above freezing, cold weather can make it more difficult to obtain a given compaction. Researches indicate that, other things being equal, the density of a compacted soil will be several percent less when placed at 32°F. than when placed at 75°F., and the difference may be sometimes as much as 10 lb./ft.^{3} (Ref. 607).

closes down the construction and marks the end of the season in a natural fashion, the supervising engineer may be faced with a difficult problem in deciding when the weather has become too cold to continue.

No definite criterion can be established for the temperature below which it is impossible to do satisfactory work. The borrow pit soil is frozen only to a relatively shallow depth, and the rate with which the unfrozen soil loses heat and freezes as it is handled during construction depends on the size of the construction surface and the rate of earth placing.

Experience in the cold mid-western part of the United States has shown that it is feasible to continue placing the fill at temperatures as low as 20 to 25°F., providing that precautions are taken to keep the construction surface active. At some dams work has been carried out 24 hours a day and 7 days a week in order to avoid shutting down during cold periods. At temperatures much below 20°F., equipment operation becomes more awkward and the water in the soil commences to freeze sufficiently so that it is very difficult to compact it to the densities desired.

At the Brownlee Dam in Idaho in 1958, it was found possible to continue construction of the clay core when the temperature dropped as low as 10°F. On this job the temperatures of the soil were actually measured in the borrow pits and in the compacted embankment (Ref 331), and it was found that with rapid work the heat loss was not so great as to cause the temperature to drop below freezing.

The Kenney Dam in British Columbia was constructed in a single season, and the weather became cold before the rolled-earth core was completed. In order to avoid the necessity for completing a small portion of the dam in the second year and keeping the construction plant at the site over the winter, tests were made to determine the influence of adding salt (sodium chloride) to the soil in the borrow pit. These tests indicated that the addition of about 1% of salt by weight would allow the material to be compacted satisfactorily at much lower temperatures (Ref. 48). Salt was added to portions of the borrow material, but the weather was not as cold as anticipated and it was not necessary to use this technique for large quantities of soil.

Another problem in cold areas is how the construction surface of the partially completed embankment should be protected during the winter when the construction is shut down. A number of methods have been used with varying costs and degrees of success. Experience indicates that in most parts of the United States it is not necessary to provide any protective covering: the worst damage which occurs is a moderate loosening of the upper few inches of the completed embankment by frost. The loose surface layer is excavated in the spring, and construction continues. The depth of stripping required can best be determined by evaluating visually the condition of the upper portion of the embankment in shallow test pits.

In the coldest areas of the United States, where the embankment remains frozen to a depth of several feet in spring after the weather is warm enough for construction again, it has been found desirable to protect the construction surface during the winter with a layer of several feet of loose soil. In very cold areas more elaborate means of protection have been used, though the necessity is problematical.[1]

11.4 CONSTRUCTION PROCEDURES: PERVIOUS EMBANKMENT SECTIONS

11.4a Construction Control

DENSITY CONTROL

For pervious embankment sections consisting of sand, gravel, or sand-gravel mixtures, less laboratory testing is usually carried out to supplement the primary visual construction control than for impervious embankment sections. The strength of compacted sands and gravels does not vary greatly with small changes in the density, and good compaction can be obtained without the close water content control necessary for impervious soils. Because of these facts and because tests on pervious materials (especially coarse gravels) are more difficult to make and the results are less accurate, many engineers do not use density control tests for pervious embankment sections consisting primarily of gravels and cobbles, even for major dams. Other engineers require only occasional tests to provide a quantitative index and record of the degree of compaction being obtained. On

[1] At the Tustervatn Dam in Norway, in which the core was constructed of fine silt with the "wet compaction" method, the construction surface was kept warm during the winter with electric heating cables covered with canvas and a layer of loose earth. Also in Norway at the Slottmoberget Dam, the surface of the earth core was ponded to a depth of 6 ft. with water retained by dikes. In the spring the upper 5 ft. or so of water had frozen, but no freezing had occurred in the rolled earth core. This last method was considered especially suitable and practical (Ref. 202).

the other hand, density control tests are usually carried out for pervious embankment sections consisting primarily of sand size particles, if only because the tests are easy to perform.

Pervious materials are best compacted with an excess of water and vibration; for these materials there is no "optimum" water content at which a maximum density is obtained, as in the case of impervious soils, and different methods of compaction control must be used.

USBR Relative Density Tests. The most common density control test is the "relative density" test with standard procedures as adopted by the Bureau of Reclamation.[1] The relative density of the soil at any given state of compaction is defined as follows:

$$\text{R.D. } (\%) = \frac{e_{max} - e}{e_{max} - e_{min}}$$

where e_{max} = void ratio of the soil in its loosest possible state
e_{min} = void ratio of the soil in its densest possible state
e = void ratio of the soil in the embankment

In terms of dry density, this equation becomes:

$$\text{R.D. } (\%) = \frac{D_{max}(D - D_{min})}{D(D_{max} - D_{min})} 100$$

where D_{max} = the dry density (lb./ft.^{3}) of the soil in its densest possible state
D_{min} = the dry density of the soil in its loosest possible state
D = the dry density of the soil in the embankment as measured in the field density test

The maximum dry density (D_{max}) is obtained by vibrating the soil in a completely saturated state in a large laboratory container with a small hand vibrator. The minimum density (D_{min}) is obtained by pouring the oven-dried material into a container in such a manner as to result in the least possible densification. The test is used for material with maximum particle size up to 3 in.

The coarser the embankment material, the larger is the field density test necessary to obtain reasonable accuracy and the more work required. For very coarse materials the volume of the large density holes has been measured satisfactorily with water by using a thin plastic sheet to act as an impervious liner, Fig. 11.4:1.

The USBR requires that the pervious sections of their dams be compacted to a minimum relative density of 70%. This criterion was

[1] See test procedure, USBR Earth Manual (Ref. 274).

'ig. 11.4:1 *Field density test in coarse soils using water and plastic sheet lining.*

elected as the density which could be achieved practicably with ractor compaction and at which the embankment section would have hear strength at least equal to the value assumed by the designer. 'he USBR test procedures and the minimum criterion of 70% relative lensity are satisfactory to assure a reasonably uniform embankment vith no excessively loose zones, and consequently they have been .dopted by many private engineers.

'rovidence Vibrated Density Test. The Providence vibrated density est, which was developed at the Providence District of the U.S. Army Corps of Engineers in 1942, is an alternate test for density control of ohesionless embankment sections and has been used on a number of najor earth dams.[1] It is similar to the USBR test except that the maxinum densities are obtained in a container in which the soil is loaded ısing a heavy spring-activated piston with a force of 1,000 lb. Use of the Providence vibrated density apparatus has several advantages: he maximum density is obtained with less work, the results can be eproduced more reliably, and slightly higher maximum densities are

Test details and drawings of the apparatus are given in Ref. 423.

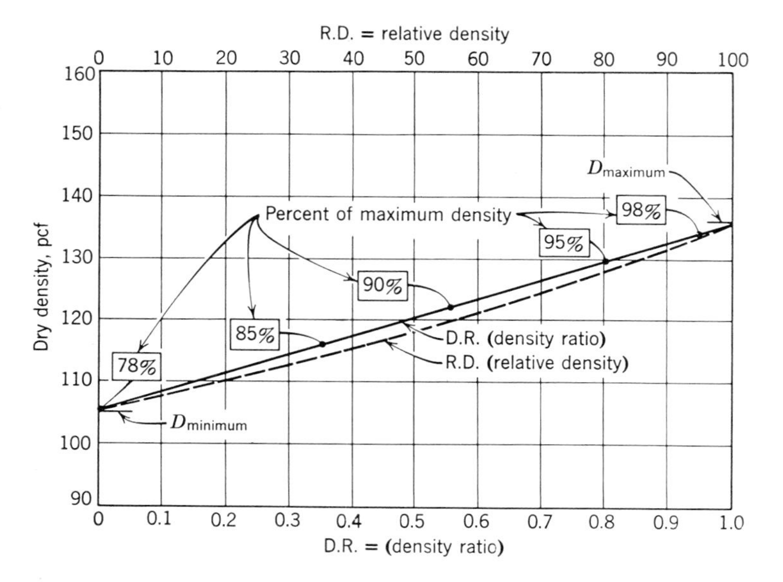

Fig. 11.4:2 Relationship between relative density and density ratio for typic gravelly sand (after Lane, Ref. 423).

achieved. The minimum density is found by pouring oven-dried mate rial into a container in a careful manner similar to that used in th USBR procedure. Gravel with a maximum size of 2½ in. can b tested.

The results of the test can be plotted in the form of relative density as is done by the USBR, or in the form of a slightly different paramete devised by the Army Engineers. Called the "density ratio," it is de fined as follows:

$$\text{D.R.} = \frac{D - D_{\text{min}}}{D_{\text{max}} - D_{\text{min}}}$$

The density ratio is nearly the same ratio as the relative density an has a slight advantage in that the formula is easier to compute.

The relationship between the two terms for a typical gravelly san at various densities is shown in Fig. 11.4:2. In this chart the densit ratio plots as a straight line and the relative density as a slight curve The variation in relative density can be plotted as a straight line o a similar chart in which the vertical density scale is logarithmic.

Figure 11.4:2 also shows the dry density in terms of "percent o maximum density" and demonstrates that increasing this percentag from 90 to 95 is roughly equivalent to changing the relative density from 60 to 80%. Thus it can be seen that the numerical value of the relative density (or density ratio) is more sensitive to density changes (in pounds per cubic foot) than the percent of maximum density, and consequently is easier to use in the field as a criterion for density control.

PLATE BEARING TESTS

Controlling pervious embankment sections with relative density tests requires considerable laboratory work and is not very precise, especially for very coarse soils. One of the most promising alternate methods which has been developed is the plate bearing test, which is based on the fact that the settlement of a loaded plate or footing resting on clean sands or gravels depends primarily on the relative density and only to a small degree on the gradation and other factors (Ref. 126, p. 296).

If plate bearing tests are performed on sections of the embankment surface which are compacted to varying relative densities, a family of load-settlement curves of the form shown in Fig. 11.4:3 can be estab-

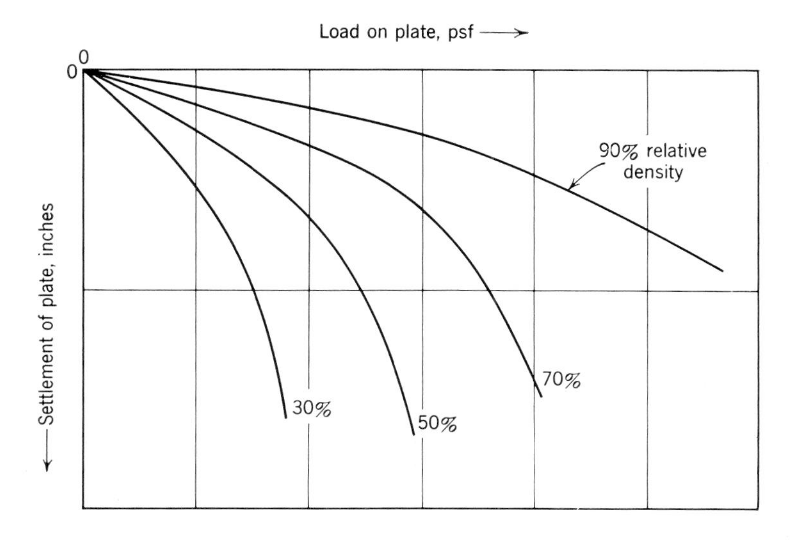

Fig. 11.4:3 Family of load-settlement curves which can be used for estimating relative density of clean sand and gravel embankment sections from plate bearing tests.

lished. Subsequently, the relative density of an embankment section on which a plate bearing test is made can then be estimated by interpolation. This method of control has not yet been widely used, but since it is faster and easier than relative density tests, it will probably be adopted in the future for coarse granular embankments.[1]

CONTROL OF EMBANKMENT PERMEABILITY

Control of the permeability of sand and gravel sections during construction may be both more important and more difficult than control of the density. As the result of problems inherent in exploring sand and gravel deposits (Sec. 4.2), pervious embankment sections are often found during construction to be less pervious than anticipated. A small difference in the quantity of fines in a well-graded sand or gravel can have a great effect on the permeability of the compacted embankment.

Basic control of permeability is accomplished by visual inspection of the fine content and of the ability of the construction surface to absorb the sluicing water added during compaction. At dams where permeability is critical, laboratory gradation tests are performed as a guide to visual control. In addition, it has often been necessary to perform frequent permeability tests in the field laboratory as a routine part of construction control. These tests are sometimes made on unloaded specimens and sometimes on compacted specimens which are loaded by weights or springs, Fig. 11.4:4. Experience indicates that the pressure applied to the sample influences the results for coarse granular soils very little.

At a number of dams where the pervious embankment sections have been constructed of coarse sand-gravel mixtures with fines or dirty rock, it has been found that laboratory tests give an inadequate index of the overall permeability of the compacted embankment because the distribution of the fines and the segregation in the embankment are different from the laboratory specimen. In critical cases, large-scale permeability tests have been performed by ponding water in holes or between dikes on the construction surface of the pervious section, measuring the rate of downward seepage, and computing the coef-

[1] Plate bearing tests were used successfully by French engineers for the control of the 250-ft.-high gravel-fill Iril-Emda Dam in Algeria, which was completed in 1954 (Ref. 434). The main part of the embankment was a well-graded sandy gravel with cobbles up to about 12 in. The plate bearing tests were made by jacking against the chassis of a loaded truck.

[2] Some examples of trouble from this cause are given in Sec. 1.4b.

Fig. 11.4:4 Large-size field permeability testing apparatus for construction tests of coarse pervious soils. (Courtesy USBR)

ficient of permeability from flow nets, Fig. 11.4:5. Tests of this type have also been used during the design stage for embankment test sections where it was considered that laboratory permeability tests were not reliable (Refs. 113, 395).

11.4b Construction Problems Caused by Fines

Although it is usually easier to compact pervious embankment sections than the impervious core, the ease of compaction decreases

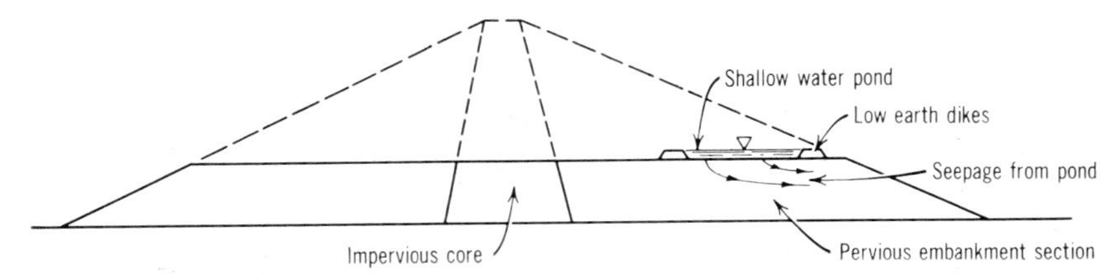

Fig. 11.4:5 Field test used to estimate average permeability of pervious embankment zones during construction.

rapidly with a small decrease in permeability. This is particularly true for pervious soils which are excavated with draglines from borrow pits below the water table. Even though such a material can be used for relatively pervious embankment sections, the fines in it may retain enough water to make the construction surface too soft and sloppy to support trucks.

During construction at a number of major dams, considerable unexpected trouble of this type has developed with relatively "clean" materials excavated below the water table. While the experience on the subject is not great, it appears that construction problems should be anticipated if sand and gravel deposits below the water table contain more than about 4% by weight of fines passing the No. 200 sieve. Experiences at three dams are described below.

WANSHIP DAM

During construction of the Wanship Dam in Utah (1957), sand and gravel for the pervious zones were first excavated from below the water table (Ref. 147). Because of the silty fines, (3 to 5% finer than the No. 200 sieve) the material was not pervious enough for the water on the fill to drain out rapidly, and the construction surface became so soft that work was halted, Fig. 11.4:6. Subsequently it was necessary to unwater the borrow pit partially and to limit the rate of construction and the amount of compaction in order to prevent the fines from working to the surface. As a result of this experience, the engineers of the USBR have adopted the policy of informing contractors on similar jobs that difficulty with the compaction of wet pervious soils which have small contents of silty fines may necessitate the draining of the material in stockpiles.

Fig. 11.4:6 Excessively soft construction surface of zone from sandy gravel excavated below the water table, Wanship Dam, Utah (1957).

TUTTLE CREEK DAM

A similar problem which developed during the construction of the large Tuttle Creek Dam in Kansas (1956) was not only handled satisfactorily but also turned to economic advantage (Refs. 46, 420). A sizable zone of the dam was to be constructed of medium to coarse sand which contained 5 to 15% of silty fines passing the No. 200 sieve. The material was excavated with draglines and loaded directly into trucks from a borrow pit in the river valley below the water table. Trouble soon developed because the trucks which were hauling large quantities of water were bogging down and had to be pulled with tractors out of the borrow pit and off the soft construction surface. Moreover, the sand was too soft and wet to be compacted satisfactorily with either tractors or vibratory rollers.

A small dredge was designed and constructed at a cost of about $250,000, and a 20-in. pipeline was laid to the dam surface. The dredged sand could then be discharged on the construction surface while the silty fines, kept in suspension in the rapidly moving water, were wasted through a 36-in. vertical steel pipe which was constructed into the dam embankment, Figs. 11.4:7, 11.4:8. This procedure was so successful in separating the silt from the sand that the average embankment material contained less than 2% finer than the No. 200 sieve.

Fig. 11.4:7 Dredged sand being discharged on embankment construction surface, Tuttle Creek Dam (1958).

Fig. 11.4:8 A 36-in. vertical steel pipe used in embankment to carry away dredging water and silt, Tuttle Creek Dam (1958).

The water-deposited sand was found to be adequately dense without any further compaction, and field density tests indicated that it was only loosened by the travel of a tractor. Consequently tractor work was kept to a minimum; the sand was distributed over the embankment surface by moving the discharge end of the pipe. About 7,000,000 yd.^{3} of sand were dredged at rates up to 15,000 yd.^{3}/day. It was estimated that the dredged sand cost roughly 30¢/yd.^{3} in place, whereas the previous unsatisfactory construction procedure had cost about 60¢/yd.^{3}

HILLS CREEK DAM

Similar construction difficulties which developed at Hills Creek Dam in Oregon (1958) indicate that both the weather and the plasticity characteristics of the fines have a large influence on the compactability of pervious embankment zones with fines. At Hills Creek the fines consisted of a halloysitic clay with liquid limit between 30 and 60%; the pervious material was a well-graded sand-gravel-cobble mixture with only 20% finer than ¼ in., and 2 to 3% finer than the No. 200 sieve. When excavated below the water table, particularly in wet weather, this mixture became so soft on the construction surface that it would not support the hauling units. In order to get the water out, the material was spread on the construction surface and allowed to drain, sometimes for several days with very little improvement. When excavated above the water table in dry weather, however, the same material could be easily compacted into a typical pervious embankment section.

In addition to the three cases cited above, several other examples have come to the authors' attention in the last few years. From these the following information can be added concerning the problem:

1. The better graded the material and the denser the finally compacted embankment, the greater the likelihood of trouble during construction.
2. Excavating with draglines below water in pits where the pervious sand and gravel is underlain by a clay or silt layer has caused similar trouble since it is almost impossible to keep the pervious material free of contamination.
3. In several cases, stockpiling dirty sand and gravel excavated below water and allowing it to drain for only a few days has been sufficient to allow it to be placed without trouble.

4. At one dam the problem was solved well by bringing in and dumping on the construction surface, alternate piles of wet material excavated below water, and dry material. When blended together by spreading into a layer with a bulldozer, the resulting mixture could be compacted adequately in the specified fashion.

11.5 CONSTRUCTION PROCEDURES: ROCKFILL EMBANKMENT SECTIONS

Because the construction of rockfill embankment sections is well suited to modern excavating and hauling equipment, the cost of rockfill as well as earthfill is becoming increasingly favorable relative to the cost of concrete. Many recent large dams are earth-rock dams, which vary in composition from earth embankments with small quantities of rock in the toe sections or in thin zones on the outer slopes, to rock embankments with earth used only in the thin impervious cores.

Rockfill sections can be constructed of almost any kind of rock in the range between uniform large, hard rocks, which make very pervious embankments, and soft sedimentary rocks in which the individual fragments are broken up by heavy equipment and tightly compacted like a soil. One of the major anachronisms in the field of embankment design is the lack of an adequate rock classification system; there are no well-established criteria for the construction methods best suited to the various rock types. Soft rocks in particular are troublesome because they break down in unpredictable ways during construction. However, even for hard rocks we have no clear understanding of the best method of construction as a function of average rock size or content of fines.

11.5a Hard Rock

In most hard rock formations it is possible to vary the average size of rock and content of fines within reasonable limits by varying the quarrying procedures. In the design stage, however, it can be extremely difficult to predict how the rock will break from examinations of cores from borings or even from small test blasts of exposed rock surfaces. Experienced engineers and geologists have made many serious errors in evaluating the type of rock which will be obtained from a proposed quarry. The most frequent trouble occurs when the quarried material is finer or contains more quarry fines and dust than had been anticipated. During construction at a number of large dams, it has been necessary to make major changes in the proposed methods of rockfill placing and, in some extreme cases, to redesign the dam completely.

Although rockfill sections are constructed of rock from such sources as required excavations (especially spillways), natural rock talus slopes, and the oversized material taken out of soil deposits in screening plants (Sec. 1.1*a*), the rock for the great majority of earth-rock dams is obtained from quarries especially opened for the purpose. Rock is quarried either by the "bench method" with vertical holes or by "coyote holing." The size and gradation of the rock can be controlled much more reliably when vertical hole shooting is used; coyote hole blasting, which more frequently results in an excess of fines and of oversize rocks requiring secondary blasting, is best adapted to hard rock formations with a system of closely spaced internal cracks which will assure that rock of manageable size will be obtained. As well as possible, the engineer should decide in advance whether he will allow coyote hole blasting at the job. A number of unfortunate incidents have occurred because, since the core borings and test blasts indicated that good rock was available, the engineer did not specify the method of quarrying to be used. Subsequently the contractor elected to blast large quantities of rock by coyote holing, and the result was hundreds of thousands or even millions of tons of rock containing so much dust and fines that it was of doubtful value for construction, or an excessive quantity of big blocks which had to be drilled and shot independently.

Construction methods for sections of hard rock have varied widely depending on the type of rock, the maximum tolerable embankment settlement, and the personal preference of the engineer—and opinion and practice are still changing. In Europe recently, the rock has often been placed in thin layers and compacted with heavy vibrators of various types or with vibrating rollers (see, for example, Refs. 201, 186, 452). In North Africa, several major dams have been wholly constructed as "placed rock," cranes being used to space the larger rocks and fill the interstices with increasingly smaller ones (Ref. 220). In Germany, for more than 30 years, rockfill dams have been compacted with heavy steel tamping plates, Fig. 9.4:4. In the United States, pervious rockfill sections are almost always constructed by one of two methods: (1) placing in layers 2 to 6 ft. thick and compacting with the travel of the hauling equipment alone or with rollers, or (2) dumping in high, sluiced lifts. At present both methods are used with approximately equal frequency.

The practice of placing rockfill sections in thin lifts is a direct extension of the procedures for constructing earth embankments. Some of the earliest earth-rock dams built in the United States were constructed in layers several feet in thickness; however, between 1920 and 1945 relatively few dams were built in this manner for reasons which are not known by the authors. In recent years the practice has been again in vogue and has been used for some of the highest modern dams.

The modern practice of building rockfill embankment sections by dumping the rock in high lifts can probably be said to have originated with the construction of reservoirs for hydraulic sluicing in the gold mining days in California about 100 years ago. The technique was subsequently used for hydroelectric developments in the Sierra Nevada Mountains in California. The popularity of this type of construction has continued and a considerable number of high rockfill dams with concrete faces have recently been constructed in California by the Pacific Gas and Electric Company (Refs. 262, 337). The method was first applied on a large scale for use with dams with rolled earth cores in the early 1940's by the engineers of the TVA and the Aluminum Company of America (Sec. 1.2*b*). Subsequently, dumped and sluiced rockfills placed in high lifts have been used for a considerable number of major sloping core and central core earth rockfill dams.

At a few recent large dams, the specifications have been written to allow the contractor to place the rock by either or both methods. In many quarries which ultimately give large clean rock, it is necessary also to excavate large quantities of rock which breaks into small pieces and/or contains considerable quantities of dirt and quarry fines. The poorer rock may be excavated first in the process of opening the quarry (talus or weathered rock) or it may be encountered at any time as the result of excavating zones of broken or weathered material. Economy requires that as much as possible of the excavated material be placed in the dam since it frequently costs almost as much to waste a truck load as to put it in the embankment. Since smaller and dirtier rock can be allowed in embankment sections constructed in horizontal layers than in sections constructed by dumping in high lifts, one way to handle this problem is to allow the contractor to place large clean rock by either method but to require that the smaller dirtier rock be placed in horizontal layers. In a very narrow valley it may not be practical to place rock by two methods except in limited quantities; however, where the construction surface is large the rock can be dumped in high lifts at one location and placed in horizontal layers at another.

In embankment zones where the compressibility and/or permeability of the rockfill are not critical, almost any construction method can be used safely. For example, the rockfill zone located upstream from the core of a sloping core earth-rock dam serves primarily as ballast. In this zone the post construction settlement is much less important than in the zone downstream from the core and considerably less stringent construction specifications can be allowed.

HARD ROCK PLACED IN LAYERS

At some dams where relatively large rock was placed in layers, it was dumped on the embankment surface and pushed over the advancing construction face with a bulldozer. The major advantage of this method is that the bulldozer pushes the fines into the voids in the top of each compacted layer, thereby creating a smooth working surface which minimizes abrasion to the tires of the rubber-tired equipment. When the rock contains such a quantity of fines that this procedure results in an obviously more compressible skin of fines on the surface of each layer, it is necessary to require that the rock be dumped directly in place over the edge of the advancing lift.

The thickness of layers must depend somewhat on the size of rock being placed. The layer thickness for maximum compaction and minimum segregation is not well known. In general, layer thicknesses ranging between 1.0 and 2.0 times the diameter of the maximum rock size have been used with a minimum thickness of 12 or 18 in. for material composed mainly of spalls and rock fines and a maximum of about 6 feet for coarse rock. Some engineers do not believe that it is good practice to place rock in layers of more than about 2 or 3 ft. in thickness but many high dams have been successfully constructed in thicker layers.

When being placed, the rock is usually wetted with hoses on the construction surface although some rockfill sections have been placed dry (Fig. 11.5:1). Except for hard rock containing little or no fines, relatively larger postconstruction settlements are likely if the rock is not wetted in some way. The main action of the water in reducing the settlement is in softening the fines so that they will be squeezed out from between the points of contact of the larger rocks into the voids.[1]

[1] In addition, there is some evidence that the fill tends toward greater segregation if it is not wetted. Extensive tests carried out for the rockfill section of the Goschenenalp Dam in Switzerland indicated that very dry rock segregated to a much larger extent than rock to which about 30% water had been added in the trucks before dumping (Ref. 113). The practice of adding water to dirty rock in the trucks before dumping, in lieu of sluicing on the construction surface, has been tried on a number of dams. Opinion on the desirability of the practice varies but it is undoubtedly preferable to dumping the rock dry.

(*a*)

(*b*)

Fig. 11.5:1(a) Placing hard, sound, clean basalt in 2- to 3-ft.-thick, wetted layers, McNary Dam, Oregon (1950). (b) Placing limestone with fines in 3- to 5-ft.-thick, dry layers, Pomme de Terre Dam, Missouri (1959).

A secondary reason for wetting the rock during construction results from the fact that for many rock types the compressive strength of the individual rocks is lower when they are wet than when dry. Consequently, wetting the rock causes the main portion of the settlement due to "point crushing" to occur during construction. (See an interesting discussion of this point by Terzaghi, Ref. 451, and the closing discussion of Steele and Cooke in Ref. 337.)

No general rules can be given for the amount of water which should be used for sluicing. Very little is actually necessary if the rock is almost free of fines. A volume equal to 30 to 50% of the volume of the rock should be adequate for the great majority of materials. High-pressure jets have little value except to wash excess fines from the surface of the layer.

At the 150 meter high Gepatsch Dam currently under construction in Austria (1962), the large upstream and downstream rockfill shells are being placed in two meter thick layers without the addition of water. (See Fig. 1.2:18 for embankment cross-section.) Each layer is compacted with four passes of an eight-ton smooth steel drum vibrating roller and the surface of each layer is scarified before the next layer is placed, Fig. 11.5:2.

Usually any material obtained from a hard rock quarry can be used satisfactorily by placing and compacting in layers. As discussed above, in the extreme case where all the material consists only of spalls and dust, it can be placed and compacted in 12- or 18-in. layers using the same general procedures as would be used for an embankment of coarse cohesionless soil. When the material contains both large rock and fines, it is not possible to compact the fines to the same high density as can be obtained if the large rock were not present. For an embankment of minimum compressibility, therefore, the weight of the overburden must be carried by a skeleton formed by the larger rock which are bearing on each other. In this case, the quantity of fines should not be greater than the volume of voids between the larger rocks if it is desired to make the embankment zone relatively incompressible.

At most dams in the United States, the rock has been compacted only with the hauling equipment which brings in the material. In a few cases, 50-ton rubber tire rollers have been used. In Europe, a number of dams have been compacted in layers from 3 to 6 ft. in thickness with heavy steel-drum vibrating rollers, and there is no question that this procedure results in a denser and less compressible fill. Until very recently, the only large rollers of this kind were of German manufacture. At present, during the construction of the high

(a)

(b)

(c)

Fig. 11.5:2 Rock placement at Gepatsch Dam, Austria, 1962. (A. Vincent, Civil Engineer.) *(a) Typical rock placed in 2-meter lifts. (b) Typical appearance of scarified surface of layer. (c) Eight-ton vibrating-steel-drum German roller.*

Cougar Dam in Oregon, 5- and 10-ton steel vibrating rollers of American manufacture are being used, and the U.S. Army Corps of Engineers are experimenting with the use of these heavy rollers at rockfills at several other sites. It is probable that these rollers will be used much more frequently in the future, especially for high dams.[1] In Germany, some major rockfill dams have been compacted with 3-ton steel tamping plates which are simply raised on a crane boom and dropped about 10 ft., and with other plates which have a horizontal vibrating action, Fig. 9.4:4. The principle of compacting rock in layers with horizontal vibration is undoubtedly very good, but in practice it is slower and more costly than compacting with ordinary steel drum vibrating rollers.

No definite rules can be given for the maximum tolerable quantity of fines and it is not desirable to attempt to specify a value in advance. The engineer must choose the maximum tolerable fine content on the basis of his visual evaluation of the fill. If he judges that the rock being placed has too many fines, he must require that the large rock be removed from the mixture and the fines be compacted in thin layers.

HARD ROCK DUMPED IN HIGH SLUICED LIFTS

The rock is usually hauled from the quarry in end-dump trucks, dumped down the sloping face of the construction lift, and sluiced with high pressure water jets from monitors. The fines and smaller rocks are deposited at the top of the lift and the larger rocks slide and roll down the face of the slope to the lower part of the lift.

For reasons of economy in the haul roads, the lift is usually kept as high as possible. Also it is commonly believed that an embankment section which is dumped in a single high lift will be somewhat less compressible than if it were dumped in several lower lifts for a number of reasons:

1. The fines which inevitably collect on the horizontal tops of each lift create compressible layers.
2. The greater energy of the large rocks sliding down the higher slope is more effective in compacting the embankment to a denser state.
3. The sharp points on rocks are broken off more effectively by the fall down the face of the higher lift.

[1] See, for example, Ref. 688, which describes a large, smooth steel-drum vibrating roller recently manufactured in the United States.

Actually the measurements available on dams constructed in lifts of various heights are not sufficient to allow any general comparisons between settlements of rockfills placed in high and low lifts; however, it is doubtful that there is a great difference. It is probable that the latter two actions listed above are generally of negligible importance and the first can be overcome to a large degree by thorough sluicing of the top of the lift.

For sloping core dams, lifts of 200 ft. in height or more have been used. For central core dams in which it is necessary to construct the rolled earth core at the same time as the rockfill sections, the rock is usually dumped in lifts varying in height between 20 and 40 ft. In Europe recently, several of the major dams have been constructed quite satisfactorily in 6 to 10 ft. lifts with high pressure, high volume sluicing.

The main function of the sluicing action is to wash the fines into the voids between the larger rocks. Most of the fines are deposited on the upper 20 to 30 ft. of the dumped face and the sluicing is limited to this area. The beneficial action of the sluicing water in driving the fines and smaller rocks into voids between the larger rocks is limited essentially to the construction surface and it has little or no effect on the relative positions of rocks located within the embankment a few feet behind the surface on which the jet is played.

The quantity of the water used and the attention of the operator are the most important elements of the sluicing process. The dirtier the rock, the more water needed to wash the fines into the voids. The nozzle size and pressure are important mainly to assure that adequate quantities of water can be obtained and that the water can be thrown for sufficient distances to cover the area of the face on which fines are deposited. For most rocks, quantities of water in the range between 2 and 4 times the volume of the rock being placed have been required by the specifications though for very clean rock less water would probably be adequate. Depending on the estimated quantity of water needed, specifications commonly require nozzle diameters of 2 to 3 in. and pressures of 70 to 100 p.s.i.

There is some difference of opinion concerning the best method of using the monitors.[1] Some engineers require that the monitors be mounted on long booms which are cantilevered out in front of the advancing lift in such a way that the jet can be played back against the freshly dumped rock, Figs. 11.5:3 and 11.5:4. On many dams, however, the contractor has obtained satisfactory results by mounting

[1] See Ref. 48 for a good description of the movement of rocks on the surface under the action of the jet.

Fig. 11.5:3 Use of monitors on opposite sides of advancing lift, Courtright Dam, California. (Courtesy, Pacific Gas and Electric Company)

the monitors on the top of the lift itself near the location of the dumping trucks and playing the stream of water down the slope. For embankment sections dumped in lifts of 30 ft. in height or less, the monitors have sometimes been placed at the bottom of the lift, Fig. 11.5:5.

The maximum tolerable quantity of fines depends on the size and gradation of the rock and the efficiency of the sluicing operation and can only be judged well in the field during construction. After thorough sluicing, all fines in each dumped load of rock should be washed from the sloping face of the lift into the voids between the larger rocks. If after sluicing there is still a blanket of fines covering the larger rocks on the advancing face of the lift, the rock is too dirty and should be wasted or placed in layers as a compacted fill.

At many jobs the engineer has specified the maximum quantity of fines, for example, not more than 10 or 15% by weight finer than 5 pounds. This practice has several drawbacks. First, it is practically impossible to determine the correct figure for maximum tolerable fine content in advance and, second, it is impossible from the practical standpoint to measure the quantity of fines. Consequently, the suitability of the rock becomes a matter of judgment and differences of opinion between the engineer and contractor are common. Disputes can be minimized by writing a contract which allows the engineer to reject rock on the basis of his visual judgment and in which payment for rejected rock results in a loss equally for the contractor and the owner. This situation provides an incentive to the contractor to develop quarrying procedures which give better rock and it lessens the inclination of the engineer to reject arbitrarily questionable rock.

Specifications also are frequently written to require certain minimum average rock sizes, for example, 500 or 1,000 lb. This again is very difficult to control and not necessary. Actually, there is very little information available to indicate what average size a rock should have to be satisfactory for a dumped rockfill. If the basic quality of the rock formation from which the embankment material is to be ob-

Fig. 11.5:4 Use of two monitors on single cantilever with one operator, Wishon Dam, California. (Courtesy Pacific Gas and Electric Company)

Fig. 11.5:5 Monitor mounted on crawler-type tractor used to sluice dumped rock fill at Cherry Valley Dam, California (1954).

tained is judged satisfactory and the blasting does not produce excess fines, it is not necessary to require any special gradation or average rock size. Probably, other things being equal, the larger the rock is, the less compressible will be the embankment; however, several major dams in recent years have been constructed in high dumped lifts with rock, with an average size of 6 in. or less, obtained as tunnel spoil or from quarries in which the rock broke on closely spaced jointing, Fig. 11.5:6.

THE RELATIVE ADVANTAGES OF CONSTRUCTION IN LAYERS AND HIGH LIFTS

The primary advantage to be obtained by dumping the rock in high sluiced lifts is economy. Since the rock is all dumped from only a few levels, less average haul distance and fewer hauling roads are required than for embankments constructed in layers.[1] Construction in layers,

[1] Another advantage was obtained at the Dalles Dam on the Columbia River. This was built without dewatering the foundation by using a design which had a large section from heavy dumped rock (Sec. 1.2*b*).

Fig. 11.5:6 Typical rock (tunnel spoil) dumped and sluiced in 30-ft. lifts in the downstrem half of Venemo Dam in Norway (1962). (G. Post, Civil Engineer)

on the other hand, has a number of advantages which vary in importance depending on the conditions at each site:

1. Rock with a higher percentage of fines can be placed, since there is less danger that segregated accumulations of fines will produce large zones of compressible material.
2. The embankment obtained is denser, more uniform, and less compressible (especially if the layers are compacted).
3. Rockfill sections constructed in layers can be built with any side slope inclination, whereas rockfills dumped in high lifts must be placed on the angle of repose of the material.
4. The rolled-earth core of the dam can be constructed at the same time as the rockfill section.
5. For relatively low, central-core dams, there is not enough room to place high dumped lifts of rock.
6. The large quantities of sluicing water used for high dumped lifts can interfere with the construction of the core and the foundation treatment, especially if it is necessary to work in the center of the valley with the core at a low elevation while placing rock on the abutments at higher elevations.

In practice, the choice of one construction method over another depends to a great extent on the personal preference and experience of the designing engineer. Some engineers will use dumped rockfill wherever possible and others will not consider it except under special circumstances. For example, the U.S. Army Engineers who have built dozens of dams with large rockfill sections have not used dumped rockfill since 1940 in their Mud Mountain Dam.

11.5b Field Tests on Rockfill Embankments

Because of the large size of the individual rock particles, it has not been considered practicable or necessary on most jobs to carry out any testing for control during construction.[1] The porosity (or density) can be roughly estimated and often is, although the value obtained is of little benefit to the engineer and would not be of much additional help even if it were an exact quantity. The estimate is made by measuring (1) the quantity of rock used (either by weighing it in trucks as it is delivered to the dam or by measuring the volume in the quarry) and (2) the volume of the rockfill section.

The porosity (the ratio of the volume of the voids to the total fill volume) generally varies between 18 and 35%. The numerical value depends primarily on the gradation of the rock and the percentage of the fines and only secondarily on the degree of the compactness of the fill. In general, the higher the content of fines, the higher the density of the embankment and the lower the porosity.

The most valuable field testing on rockfill sections is that which is carried out to determine the relative influence of the various types of compaction equipment and effort. Since measurements of the density are almost impossible to make, the results of tests of various types of compactors can most easily be measured by leveling the surface of the layer. A method that has proved practical is to paint the exposed points of the number of rocks on the surface of the layer being studied and to measure their elevations before and after compaction. The average difference in elevations gives an indication of the change in the density. With such field tests it is possible to study the influence of the thickness of the layer, the type and number of roller passes, and of sluicing the fill. For a given roller and layer thickness, we can deter-

[1] At a few dams, large-scale field density tests have been made by excavating a large hole in the fill and measuring the volume of the hole by surveying. For the extensive tests at Goschenenalp Dam (Ref. 113), it was found necessary to dig out about 250 yd.^{3} of rock in order to assure that the density could be measured with an accuracy of 1%.

mine the number of passes at which additional coverages no longer give any large increase in the density. However, these tests still do not give any assistance in evaluating the problem of determining the influence of the method of compaction on the compressibility of the embankment.

11.5c Compacting Hard Rock and Soil Together to Form Local Embankment Zones Which Are Both Strong and Impermeable

In a few cases, soil and rock have been compacted together in such a manner as to provide a rock skeleton and voids which are filled with impervious soil. Although the procedure has not been widely used, it has such potential value to dam designers that it is desirable to devote some space to it here.

The procedure was used to construct a 5-ft.-thick horizontal impervious blanket under the upstream gravel shell of the Rosshaupten Dam in Germany in 1953 (Ref. 70). The purpose was to provide an impervious blanket under the pervious upstream slope without the necessity for introducing a horizontal bed of soil which was relatively weaker than the foundation or the overlying embankment material, Fig. 1.4:5.

A similar rock soil mixture was studied seriously during the design for the High Aswan Dam in 1958 (Ref. 685). For these studies, rock graded between approximately 10 and 40 cm. was compacted with the Nile silt, using procedures approximately as follows. The silt was mixed to a very wet, soft consistency and spread in layers of about 40 cm. in thickness. A layer of the rock with a thickness of about 70 cm. was then dumped on top. With a 15-ton plate vibrator (about 1½ meters in diameter) which transmitted primarily horizontal vibrations, the rock was vibrated into the soft silt to form a layer with a final thickness of about 60 cm. In this fashion it was shown to be possible to construct an embankment zone in which the rocks were tightly compacted together, and the silt filled all the voids so that the finished product had high shear strength and incompressibility of the densely compacted rock and the low permeability of the wet silt. The authors believe that this procedure can be used to advantage to solve a number of design problems and will probably be used more widely in the future.[1]

[1] For example, such a soil-rock mixture might be used to construct a thin, impervious zone on the upstream face of a rockfill dam. Since it would be strong,

11.5d Embankment Sections of Soft Rock

Soft rocks such as shales, mudstones, siltstones, chalks, and badly weathered igneous and metamorphic rocks have been avoided wherever possible for use in dam embankments, and consequently there is little experience available to indicate the best methods of construction with soft rocks of various types. Yet where large quantities of these materials are available from required excavations, it may be essential to use them.

The main concern about these materials has been that they may soften appreciably with time and exposure to the air and water in the pores of the embankment. With most soft rocks, however, it is possible after some experimentation to construct an impervious embankment in which the spaces between the harder of the soft rock fragments are filled with crushed rock or soil. If there is enough fine material to fill the voids adequately and if the fines are reasonably well compacted at a good water content, no appreciable embankment softening will occur.

Such an embankment of soft rock can be constructed in two ways: (1) by using a construction method which will chew up the rock to create sufficient fines or (2) by mixing and blending the excavated rock with soil from another borrow pit.[1] Embankment sections of this type are most commonly used as random zones (Sec. 1.1*a*).

Each formation of soft rock must be handled as an individual problem, and the best method of construction should be determined by making test embankment sections. Soft rocks can be broken up in various ways:

1. By blasting in the borrow pit.
2. By ripping in the borrow pit and excavating in thin layers with scrapers.

---

impervious, and not subject to deterioration by weathering, such a zone could be used on a steep slope and should compete in cost and other considerations with membranes of concrete, asphalt, and steel. Such a zone would, however, be quite brittle and subject to cracking under imposed shearing strains and would need adequate protective filters.

[1] At the Youghiogheny Dam constructed in 1942, shale from the spillway excavation was passed through a crushing plant and then mixed with soil before compacting in the dam (Ref. 406). At the high Trinity Dam in California in 1958, Fig. 1.2:29, large quantities of weathered igneous rock were excavated with scrapers and passed through a screening and crushing plant (Ref. 695).

3. By wetting the material and allowing it to slake.
4. By rolling on the construction surface with heavy spike tooth or chisel tooth tamping rollers, Fig. 11.5:7.

An interesting example of what can be done in soft rock is given by the experience at the Briones Dam presently (1963) under construction in California, (Ref. 696). For this structure, about 5,000,000 yd.[3] of the embankment consist of a soft, fine-grained sandstone. The material is first ripped in the borrow excavation, Fig. 11.5:8, excavated with scrapers and then broken down on the construction surface with a specially developed roller into a relatively fine-grained impervious soil with a maximum size of about 4 in., Fig. 11.5:9. The new roller, which has 80-in.-diameter lead-ballasted wheels with replaceable chisel-type feet, weighs 60 tons, Fig. 11.5:10. Powered by two 350-hp. engines, the roller has a top speed of about 30 miles/hr., and has proved to be very effective in breaking down the large sandstone fragments and compacting material to a suitable density. Because of the satisfactory experience of this roller and because its speed allows it to do much more work than a conventional roller, there can be little doubt that similar, perhaps even larger, rollers will be developed and used on future dams.

Fig. 11.5:7 Spike tooth roller used to break up shale chunks at Tuttle Creek Dam (see description, Refs. 74 and 420).

Fig. 11.5:8 Ripping soft sandstone in borrow excavation at Briones Dam.

11.6 SLOPE TREATMENT AND RIPRAP

FINISHING EMBANKMENT SLOPES

Since the soil at the edge of the embankment construction surface has no confinement, it is difficult to obtain the full compaction effect of the heavy rollers on the outer few feet of the slopes. This has been a matter of concern over the years to engineers who feared that the outer slopes, because they were not being as well compacted as the interior, might soften and slough.[1] Since no serious or frequent trouble has arisen from loosely compacted outer slopes (Sec. 2.6), it is current practice to compact the material simply by running the rollers on the construction surface as far out on the slope as possible and then to smooth and compact the finished slope with the travel of a bulldozer or grader, Fig. 11.6:1. The upstream slope under the riprap and the downstream slope are finished in the same manner.

[1] A few older dams were even constructed several feet larger than the final dimensions and then cut back to the design slopes.

The only exceptions to this general statement, as discussed briefly at the end of Sec. 2.4*c*, are dams which have a steep upstream slope and no riprap on the lower elevations. For such dams, some troubles have been experienced with shallow surface sloughing on the unprotected embankment slope. These experiences tend to indicate that the

(*a*)

(*b*)

Fig. 11.5:9 Sandstone being compacted at Briones Dam. (a) Typical material as spread on the construction surface. (b) Material after compaction.

Fig. 11.5:10 New, heavy, high-speed roller developed for breaking up and compacting sandstone at Briones Dam.

weights of only a few feet of free draining rock on the surface of the slope has a large influence on stabilizing the outer skin of the embankment which is difficult to compact. Consequently, in such cases more attention should be given to compacting the embankment surface, to dressing up the slope to a uniform surface with no localized steep areas where sloughing can start, and if there is any coarse material available, as oversized cobbles from the fill, it should be placed on the upstream slope.

CONTROL OF RIPRAP GRADATION

The two main objectives in the construction of a good riprap layer are to obtain a well graded rock mixture from the source (quarry or other) and to place it on the upstream slope in a well-knit, uniform layer without segregation. Each rock source for riprap presents different construction problems, and different handling and placing techniques must be utilized to get the rock into a densely packed riprap blanket.

The best time to control the gradation of quarried rock is during the quarrying; sorting and mixing later in stockpiles or on the dam

Fig. 11.6:1 Typical procedures used for finishing embankment slopes.

surface is always less than satisfactory. The sizes and gradation of the rock obtained can be varied considerably by changes in the quarrying operation (the spacing of the blast holes and the quantity of explosives used). It is usually possible after some experimentation to shoot the rock in such a way as to produce a well-graded mixture of rock sizes meeting the specifications. In other cases, a suitable gradation can be obtained from the blasted rock by wasting some of the fines through selective excavation in the quarry or by breaking down some of the larger rocks with a wrecking ball or with secondary blasting. Coyote hole blasting seldom yields well-graded rock for riprap. The services of an experienced quarry operator are very valuable in the development of rocks for riprap from deposits which do not break properly.

Control of the gradation of the riprap to meet the specifications is almost always carried out by visual inspection. In order to calibrate the judgment of the inspector, it is very helpful to establish, at a convenient location in the quarry, a pile of rock with the desired gradation. The pile should contain 5 to 10 tons, and should be formed by measuring and picking out individual rocks and combining them in the correct proportion. By frequent reference to the appearance of the rock in this pile, the inspector can judge "by eye" the suitability of the rock being quarried. In the event of controversy with the contractor concerning the gradation of the rock riprap being used, it is not impossible or very expensive to measure the actual gradation of the rock. This can be done by dumping one or two typical truck loads on a flat area, segregating the rocks into fragments of more or less equal size and measuring them.

PLACING RIPRAP

Riprap can be placed satisfactorily either at the time the embankment is being constructed or after it has been completed. When placed during construction, the riprap layer is kept a few feet lower than the construction surface, and the rock can be dumped directly from trucks on top of the previously prepared filter layer, Fig. 11.6:2. When placed after the embankment is completed, there are a number of different methods for the operation. Frequently the rock is hauled to the crest of the dam in trucks which are then lowered down the slope on a cable fastened to a tractor winch, Fig. 11.6:3. The rock should never be allowed to drop down the slope in a chute or be pushed down the slope with a bulldozer, since these operations result in excessive segregation.[1]

One of the principal advantages of dumped riprap is that the construction requires a minimum of expensive hand work. Well-graded

[1] Where the dam is not too high and the rock is hard and not too brittle, the riprap can be dumped at the toe of the dam and pushed up the slope with a bulldozer. Fig. 11.6:4 shows hard, bulky limestone riprap being placed in this manner at Enders Dam. In this case the rock was so hard that it was not appreciably broken down by the continuous traffic of the tractors, and each pass of the tractor consolidated the riprap already placed into a more tightly interlocked layer.

Fig. 11.6:2 Placing riprap as the embankment is being constructed by dumping from the construction surface.

Fig. 11.6:3 Placing 3-ft.-thick riprap layer on upstream slope after completion of embankment, Casitas Dam, California (1959). Truck is held on slope with cable attached to tractor winch. Rock is later spread and evened with crane and orange peel bucket. (Courtesy USBR)

and bulky rock can be placed in a riprap layer by careful dumping, and almost no additional movement of the rocks is required. The poorer the gradation and the more slablike the individual rocks, the more work is required to place them in the form of a tightly knit protective blanket without large holes through which the filter material can escape. After dumping, the rocks are worked by men with bars or

Fig. 11.6:4 Placing riprap by pushing it with bulldozer up the slope from the upstream toe, Enders Dam (1949). (Courtesy USBR)

other equipment. Draglines with orange peel buckets, back hoes, and other power equipment can also be used advantageously to work the rock, Fig. 11.6:5. While many contractors would automatically assume that such work can only be done economically with power machinery and will balk at the suggestion of using hand labor, on a few recent jobs contractors have used hand labor extensively for this work (Ref. 675). Actually, with most rocks a half dozen laborers can do a surprising volume of work and can often do it cheaper and better than power machinery, even in the United States. The authors believe that some hand labor should be specified on all jobs.

(*a*)

(*b*)

(*c*)

Fig. 11.6:5(a) Placing riprap on the upstream slope of Bonny Dam, Colorado (1951), with a dragline and orange peel bucket. The rock was dumped on the slope with winched trucks travelling on the wooden mat shown, and was reworked with the crane. Note platform used to keep crane level on the slope. (Courtesy USBR). *(b) View showing "cow dozer" used to level off riprap at Cachuma Dam. The teeth fold up when the blade is used as a bull dozer. When the attachment is pulled over the riprap, as shown, the teeth extend to permit a combing action.* (Courtesy USBR) *(c) Dressing upstream slope of rockfill zone, Ball Mountain Dam, Vermont (1959), with Gradall from the construction surface.*

In addition to quarried rock, riprap is often constructed from rock screened as the oversize from earth borrow pits, from deposits of talus rock or from rock gathered as "field stones." Different types of rock create somewhat different construction problems. At Heart Butte Dam (Ref. 233), constructed in North Dakota in 1949, the rock for the riprap was obtained by collecting field stones over a wide area.[1] These ranged in size from cobbles to boulders with a volume of more than 1 yd.^{3} The rock posed two problems: (1) it was deficient in fines to fill the spaces between the larger rocks; (2) some of the rocks were too large. Both problems were solved satisfactorily by dumping the rock on the upstream slope of the dam and then breaking the large rocks with a 3,700-lb. steel-wrecking ball. The smaller rock and the spalls obtained from breaking the oversized boulders were used to fill the larger openings in the riprap blanket. In addition to breaking

[1] See also Ref. 482 for description of the "harvest" of 650,000 yd.^{3} of field boulders for the construction of the riprap blanket on the mammoth Oahe Dam on the Missouri River.

Fig. 11.6:6 Riprap of large field stones at Heart Butte Dam before reworking. Note large voids. (Courtesy USBR)

Fig. 11.6:7 Reworking slabby riprap at Heart Butte Dam with hand labor using bars. (Courtesy USBR)

Fig. 11.6:8 Final well-graded, tightly knit riprap blanket affording excellent wave protection at Heart Butte Dam. (Courtesy USBR)

the big rocks, the wrecking ball was very effectively used to pound the rocks together into a more tightly knit layer than could have been obtained by simply dumping. Fig. 11.6:6 shows the riprap blanket before the broken fines were used to fill the voids. Because of the slabby character of the rock, a considerable amount of hand work was required to provide a satisfactory riprap, Fig. 11.6:7. Figure 11.6:8 shows the completed riprap blanket.

References

1. T. Nilsson and B. Lofquist, "An Earth and Rockfill Dam on Stratified Soil. The Wet Construction Method," Fifth Congress on Large Dams, Paris (1955), vol. I, p. 403.
2. J. G. Lewis, "The Use of a Sloping Core in Earth and Rockfill Dams," Fifth Congress on Large Dams, Paris (1955), vol. I, p. 532.
3. T. A. Middlebrooks, "Earth Dam Practice in the United States," *Transactions, American Society of Civil Engineers,* Centennial Volume (1953) p. 697.
4. K. Terzaghi, "Origin and Functions of Soil Mechanics," *Transactions, American Society of Civil Engineers,* Centennial Volume (1953) p. 697.
5. W. Eggenberger, "The Goschenenalp Rockfill Dam Project, Switzerland," Third International Conference of Soil Mechanics and Foundation Engineering, Zurich (1953), vol. III, p. 296.
6. G. G. Meyerhof, "Compaction of Sands and Bearing Capacity of Piles," *Transactions, American Society of Civil Engineers,* vol. 126 (1961) part I, p. 1293.
7. A. W. Skempton and A. W. Bishop, "The Gain in Stability Due to Pore Pressure Dissipation in a Soft Clay Foundation," Fifth Congress on Large Dams, Paris (1955), vol. I, p. 613.
8. C. F. Grundy, "The Treatment by Grouting of Permeable Foundations of Dams," Fifth Congress on Large Dams, Paris (1955), vol. I, p. 647.
9. M. R. Lefoulon and M. E. Ischy, "Usine de Fessenheim 3e, Bief de Grand Canal D'Alsace, protection des fouilles par injection des alluvions," Fifth Congress on Large Dams, Paris (1955), vol. I, p. 711.
10. B. Rajcevic and M. Vercon, "Foundation de Barrage de Vlasina," Fifth Congress on Large Dams, Paris (1955), vol. I, p. 859.
11. "Texarkana Dam, Foundation and Borrow Investigation and Embankment Design," U.S. Corps of Engineers, Waterways Experiment Station Technical Memo No. 3-293 (July 1949).
12. "Review of Soils Design and Field Observations, Enid Dam, Mississippi," U.S. Corps of Engineers, Waterways Experiment Station Technical Memo No. 3-409 (October 1955).
13. E. C. Steer and G. M. Binnie, "Design and Construction of Earth Dams, Core Walls, and Diaphragms of Earth and Rockfill Dams," Fourth Congress on Large Dams, New Delhi (1951), vol. 1, p. 47.
14. K. L. Powers, J. R. Benson, and V. S. Meissner, "Asphaltic Concrete and Soil-Cement Test as Riprap Substitutes at Bonny Reservoir," *Civil Engineering* (June 1952), p. 34.
15. A. Casagrande, "Notes on the Design of Earth Dams," *Journal, Boston Society of Civil Engineers* (Oct. 1950).
16. A. Casagrande, "Soil Mechanics in the Design and Construction of Logan International Airport," *Journal, Boston Society of Civil Engineers* (April 1949).
17. F. S. Brown, "Foundation Investigations for Franklin Falls Dam," *Journal, Boston Society of Civil Engineers* (April 1941). "Contributions to Soil Mechanics, 1941–1953," *Boston Society of Civil Engineers.*

18. F. C. Walker and W. G. Holtz, "Control of Embankment Material by Laboratory Testing," *Transactions, American Society of Civil Engineers,* vol. 118 (1953), p. 1.
19. G. Post and P. Londe, *Les Barrages en terre compactée—pratiques Americaines,* Gauthier-Villars, Paris (1953).
20. "Review of Soils Design, Construction, and Performance Observations, Tom Jenkins Dam, Ohio," U.S. Waterways Experiment Station Technical Report No. 3-474 (March 1958).
21. C. E. Blee and A. A. Meyer, "Measurements of Settlements at Certain Dams on the TVA System and Assumptions for Earthquake Loadings for Dams in the TVA Area," Fifth Congress on Large Dams, Paris (1955), vol. III, p. 141.
22. F. C. Walker, "Experience in the Measurement of Consolidation and Pore Pressure of Earth Dams," Third Congress on Large Dams, Stockholm (1948), vol. III, Report 58.
23. J. B. Cooke, "Wishon and Courtright Rockfill Dams," *Transactions, American Society of Civil Engineers,* vol. 125 (1960), part II, p. 587.
24. L. H. Gomes Fernandes, et al., "The Paradela Concrete Face Rockfill Dam," *Transactions, American Society of Civil Engineers,* vol. 125 (1960), part II, p. 365.
25. W. J. Weyerman, "Paradela Rockfill Dam Foundation Treatment," *Transactions, American Society of Civil Engineers,* vol. 125 (1960), part II, p. 419.
26. H. E. Lloyd, O. L. Moore, and W. F. Getts, "Construction and Performance of Cherry Valley Central Core Rockfill Dam," *Transactions, American Society of Civil Engineers,* vol. 125 (1960), part II, p. 336.
27. "Review of Soils Design, Construction, and Prototype Analysis, Blakely Mountain Dam, Arkansas," U.S. Corps of Engineers, Waterways Experiment Station Technical Memo No. 3-439 (October 1956).
28. R. E. Davis, "Earthquake Tests on Models of Rockfill Dams," unpublished report to International Engineering Company, Inc., San Francisco, California, February 1952.
29. J. B. Snethlage, F. H. Schneidenhelm, and A. N. Vanderlip, "Rockfill Dams—Review and Statistics," *Transactions, American Society of Civil Engineers,* vol. 125 (1960), part II, p. 678.
30. I. C. Steele and J. B. Cooke, "Salt Springs and Lower Bear River Concrete Face Rockfill Dams," *Transactions, American Society of Civil Engineers,* vol. 125 (1960), part II, p. 74.
31. "Anderson Ranch Dam—Technical Record of Design and Construction," U.S. Bureau of Reclamation, Denver (1956).
32. "Jamestown Dam—Technical Record of Design and Construction," U.S. Bureau of Reclamation, Denver (1956).
33. "Trenton Dam—Technical Record of Design and Construction," U.S. Bureau of Reclamation, Denver (1956).
34. W. P. Creager, "Design and Maintenance of Earth Dams," American Water Works Association, vol. 31, No. 8, (August 1939).
35. *Treatise on Dams,* Design and Construction, U.S. Bureau of Reclamation Manual, Chap. 8 (Earth Dams), (October 10, 1950).
36. J. A. Fraps, "Collapse of Littlefield Dam," *Western Construction News* (November 25, 1929).
37. C. H. Lee, "Selection of Materials for Rolled-Fill Earth Dams," *Transactions, American Society of Civil Engineers,* vol. 103 (1938).

38. "Medicine Creek Dam—Technical Record of Design and Construction," U.S. Bureau of Reclamation, Denver (1955).
39. H. J. Gibbs and W. G. Holtz, "Research on Determining the Density of Sands by Spoon Penetration Tests," Fourth International Conference on Soil Mechanics and Foundation Engineering, London (1957), vol. I, p. 38.
40. "San Simon Silt Barrier, Arizona," Design Memorandum, U.S. Bureau of Reclamation, Denver (October 1957).
41. "O'Sullivan Dam—Technical Record of Design and Construction," U.S. Bureau of Reclamation, Denver (1954).
42. "Enders Dam—Technical Record of Design and Construction," U.S. Bureau of Reclamation, Denver (1958).
43. H. P. Bunger, "Constructing Steel Faced Storage Dams," *Western Construction News* (February, 1935), p. 36.
44. A. W. Bishop, *Hydro-Electric Engineering Practice,* Chapter IX, "Embankment Dams" (Handbook on Hydro-Electric Practice) Blackie and Sons, Ltd., London (1958).
45. O. E. Hovey, "Steel Dams," American Institute of Steel Construction (1935).
46. "Dredge Rescues Big Dam Job," (Tuttle Creek Dam) *Construction Methods and Equipment* (February 1957), p. 91.
47. J. W. Hilf, "Use of Materials from Structural Excavations," *Journal, Soil Mechanics and Foundation Division, American Society of Civil Engineers,* vol. 83 (1957), Paper 1207.
48. D. J. Bleifuss and James P. Hawke, "Rockfill Dam Design and Construction Problems," *Transactions, American Society of Civil Engineers,* vol. 125 (1960), part II, p. 275.
49. J. B. Hayes, "Gunite Blanket Improves Foundation Grouting for Earth Dams on Rock," *Civil Engineering* (1953).
50. "The Upper Holston Projects," Technical Report No. 14, Tennessee Valley Authority (1957).
51. "Earthfill in Wet Weather," *Engineering News Record* (August 7, 1952), p. 49.
52. "Civil and Structural Design," Technical Report No. 24, vol. I, Tennessee Valley Authority (1952).
53. F. H. Kellogg, "Clay Grouting at Madden Reservoir," *Engineering News-Record,* vol. 109 (1932).
54. J. A. Horn, "Computer Analysis of Slope Stability," *Transactions, American Society of Civil Engineers,* vol. 126 (1961), part 1, p. 850.
55. "Report on Slope Protection Methods for Earth Dams in Missouri River Basin," Office of Division Engineer, Missouri River Division, Corps of Engineers, Omaha, Nebraska (December 1, 1946) (unpublished).
56. P. T. Bennett and R. A. Barron, "Design Data for Partially Penetrating Relief Wells," *Proceedings,* Fourth International Conference on Soil Mechanics and Foundation Engineering, London (1957), vol. II, p. 282.
57. "Report on the Slide of a Portion of the Upstream Face of the Fort Peck Dam," *U.S. Government Printing Office,* Washington (1939).
58. A. W. Bishop, "The Stability of Earth Dams," Thesis submitted to the Imperial College (London University) for the degree of Doctor of Philosophy (May 1952).
59. P. Habib, "Study of Water Circulation in a Permeable Soil by means of Three Dimensional Electrical Analogy," (In French) *Proceedings,* Third

International Conference on Soil Mechanics and Foundation Engineering, Zurich (1953), vol. II, p. 250.
60. T. W. Lambe, "The Engineering Behavior of Compacted Clays," *Transactions, American Society of Civil Engineers,* vol. 125 (1960), p. 718.
61. T. W. Lambe, "The Structure of Compacted Clay," *Transactions, American Society of Civil Engineers,* vol. 125 (1960), p. 682.
62. A. Casagrande and S. D. Wilson, "Prestress Induced in Consolidated Quick Triaxial Tests," Third International Conference on Soil Mechanics and Foundation Engineering, Zurich (1953), vol. I, p. 106.
63. A. W. Bishop, "The Use of the Slip Circle in the Stability Analysis of Slopes," *Geotechnique,* vol. 5, p. 7.
64. M. Bar-Shany, G. Korlath, and J. G. Zeitlin, "The Use of Fat Clay in Dam Construction in Israel," Fourth International Conference on Soil Mechanics and Foundation Engineering, London (1957), vol. 2, p. 273.
65. M. Vercon and B. Rajcevic, "Observation des contraintes et des deformations dans quelque barrages en Yougoslavie," Sixth Congress on Large Dams, New York (1958), Question 21, Report 136.
66. E. E. Nonveiller, "Compaction Methods and Moisture Content for Materials Used in the Construction of Earth Core and Supporting Fill for Earth and Rockfill Dams," *Transactions,* Sixth Congress on Large Dams, New York, (1958), General Report on Question No. 22.
67. C. M. Roberts, "The Quoich Rockfill Dam," Sixth Congress on Large Dams, New York (1958), Report 6, Question 22.
68. H. H. Dixon, "Moisture Control and Compaction Methods Used During Construction of the Sasumua Dam, Kenya," Sixth Congress on Large Dams, New York (1958), Report 10, Question 22.
69. F. Treiber, "Measurements and Observations on Rosshaupton Dam," Sixth Congress on Large Dams, New York (1958), Report 5, Question 21.
70. F. Treiber, "Compaction Methods Adopted for the Construction of Rosshaupton Dam, Their Effectiveness and the Behavior of the Impervious Loam Core," Sixth Congress on Large Dams, New York (1958), Report 8, Question 22.
71. F. Kawakomi, "Compaction Methods and Field Moisture Content of the Earth Core for Earth Dams Using Moist Materials," Sixth Congress on Large Dams, New York, (1958) Report 28, Question No. 22.
72. A. L. Little, "Compaction and Pore Water Measurements on Some Recent Earth Dams," Sixth Congress on Large Dams, New York (1958), Report 42, Question No. 22.
73. P. A. Scott, "Observations on an Experimental Clay Embankment to Assist the Design of the Impermeable Layer of a Large Rockfill Dam," Sixth Congress on Large Dams, New York (1958), Report 57, Question 22.
74. P. T. Bennett, "Materials and Compaction Methods—Missouri Basin Dams," Sixth Congress on Large Dams, New York (1958), Report 95, Question No. 22.
75. M. M. Maslov, "Problem of High Earth Dam Density in Condition of Seismic Activity," Sixth Congress on Large Dams, New York (1958), Report 115, Question No. 22.
76. L. Bernell, "Water Content and its Effect on Settlements in Earth Dams," Sixth Congress on Large Dams, New York, (1958), Report 117, Question 22.
77. E. R. de Luccia, "Earth Dam Construction in a Wet Climate," Sixth Congress on Large Dams, New York (1958), Report 128, Question 22.

78. R. Maigre and J. Barge, "Compaction of Materials for the Construction of the Watertight Core of Serre-Ponçon Earth Dam," Sixth Congress on Large Dams, New York (1958), Report 132, Question 22.
79. T. Abou-Wafa and H. Said, "Compaction Tests of Dune Sand Fill Under Water by Deep Internal Vibration (Aswan Dam)," Sixth Congress on Large Dams, New York (1958), Report 138, Question 22.
80. H. Moser and F. Neumaier, "Determination of Seepage Flow Under Rosshaupton Dam by Means of Radioactive Isotopes," Sixth Congress on Large Dams, New York (1958), Communication 10.
81. "Review of Soils Design, Construction, and Performance Observations, Benbrook Dam, Texas," U.S. Corps of Engineers, Waterways Experiment Station Technical Memo No. 3-452 (March 1957).
82. H. Press, "Heightening of Dams," Sixth Congress on Large Dams, New York (1958), Report 3, Question 20.
83. J. Mencl, "Compaction of Sandy Soils in Upstream Slopes of Earth Dams in Seismic Regions," Sixth Congress on Large Dams, New York (1958).
84. G. R. Scott, "Montgomery Dam—A Rockfill Structure," Sixth Congress on Large Dams, New York (1958), Report 92, Question 20.
85. L. Bernell, "Determination of Pore Pressures in Earth Dams During Construction," Sixth Congress on Large Dams, New York (1958), Report 116, Question 21.
86. L. Bernell and R. Nilsson, "Electrical Analogy Equipment for Solving Non-Stationary, Two-Dimensional Flow Problems," Fourth International Conference on Soil Mechanics and Foundation Engineering, London (1957).
87. "Review of Soils Design, Construction, and Performance Observations, John H. Kerr Project (Bugs Island and Island Creek Dams), Virginia," U.S. Army Corps of Engineers, Waterways Experiment Station Technical Memo No. 3-464 (August 1957).
88. J. Lowe, III, "Earth Dams—Section 5 in Handbook of Applied Hydraulics," (2nd Ed.) (A. P. Davis, Ed.), McGraw-Hill Book Co., (1952).
89. B. Fruhauf, "Wet Vibration Puts Strength in Sand," *Engineering News Record* (June 23, 1949).
90. P. T. Bennett, "The Effects of Blankets on Seepage Through Porous Foundations," *Transactions, American Society of Civil Engineers,* vol. III (1946), p. 215.
91. P. D. Evdokimov, "Practice of Geotechnic Research of Certain Soviet Hydraulic Power Plants Carried Out by V.N.I.G.," Sixth Congress on Large Dams, New York (1958).
92. P. D. Evdokimov, "Investigation of Foundation Strength and Shearing Resistance of Hydraulic Structures on Soft Soils and Rocks," Sixth International Congress on Large Dams, New York (1958).
93. K. Terzaghi and T. M. Leps, "Design and Performance of Vermillion Dam, California," *Transactions, American Society of Civil Engineers,* vol. 125 (1960), p. 63.
94. E. M. Fucik and R. F. Edbrooke, "Design and Construction of the Ambuklao Rock Fill Dam," *Transactions, American Society of Civil Engineers,* vol. 125 (1960), p. 1207.
95. J. L. Sherard, "Influence of Soil Properties and Construction Methods on the Performance of Homogeneous Earth Dams," U.S. Bureau of Reclamation Technical Memo No. 645, Denver (January, 1953).

96. E. M. Fucik, "Petenwell Hydroelectric Project," *Transactions, American Society of Civil Engineers,* vol. 117 (1952), p. 528.
97. E. Habetha, "The Use of Radioactive Isotopes to Determine Permeability in Connection with Dam Construction," Seventh Congress on Large Dams, Rome (1961), Report 7, Question 25.
98. F. C. Walker, "The Design of Earth Dams for Pervious Foundations," Third International Conference on Soil Mechanics and Foundation Engineering, Zurich (1953), vol. II, p. 294.
99. F. J. Davis, "Quality Control of Earth Embankment," Third International Conference on Soil Mechanics and Foundation Engineering, Zurich (1953), vol. I, p. 218.
100. J. W. Hilf, "Compacting Earth Dams With Heavy Tamping Rollers," *Transactions, American Society of Civil Engineers,* vol. 124 (1959), p. 409.
101. O. L. Rice and H. G. Arthur, "The Most Recent Methods Developed to Avoid Piping or Blow-Outs in Dams," Third Congress on Large Dams, Stockholm (1948).
102. K. Keil, *Der Dammbau,* Springer, Berlin (1951) (German reference book on dams).
103. A. J. Ackerman, F. T. Matthias and G. O. Vogan, "Brazil's Paraiba-Pirai Diversion Project Includes World's Largest Underground Power Station," *Civil Engineering* (November 1951), p. 652.
104. M. C. Hinderlider, "The Failure of the Apishapa Dam and Lessons to be Learned Therefrom," *Colorado Society of Engineers Bulletin,* vol. 7 (September 1923), p. 4.
105. Engineering News Record, Vol. 91, 1923, pp. 357, 418, 422, 900 and 1029.
106. B. Lofquist, Oral discussion of Question 18, Fifth Congress on Large Dams, Paris (1955), vol. III, p. 22.
107. J. Salva, "Methodes de compactage et teneur en eau des materiaux Utilisés pour la construction de la Dique en Terre du Sarno," Sixth Congress on Large Dams, New York (1958), Question No. 22, Report 100.
108. A. W. Bishop, "Some Factors Controlling Pore Pressures Set Up During the Construction of Earth Dams," Fourth International Conference on Soil Mechanics and Foundation Engineering, London (1957), vol. II, p. 297.
109. R. Peterson, N. L. Iverson and P. J. Rivard, "Studies of Several Dam Failures on Clay Foundations," Fourth International Conference on Soil Mechanics and Foundation Engineering, London (1957), vol. II, p. 348.
110. A. Casagrande and S. D. Wilson, "Effects of Stress History on the Strength of Clays," Soil Mechanics Series, No. 43 (1953), Harvard University.
111. R. Schiltknecht and H. Bickel, "Control Measurements at the Castilletto Earth Dam," Fourth International Conference on Soil Mechanics and Foundation Engineering, London (1957), vol. II, p. 373.
112. J. Zeller and R. Wullimann, "The Shear Strength of the Shell Materials for the Goschenenalp Dam, Switzerland," Fourth International Conference on Soil Mechanics and Foundation Engineering (London), 1957, vol. II, p. 399.
113. J. Zeller and H. Zeindler, "Test Fills with Coarse Shell Materials for Goschenenalp Dam," Fourth International Conference on Soil Mechanics and Foundation Engineering, London (1957), vol. II, p. 405.
114. R. A. Barron, "The Efficacy of Toe Drains in Controlling Seepage Uplift in Layered, Pervious Foundations," Third International Conference on Soil Mechanics and Foundation Engineering, Zurich (1953), vol. II, p. 195.

115. V. D. Harty, "Slide in Fort Henry Embankment, River Shannon, Ireland," Third International Conference on Soil Mechanics and Foundation Engineering, Zurich (1953), vol. II, p. 255.
116. J. L. Jaspar and A. S. Ringheim, "Steel Sheet Piling Studies," Third International Conference on Soil Mechanics and Foundation Engineering, Zurich (1953), vol. II, p. 264.
117. R. Peterson and N. L. Iverson, "Study of Several Low Earth Dam Failures," Third International Conference on Soil Mechanics and Foundation Engineering, Zurich (1953), vol. II, p. 273.
118. C. Mallet and J. Paquant, *Les barrages en terre,* Eyrolles, Paris (1951) (French reference book).
119. J. A. Banks, "Construction of Muirhead Reservoir, Scotland," Second International Conference on Soil Mechanics and Foundation Engineering, Rotterdam (1948), vol. II, p. 24.
120. C. I. Mansur and W. I. Perret, "Efficacy of Partial Cutoffs for Controlling Seepage Beneath Dams and Levees Constructed on Pervious Foundations," Second International Conference on Soil Mechanics and Foundation Engineering, Rotterdam (1948), vol. V, p. 299.
121. J. G. Patrick, "Stability Analysis by Application of the Elastic Theory," Second International Conference on Soil Mechanics and Foundation Engineering, Rotterdam (1948), vol. VII, p. 102.
122. G. Wetter, "Addition of Bentonite to the Impervious Core of San Valentino Dam," Second International Conference on Soil Mechanics and Foundation Engineering, Rotterdam (1948), vol. IV, p. 273.
123. L. F. Cooling and H. Q. Golder, "The Analysis of the Failure of an Earth Dam During Construction," *Journal, Institution of Civil Engineers* (London), vol. 19 (1942), p. 38.
124. A. W. Skempton and H. Q. Golder, "Practical Examples of the $\phi = 0$ Analysis of Stability of Clays," Second International Conference on Soil Mechanics and Foundation Engineering, Rotterdam (1948), vol. IV, p. 63.
125. K. Terzaghi, "Mechanism of Landslides," Berkey Volume, *Geological Society of America* (November 1950), p. 83.
126. Terzaghi and Peck, *Soil Mechanics in Engineering Practice,* John Wiley & Sons (1948).
127. "Foundation Failure Causes Slump in Big Dike at Hartford, Conn.," *Engineering News Record,* vol. 127 (1941), p. 142.
128. "Foundation of Marshall Creek Earth Dam Fails," *Engineering News Record,* vol. 119 (1937), p. 532.
129. "Why Marshall Dam Failed," *Engineering News Record,* vol. 120 (1938), p. 431.
130. "Uncompleted Lafayette Rolled-Fill Earth Dam Damaged by Movement," *Engineering News Record,* vol. 101 (1928), p. 483.
131. A. K. B. Lyman, "Compaction of Cohesionless Foundation Soils by Explosives," *Transactions, American Society of Civil Engineers,* vol. 107, (1942). p. 1330.
132. T. T. Knappen, "Calculation of the Stability of Earth Dams," Second Congress on Large Dams, Washington, D.C. (1936), vol. 4, p. 505.
133. T. A. Middlebrooks, "Fort Peck Slide," *Transactions, American Society of Civil Engineers* (1942).
134. Justin, Creager and Hinds, *Engineering for Dams,* John Wiley & Sons, Vol. III (June 1947).

135. J. L. Savage, "Report on Slide of Upstream Slope of Belle Fourche Dam," unpublished U.S. Bureau of Reclamation Report (August 20, 1931).
136. "Small Earth Dam Fails" (Fruit Grower's Dam), *Engineering News Record,* vol. 118 (1937), p. 932.
137. Records of Inspection of Costilla Dam, New Mexico State Engineer.
138. E. E. Esmiol, "Seismic Stability of Earth Dams," U.S. Bureau of Reclamation Technical Memorandum No. 641, Denver (April 1951).
139. F. H. Tibbetts, "Earthquake Proof Earth Dams" (Coyote Dam), *Engineering News Record* (July 2, 1936).
140. H. B. Seed and C. K. Chan, "Structure and Strength Characteristics of Compacted Clays," *Transactions, American Society of Civil Engineers,* vol. 126 (1961), part I, p. 1343.
141. H. Schussler, *The Water Supply of San Francisco,* Herman B. Brown Press (1906).
142. J. F. Dawson, "Control of Damage by Muskrats to Earth Structures," Thesis, University of Saskatchewan (1950).
143. *Engineering News,* vol. 60, p. 548 (1911).
144. M. M. O'Shaughnessy, Letter to Engineering News Record (July 9, 1925).
145. Records of Dam Inspection, California State Department of Water Resources, Sacramento, California.
146. K. V. Steinbrugge and D. F. Moran, "An Engineering Study of the Southern California Earthquake of July 21, 1952, and its Aftereffects," *Bulletin of the Seismological Society of America,* vol. 44, No. 2B (1954).
147. F. C. Walker, "Development of Earth Dam Design in the Bureau of Reclamation," *U.S. Bureau of Reclamation Publication* (August 1958). (For distribution to the Sixth World Congress on Large Dams, New York, 1958).
148. H. H. Jewell, "Rock Riprap Replaces Porous Concrete Slope Protection at Santee-Cooper Project," *Civil Engineering* (January 1948), p. 14.
149. Records of the North Sterling Irrigation District, Sterling, Colorado.
150. H. H. Jewell, "Protecting Upstream Slope of Kingsley Dam," *Civil Engineering* (November 1945), p. 493.
151. Records of the New Mexico State Engineer, Santa Fe, New Mexico.
152. "The Throttle Dam, Raton, New Mexico," *Engineering News Record,* vol. 72 (1914), p. 72.
153. J. G. Lewis, "The Sloping Core Principle for Earth and Rock Fill Dams," Doctor's Thesis, University of London (1953).
154. B. M. Jones and O. N. Floyd, "Further Report on Apishapa Dam Failure," *Engineering News Record,* vol. 91 (September 13, 1923), p. 422.
155. J. E. Field, "Failure of Apishapa Earth Dam in Colorado," *Engineering News Record,* vol. 91 (September 13, 1923), p. 418.
156. J. D. Justin, "The Design of Earth Dams," *Transactions, American Society of Civil Engineers,* vol. 87 (1924), p. 1.
157. J. P. Gould, "Compression Characteristics of Rolled Fill Materials in Earth Dams," Technical Memo 648, U.S. Bureau of Reclamation, Denver (March 1954).
158. J. P. Gould, "The Compressibility of Rolled Fill Materials Determined From Field Observations," Third International Conference on Soil Mechanics and Foundation Engineering, Zurich (1953), vol. II, p. 239.
159. J. G. Patrick, "Swift Dam Design," Paper presented at the American Society of Civil Engineers' Convention in Portland, Oregon (June 24, 1958).

160. H. H. Burke, "Swift Dam Construction," Paper presented at the American Society of Civil Engineers' Convention in Portland, Oregon (June 24, 1958).
161. R. Peterson, "Design and Construction of Earth Dams in Western Canada," *Journal, Engineering Institute of Canada* (February 1957).
162. A. Weiss, "Construction Technique of Passing Floods Over Earth Dams," *Transactions, American Society of Civil Engineers,* vol. 116 (1951), p. 1158.
163. W. L. Lowe-Brown, "British Practice in Dam Foundations," Third Congress on Large Dams, Stockholm (1948), vol. I.
164. P. T. Bennett, "Notes on Embankment Design," Fourth Congress on Large Dams, New Delhi (1951), vol. I, p. 223.
165. W. L. Shannon, "Measurements of Lateral Movements in Soils," *Proceedings,* Ninth Canadian Soil Mechanics Conference, Ottawa (December 1955), p. 75.
166. H. P. Hill, "The Ladybower Reservoir," *Journal, Institution of Water Engineers* (England), vol. 3 (1949), p. 414.
167. C. M. Clark, "Caissons Play New Part in Design of Large Dam" (Merriman Dam), *Water Works Engineering,* vol. 93 (1940), p. 1144.
168. E. Reinius, "The Stability of the Slopes of Earth Dams," *Geotechnique,* vol. V (1955), p. 181.
169. A. E. Bretting, Comments of General Reporter, "Stability and Deformations of Earth Constructions," Second International Conference on Soil Mechanics and Foundation Engineering, Rotterdam (1943), vol. 6, p. 92.
170. "Swiss Dam Taps Separate Streams for Power," *Engineering News Record* (May 16, 1957), p. 82.
171. "Earth Fill Dam Built on Frozen Ground," *Engineering News Record* (February 5, 1948), p. 70.
172. "Review of Slope Protection Methods," Report of the Subcommittee on Slope Protection of the Committee on Earth Dams of the Soil Mechanics and Foundations Division, *Proceedings, American Society of Civil Engineers,* vol. 74 (1948), p. 845.
173. C. W. O. Turner, Oral Discussion of Question No. 13, Fourth Congress on Large Dams, New Delhi (1951), vol. I.
174. L. F. Harza, Oral Discussion of Question No. 13, Fourth Congress on Large Dams, New Delhi (1951), vol. I.
175. K. Sain, Oral Discussion of Question No. 13, Fourth Congress on Large Dams, New Delhi (1951), vol. I.
176. E. G. H. Westerberg, Oral Discussion of Question No. 13, Fourth Congress on Large Dams, New Delhi, (1951), vol. I.
177. C. E. Blee, Oral Discussion of Question No. 13, Fourth Congress on Large Dams, New Delhi (1951), vol. I.
178. B. Hellstrom, Oral Discussion of Question No. 13, Fourth Congress on Large Dams (New Delhi), 1951, vol. I.
179. K. S. Lane, Oral Discussion of Question 18, Fifth Congress on Large Dams, Paris (1955), vol. III, p. 34.
180. A. W. Bishop, Oral Discussion on Question 18, Fifth Congress on Large Dams, Paris (1955), vol. III, p. 36.
181. J. K. Hunter and H. G. Keefe, "Special Problems Relating to the Construction of Dams in Active Volcanic Country," Fifth Congress on Large Dams, Paris (1955), vol. III, p. 511.

182. W. W. Daehn, "Behavior of a Rolled Earth Dam Constructed on a Compressible Foundation (Fresno Dam)," Fifth Congress on Large Dams, Paris (1955), vol. III, p. 171.
183. K. S. Lane, "Designing for Foundation Movements at Garrison Dam," Fifth Congress on Large Dams, Paris (1955), vol. III, p. 193.
184. A. de Palma Carlos, "Valeurs prevues et valeurs observées pour les tassements des enrochements du Barrage Salazar pendant et après sa construction," Fifth Congress on Large Dams, Paris (1955), vol. III, p. 245.
185. H. Mohle, "Settlings and Displacements at the Bever Dam of the Wupperverband," Fifth Congress on Large Dams, Paris (1955), vol. III, p. 323.
186. B. Hellstrom, "Compaction of a Rock Fill Dam (Nissastrom Dam)," Fifth Congress on Large Dams, Paris (1955), vol. III, p. 331.
187. J. Guilhamon and H. Castelnan, "Tassement des materiaux constutifs et de la foundation (Barrage de L'Iril-Emda)," Fifth Congress on Large Dams, Paris (1955), vol. III, p. 389.
188. G. Westerberg, G. Pira and J. Hagrup, "Description of Some Swedish Earth and Rockfill Dams With Concrete Core Walls and Measurements of the Movements and Pressure in the Filling Material and the Core Walls," Fourth Congress on Large Dams, New Delhi (1951), vol. I, p. 67.
189. B. Lofquist, "Calculating a Concrete Wall," Fourth Congress on Large Dams, New Delhi (1951), vol. I, p. 111.
190. V. M. Jevdjevic and B. M. Rajcevic, "Some Special Features of Designs and Constructions for Earth Dams and Rockfill Dams in Yugoslavia," Fourth Congress on Large Dams, New Delhi (1951), vol. I, p. 136.
191. L. F. Harza, "Rock Fill Dams," Fourth Congress on Large Dams, New Delhi (1951), vol. I, p. 153.
192. C. E. Blee and R. M. Riegel, "Rock Fill Dams," Fourth Congress on Large Dams, New Delhi (1951), vol. I, p. 189.
193. G. E. Bertram, "Slope Protection For Earth Dams," Fourth Congress on Large Dams, New Delhi (1951), vol. I, p. 209.
194. R. W. Bock and W. E. Collins, "Design and Construction of Bureau of Reclamation Earth Dams in the Missouri Basin," Fourth Congress on Large Dams, New Delhi (1951), vol. I, p. 247.
195. A. Cathabard, "Le Barrage de Lavand-Gelade," Fourth Congress on Large Dams, New Delhi (1951), vol. I, p. 413.
196. M. L. Dickinson, Discussion of Ref. 93.
197. W. F. Van Asbeck, "Bitumen in Hydraulic Engineering," Shell Petroleum Company, Ltd., London (1955).
198. L. Bjerrum, Comments of General Reporter, "Stability and Deformations of Slopes and Earth Dams," Third International Conference on Soil Mechanics and Foundation Engineering, Zurich (1953), vol. III, p. 209.
199. A. L. Little and V. E. Price, "The Use of an Electronic Computer for Slope Stability Analysis," *Geotechnique,* vol. VIII (September 1958), p. 118.
200. G. A. R. Sheppard and L. B. Aylen, "The USK Scheme for the Water Supply of Swensea," *Proceedings, Institution of Civil Engineers* (London), vol. 7 (June 1957), p. 246.
201. H. Press, *Talsperren,* Verlag Wilhelm Ernest & Sohn, Berlin (1958). (German reference book.)
202. B. Kjaernsli, "Some Notes on Earth and Rock Fill Dams in Norway," Norwegian Geotechnical Institute, September 1962.

203. F. C. Walker, Comments of General Reporter, "Earth Dams, Slopes and Open Excavations," Fourth International Conference on Soil Mechanics and Foundation Engineering, London (1957), vol. II, p. 463.
204. C. C. Delal, "Drains Under Dams," *Engineering News-Record,* vol. 95 (1925), p. 483.
205. B. Bassell, "Earth Dams," *Engineering News Publishing Company,* (1907).
206. L. M. Ellsperman, "Laboratory Testing and Construction of Asphaltic Concrete Dam Facings," *Proceedings,* Conference on the Use of Asphalt in Hydraulic Structures, Bakersfield, California (May 24, 1961). The Asphalt Institute Pacific Coast Division, Berkeley, California.
207. J. D. Schuyler, *Reservoirs for Irrigation, Water Power, and Domestic Water Supply,* John Wiley and Sons (1905).
208. J. Toran, Discussion of General Reporter "Raising of Dams," Sixth Congress on Large Dams, New York, (1958), Question No. 20.
209. R. R. Proctor, "The Design and Construction of Rolled Earth Dams," *Engineering News-Record, vol.* 111, (1933), pp. 245, 286, 348, 372.
210. P. Forchheimer, "Über die Ergiebigkeit von Brunnenanlagen und Sickerschlitzen," *Zeitschrift, Architekten-und Ingenier-Verein,* Hannover (1886), vol. 32, No. 7, pp. 539–64.
211. A. Casagrande, "Seepage Through Dams," Contributions to Soil Mechanics 1925–1940, *Boston Society of Civil Engineers* (1940), p. 295.
212. K. Terzaghi, *Theoretical Soil Mechanics,* John Wiley & Sons, (1943).
213. D. W. Taylor, *Fundamentals of Soil Mechanics,* John Wiley & Sons (1948).
214. H. R. Cedergren, "Use of Flow Net In Earth Dam and Levee Design," Second International Conference on Soil Mechanics and Foundation Engineering, Rotterdam, (1948), vol. 5, p. 293.
215. P. L. Capper and W. F. Cassie, *The Mechanics of Engineering Soils,* McGraw-Hill Book Company, (1949), p. 238.
216. A. de Palma Carlos, Discussion of paper, "Rockfill Dams: The Paradela Concrete Face Dam," *Transactions, American Society of Civil Engineers,* vol. 125 (1960), part II, p. 397.
217. T. Nilsson, "Bigger and Better Dams—The Swedish Contribution," *Engineering News-Record* (September 11, 1958), p. 60.
218. A. Coyne, "Bigger and Better Dams—The French Contribution," *Engineering News-Record* (September 11, 1958), p. 42.
219. C. Cabanius and R. Maigre, "The Serre-Ponçon Dam," *Travaux,* Paris (August 1958), p. 43.
220. J. Thevenin, "The Ghrib Dam," *Travaux,* Paris (August, 1958), p. 141.
221. R. J. Pope, "The Dalles Closure Dam," *Transactions, American Society of Civil Engineers,* vol. 125 (1960), part II, p. 473.
222. H. W. McFarlane, "The Beechwood Earth-fill Dam," *Proceeding,* Thirteenth Canadian Soil Mechanics Conference, National Research Council Technical Memo 63, Ottawa (June 1960).
223. T. F. Thompson, "Foundation Treatment for Earth Dams on Rock," *Proceedings, American Society of Civil Engineers,* Separate No. 548 (November 1954).
224. W. G. Holtz and H. J. Gibbs, "Triaxial Shear Tests on Pervious Gravelly Soils," *Journal, Soil Mechanics and Foundation Division, American Society of Civil Engineers,* Paper 867 (January 1956).

225. E. E. Esmiol, "Seepage Through Foundations Containing Discontinuities," *Journal, Soil Mechanics and Foundation Division, American Society of Civil Engineers,* Paper No. 1143 (January 1957).
226. C. I. Mansur, "Laboratory and In-Situ Permeability of Sands," *Transactions, American Society of Civil Engineers,* vol. 123 (1958), p. 868.
227. V. L. Minear, "General Aspects of Cement Grouting of Rock," *Journal, Soil Mechanics and Foundation Division, American Society of Civil Engineers,* Paper 1145 (January 1957).
228. C. M. Duke, "Effects of Ground on the Destructiveness of Large Earthquakes," *Journal, Soil Mechanics and Foundation Division, American Society of Civil Engineers,* Paper 1730 (August 1958).
229. L. G. Feil, "Unusual Design Problems-Harlan County Dam," *Transactions, American Society of Civil Engineers,* vol. 114 (1949), p. 321.
230. A. Mayer, "French Practice in Grouting Alluvial Materials," *Journal, Soil Mechanics and Foundations Division, American Society of Civil Engineers,* (February 1958).
231. "Pressure Grouting," U.S. Bureau of Reclamation Technical Memorandum No. 646, Denver (June 1957).
232. D. H. Trollope, Comments of General Reporter, "Earth Dams, Slopes and Open Excavations," Fifth International Conference on Soil Mechanics and Foundation Engineering, Paris (1961), vol. II, p. 859.
233. "Heart Butte Dam—Technical Record of Design and Construction," U.S. Bureau of Reclamation, Denver (1952).
234. S. T. Thorfinnson, "A Large Scale Field Test on a Bentonite Seam," *Proceedings, American Society of Civil Engineers,* Separate 549 (November 1954).
235. L. M. Ellsperman and R. E. Burnett, "Asphalt 'Ripraps' Earth Dam (Glen Anne Dam)," *Western Construction,* (September 1953).
236. E. B. Burwell, Jr., "Cement and Clay Grouting of Foundations—Practice of the Corps of Engineers," *Journal, Soil Mechanics and Foundations Division, American Society of Civil Engineers,* Paper 1551 (February 1958).
237. G. A. Kravetz, "Cement and Clay Grouting of Foundations—The Use of Clay in Pressure Grouting," *Journal, Soil Mechanics and Foundations Division, American Society of Civil Engineers,* Paper 1546 (February 1958).
238. J. P. Elston, "Cement and Clay Grouting for Foundations—Suggested Specifications for Pressure Grouting," *Journal, Soil Mechanics and Foundations Division, American Society of Civil Engineers,* Paper No. 1548 (February 1958).
239. G. K. Leonard and L. F. Grant, "Cement and Clay Grouting of Foundations —Experience of TVA with clay-cement and related Grouts," *Journal, Soil Mechanics and Foundations Division, American Society of Civil Engineers,* Paper No. 1552 (February 1958).
240. T. B. Kennedy, "Pressure Grouting Fine Fissures," *Journal, Soil Mechanics and Foundations Division, American Society of Civil Engineers,* Paper 1731 (August 1958).
241. E. B. Burwell and R. H. Nesbitt, "The NX Bore-hole Camera," *Mining Engineering* (August 1954), p. 805.
242. A. W. Simonds, "Cement and Clay Grouting of Foundations—Present Status of Pressure Grouting Foundations," *Journal, Soil Mechanics and Foundations Division, American Society of Civil Engineers,* Paper 1544, (February 1958).

243. R. W. Clough and D. Pirtz, "Earthquake Resistance of Rock-Fill Dams," *Transactions, American Society of Civil Engineers,* vol. 123 (1958), p. 792.
244. H. B. Seed, "The Strength of Soils Under Earthquake Loading Conditions," A report prepared for the Department of Water Resources, State of California, November (1958).
245. H. B. Seed and R. Lundgren, "Investigation of the Effect of Transient Loading on the Strength and Deformation Characteristics of Saturated Sands," *Proceedings, American Society of Testing Materials,* vol. 54 (1954).
246. A. Casagrande and W. L. Shannon, "Research on Stress-Deformation and Strength Characteristics of Soils and Soft Rocks Under Transient Loading," Harvard University Soil Mechanics Series, No. 31 (1948).
247. C. A. Heiland, "Geophysical Investigations Concerning the Seismic Resistance of Earth Dams," *Transactions, American Institute of Mining and Metallurgical Engineers,* vol. 138, (1940), p. 350.
248. M. Hatanaka, "Fundamental Considerations on the Earthquake Resistant Properties of the Earth Dam," Disaster Prevention Research Institute, Bulletin No. 11, Kyoto University, Kyoto, Japan (December 1955).
249. S. K. Guha and G. Ram, "Earthquake Forces on Earth Dam," *Journal, Central Board of Irrigation and Power,* Simla, India (July 1951).
250. P. Byerly, "Earthquake Origin and Propagation," *Proceedings,* Symposium on Earthquake and Blast Effects on Structures" Earthquake Engineering Research Institute, Berkeley, California (September 1952), p. 1.
251. F. Neumann, "Some Generalized Concepts of Earthquake Motion," *Proceedings,* Symposium on Earthquake and Blast Effects on Structures, Earthquake Engineering Research Institute, Berkeley, California (September 1952), p. 8.
252. A. L. Miller, "Earthquake Lessons From the Pacific North West," *Proceedings,* Symposium on Earthquake Blast Effects on Structures, Earthquake Engineering Research Institute, Berkeley, California (September 1952), p. 212.
253. F. W. Furkert, "The Effect of Earthquakes on Engineering Structures," *Proceedings, Institution of Civil Engineers,* Paper 4872, vol. 236 (1932–1933), p. 344.
254. F. R. Melville and R. LeGeyt, "The Pegu Yomas Scheme for the Water Supply of Rangoon," *Proceedings, Institution of Civil Engineers,* Paper 5595, vol. 30 (1947–1948), p. 59.
255. "Water Supply of Rangoon, Burma," *The Engineer,* vol. 169 (1940), pp. 488, 496, 508.
256. F. Neumann, *Earthquake Intensity and Related Ground Movement,* University of Washington Press, Seattle, (1954).
257. G. O. Louderback, "Faults and Engineering Geology," *The Geological Society of America,* The Berkey Volume, (November 1950), p. 125.
258. L. D. Leet, *Earth Waves,* Harvard University Press, (1950).
259. G. O. Louderback, "Central California Earthquakes of the 1830's," Bulletin, Seismological Society of America, vol. 37 (1947).
260. F. W. Housner, R. R. Martel and J. L. Alford, "Spectrum Analysis of Strong-Motion Earthquakes," Bulletin, Seismological Society of America, vol. 43 (1953), p. 97.
261. G. O. Louderback, "Characteristics of Active Faults in the Central Coast Ranges of California with Application to the Safety of Dams," Bulletin, Seismological Society of America, vol. 27 (1937), pp. 1–27.

262. J. D. Galloway, "The Design of Rockfill Dams," *Transactions, American Society of Civil Engineers*, vol. 104 (1939), p. 84.
263. "Outlet Tunnel Serves Multiple Purposes at Rector Creek Dam," *Western Construction News* (July 1946), p. 79.
264. F. A. Noetzli, "Core Wall in Rockfill Dam Tilts When Reservoir Fills," *Engineering News-Record*, vol. 109 (1932), p. 529.
265. R. R. Martel, "Effect of Earthquakes on Earth Dams," *The Military Engineer*, vol. 30 (1938), p. 359.
266. L. V. Branch, "Earth Dam Has Screened Gravel Core and Pipe Drains," *Engineering News-Record*, vol. 80 (1918), p. 366.
267. N. Mononobe, A. Takata and M. Matamura, "Seismic Stability of the Earth Dam," Second Congress on Large Dams, Washington, D.C., vol. IV (1936), p. 435.
268. H. O. Wood and F. Neumann, "Modified Mercalli Intensity Scale of 1931," Bulletin, Seismological Society of America, vol. 21, No. 4 (December 1931).
269. A. W. Anderson, et al., "Lateral Forces of Earthquake and Wind," *Transactions, American Society of Civil Engineers*, vol. 117 (1952), p. 716.
270. A. T. Blythe and G. S. Williams, Exchange of Letters to the Editor, *Engineering News-Record*, vol. 94, (1925), p. 1070.
271. F. S. Brown, "Service Behavior of Blankets as a Method of Sealing Dams," Seventh Congress on Large Dams, Rome (1961), Question 27, Report 67.
272. "Soil Mechanics Design—Stability of Slopes and Foundations," Engineering Manual, Civil Works Construction, U.S. Corps of Engineers, Part CXIX (February, 1952) Chapter 2.
273. D. W. Taylor, "Stability of Earth Slopes," *Journal, Boston Society of Civil Engineers* (July 1937).
274. *Earth Manual* (First Edition), U.S. Bureau of Reclamation, Denver (July 1960).
275. K. Terzaghi, "Mechanics of Shear Failures in Clay Slopes," *Public Roads*, vol. 10 (1929), p. 177.
276. D. R. May, "Application of the Planimeter to the Swedish Method of Analyzing the Stability of Earth Slopes," Second Congress on Large Dams, Washington, D.C. (1936), vol. IV, p. 540.
277. A. Mayer, "Characteristics of Materials Used in Earth Dam Construction—Stability of Earth Dams in ases of Reseervoir Discharge," Second Congress on Large Dams, Washington, D.C. (1936), vol. IV, pp. 295–331.
278. J. D. Justin, *Earth Dam Projects*, John Wiley & Sons, (1932).
279. K. Terzaghi, (Leading a discussion) "Investigation of the Characteristics of Soils with Respect to Their Suitability for the Construction of Earth Dams," First Congress on Large Dams (1933), vol. III, p. 321–3.
280. "Statens Jarnvagers Geotekniska Commission," *Slutbeton Kande*, vol. 31 (May 1922).
281. K. Terzaghi, "Die Prüfung von Baumaterialien für gewalzte Erddämme," First Congress on Large Dams, (1933), vol. III, p. 69–106.
282. D. B. Gumensky, "Earthquakes and Earthquake-Resistant Design," in *American Civil Engineering Practice* (Edited by R. W. Abbett), vol. III, John Wiley & Sons, (1957), p. 34–01.
283. "Le Basin d'accumulation d'eau de Viandem au Luxembourg," *Bitume Actualités*, Paris (July 1962), p. 19.
284. P. C. Varghese, "Investigations of a New Procedure for Analyzing The Stability of Slopes," Master's Thesis, Harvard University (June 1949).

285. W. A. Brown, "Earth Dam Stability Analysis," Technical Memorandum (April 18, 1952), Supervision of Dams, California State Division of Water Resources, Sacramento, California.
286. C. A. Coulomb, "Essai sur une application des Règles des maximus et minimus a quelques problèmes de statique relatifs a l'architecture," *Mem. Acad. Roy. Pres. Divers Savants,* vol. 7, Paris (1776).
287. W. J. M. Rankine, "On the Stability of Loose Earth," *Transactions of the Royal Philosophic Society,* vol. 147, London (1857).
288. K. E. Pettersson, "Kajraset i Goteborg den 5te Mars 1916," *Teknisk Tidsskeift,* V. U. vol. 30 Stockholm (1916), pp. 281–287.
289. S. Hultin, "Grufyllnader for kajbyggnader," *Teknisk Tidsskeift,* V. U., vol. 31, Stockholm (1916), p. 292.
290. W. Kjellman, "Do Slip Surfaces Exist," Geotechnique, vol. V, London (1955), p. 18.
291. K. E. Pettersson, "The Early History of Circular Sliding Surfaces," *Geotechnique,* vol. V, London, (1955), p. 275.
292. W. Fellenius, *Erdstätische Berechnungen mit Reibung und Kohäsion,* Ernst, Berlin, 1927.
293. W. Fellenius, "Calculation of the Stability of Earth Dams," Second Congress on Large Dams, Washington, D.C. (1936), vol. 4, p. 445.
294. A. Collin, "Landslides in Clays," University of Toronto Press (1956). (Translated from the French work dated 1846 by W. R. Schriever.)
295. N. Janbu, "Stability Analysis of Slopes with Dimensionless Parameters," Soil Mechanics Series, No. 46, Harvard University (January 1954).
296. K. O. Strenge, "Noxon Rapids Dam," *Civil Engineering* (July 1959).
297. H. K. Fairbanks and R. A. Sutherland, "Design Features for Safety and Economy for Wide Valleys," Seventh Congress on Large Dams, Rome (1961), Question 26, Report 61.
298. D. W. Taylor, Paper Presented to the American Society of Civil Engineers Convention, New York (January 1949).
299. H. Krey, *Erddruck, Erdurterstand, und Tragfähigkeit des Baugrundes,* Ernst, Berlin, 1936.
300. J. P. Gould, "Construction Pore Pressures in Rolled Earth Dams," Technical Memorandum 650, Bureau of Reclamation, Denver, Colorado (April 1959).
301. W. J. Turnbull and W. G. Shockley, "Compaction of Earth Dams in the Corps of Engineers, U.S. Army," Sixth Congress on Large Dams, New York (1958), Report 96, Question 22.
302. J. W. Hilf, "Estimating Construction Pore Pressures in Rolled Earth Dams," Second International Conference on Soil Mechanics and Foundation Engineering, Rotterdam (1948), vol. III, p. 234.
303. E. E. Nonveiller, "Pore Pressures in the Lokvarka Dam," Fourth International Conference on Soil Mechanics and Foundation Engineering, London (1957), vol. 2, p. 341.
304. Karl Terzaghi, "The Design and Performance of the Sasumua Dam," Paper No. 6252, *Proceedings, Institution of Civil Engineers, London* (April 1958), p. 369.
305. H. H. Dixon, G. A. Edington, and E. P. Fitzgerald, "The Chama-Sasumua Water Supply System of Nairobi," *Proceedings, Institution of Civil Engineers,* London (April 1959).

306. C. J. Hoffman, "Spillway Considerations in the Heightening of Existing Dams," Sixth Congress on Large Dams, New York (1958), Question 20, Report 91.
307. A. Aastrup and S. Sallstrom, "Bergeforsen—A Swedish Power Plant Built on Non-Resistant Rock," Seventh Congress on Large Dams, Rome (1961), Question 25, Report 69.
308. J. A. H. Brahtz, C. N. Zanger, and J. R. Bruggeman, "Notes on Analytical Soil Mechanics," Technical Memo No. 592, U.S. Bureau of Reclamation, Denver, Colorado (1939).
309. H. V. Klaas and W. H. Stewart, "Willamette Basin—Large-Scale Flood Control Plan," Sixth Congress on Large Dams, U.S. Committee Souvenir Publication (1958), p. 137.
310. "El Vado Dam Rehabilitation," Specifications and Drawings No. DC-4203, U.S. Bureau of Reclamation (1954).
311. A. W. Bishop and D. Henkel, *The Measurement of Soil Properties in the Triaxial Test,* Edward Arnold, Ltd., (1957).
312. T. A. Middlebrooks, "Foundation Investigation of Fort Peck Dam Closure Section," International Conference on Soil Mechanics and Foundation Engineering (1936), vol. 1, p. 135.
313. L. Jurgenson, "The Application of Theories of Elasticity and Plasticity to Foundation Problems," *Journal, Boston Society of Civil Engineers,* vol. 21 (1934), p. 206.
314. T. A. Middlebrooks, "Seepage Control for Large Dams," Third Congress on Large Dams, Stockholm (1948).
315. T. A. Middlebrooks and G. E. Bertram, "Corps of Engineers Practice of Earth Dam Design and Construction," Second International Conference on Soil Mechanics and Foundation Engineering, Rotterdam (1948), vol. 7, p. 100.
316. R. R. Proctor, "Earth Dam Design, Construction and Performance as Practiced by the Water System of the Dept. of Water and Power, City of Los Angeles," Second International Conference on Soil Mechanics and Foundation Engineering, Rotterdam (1948), vol. 5, p. 81.
317. R. E. Glover, H. J. Gibbs, and W. W. Daehn, "Deformability of Earth Materials and its Effect on the Stability of Earth Dams Following a Rapid Drawdown," Second International Conference on Soil Mechanics and Foundation Engineering, Rotterdam (1948), vol. 5, p. 77.
318. F. C. Walker and W. W. Daehn, "Ten Years of Pore Pressure Measurements," Second International Conference on Soil Mechanics and Foundation Engineering, Rotterdam, 1948, vol. 3, p. 245.
319. F. W. Kellogg, "Investigation of Drainage Rates Affecting the Stability of Earth Dams," *Transactions, American Society of Civil Engineers,* vol. 113 (1948), p. 1261.
320. J. Lowe and L. Karafiath, "Stability of Earth Dams Upon Drawdown," First Pan American Conference on Soil Mechanics and Foundation Engineering, Mexico City (1959), vol. II, p. 537.
321. G. F. Sudman, "Kajakai Central Core Dam, Afganistan," *Transactions, American Society of Civil Engineers,* vol. 125 (1960), part II, p. 301.
322. W. A. Clevenger, discussion of "A Rapid Method of Construction Control for Embankments of Cohesive Soil," Conference on Soils for Engineering Purposes, Mexico, *American Society for Testing Materials Special Technical Publication,* No. 232 (December 1957), p. 451.

323. H. L. Drake, et al., "McNary Dam—Design From Technical Considerations," *Proceedings, American Society of Civil Engineers* (September 1958).
324. K. Terzaghi, "Retaining-Wall Design for Fifteen-Mile Falls Dam," *Engineering News-Record* (May 17, 1934).
325. H. R. Cedergren, Discussion of Paper, "Investigation of Drainage Rates Affecting Stability of Earth Dams," Transactions, *American Society of Civil Engineers,* vol. 113 (1948), p. 1285.
326. W. L. Shannon, Discussion of Paper "Investigation of Drainage Rates Affecting Stability of Earth Dams," *Transactions, American Society of Civil Engineers,* vol. 113 (1948), p. 1302.
327. T. W. Lambe, Discussion of Paper, "Investigation of Drainage Rates Affecting Stability of Earth Dams," *Transactions, American Society of Civil Engi*neers, vol. 113 (1948), p. 1294.
328. W. G. Huber, "Kenney and Cheakamus Dams," *Transactions, American Society of Civil Engineers,* vol. 125 (1960), part II, p. 255.
329. L. A. Schmidt, "Performance and Maintenance of Dix River Dam," *Transactions, American Society of Civil Engineers,* vol. 125 (1960), part II, p. 1.
330. P. A. Baumann, "Cogswell and San Gabriel Dams," *Transactions, American Society of Civil Engineers,* vol. 125 (1960), part II, p. 29.
331. T. Mundal, "Brownlee Sloping Core Dam," *Transactions, American Society of Civil Engineers,* vol. 125 (1960), part II, p. 520.
332. G. K. Leonard and O. H. Raine, "Performance of TVA Central Core Dams," *Transactions, American Society of Civil Engineers,* vol. 125 (1960), part II, p. 190.
333. F. W. Patterson and D. H. MacDonald, "The Bersimis Sloping Core Dams," *Transactions, American Society of Civil Engineers,* vol. 125 (1960), part II, p. 488.
334. J. P. Growdon, "Nantahala Sloping Core Dam," *Transactions, American Society of Civil Engineers,* vol. 125 (1960), part II, p. 160.
335. J. P. Growdon, "Dams With Sloping Earth Cores," *Transactions, American Society of Civil Engineers,* vol. 125 (1960), part II, p. 207.
336. J. P. Growdon, "Performance of Seven Sloping Core Dams," *Transactions, American Society of Civil Engineers,* vol. 125 (1960), part II, p. 237.
337. I. C. Steele and J. B. Cooke, "Salt Springs and Lower Bear River Concrete Face Dams," *Transactions, American Society of Civil Engineers,* vol. 125 (1960), part II, p. 74.
338. C. P. Vetter, Oral Discussion of Question 18, Fifth Congress on Large Dams, Paris (1955), vol. III.
339. C. Y. Li, "Construction Pore Pressures in Quebradona Earth Dam," *Journal, Soil Mechanics and Foundations Division, American Society of Civil Engineers* (October 1959), p. 43.
340. J. L. Sherard, Discussion of Paper, "Tractive Resistance of Cohesive Soils," by I. S. Dunn, *Journal, Soil Mechanics and Foundation Division, American Society of Civil Engineers* (December 1959).
341. "Slope Protection for Earth Dams," U.S. Army Corps of Engineers, Waterways Experiment Station, Vicksburg, Mississippi (1949).
342. H. G. Arthur, "Design of Earth Embankment—Falcon Dam and Power Plant," U.S. Bureau of Reclamation Technical Memorandum No. 642, Denver, Colorado (August 1951).
343. A. Hedar, "Design of Rock-Fill Breakwaters," *Proceedings, Minnesota International Hydraulics Convention* (September 1953), p. 241.

344. "Presas de Tierra y de Enrocamiento en Mexico," Secretario de Recursos Hidraulicos, Contribution to the First Pan American Congress of Soil Mechanics, Mexico (1959).
345. "Bonny Dam," Technical Record of Design and Construction, U.S. Bureau of Reclamation, Denver, Colorado (1954).
346. "Platoro Dam," Technical Record of Design and Construction, U.S. Bureau of Reclamation, Denver, Colorado (1954).
347. "Long Lake Dam and Main Canal," Technical Record of Design and Construction, U.S. Bureau of Reclamation, Denver, Colorado (1955).
348. "Cedar Bluff Dam," Technical Record of Design and Construction, U.S. Bureau of Reclamation, Denver, Colorado (1955).
349. "Boysen Dam and Power Plant," Technical Record of Design and Construction, U.S. Bureau of Reclamation, Denver, Colorado (1957).
350. "Trenton Dam," Technical Record of Design and Construction, U.S. Bureau of Reclamation, Denver, Colorado (1957).
351. "Davis Dam and Power Plant," Technical Record of Design and Construction, U.S. Bureau of Reclamation, Denver, Colorado (1957).
352. "Slope Protection," Civil Works Engineer Bulletin, 52-15 (June 2, 1952), Office of Chief of Engineers, U.S. Corps of Engineers, Washington, D.C.
353. G. W. Lamb, "Steel Dam in Good Shape After 50 Years," *Steel Construction Digest,* vol. 7, No. 2 (April 1950).
354. C. Marcello, Article on Construction of Khram Dam (in Italian), *L'Energia Elettrica,* No. 2 (1936), p. 79.
355. A. N. Vanderlip, F. W. Scheidenhelm, and J. B. Snethlage, "Laboratory Investigation of Asphaltic Concrete, Montgomery Dam, Colorado," Proceedings, Association of Asphalt Paving Technologists, Montreal (February 1958), p. 383.
356. C. W. Lauritzen, F. W. Haws and Allan S. Humphreys, "Plastic Film for Controlling Seepage Losses in Farm Reservoirs," Bulletin 391, Utah State College Division of Agricultural Sciences (July 1956).
357. "Nuclear Soil Testing," *Contractors and Engineers* (April 1962), p. 76.
358. F. W. Scheidenhelm, J. B. Snethlage and A. N. Vanderlip, "Montgomery Dam—Rockfill With Asphalt Concrete Deck," *Transactions, American Society of Civil Engineers,* vol. 125 (1960), part II, p. 431.
359. A. J. Cleary, "The Calaveras Dam, California, The Highest Earth Dam," *Engineering News-Record,* vol. 72 (1914), p. 692.
360. "Necaxa Dam Fails During Construction," *Engineering News-Record,* vol. 62 (1909), p. 1.
361. "Hydraulic Fill Dam Fails Disasterously," *Engineering News-Record,* vol. 104, p. 869.
362. "Earth Embankments," Manual, U.S. Corps of Engineers, EM1110-2-2300 (April 1, 1959).
363. W. J. Turnbull, "Trench Digs and Mixes Cutoff Materials," *Engineering News-Record,* vol. 122 (1939), p. 63.
364. A. da Morais and A. da Palma Carlos, "Design and Construction of Earthfill and Rockfill Dams and Their Waterproofing Blankets. Salazar (Pego do Altor) and Vale do Gaio Dams," Communication C 7, Fourth Congress on Large Dams, New Delhi (1951), vol. 4, p. 453.
365. C. Marcello, "Gravel Fill Dam," Communication No. 12, Fourth Congress on Large Dams, New Delhi (1951), vol. 4, p. 517.

366. R. T. Logeman, "Steel Dams," *Journal, Western Society of Engineers*, vol. 45 (1940), p. 313.
367. M. Duffaut et al, "Resistance des barrages au glissement au contact béton-rocher et sur une réprise de bétonnage," Seventh Congress on Large Dams, Rome (1961), Question 26, Report 43.
368. A. Casagrande, "Characteristics of Cohesionless Soils Affecting the Stability of Slopes and Earth Fills," Contributions to Soil Mechanics, 1925–1940, *Boston Society of Civil Engineers* (1940), p. 257.
369. A. P. Geuss, "Box Canyon Hydroelectric Project," *Journal, Power Division, American Society of Civil Engineers* (June 1958).
370. W. A. Clevenger, "Experiences with Loess as a Foundation Material," *Transactions, American Society of Civil Engineers,* vol. 123 (1958).
371. J. L. Sherard, "A Steel-Faced Rockfill Dam," *Transactions, American Society of Civil Engineers,* vol. 125 (1960), part II, p. 465.
372. "Bucks Dam, California," *Engineering News-Record,* vol. 100 (1928), p. 140.
373. H. Lauffer and W. Schober, "Investigations for the Earth Core of the Gepatsch Rockfill Dam with a Height of 500 ft," Seventh Congress on Large Dams, Rome (1961), Question 27, Report 92.
374. C. P. Segar, "Steel Used Extensively in Building El Vado Dam," *Engineering News-Record* (1935), p. 211.
375. G. R. Scott, Discussion to Ref. 358.
376. Ohde, "Zur Static der Staudämme mit Betonkern," *Mitteilungen des Institutes für Wasserbau,* University of Berlin (1951), No. 38.
377. F. T. Crowe, "Reclamation Service Building Highest Earth Dam," *Engineering News-Record,* vol. 87, (December 1, 1921), p. 890.
378. F. T. Crowe, "Building the Tieton Dam," *Pacific Builder and Engineer,* vol. 29 (January 26, 1923), p. 7.
379. A. da Palma Carlos and R. da Silva Sanches, "Aims and Conditions of the Utilizaiton of Bituminous Products in Some Portuguese Dams," Seventh Congress on Large Dams, Rome (1961), Question 27, Report 105.
380. M. A. Mason, "The Problem of Wave Action on Earth Slopes," *Transactions, American Society of Civil Engineers,* vol. 116 (1951), p. 1398.
381. A. Casagrande, "Classification and Identification of Soils," *Transactions, American Society of Civil Engineers,* vol. 113 (1948), p. 901.
382. "Unified Soil Classification System," Technical Memorandum No. 3-357, Corps of Engineers, U.S. Army, Waterways, Experiment Station, Vicksburg, Miss. (March 1953).
383. N. N. Ambraseys, "On The Shear Response of a Two-Dimensional Truncated Wedge Subjected to an Arbitrary Disturbance," *Bulletin, Geological Society of America,* vol. 50, No. 1, p. 45 (Jan. 1960).
384. V. F. B. de Mello and P. T. da Cruz, "Some Quantitative Investigations on Curtain Grouting in Rock Foundations of Earth Dams." First Pan American Congress of Soil Mechanics and Foundation Engineering, Mexico (September 1959).
385. C. J. Ullrich, "Earth and Rockfill Dam for Irrigation," *Engineering News-Record,* vol. 100 (1928), p. 864, 872.
386. E. C. Murphy, "Failures of Lake Avalon Dam Near Carlsbad, New Mexico," *Engineering News,* vol. 54 (1905), p. 9.
387. G. E. Bertram, "An Experimental Investigation of Protective Filters," Harvard University Soil Mechanics Series, No. 7 (January 1940).

388. "Seepage Control—Soil Mechanics Design," Engineering Manual for Civil Works Construction, Part CXIX, Chapter 1 (February 1952), U.S. Army Corps of Engineers.
389. F. P. Sterns, "Discussion on the Bohio Dam," *Transactions, American Society of Civil Engineers,* vol. 48 (1902), p. 272–3.
390. "Investigation of Filter Requirements for Underdrains," U.S. Corps of Engineers, Waterways Experiment Station Technical Memorandum No. 183-1 (December 1941).
391. K. P. Karpoff, "The Use of Laboratory Tests to Develop Design Criteria for Protective Filters," *Proceedings, American Society for Testing Materials,* vol. 55 (1955), p. 1183.
392. N. R. Twelker, "Analysis of Seepage in Pervious Abutments of Dams," Fourth International Conference on Soil Mechanics and Foundation Engineering, London (1957), vol. II, p. 389.
393. W. G. Holtz, "The Determination of Limits for Control of Placement Moisture on High Rolled Earth Dams," *Proceedings, American Society for Testing Materials,* vol. 48 (1948), p. 1240.
394. H. W. Koenig, "Die Biggetalsperre," *Das Gas und Wasserfach,* vol. 26 (1962), p. 661.
395. G. F. Sowers and C. E. Gove, "Large Scale Preconstruction Tests of Embankment Materials for an Earth-Rockfill Dam," Fifth International Conference on Soil Mechanics and Foundation Engineering, Paris (1961).
396. F. C. Walker and W. G. Harber, "Design of the Trinity Dam, an Earthfill Structure 537 Feet High," Fifth International Conference on Soil Mechanics and Foundation Engineering, Paris (1961).
397. E. Meyer-Peter, "Soil Mechanics and Foundation Problems of the Marmorera Earth Dam," Third International Conference on Soil Mechanics and Foundation Engineering, Zurich, Switzerland (1953).
398. K. Noyan and T. Acatay, "The Construction of the Impervious Core and Supporting Fill at Hirfanli Dam," Sixth Congress on Large Dams, New York (1959), Report No. 22, Question 22.
399. Dan K. Heiple, "Earth Moving—an Art and a Science," Construction Methods Reprint, McGraw-Hill Publishing Company (1952).
400. W. M. Aldous and H. W. Wills, "Field Equipment for Vibratory Compaction of Soils and Base Courses," Civil Aeronautics Administration, Technical Development Report No. 186 (November 1952).
401. D. L. Townsend, "The Performance and Efficiency of Standard Compacting Equipment," Ontario Joint Highway Research Programme, Queens University, Kingston, Ontario, Report No. 6 (January 1959).
402. W. J. Turnbull and C. R. Foster, "Stabilization of Materials by Compaction," *Transactions, American Society of Civil Engineers,* vol. 123 (1958), p. 1.
403. "Dam Slide Laid to Rock Fault," *Engineering News-Record* (January 7, 1962), p. 22.
404. F. H. P. Williams and D. J. MacLean, "The Compaction of Soil—A Study of the Performance of Plant," Road Research Technical Paper No. 17, Department of Scientific and Industrial Research, Road Research Laboratory (England) (1950).
405. J. W. Hilf, "A Rapid Method of Construction Control for Embankments of Cohesive Soil," Conference on Soils for Engineering Purposes, Mexico City (December 1957), *American Society of Testing Materials Special Publication* STP No. 232.

406. "Belt Delivers Fill for Earth Dam," (Youghiogheny Dam), *Engineering News-Record* (December 3, 1942).
407. "Making the Fill at Dorena Dam," *Engineering News-Record* (November 3, 1949), p. 24.
408. R. R. Phillippe, "Compaction of Cohesive Soils," Progress Report of the Subcommittee on the Consolidation in Earth Dams and Their Foundations, *Proceedings, American Society of Civil Engineers,* Paper 48 (December 1950).
409. "Compaction Studies on Clayey Sands," Report No. 1, Soil Compaction Investigation, Technical Memo 3-271, Waterways Experiment Station, Vicksburg, Miss. (April 1949).
410. "Effect on Soil Compaction of Tire Pressure and Number of Coverages of Rubber-Tired Rollers and Foot-Contact Pressure of Sheepsfoot Rollers," Report No. 7, Soil Compaction Investigation, Technical Memo 3-271, Waterways Experiment Station, Vicksburg, Miss. (June 1956).
411. "Effect of Lift Thickness and Tire Pressure," Report No. 8, Soil Compaction Investigation, Technical Memo 3-271, Waterways Experiment Station, Vicksburg, Miss. (October 1957).
412. C. Y. Li, "Basic Concepts on the Compaction of Soil," *Journal, Soil Mechanics and Foundations Division, American Society of Civil Engineers* (January 1956).
413. S. J. Johnson and W. G. Shockley, "Field Penetration Tests for Selection of Sheepsfoot Rollers," *Proceedings, American Society of Civil Engineers,* Paper 363, vol. 79 (1953).
414. W. G. Holtz and C. A. Lowitz, "Compaction Characteristics of Gravelly Soils," Conference on Soils for Engineering Purposes, University of Mexico (December 1957), *Special Technical Publication No. 232, American Society for Testing Materials,* p. 67.
415. W. J. Turnbull and C. R. Foster, "Effect of Tire Pressures and Lift Thicknesses on Comparison of Soil with Rubber-tired Rollers," Conference on Soils for Engineering Purposes, University of Mexico (December 1957), *Special Technical Publication No. 232, American Society for Testing Materials,* p. 102.
416. S. D. Wilson, "Effect of Compaction on Soil Properties," Proceedings, Conference on Soil Stabilization, Massachusetts Institute of Technology (June 1952), p. 149.
417. J. F. Haley, Discussion of several papers, Proceedings, Conference on Soil Stabilization, Massachusetts Institute of Technology (June 1952), p. 264.
418. "Earth Movers Set Fast Pace on Dam Job—Navajo Dam," *Construction Methods and Equipment* (August 1959).
419. P. Baumann, "Design and Construction of San Gabriel Dam No. 1," *Transactions, American Society of Civil Engineers,* vol. 107 (1942), p. 1607.
420. K. S. Lane and R. G. Fehrman, "Tuttle Creek—A Dam of Rolled Shale and Dredged Sand," A paper presented at the Reno Convention, American Society of Civil Engineers (June 1960).
421. W. W. Daehn and E. E. Esmiol, "Field Measurements at Willard Dam Test Embankment," Paper presented at the New York Convention of the American Society of Civil Engineers (October 1957).
422. J. W. Hilf, "A Rapid Method of Construction Control for Embankments of Cohesive Soils," U.S. Bureau of Reclamation Engineering Monograph No. 26, Denver, Colorado (October 1959).

423. K. S. Lane, "Providence Vibrated Density Test," Second International Conference on Soil Mechanics and Foundation Engineering, Rotterdam (1948), vol. 4, p. 243.
424. "Design of Small Dams," U.S. Bureau of Reclamation, U.S. Government Printing Office, Washington, D.C. First Edition (1960).
425. "Cachuma Dam," Technical Record of Design and Construction, U.S. Bureau of Reclamation, Denver, Colorado (July 1959).
426. "Colorado-Big Thompson Project," Technical Record of Design and Construction, vol. 11, U.S. Bureau of Reclamation, Denver, Colorado (April 1957).
427. V. Lanner, discussion of Ref. 328, *Transactions, American Society of Civil Engineers,* vol. 125 (1960), part II, p. 564.
428. K. L. Powers, "Test Section for Embankment Protection—St. Francis Unit, Upper Republican Division, Missouri River Basin Project," Final Report (June 1951) (unpublished U.S. Bureau of Reclamation Report).
429. "Três Marias Dam, Report of Laboratory and Field Investigations Related to the Choice of the Compaction and Moisture Control Method," unpublished Report of the Centrais Electrica de Minas Gerais, S.A. (Brazil) (January 1960).
430. R. L. Boyce, "Final Construction Report on Lovewell Dam," unpublished U.S. Bureau of Reclamation Report, March 20 (1958).
431. Lohr, "Problems Encountered in the Construction of Fill Dams with Bituminous Sealing Elements," Seventh Congress on Large Dams, Rome (1961), Question 27, Report 77.
432. F. C. Walker, "The Use of Residual Soils in Earth and Rockfill Dams," Proceedings, First Panamerican Conference on Soil Mechanics and Foundation Engineering, Mexico (1959), vol. II, p. 589.
433. E. Tamez, "Some Factors Affecting the Dynamic Compaction Test," Conference on Soils for Engineering Purposes, University of Mexico (December 1957), *American Society for Testing Materials Special Technical Publication No. 232,* p. 62.
434. J. Lordet, "Hydro-Electric Development of the Oued Agrioun," Fifth International Congress on Large Dams, Paris (1955), Supplement to *Travaux,* p. 190.
435. N. N. Ambraseys, "On the Seismic Behavior of Earth Dams," Proceedings, Second World Conference on Earthquake Engineering (Japan, 1960), vol. I, p. 331.
436. D. C. Henny, "Dams on the San Andreas Fault—Letter to the Editor," *Engineering News-Record,* vol. 109, (September 29, 1932), p. 385.
437. M. Shebusawa, "Déscription des dégates causes aux installations électriques au Japon par le grand tremblement de terre du 1 Septembre, 1923," International Conference on Large Electrical Installations, Paris, (1925), vol. 1, p. 1191.
438. S. D. Wilson, "Small Soil Compaction Apparatus Duplicates Field Results Closely," *Engineering News-Record* (November 2, 1950).
439. "Hills Creek Reservoir—Embankment Design Review," unpublished Report of Portland District, Corps of Engineers, U.S. Army (June 15, 1959).
440. H. J. McKeever, "Construction Methods for Goschenen Dam," *World Construction* (September 1960), p. 16.
441. H. B. Seed, J. K. Mitchell, and C. K. Chan, "The Strength of Compacted Cohesive Soils," Paper presented to the American Society of Civil Engineers

Research Conference on the Shear Strength of Cohesive Soils, Boulder, Colorado (June 1960).

442. B. Lofquist, Discussion of "Stability and Deformations of Slopes and Earth Dams," Third International Conference on Soil Mechanics and Foundation Engineering, Zurich, Switzerland, (1953), vol. III, p. 214.
443. John L. Kurantz, "Measurement of Moisture and Density in Soils by the Nuclear Method," Paper presented to the American Society of Testing Materials Symposium on Radioisotope Test Methods, San Francisco, California, (October 13), 1959.
444. R. Horonjeff and D. F. Javette, "Neutron and Gamma-Ray Methods for Measuring Moisture Content and Density to Control Field Compaction," Bulletin No. 122, Soil Testing Methods, Highway Research Board, Washington, D.C. (1956), p. 23.
445. F. Reinhold, "Soil Cement Methods for Residential Streets in Western Germany," *World Construction*, (March–April 1955), p. 41.
446. "Três Marias Dam," *Water Power*, part I (December 1959); part II (January 1960); part III (February 1960).
447. J. F. Redus, "A Study of In-Place Density Determinations for Base Courses and Soils," Highway Research Board Bulletin 159, Washington, D.C. (1957), p. 24.
448. H. W. Humphres, "A Method for Controlling Compaction of Granular Materials," Highway Research Board Bulletin 159, Washington, D.C. (1957).
449. W. A. Brown, Discussion of Ref. 412, *Journal, Soil Mechanics and Foundations Division, American Society of Civil Engineers* (July 1956), p. 1028.
450. "Closure Dam—The Dalles Project," Design Memo No. 14, Unpublished Report, Portland, Oregon District, U.S. Corps of Engineers (February 24, 1951).
451. K. Terzaghi, Discussion of Ref. 337.
452. "Republic Bore Hole Color Camera," Bulletin 1161, Brochure of the Republic Company, 611 South Snelling Ave., St. Paul, Minnesota, U.S.A.
453. W. F. Swiger, "Design and Construction of Grouted Cutoff—Rocky Reach Hydro-Electric Power Project," *Journal, Soil Mechanics and Foundations Division, American Society of Civil Engineers* (April 1961), p. 95.
454. "Slurry Test Experimentation at The Wanapum Site," Unpublished Report to the Public Utility District No. 2 of Grant County, Ephrata, Washington by Harza Engineering Company, Chicago (1958).
455. D. A. Werblin, "Installation and Operation of Dewatering Systems," *Journal, Soil Mechanics and Foundations Division, American Society of Civil Engineers* (February 1960).
456. K. Terzaghi, "Storage Dam Founded on Land Slide Debris (Cheakamus Dam,)" *Journal, Boston Society of Civil Engineers* (January 1960), p. 64.
457. K. Terzaghi, "Fifty Years of Subsoil Exploration," Third International Conference on Soil Mechanics and Foundation Engineering, Switzerland (1953), vol. III, p. 227.
458. L. R. Wenzel, "Methods for Determining Permeability of Water Bearing Materials," U.S. Department of Interior, Water Supply Paper No. 887 (1942), p. 1.
459. G. Plantema, "Electrical Pore Water Pressure Cells: Some Designs and Experiences," Third International Conference on Soil Mechanics and Foundation Engineering, Switzerland (1953), vol. 1, p. 279.

460. "Core Wall of 31 Caissons Sunk Under Air (Quabbin Dam)," *Engineering News-Record,* vol. III (1933), p. 215.
461. G. McIldowie, "The Construction of the Silent Valley Reservoir," Paper No. 4947, *Institute of Civil Engineers* (England), vol. 239, (1934–1935), p. 465.
462. T. A. Middlebrooks and W. H. Jervis, "Relief Wells for Dams and Levies," *Transactions, American Society of Civil Engineers,* vol. 112 (1947), p. 1321.
463. J. G. Patrick, "Under-Seepage Control Methods on Corps of Engineers Projects," Fifth Congress on Large Dams, Paris (1955), vol. 1, p. 297.
464. W. J. Turnbull, and C. I. Mansur, "Investigation of Underseepage-Mississippi River Levees," *Transactions, American Society of Civil Engineers,* vol. 126 (1961), part I, p. 1429.
465. W. J. Turnbull and C. I. Mansur, "Design of Underseepage Control Measures for Dams and Levees," *Transactions, American Society of Civil Engineers,* vol. 126 (1961), part I, p. 1486.
466. W. J. Turnbull and C. I. Mansur, "Construction and Maintenance of Underseepage Control Measures," *Transactions, American Society of Civil Engineers,* vol. 126 (1961), part I, p. 1540.
467. W. J. Turnbull and C. I. Mansur, "Relief Well System for Dams and Levees," *Transactions, American Society of Civil Engineers,* vol. 119 (1954), p. 842.
468. R. A. Barron, "The Effect of a Slightly Pervious Top Blanket on The Performance of Relief Wells," Second International Conference on Soil Mechanics and Foundation Engineering, Rotterdam (1948), vol. 4, p. 324.
469. F. C. Walker, "Experience in The Evaluation and Treatment of Seepage from Operating Reservoirs, Fifth Congress on Large Dams, Paris (1955), vol. 1, p. 285.
470. H. Cambefort, "Parafouilles speciaux en terrains permeables," Fifth Congress on Large Dams, Paris (1955), vol. 1, p. 883.
471. N. L. Liver, E. C. Mardoff and J. C. King, "Mixed-in-Place Piles Form Jetty to Control Beach Erosion," *Civil Engineering* (March 1954).
472. P. Franke, "The Concrete-Pile Cutoff Wall According to The ICOS-Veder Patent," Translated from the German (*Die Wasserwirschaft,* vol. 44, No. 3, p. 60) by the Army Engineers Research Center, Vicksburg, Mississippi, Translation No. 54-6 (September 1954).
473. "The I.C.O.S. Processes-Monographs of Characteristic Works," Brochure of the construction firm Impress Construzioni Opere Specializzate Milan, Italy (1959).
474. E. E. Esmiol, "Impervious Soils Used in Rolled Earth Dams," U.S. Bureau of Reclamation Technical Memo 649, Denver (1954).
475. J. Kell, "Pre-treatment of Gravel for Compressed Air Tunnelling under the River Thames at Dartford," *Chartered Civil Engineers* (London) (March 1957).
476. "River Gravel Solidified," *Engineering News-Record,* (November 14, 1957).
477. "Clay Grouting in Alluvial Sands and Gravels," Brochure of The Rodio Group, Zurich, Switzerland (1957).
478. E. Ischy, "The Lac Noir Dam Consolidation and Water Proofing," Second International Conference on Soil Mechanics and Foundation Engineering, Rotterdam (1948), vol. 5, p. 270.
479. R. Maigre, "Réalization par injéction d'un ecran impermeable en material alluvionnaire," Fifth Congress on Large Dams, Paris (1955), vol. 1, p. 755.
480. E. W. Lane, "Security from Underseepage-Masonry Dams on Earth Founda-

tions," *Transactions, American Society of Civil Engineers*, vol. 100 (1935) p. 1257.

481. J. B. Cooke, Discussion of Ref. 328, *Transactions, American Society of Civil Engineers*, vol. 125 (1960), part II, p. 575.
482. "How to Rap Up a Riprap Job," *Construction Methods and Equipment* (June 1960), p. 135.
483. M. D. Morris, "Earth Compaction—How to Achieve Better Results at Less Cost," *Construction Methods and Equipment* (June 1960), p. 160.
484. Allen S. Cary, "Performance of Mud Mountain Dam," *Transactions, American Society of Civil Engineers*, vol. 125 (1960), part II, p. 183.
485. N. N. Ambraseys, "The Seismic Stability of Earth Dams," Proceedings, Second World Conference on Earthquake Engineering, Japan (1960), vol. II, p. 1345.
486. L. A. Queiroz, "Compressible Foundation at Três Marias Earth Dam," Proceedings, First Panamerican Conference on Soil Mechanics and Foundation Engineering, Mexico (1959), vol. II, p. 763.
487. E. Nonveiller, "The Stability of Slopes of Dams Composed of Heterogeneous Materials," Third International Conference in Soil Mechanics and Foundation Engineering, Zurich (1953), vol. 2, p. 268.
488. R. Chadeisson, "Continuous Diaphragm Cast In-Situ," (in French), Fifth International Conference on Soil Mechanics and Foundation Engineering, Paris (1961), vol. II, p. 563.
489. K. A. Linell and H. F. Shea, "Strength and Deformation Characteristics of Various Glacial Tills in New England," American Society of Civil Engineers Conference on the Shear Strength of Cohesive Soils, Boulder, Colorado (June 1960).
490. "High Earth Dam Plugs Narrow Canyon For Power (Mammoth Pool Dam)," *Engineering News-Record* (April 7, 1960), p. 44.
491. J. C. King and E. C. W. Bush, "Grouting of Granular Materials," *Journal, Soil Mechanics and Foundations Division, American Society of Civil Engineers*, (April 1961).
492. W. G. Shockley and R. G. Ahlvin, "Non-Uniform Conditions in Triaxial Test Specimens," Paper presented to the American Society of Civil Engineers Conference on the Shear Strength of Cohesive Soils, Boulder, Colorado (June 1960).
493. C. N. Zanger, "Theory and Problems of Water Percolation," U.S. Bureau of Reclamation, Engineering Monograph No. 8, Denver, Colorado (April 1953).
494. H. Darcy, *Les fontaines publiques de la ville de Dijon*, Dalmont, Paris (1856).
495. M. Muskat, *The Flow of Homogeneous Fluids Through Porous Media*, J. W. Edwards (1946).
496. P. Thurber, "Design of Cougar Central Core Dam," *Transactions, American Society of Civil Engineers*, vol. 125 (1960), part II, p. 660.
497. L. F. Harza, "Uplift and Seepage Under Dams on Sand," *Transactions, American Society of Civil Engineers*, vol. 100 (1935), p. 1362.
498. R. V. Southwell, *Relaxation Methods in Theoretical Physics*, Oxford Press, 1946.
499. S. T. Yang, "Seepage Toward a Well Analyzed by the Relaxation Method," Doctor's Thesis, Harvard University (1949).
500. F. S. Shaw and R. V. Southwell, "Relaxation Methods Applied to Engineering Problems (Problems Relating to Percolation of Fluid Through Porous

Materials)," *Proceedings, Royal Society (London)*, Series A, vol. 178 (1941), p. 1.

501. J. McNamee, "Seepage Into a Sheeted Excavation," *Geotechnique* (London), vol. 4 (1949), p. 227.
502. A. I. Kashef et al., "Numerical Solutions of Steady State and Transient Flow Problems," Engineering Experiment Station Bulletin No. 117, Purdue University (July 1952).
503. "Pressure cells for Field Use," Bulletin No. 40, Waterways Experiment Station, U.S. Army Corps of Engineers, Vicksburg, Mississippi (January 1955).
504. M. J. Hvorslev, "Time Lag and Soil Permeability in Ground Water Observations," Bulletin No. 36, Waterways Experiment Station, U.S. Corps of Engineers, Vicksburg, Mississippi (April 1951).
505. D. Finzi, "Foundation Settlement and Pore Pressure Observations at San Valentino Earth Dam During the First Five Years of Operation," Fourth International Conference on Soil Mechanics and Foundation Engineering, London (1957), vol. 2, p. 305.
506. A. D. N. Penman, "A Field Piezometer Apparatus," *Geotechnique,* vol. 6, No. 2 (June 1956), p. 57.
507. J. W. Hilf, "An Investigation of Pore Water Pressure in Compacted Cohesive Soils," University of Colorado Doctor's Thesis Published as Technical Memorandum No. 654, U.S. Bureau of Reclamation, Denver, Colorado (October 1956).
508. *Engineering Manual for Civil Works Construction,* part CXXIII, U.S. Army Engineers (June 1954).
509. J. D. Parsons and S. D. Wilson, "Safe Loads on Dog-Leg Piles," *Transactions, American Society of Civil Engineers,* vol. 121 (1956), p. 695.
510. "Subsurface Investigation—Geophysical Explorations," U.S. Army Corps of Engineers, *Engineering Manual for Civil Works Construction,* part CXVIII, Chapter 2, (September 1948).
511. R. E. Davis and Associates, "Model Study of Stability of Portage Mountain Dam During Earthquakes," Unpublished Report to International Power and Engineering Consultants, Ltd., Berkeley, California (November 1960).
512. R. Peterson, Oral Discussion on "Stability and Deformations of Slopes," Third International Conference on Soil Mechanics and Foundation Engineering, Zurich (1953), vol. 2, p. 211.
513. R. Peterson, "Summary of Data Extracted From Reports on Boundary Dam Pertaining to Sand Drains in the Foundation," Unpublished Report of the Prairie Farm Rehabilitation Administration, Saskatoon, Saskatchewan, (January 19, 1959).
514. R. V. Whitman, "Testing Soils With Transient Loads," Conference on Soils For Engineering Purposes, *American Society for Testing Materials Special Technical Publication No. 232* (December 1957), p. 514.
515. D. Croney and J. C. Jacobs, "The Rapid Measurement of Soil Moisture Content in the Field," *Roads and Road Construction* (London) (July 1951), p. 191.
516. A. W. Skempton, "The $\phi = 0$ Analysis of Stability and its Theoretical Basis," Second International Conference on Soil Mechanics and Foundation Engineering, Rotterdam (1948), vol. 1, p. 72.
517. E. D'Appolonia, "Loose Sands—Their Compaction by Vibroflotation," *American Society for Testing Materials Publication No. 156* (1953), p. 138.

518. "Permeability Tests Using Drill Holes and Wells," and "Comments Regarding Equipment and Methods of Conducting Permeability Tests in Drill Holes," (Bound in same volume), Geology Report No. G-37, U.S. Bureau of Reclamation, Denver, Colorado (January 3, 1951).
519. R. Peterson et al., "Limitations of Laboratory Shear Strength in Evaluating Stability of Highly Plastic Clays," Research Conference on Shear Strength of Cohesive Soils, American Society of Civil Engineers, Boulder, Colorado (June 1960), p. 765.
520. D. H. Trollope, "The Systematic Arching Theory Applied to the Stability Analysis of Embankments," Fourth International Conference on Soil Mechanics and Foundation Engineering, London (1957), vol. 2, p. 382.
521. A. Casagrande, "An Unsolved Problem of Embankment Stability on Soft Ground," Proceedings, First Pan-American Conference on Soil Mechanics, Mexico City, (September 1959), vol. II, p. 721.
522. A. W. Bishop and L. Bjerrum, "The Relevance of the Triaxial Test to the Solution of Stability Problems," Research Conference on Shear Strength of Cohesive Soils American Society of Civil Engineers, Boulder, Colorado (June 1960), p. 437.
523. N. Janbu, "Earth Pressure and Bearing Capacity by the Generalized Procedure of Slices," Fourth International Conference on Soil Mechanics and Foundation Engineering, London (1959), vol. 2, p. 207.
524. A. Casagrande and R. C. Hirschfeld, "Investigation of Stress-Deformation and Strength Characteristics of Compacted Clays," Harvard University Soil Mechanics Series, No. 61.
525. A. W. Bishop et al., "Factors Controlling the Strength of Partly Saturated Cohesive Soils," Research Conference on Shear Strength of Cohesive Soils, American Society of Civil Engineers, Boulder, Colorado (June 1960), p. 503.
526. H. J. Gibbs et al., "Shear Strength of Cohesive Soils," Research Conference on Shear Strength of Cohesive Soils, American Society of Civil Engineers, Boulder, Colorado (June 1960), p. 33.
527. J. Lowe and L. Karafiath, "Effect of Anisotropic Consolidation on the Undrained Shear Strength of Compacted Clays," Research Conference on Shear Strength of Cohesive Soils, American Society of Civil Engineers, Boulder, Colorado (June 1960), p. 837.
528. A. Anderson et al., "Triaxial Equipment Developed at the Norwegian Geotechnical Institute," Publication No. 21, Norwegian Geotechnical Institute, Oslo (1957).
529. A. W. Bishop, "The Measurement of Pore Pressure in the Triaxial Test," Proceedings, Conference on Pore Pressure and Suction in Soils, Butterworth, London (1960).
530. D. W. Taylor, "Pressure Distribution Theories, Earth Pressure Cell Investigations and Pressure Distribution Data," U.S. Army Waterways Experiment Station Publication, Vicksburg, Mississippi (1947).
531. A. W. Bishop et al., "Pore Pressure Observations at Selset Dam," Pore Pressure and Suction in Soil Conference, Butterworth, London (1960).
532. A. Caquot and J. Kerisel, *Traité de mechanique des sols,* Gauthier-Villars, Paris (1949).
533. O. K. Frohlich, "General Theory of the Stability of Slopes," *Geotechnique,* vol. 5 (March 1955), p. 37.
534. A. F. Samsioe, "Stresses in Downstream Part of an Earth or Rockfill Dam," *Geotechnique,* vol. 5 (1955).

535. J. W. Libby, "Furnas Hydroelectric Project," *Transactions, American Society of Civil Engineers,* vol. 126 (1961), part III, p. 94.
536. K. S. Lane, "Field Slope Charts for Stability Studies," Fifth International Conference on Soil Mechanics and Foundation Engineering, Paris (1961), vol. II, p. 651.
537. N. Janbu, "Application of Composite Slip Surfaces for Stability Analysis," Proceedings, European Conference on Stability of Slopes, Sweden, (September 1954), vol. 3, p. 43.
538. R. A. Sevaldson, "The Slide at Lodalen, October 6, 1954," *Geotechnique,* vol. 6 (1956), p. 167.
539. G. Plantema, "A Soil Pressure Cell and Calibration Equipment," Third International Conference on Soil Mechanics and Foundation Engineering, Zurich, (1953), vol. 1, p. 283.
540. J. Paton and N. G. Semple, "Investigation of the Stability of an Earth Dam Subject to Rapid Drawdown Including Details of Pore Pressures Recorded During a Controlled Drawdown Test," Conference on Pore Pressure and Suction in Soil, London (1960).
541. D. J. Bazett, Discussion of "Drawdown Pore Pressures Measured in a Small Dike for a Storage Reservoir at Niagara, Ontario," Conference on Pore Pressures and Suction in Soil, London (1960).
542. B. Kjaernsli and I. Torblaa, "Compaction of Morraine in Three Foot Layers," Seventh Congress on Large Dams, Rome (1961), Report 81, Question 27.
543. C. Y. Li, "Columbia's Troneras Dam," *Civil Engineering* (February 1963), p. 37.
544. J. Thevenin, "Les barrages Algeriens avec masque amont en béton bitumineux," *Travaux* (November 1960), p. 690.
545. M. Lang, "Les proprietés des bitumes et mélanges bitumineux utilisés dans les revetements de canaux et masques de barrages," *Travaux* (November 1960), p. 661.
546. "Colloque sur les masques d'etanchéité des barrage en terre et en enrochements," *Construction* (October 1960).
547. D. J. Dayton, "Asphalt Linings for Filtered Water Reservoirs," Proceedings, Conference on the Use of Asphalt in Hydraulic Structures, Bakersfield, California (May 24, 1961), The Asphalt Institute Pacific Coast Division, Berkeley, California.
548. J. Salva, "The Sarno Dam," Fifth International Congress on Large Dams, Paris (1955), Supplement to *Travaux,* No. 247, p. 179.
549. R. Vermeringer, "Constitution d'un masque souple et raccordement aux ouvrages rigides (Barrage de L'Iril Emda)," Fifth Congress on Large Dams, Paris (1955), vol. IV, communication 22, p. 951.
550. H. Castenow and Y. Haller, "Barrage L'Iril Emda," *La Technique Moderne, Construction,* (September 1954).
551. J. P. Gould and P. C. Rutledge, "Report on Investigation of Deformations in Foundations of Earth Embankments Containing Concrete Pressure Pipe Conduits," Report to the American Concrete Pressure Pipe Association (limited private printing) (September 1960).
552. H. Haas, "Technische Eigenschaten von Asphaltbeton für Wasserbauzwecke," *Die Bautechnik,* (September 1959), p. 339.
553. P. Schmitz, "Die Genkeltalsperre," *Bitumen* (1953), p. 204.

554. H. W. Koenig, "Der Neubau der Hennetalsperre des Ruhrtalsperrenvereins," Publication of the Westdeutschen Wasserwerschaftsverbandes E. V., Essen, (1957).
555. W. Lorenz, "Der Staudamn am Sylvenstein mit Dichtungsschürze," *Die Bautechnik* (June 1958), p. 205.
556. J. C. Ott, "La construction du barrage de Bou-Hanifia," Bulletin Technique de la Suisse Romande, (February, 1944), p. 25.
557. J. C. Ott, "Comportment du barrage de Bou-Hanifia," Bulletin Technique de la Suisse Romande (February 1955), p. 57.
558. S. F. Gizienski, et al., "Application of Electrical Analogy to Specific Problems in Civil Engineering Hydraulics," Paper presented at the Hydraulics Division Conference, American Society of Civil Engineers, Jackson, Mississippi (November 1950).
559. A. D. M. Penman, "A Study of the Response Time of Various Types of Piezometer," Proceedings of the Conference on Pore Pressure and Suction in Soils, London (March 1960), p. 53.
560. "Acoustical Examination of the Embankment of Serre-Ponçon Dam with Telemac Monometric Cells," Paper published in the form of a brochure by the manufacturer of the French acoustical piezometer, Telemac, 19, rue Alphonse de Neuville, Paris 17, France.
561. H. A. Sikso, "Chalk Used as a Construction Material at Gavin's Point Dam," *Civil Engineering*, (September 1957), p. 612.
562. S. T. Thorfinnson, "Underseepage Control at Fort Randall Dam," *Transactions, American Society of Civil Engineers*, vol. 125 (1960), p. 792.
563. J. L. Sherard, "A Review of Experience with Asphaltic Concrete Impervious Membranes on the Upstream Slope of Earth and Rockfill Dams," Proceedings, Conference on the Use of Asphalt in Hydraulic Structures, Bakersfield, California, (May 24), 1961.
564. E. J. Zegarra, Discussion of paper "Stabilization of Materials by Compaction," by W. J. Turnbull and C. R. Foster, *Transactions, American Society of Civil Engineers*, vol. 123 (1958), p. 20.
565. "Drilled Piles form Cutoff Trench Under Dam," *Civil Engineering*, vol. 23 (1953), p. 126.
566. R. F. Walter, "The Belle Fourche Dam," *Engineering Record*, vol. 53 (1906), p. 307.
567. O. T. Ready, "Construction of Belle Fourche Dam," *Engineering Record*, vol. 61 (1910), p. 466.
568. "Steel Dam," *Engineering News* (May 12, 1898), p. 299.
569. K. S. Lane and P. E. Wohlt, "Performance of Sheet Piling and Blankets for Sealing Missouri River Reservoirs," Seventh Congress on Large Dams, Rome (1961), Report 65, Question No. 27.
570. K. R. Barney, "Madison Canyon Slide," Paper presented at the joint meeting of The American Society of Civil Engineers and the U.S. Committee on Large Dams, Reno, Nevada (June 1960).
571. R. M. Ball, "The Madison Earthquake and its Effects," Paper presented at the joint meeting of The American Society of Civil Engineers and the U.S. Committee on Large Dams, Reno, Nevada (June 1960).
572. J. L. Sherard, "A Report on the Damage to Hebgen Dam in the West Yellowstone Earthquake," *Engineering News-Record* (August 17, 1959).
573. C. H. Kirk, "An Engineering Appraisal of the August 1959 Earthquake Damage to Hebgen Dam," Paper presented to the joint meeting of the

American Society of Civil Engineers and the U.S. Committee on Large Dams, Reno, Nevada (June 1960).
574. E. Reinius, "The Stability of the Upstream Slope of Earth Dams," Meddelanden Statens Kommitte for Bygghadsforkning, No. 12 (1948) Published Doctor's Thesis, Royal Institute of Technology, Stockholm (in English).
575. H. B. Seed and C. K. Chan, "Thixotropic Characteristics of Compacted Clays," *Transactions, American Society of Civil Engineers,* vol. 124 (1959), p. 894.
576. A. L. Grandi, et al., "Earth Dams in Argentina," Fifth International Conference on Soil Mechanics and Foundation Engineering, Paris (1961), vol. II, p. 613.
577. E. Nonveiller and P. Agnosti, "Stresses and Deformations in Cores of Rockfill Dams," Fifth International Conference on Soil Mechanics and Foundation Engineering, Paris (1961), vol. II.
578. J. B. Hays, "Deep Solution Channel, Kentucky Dam," *Transactions, American Society of Civil Engineers,* vol. 105 (1940), p. 948.
579. J. S. Lewis, "Large Core Drills at Chickamauga Dam," *Transactions, American Society of Civil Engineers,* vol. 105 (1940), p. 850.
580. J. S. Lewis et al., "Foundation Experiences, Tennessee Valley Authority—A Symposium," *Transactions, American Society of Civil Engineers,* vol. 106 (1941), p. 685.
581. M. Lugeon, *Barrage et geologie,* F. Rouge and Cie, Lausanne (1932).
582. W. J. Mead, "Engineering Geology of Dam Sites," Second Congress on Large Dams, Washington (1936), vol. 4, p. 171.
583. "Geology and Foundation Treatment," Tennessee Valley Authority Projects, Technical Report No. 22, U.S. Government Printing Office, Washington, D.C. (1949).
584. "Review of Soils Design, Construction and Performance—Harlan County Dam, Nebraska," U.S. Waterways Experiment Station Technical Report No. 3-501 (April 1959).
585. "Review of Soils Design, Construction and Performance of Wister Dam, Oklahoma," U.S. Waterways Experiment Station Technical Report No. 3-508 (June 1959).
586. "Soils Design, Construction and Performance Observations, Lookout Dam, Oregon," U.S. Waterways Experiment Station Technical Report No. 3-502, (May 1959).
587. "Neversink Dam, Grouting the Cutoff Wall," The Delaware Water Supply News, Board of Water Supply of the City of New York, April 1 (1949).
588. Marten and Dronhin, "Les masques en béton armé et en beton bitumineux des barrage en enrochements," Second Congress on Large Dams, Washington (1936), vol. 5, p. 225.
589. V. Thevenin, "Observations fortes au barrage du Ghrib sur le viellissement des masques a liant plastique" *Terre et Eaux,* No. 30 (1957).
590. D. R. Schuster, "Dammbau und Asphaltbeton—Dichtung der Hennetalsperre," *Bitumen,* Nos. 8 and 9, Hamburg (1957).
591. W. F. Van Asbeck, "The Use of Asphaltic Bitumen for Sealing Earth and Rockfill Dams," Seventh Congress on Large Dams, Rome (1961), Question No. 25, Report No. 13.
592. T. Abou Wafa, "Field Tests for Grouting Nile Alluvials Under the Aswan High Dam," Seventh Congress on Large Dams, Rome (1961), Question No. 25, Report No. 13.

593. J. Thevenin, "Barrage du Ghrib—observations et études sur le viellissement due masque en beton bitumineux," Seventh Congress on Large Dams, Rome (1961), Question 27, Report No. 36.
594. F. C. Walker and L. M. Ellsperman, "Factors Affecting the Use of Asphaltic Diaphragms in Earth and Rockfill Dams in the Western United States," Seventh Congress on Large Dams, Rome (1961), Question No. 27, Report 66.
595. H. W. Koenig, "Waterproofing of Earth-fill and Rock-fill Dams with Bitumen and Other Materials," Seventh Congress on Large Dams, Rome (1961), Question No. 27, Report No. 80.
596. L. Hobst, "The Sealing of Rockfill and Earth Dams by Precast Elements and P.V.C. Liners," Seventh Congress On Large Dams, Rome (1961), Question No. 27, Report No. 97.
597. G. Schnitter and J. Zeller, "Geotechnical Investigations of Mixture of Bitumen, Clay or Bentonite with Sandy Gravel," Seventh Congress on Large Dams, Rome (1961), Question No. 27, Report No. 38.
598. M. Vercon and M. Markovic, "Masque d'etancheité en béton bitumineux, du barrage de Rodoina," Seventh Congress on Large Dams (Rome), 1961, Question No. 27, Report No. 44.
599. E. Nonveiller and M. Hobekovic, "Properties of Clay-Cement Suspensions for Grouting," Seventh Congress on Large Dams (Rome), 1961, Question No. 25, Report No. 86.
600. C. F. Ripley and K. L. Lee, "Sliding Friction Tests on Sedimentary Rock Specimens," Seventh Congress on Large Dams, Rome (1961), Communication No. 8.
601. P. H. Dunaway, "New Tools for Earth-Dam Construction," *Civil Engineering* (September 1960), p. 35.
602. "Bentonite Slurry Stabilizes Trench-Keeps out Ground Water," *Engineering News-Record,* (February 11, 1960), p. 42.
603. "Wanapum Dam is off to a Fast Start," *Western Construction* (September 1960), p. 35.
604. H. Muhs, "Measurement of Pore Water Pressure in Open Ground—Particularly in Earth Dams," (in German), *Baumaschine und Bautechnik,* October, 1954 and November, 1954.
605. A. W. Johnson and J. R. Sallberg, "Factors that Influence Field Compaction of Soils," Highway Research Board Bulletin 272, Washington, D.C. (1960).
606. C. A. Hogentogler and E. A. Willis, "Stabilized Soil Roads," *Public Roads* (May 1936), p. 45.
607. D. J. Belcher, "A Field Investigation of Low Cost Stabilized Roads," Purdue University of Engineering Experiment Station Research Series, No. 81 (April 1941).
608. J. Barry Cooke, Discussion on Question 27 (concerning need for measurements and dissemination of information concerning performance of rock-fill dams), Proceedings, Seventh Congress on Large Dams, Rome (1961).
609. W. G. Shockley, et al., "Investigations with Rotary Cone Penetrometer," Fifth International Conference on Soil Mechanics and Foundation Engineering, Paris (1961), Vol. 1.
610. A. Dvorak and P. Peter, "Field Tests on Soils and Rocks," Fifth International Conference on Soil Mechanics and Foundation Engineering, Paris (1961), vol. I, page 453.

611. C. H. Pigot and I. D. MacKensie, "Carillon Foundation Studies," *The Engineering Journal* (Canada) (October 1961), p. 65.
612. A. E. Niederhoff, "Field Tests on a Shale Foundation," *Proceedings, American Society of Civil Engineers,* vol. 65, (1939), p. 1239.
613. J. R. Roundtree, Discussion of Ref. 612 with data on field shear tests at Watts Bar Dam, *Proceedings, American Society of Civil Engineers,* vol. 66 (1940), p. 1217.
614. A. Coyne, "New Dam Techniques," *Proceedings, Institution of Civil Engineers* (London) (November 1959), p. 285.
615. M. Guelton, et al., "The Serre-Ponçon Dam—Conception of the Project," *Travaux* (May 1961), p. 58 (in English).
616. O. de Schnakenbourg, "The Serre-Ponçon Dam—Control of the Execution of the Dam," *Travaux* (May 1961), p. 76 (in English).
617. J. Barge, "The Serre-Ponçon Dam—Execution of the Works," *Travaux* (May 1961), p. 87 (in English).
618. A. Casagrande, "Control of Seepage Through Foundations and Abutments of Dams," *Geotechnique* (London) (September 1961), p. 161.
619. "Stability of Earth and Rockfill Dams," U.S. Army Corps of Engineers Manual EM 1110-2-1902 (December 27, 1960).
620. K. Terzaghi, "Does Foundation Technology Really Lag?" *Engineering News-Record* (February 15, 1962).
621. K. Terzaghi, "Effect of Minor Geologic Details on the Safety of Dams," Bulletin, American Institute of Mining Engineers, Technical Publication 215 (1929) Class I, Mining Geology, No. 26, p. 31.
622. E. Huni, Discussion of paper, "Grouting of Granular Materials" *Journal, Soil Mechanics and Foundations Division, American Society of Civil Engineers* (December 1961), p. 53.
623. R. D. Wyckoff and D. W. Reed, "Electrical Conduction Models for the Solution of Water Seepage Problems," *Physics,* vol. 6 (December 1936).
624. M. E. Harr, *Ground Water and Seepage,* McGraw-Hill Book Co. (1962).
625. E. Reinius, "The Stability of the Downstream Part of Earth Dams," Bulletin No. 59 of the Division of Hydraulics at the Royal Institute of Technology, (in English) Stockholm (1961).
626. J. Narain, "Flexibility of Compacted Clays," Ph.D. Thesis, Purdue University (January 1962).
627. J. M. Bird, "Uncertainties in Earth Dam Design," *Proceedings, American Society of Civil Engineers,* vol. 87, No. SM 3, part 1 (June 1961).
628. R. J. Marsal, "Earth Dams in Mexico," Proceedings, First Pan-American Conference on Soil Mechanics and Foundation Engineering, Mexico (1960), vol 3, p. 1293.
629. E. Tamez and G. Springall, "The Use of Soils as Construction Materials for Earth Dams," Proceedings, First Pan-American Conference on Soil Mechanics and Foundation Engineering, Mexico (1960), vol. 3, p. 1269.
630. K. Terzaghi, "Der Grundbruch an Staumauern und seine Verhütung," Die *Wasserkraft* (1922), p. 445.
631. K. Terzaghi, *Erdbaumechanik,* F. Deuticke, Vienna (1925).
632. H. Cambefort, *Forages et sondages,* Eyrolles, Paris (1955).
633. M. V. Pardal, "La technica de las inyecciones y su applicacion a los embalses," *Revista de Obras Publicas,* Madrid (1960).
634. G. Tschebotarioff, *Soil Mechanics, Foundations, and Earth Structures,* McGraw-Hill Book Co. (1956).

635. K. Terzaghi, "Soil Studies for the Granville Dam at Westfield, Massachusetts," *Journal, New England Water Works Association,* vol. 43 (1929), p. 191.
636. J. A. Haine and J. W. Hilf, Discussion of Ref. 381, *Transactions, American Society of Civil Engineers,* vol. 113 (1948), p. 941.
637. T. T. Knappen and R. R. Phillippe, "Practical Soil Mechanics at Muskingum III," *Engineering News-Record* (April 23, 1936), p. 595.
638. K. Terzaghi, "Past and Future of Applied Soil Mechanics," *Journal, Boston Society of Civil Engineers* (April 1961), p. 110.
639. L. F. Grant and L. A. Schmidt, "Grouting Deep Solution Channels, Under an Earth Fill Dam," *Journal, Soil Mechanics and Foundations Division, American Society of Civil Engineers* (October 1958).
640. R. Pfister, "Die baulichen Anlagen des Kraftwerkes Goshenen," (The Civil Construction Features of the Goschenen Power Project), *Die Wirtschaft* (July 1959), p. 35.
641. J. D. Hodgson et al., "Rapid Determination of Moisture Content and Its Influence on Earth Dam Control," Proceedings, Second Australia-New Zealand Conference on Soil Mechanics and Foundation Engineering, The New Zealand Institution of Civil Engineers (January 1956), p. 101.
642. J. K. Wilkins, "Flow of Water through Rockfill and Its Application to the Design of Dams," Proceedings, Second Australia-New Zealand Conference on Soil Mechanics and Foundation Engineering, The New Zealand Institution of Civil Engineers, (January 1956), p. 141.
643. "Dam Trench Drained from Below," *Engineering News-Record* (October 23, 1958), p. 47.
644. A. Hazen and L. Metcalf, "Middle Section of Upstream Side of Calaveras Dam Slips Into Reservoir," *Engineering News-Record,* vol. 80, (1918), p. 679.
645. L. A. Schmidt, "Flowing Water in Underground Channels, Hales Bar Dam, Tennessee," *Transactions, American Society of Civil Engineers,* vol. 110 (1945), p. 965.
646. A. V. Lynn and R. F. Rhoades, "Foundation Exploration at Kentucky Dam Site," *Engineering News-Record,* vol. 125 (1940), p. 70.
647. K. Bryan, "Geology of Reservoir and Dam Sites," U.S. Geological Survey Water Supply, Paper 597 (1928).
648. K. Terzaghi, Letter to Editor, *Geotechnique,* London, (March 1962). Discussing Reference 618.
649. T. Silvestri, "Determinazione Sperimentale de Resistenza Meccania del Materiale Constituente el Corpo di uno diga del tip 'Rockfill'," *Geotechnica,* vol. 8, pp. 186–191.
650. K. L. Rao, "Exit Gradients on Structures with Permeable Foundations," Fifth Congress on Large Dams, Paris (1955), vol. 1, p. 675.
651. "Papua to Get Steel-faced Dam," *Engineering News-Record,* (May 1962), p. 46.
652. D. W. Barr and R. W. Rosene, "Loose Rock Spillway for Low-Head Dam," *Civil Engineering* (April 1958), p. 54.
653. D. W. Barr, "Partial Failure of Loose-Rock Spillway," *Civil Engineering* (October 1959), p. 60.
654. K. Terzaghi, "Dam Foundation on Sheeted Granite," *Geotechnique,* London (September 1962), p. 199.
655. G. L. Counts, "Earth Dam Construction at Clark Hill Dam," Proceedings of Conference on Foundations and Embankment Construction held at Kansas

City, Mo., May 1949; published by The Waterways Experiment Station, Vicksburg, Mississippi (March 1950).

656. J. R. Stephens, "Problems in the Construction of the Grenada Dam," Proceedings of Conference on Foundations and Embankment Construction held at Kansas City, Mo., May 1949; published by the Waterways Experiment Station, Vicksburg, Mississippi (March 1950).
657. "Basic Data Report on Test Fills of Proposed Embankment Materials for Oroville Dam," State of California Department of Water Resources Report (December 1961).
658. H. Schwegler, "Dammbauten in Syrien," *Schweizerische Bauzeitung*, Zurich (May 31, 1962), p. 374.
659. J. L. Serafim and J. J. B. Lopes, "In-Situ Shear Tests and Triaxial Tests of Foundation Rocks for Concrete Dams," Proceedings, Fifth International Conference on Soil Mechanics and Foundation Engineering, Paris (1961) vol. I, p. 533.
660. J. D. Freeman, "Experiences in Construction of Earth Dams in the Tulsa District," Proceedings of Conference on Foundations and Embankment Construction held at Kansas City, Mo., May 1949; published by The Waterways Experiment Station, Vicksburg, Mississippi (March 1950).
661. D. D. Leslie, "General Considerations on Roller Specifications for Earth Dams," Proceedings of Conference on Foundations and Embankment Construction held at Kansas City, Mo., May 1949; published by The Waterways Experiment Station, Vicksburg, Mississippi (March 1950).
662. J. M. Chauvin, "Le Barrage de la Cheffia," *Le Moniteur des Travaux Public et du Batiment*, Paris, No. 53 (December 31, 1960).
663. J. M. Chauvin, "The Cheffia Dam on the Oued Bou-Namoussa," *Travaux*, Paris, English Version (May 1961), p. 225.
664. A. C. Meigh and I. K. Nixon, "Comparison of In-Situ Tests for Granular Soils," Fifth International Conference on Soil Mechanics and Foundation Engineering, Paris (1961), vol. 1, p. 499.
665. K. S. Lane, comments on "Foundation Reports on Concrete Dam Construction," Proceedings of Conference on Foundations and Embankment Construction held at Kansas City, Missouri, May 1949; published by The Waterways Experiment Station, Vicksburg, Mississippi (March 1950).
666. "A First: Soil Cement for Dam Facing," *Engineering News-Record* (April 6, 1961).
667. J. Destenay and Y. LeMay, "Les masques d'etanchéité des barrage en enrochments construits par electricité de France," Seventh Congress on Large Dams, Rome (1961), Report No. 19, Question No. 27.
668. C. Schaerer, "Le comportement des diques en terre pendant leur construction et durant l'exploitation de l'aménagement," Barrages en Suisse, Septième Congres International des Grands Barrages, Rome (1961), Numéro Spécial de la Revue Mensuelle Suisse-Cours D'Eau et Énergie, p. 164.
669. "Big Wheel Keeps Abiquiu Dam Rolling," *Engineering News-Record*, (July 19, 1962), p. 28.
670. A. A. Nitchiporovitch and A. A. Sidorov, "Anti-Seepage Installations in Earth and Rock-fill Dams Based on Experience in the U.S.S.R.," Seventh Congress on Large Dams, Rome (1961), Report No. 123, Question No. 27.
671. G. E. Blight, "Controlling Earth Dam Compaction Under Arid Conditions," *Civil Engineering* (August 1962), p. 54.

672. Charles E. Hall, "Compacting a Dam Foundation by Blasting," *Journal of the Soil Mechanics Division, American Society of Civil Engineers,* (June 1962).
673. G. E. Bertram, "Sealing of Earth and Rockfill Dams with Bitumen and other Materials," Summary of General Reporter, Question No. 27, Seventh Congress on Large Dams, Rome (1961).
674. G. E. Blight, "The Utilization of Soil Suction in the Design of Earth Dam Embankments," Proceedings, 3rd Regional Conference for Africa on Soil Mechanics and Foundation Engineering, Salisbury, (1963), vol. I, p. 140.
675. L. G. Elser and D. E. Schuster, "Pomme de Terre Dam of Earth and Rockfill," Paper submitted to *Journal, Soil Mechanics and Foundation Division, American Society of Civil Engineers* (October 1962).
676. R. P. West, "Waco Dam Slide: Its Cause and Correction," *Engineering News-Record* (August 2, 1962), p. 34.
677. "Three Phases of Mission Dam," *Engineering News-Record,* (June 1962).
678. J. R. Anderson, "Vinyl Film Liner for Earth-Fill Reservoirs," *Civil Engineering* (June 1960), p. 42.
679. H. Muhs and D. Campbell-Allen, "A Laboratory Examination of an Electrical Pore Pressure Gage for Use in Earth Dams," *The Journal, Institution of Engineers,* Australia (September 1955), p. 241.
680. E. Krebs, "Television Bore Hole Telescope FB 400," Brochure published by The Eastman International Company, Hanover, Germany (1961).
681. T. A. Middlebrooks, "Recommended Procedures for Embankment Construction and Foundation Preparation," Proceedings of Conference on Foundations and Embankment Construction held at Kansas City, Mo., May, 1949; published by The Waterways Experiment Station, Vicksburg, Mississippi (March 1950).
682. W. G. Holtz, discussion of General Report on Earth Dams, Slopes and Excavations, Fifth International Conference on Soil Mechanics and Foundation Engineering, Paris (1961) vol. III, p. 343.
683. J. A. Leadabrand, "Bonny Dam Experimental Project," Fifth International Conference on Soil and Foundation Engineering, Paris, (1961), vol. III, p. 363.
684. M. V. Mencl, discussion of General Report on Techniques of Field Measurement and Sampling, Fifth International Conference of Soil Mechanics and Foundation Engineering, vol. III, p. 165.
685. "Investigations for the Construction of the High Dam at Aswan performed for the Sadd el-Aali Authority in Cairo, (1957–8)," Unpublished Report of Johann Keller G. M. B. H., Frankfurt, Germany.
686. C. H. Mallet and J. Pacquant, "Moyens et dispositifs de mesure des déformations des barrages en terre et leur assises," Fourth Congress on Large Dams, New Delhi (1951), vol. I, p. 303.
687. J. K. Wilkins, "The Stability of Rockfill Dams Subjected to Water Flowing Over Them," unpublished paper printed by The Hydro-Electric Commission of Tasmania (1963). (Soon to be published in an Australian journal.)
688. "Bros Announces Vibrating Roller," *Engineering News Record* (November 15, 1962), p. 64.
689. W. G. Holtz and F. C. Walker, "Soil-Cement as Slope Protection for Earth Dams," *Journal, Soil Mechanics and Foundations Division,* American Society of Civil Engineers (December 1962).

690. "New Scrapers Haul Big Loads for Earthfill Dam (Glendola Dam)," *Engineers and Contractors* (July 1962), p. 26.
691. "Blasting Compacts Sand Fill," *Construction Methods and Equipment* (May 1960).
692. W. L. Shannon et al., "Field Measurements," Chapter 13, *Foundation Engineering* (Edited by G. A. Leonards), McGraw-Hill Book Company, New York, 1962.
693. G. W. Housner, "Vibrations of Structures Induced by Seismic Waves," in *Shock and Vibration Handbook*, Vol. 3, Chapter 50, McGraw-Hill Book Company, New York, 1961.
694. "Dam in Jam Gets Plastic Surgery," *Engineering News Record* (February 14), 1963.
695. "Conveyor Moves Fill Down a Mountain," *Construction Methods and Equipment* (November 1958), p. 84.
696. "Triple-threat rigs at Briones Dam," *Western Construction* (February 1963), p. 58.
697. H. B. Seed and R. W. Clough, "Earthquake Resistance of Sloping Core Dams," *Journal, Soil Mechanics and Foundations Division*, American Society of Civil Engineers (February 1963), p. 209.
698. W. K. Cloud and D. S. Carder, "The Strong-Motion Program of the U.S. Coast and Geodetic Survey," World Conference on Earthquake Engineering, Berkeley, California (June 1956).
699. G. I. Davey, "Rockfill Dams at Mary Kathleen and Mount Isa," *Journal, Institution of Engineers* (Australia) (December 1960).
700. J. B. Frazer, "A Steel Faced Rockfill Dam for Papua," *Transactions, Institution of Engineers* (Australia) (September 1962).
701. N. N. Ambraseys, "Earthquake Engineering Reference Index," Report prepared for the British National Section of the International Association for Earthquake Engineering, London (March 1963).
702. N. Morgenstern, "The Limit Equilibrium Method of Slope Stability Analysis," Ph.D. Thesis, University of London (1963).
703. J. Salva, "Méthodes pour la mesure des tassements d'un barrage souple," Fifth International Congress on Large Dams, Paris (1955), vol. II, p. 367.
704. Moran, Proctor, Meuser and Rutledge, "Study of Effective Use of Coarse Grained Soils in Construction of Earthfill Dams," Report prepared for the Soil Conservation Service, U.S. Department of Agriculture (published in limited quantity) (July 1962).
705. C. K. Willey, "Wanapum Hydroelectric Development," *Civil Engineering* (September 1960), p. 65.
706. L. Bjerrum, "Pore Pressure and Settlement Observations at Tustervatn Dam," Internal Report, Norwegian Geotechnical Institute (April 27, 1962).
707. T. Kallstenius and W. Bergan, "In Situ Determination of Horizontal Ground Movements," Fifth International Conference on Soil Mechanics and Foundation Engineering (Paris 1961), vol. I, p. 481.
708. B. Kjaernsli and I. Torblaa, "Asfalt pä fyllingsdammer" (in Norwegian with English Summary), *Publication No. 48*, Norwegian Geotechnical Institute, Oslo (1962).
709. "Anvisningar för utförande och kontroll av jorddammer," (Field Construction Manual published in Swedish). Swedish State Power Board, Stockholm (March 1958).

Index